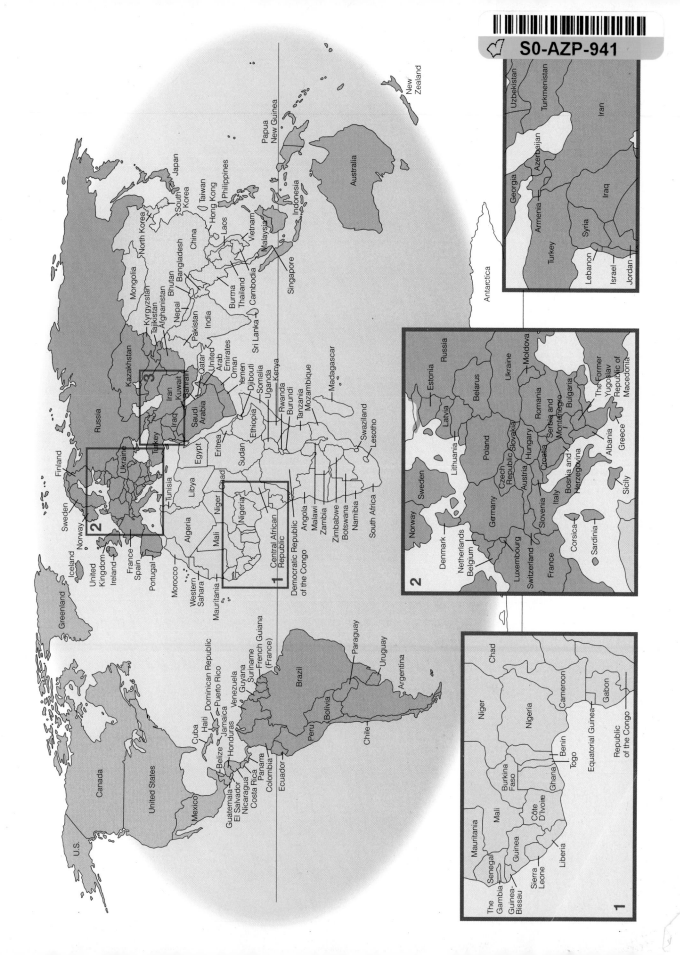

S0-AZP-941

SEVENTH EDITION

Essentials of Sociology

David B. Brinkerhoff
University of Nebraska, Lincoln

Lynn K. White
University of Nebraska, Lincoln

Suzanne T. Ortega
University of Washington

Rose Weitz
Arizona State University

WADSWORTH
CENGAGE Learning™

Australia • Brazil • Japan • Korea • Mexico • Singapore • Spain • United Kingdom • United States

Essentials of Sociology, Seventh Edition
David B. Brinkerhoff, Lynn K. White,
Suzanne T. Ortega, Rose Weitz

Acquisitions Editor, Sociology: Chris Caldeira

Assistant Editor: Christina Ho

Editorial Assistant: Katia Krukowski

Technology Project Manager: Dave Lionetti

Marketing Manager: Michelle Williams

Marketing Communications Manager:
    Linda Yip

Project Manager, Editorial Production:
    Cheri Palmer

Creative Director: Rob Hugel

Art Director: John Walker

Print Buyer: Becky Cross

Permissions Editor: Roberta Broyer

Production Service: Anne Williams, Graphic
    World Publishing Services

Text Designer: Roy Neuhaus

Photo Researcher: Terri Wright

Illustrator: Graphic World Illustration Studio

Cover Designer: Yvo Riezebos Design

Cover Image: Franz Marc Frei/Corbis

Compositor: Graphic World Inc.

For product information and technology assistance, contact us at
Cengage Learning Customer & Sales Support, 1-800-354-9706

For permission to use material from this text or product, submit all requests online at **cengage.com/permissions**
Further permissions questions can be emailed to
permissionrequest@cengage.com

Library of Congress Control Number: 2006939504

ISBN-13: 978-0-495-09636-8

ISBN-10: 0-495-09636-9

Wadsworth
10 Davis Drive
Belmont, CA 94002-3098
USA

Cengage Learning is a leading provider of customized learning solutions with office locations around the globe, including Singapore, the United Kingdom, Australia, Mexico, Brazil and Japan. Locate your local office at:
international.cengage.com/region

Cengage Learning products are represented in Canada by
Nelson Education, Ltd.

For your course and learning solutions, visit **academic.cengage.com**

Purchase any of our products at your local college store or at our preferred online store **www.ichapters.com**

Printed in the United States of America
2 3 4 5 6 7 11 10 09 08

Brief Contents

Contents

CHAPTER 3   Socialization   54

Preface

Sociology deals with the major issues—both macro and micro—that confront our planet, our nation, and our lives. At the micro level, sociology explores the substance of ordinary life—getting a job, getting (and staying) married, caring for children, and caring for aging parents. At the macro level, sociologists grapple with the crucial national and international problems of our times, including homelessness, health care reform, environmental degradation, poverty, and international conflict and dependence. An introductory textbook offers the opportunity to help students understand both of these levels and the connections between them. In this textbook, we hope to give students a sociological framework that will help them better understand how social structures shape both their personal lives and the larger social world that surrounds them.

This seventh edition of *Essentials of Sociology* has the same goal as the preceding editions: to provide a concise, balanced introduction to the field of sociology. Like the first six editions, the seventh edition provides a careful blend of theory and the latest research combined with a series of examples, case studies, and applications that will help students develop the sociological imagination. This edition also continues our tradition of providing extensive coverage of **diversity issues**, both in the body of the chapters and in the **Focus On** boxes and **Intersections.**

Major Changes in the Seventh Edition

As our readers have come to expect, this edition of *Essentials of Sociology* has been revised to take into account recent developments in the field as well as "hot" new topics such as terrorism, Hurricane Katrina, the ethnic conflict in Darfur, and the privatization of the U.S. economy. We have carefully reviewed all the major journals and many of the specialty journals to provide students with the most current findings. Figures and graphs have been revised to incorporate the very latest data available.

A major change in this edition is a full-length chapter on health and health care, which includes discussion of sociological theories of illness, underlying social causes of illness and mortality, and the nature of the U.S. health care system, among other topics. More new topics featured in this edition include the "Wal-Mart Effect" (Chapter 13), the rise of fundamentalism (Chapter 13), and sociological explanations for upward mobility (Chapter 7). This edition also includes new highlighted boxes on media and culture, offering sociological analysis of such topics as the "karaoke class wars" and the "Culture of Fear."

This edition has been streamlined to make it particularly useful for those who wish to supplement their text with readers, literature, or activities, or for instructors who teach short academic terms and have difficulty covering the material contained in a more comprehensive text. In condensing the field, we have retained all the central concepts, theories, and research findings. We also offer many examples to help students grasp this basic information and to keep the book engaging and readable.

Features of the Seventh Edition

The seventh edition of *Essentials of Sociology* retains all the pedagogical features that have made it successful. Each chapter contains at least two high-interest boxed inserts. Clearly identified concepts, **Concept Summaries,** and chapter summaries continue to aid students in mastering the material. As in the sixth edition, discussion of the three major theoretical perspectives in sociology appears near the beginning of most chapters, and **Thinking Critically** questions appear at the end of each chapter. These questions challenge students to apply sociological concepts and theory to problems relevant to their immediate lives and can be used for group discussion or individual writing assignments.

This edition also continues our tradition of incorporating cross-national examples throughout the text. Chapter 8, for instance, includes a case study on racial prejudice and violence in Darfur, and Chapter 10 explores declining life expectancies in the former Soviet Union. In Chapter 4, the **Focus on a Global Perspective** box features an analysis of impression management among Australian aborigines and the **Focus On** box in Chapter 5 assesses the strength of weak ties in China and Singapore.

As in the previous edition, this edition continues to offer **Connections** boxes in each chapter to clarify concepts and to help students see how the concepts apply to their own lives. Four types of connections are presented in the book: examples, personal applications for students, historical notes, and social policy applications.

Also as in the previous edition, end-of-chapter sections titled **Where This Leaves Us** tie chapter concepts together and show how the material relates to the larger themes, problems, and issues discussed in the chapter. These sections will help students improve their grasp of the main points in the chapter.

Focus On Boxes

A boxed insert in each chapter introduces provocative and interesting issues. These **Focus On** boxes fall into four different formats: Focus on American Diversity, Focus on a Global Perspective, Focus on Technology, and the all new Focus on the Environment feature. To demonstrate to students the importance of understanding the increasing diversity of American society, the **Focus on American Diversity** boxes examine issues such as the measurement of IQ, gay and lesbian families, and gender differences in mathematics. The **Focus on a Global Perspective** series introduces students to a comparative approach to social issues and social science research, and it deals with topics such as women and development, international migration, and conducting survey research in Nigeria. The **Focus on Media and Culture** series explores from a sociological perspective a wide range of topics students should find especially interesting, including how the mass media have spread a "culture of fear" and what the blacklist against the Dixie Chicks tells us about who controls the media.

Intersections

This popular feature continues in the current edition. Located between chapters throughout the text, essays are designed to help students understand the connections between concepts and topics that have been treated separately in preceding chapters. The **Intersections** demonstrate how institutions, social processes, culture, and systems of inequality often reinforce each other. After the first four chapters, for instance, Intersections asks students to think about how culture and social structure affect each other and how each in turn affects a society's chances of succeeding or failing. After Chapter 13, students are challenged to think further about the relationships between values and institutions by considering how the political process fosters the appearance of what some scholars have called the *culture wars*. This effective feature helps students understand the connections between important sociological concepts.

Concept Learning Aids

Learning new concepts and new vocabulary is vital to developing a new perspective. In *Essentials of Sociology*, this learning is facilitated in four ways: (1) when new terms and concepts first appear in the text, they are printed in boldface type, and complete definitions are set out clearly in the margin; (2) when several related concepts are introduced (for example, pluralist, power elite, and state autonomy models of American government), a Concept Summary is included to summarize the definitions, give examples, and clarify differences; (3) a glossary is included in the back of the book for handy reference; (4) Thinking Critically questions encourage students to make concepts and terms a working part of their vocabulary by using them to discuss a problem of personal or social relevance.

Chapter Summaries

A short point-by-point summary lists the chief points made in each chapter. These summaries will aid beginning students as they study the text and help them distinguish the central concepts from the supporting points.

Supplements for the Seventh Edition

For Instructors

Instructor's Edition

The Instructor's Edition contains a visual walk-through of the text that provides an overview of the key features, themes, and available supplements.

Instructor's Resource Manual with Test Bank

This supplement offers the instructor chapter outlines, class projects and assignments, discussion and lecture topics, Internet and InfoTrac® College Edition Activities, as well as personal applications and journal topics for each chapter. Unique to this supplement is the **Student Data Set** (SDS), which is based on a class questionnaire that enables instructors to directly link students to the data and relationships examined in the main text. The manual includes suggestions for lectures, discussions, and class activities involving the Student Data Set. The test bank contains 100 multiple-choice questions. Each multiple-choice item has the question type (factual, applied, or conceptual) indicated. Also included are 15 true/false questions and 10 short-answer questions with answers and page references along with 10 essay questions for each chapter. All questions are labeled as new, modified, or pickup so instructors know if the question is new to this edition of the test bank, modified but picked up from the previous edition of the test bank, or picked up straight from the previous edition of the test bank. A table of contents for the *ABC News* Sociology Video Series and concise user guides for CengageNow™, InfoTrac College Edition, WebTutor™, and Extension: Wadsworth's Sociology Reader Collection are also included.

Multimedia Manager with Instructor's Resource CD-ROM

With this one-stop digital library and presentation tool, instructors can assemble, edit, and present custom lectures with ease. The Multimedia Manager contains figures, tables, graphs, and maps from this text; pre-assembled Microsoft® PowerPoint® lecture slides; video clips from DALLAS TeleLearning; ShowCase presentational software; tips for teaching; the instructor's manual; and more.

ExamView® (Windows/Macintosh)

Create, deliver, and customize printed and online tests and study guides in minutes with this easy-to-use assessment and tutorial system. ExamView includes a Quick Test Wizard and an Online Test Wizard to guide instructors step by step through the process of creating tests. The test appears on screen exactly as it will print or display online. Using ExamView's complete word processing capabilities, instructors can enter an unlimited number of new questions or edit questions included with ExamView.

Introduction to Sociology 2008 Transparency Masters

A set of black and white transparency masters consisting of tables and figures from Wadsworth's introductory sociology texts is available to help prepare lecture presentations. Free to qualified adopters.

ABC® News: Sociology Videos, Volumes I–II

Launch your lectures with exciting video clips from the award-winning news coverage of ABC. Addressing topics covered in a typical course, these videos are divided into short segments—perfect for introducing key concepts in contexts relevant to students' lives.

Wadsworth's Lecture Launchers for Introductory Sociology

An exclusive offering jointly created by Cengage Learning and DALLAS TeleLearning, this video contains a collection of video highlights taken from the *Exploring Society: An Introduction to Sociology Telecourse* (formerly *The Sociological Imagination*). Each 3- to 6-minute long video segment has been specially chosen to enhance and enliven class lectures and discussions of 20 key topics covered in the introduction to sociology course. Accompanying the video is a brief written description of each clip, along with suggested discussion questions to help effectively incorporate the material into the classroom. Available on VHS or DVD.

Sociology: Core Concepts Video

Another exclusive offering jointly created by Cengage Learning and DALLAS TeleLearning, this video contains a collection of video highlights taken from the

Exploring Society: An Introduction to Sociology Telecourse (formerly *The Sociological Imagination*). Each 15- to 20-minute video segment will enhance student learning of the essential concepts in the introductory course and can be used to initiate class lectures, discussion, and review. The video covers topics such as the sociological imagination, stratification, race and ethnic relations, social change, and more. Available on VHS or DVD.

Wadsworth Sociology Video Library

Bring sociological concepts to life with videos from Wadsworth's Sociology Video Library, which includes thought provoking offerings from Films for Humanities, as well as other excellent educational video sources. This extensive collection illustrates important sociological concepts covered in many sociology courses.

Extension: Wadsworth's Sociology Reader Collection

Create your own customized reader for your sociology class, drawing from dozens of classic and contemporary articles found on the exclusive Cengage Learning TextChoice database. Using the TextChoice website (http://www.TextChoice.com), you can preview articles, select your content, and add your own original material. TextChoice will then produce your materials as a printed supplementary reader for your class.

Turnitin™ Online Originality Checker

This online "originality checker" is a simple solution for professors who want to put a strong deterrent against plagiarism into place and make sure their students are employing proper research techniques. Students upload their papers to their professor's personalized website and within seconds, the paper is checked against three databases—a constantly updated archive of over 4.5 billion web pages; a collection of millions of published works, including a number of Cengage Learning Higher Education texts; and the millions of student papers already submitted to Turnitin. For each paper submitted, the professor receives a customized report that documents any text matches found in Turnitin's databases. At a glance, the professor can see if the student has used proper research and citation skills, or if he or she has simply copied the material from a source and pasted it into the paper without giving credit where credit was due. Our exclusive deal with iParadigms, the producers of Turnitin, gives instructors the ability to package Turnitin with the *Essentials of Sociology*, Seventh Edition, Cengage Learning textbook. Please consult with your Cengage Learning sales representative to find out more!

For Students

WebTutor™ Toolbox on WebCT and Blackboard

WebTutor Toolbox combines easy-to-use course management tools with content from this text's rich companion website. Ready to use as soon as you log on—or, customize WebTutor ToolBox with weblinks, images, and other resources.

InfoTrac® College Edition with InfoMarks®

Available as a free option with newly purchased texts, InfoTrac College Edition gives instructors and students 4 months of free access to an extensive online database of reliable, full-length articles (not just abstracts) from thousands of scholarly and popular publications going back as much as 22 years. Among the journals available 24/7 are *American Journal of Sociology, Social Forces, Social Research,* and *Sociology.* InfoTrac College Edition now also comes with InfoMarks, a tool that allows you to save your search parameters, as well as save links to specific articles. (Available to North American college and university students only; journals are subject to change.)

Introduction to Sociology Group Activities Workbook

This supplement by Lori Ann Fowler of Tarrant County College contains both in- and out-of-class group activities (utilizing resources such as MicroCase® Online Data exercises from Wadsworth's Online Sociology Resource Center) that students can tear out and turn in to the instructor once complete. Also included are ideas for video clips to anchor group discussions, maps, case studies, group quizzes, ethical debates, group questions, group project topics, and ideas for outside readings for students to base group discussions on. Both a workbook for students and a repository of ideas, instructors can use this guide to get ideas for any Introductory Sociology class.

Companion Website for *Essentials of Sociology,* Seventh Edition (academic.cengage.com/sociology/brinkerhoff)

The book's companion site includes chapter-specific resources for instructors and students. For instructors, the site offers a password-protected instructor's manual, Microsoft PowerPoint presentation slides, and more. For students, there is a multitude of text-specific study aids, including the following:

- Tutorial practice quizzes that can be scored and e-mailed to the instructor
- Glossary
- Flash cards
- Crossword puzzles
- Weblinks

Researching *Sociology on the Internet,* Third Edition

Prepared by D. R. Wilson of Houston Baptist University, this guide is designed to assist sociology students with doing research on the Internet. Part One contains the general information necessary to get started and answers questions about security, the type of sociology material available on the Internet, the information that is reliable and the sites that are not, the best ways to find research, and the best links to take students where they want to go. Part Two looks at each main topic in the area of sociology and refers students to sites where they can obtain the most enlightening research and information.

Acknowledgments

As with the earlier editions, in preparing this edition we have accumulated many debts. We are especially grateful for the generous advice of our colleagues around the country who shared their expertise and advised us on our own forays into their substantive areas. Special thanks for helping with this edition go to Gray Cavendar, Susan E. Chase, Kirsten A. Dellinger, Jennifer Glick, Nancy Jurik, Miliann Kang, Jill Kiecolt, Patrician Yancey Martin, Cecilia Menjívar, Milagros Peña, and Ann Tickamyer.

Special thanks also go to the people at West and Wadsworth Publishing, including Clyde Perlee, who first prompted us to become authors, and Denise Simon, who was generous with encouragement and advice. Eve Howard has continued to be helpful over the years, and Bob Jucha was truly essential to this edition of *Essentials*. Also, many thanks to Katia Krukowski, who came to the rescue whenever needed. At all levels, the people at Wadsworth have been a pleasure to work with and helped us make our book the best possible, while leaving the substance and direction of the book in our hands.

We would like to express our gratitude to those people who reviewed the manuscript for us:

G. Thomas Behler, Ferris State University
Thomas P. Egan, Western Kentucky University
M. D. Litonjua, College of Mount St. Joseph
Mel Moore, University of Northern Colorado
Lisa Rashotte, University of North Carolina at Charlotte
David L. Strickland, East Georgia College
Allison L. Vetter, University of Central Arkansas
Leslie T. C. Wang, The University of Toledo

Once again, we thank those people who reviewed the manuscript for previous editions of *Sociology* and *Essentials of Sociology*. Their suggestions and comments made a substantial contribution to the project:

Margaret Abraham, Hofstra University, New York
Paul J. Baker, Illinois State University
Robert Benford, University of Nebraska
Susan Blackwell, Delgado Community College
Tim Brezina, Tulane University
Marie Butler, Oxnard College
John K. Cochran, Wichita State University, Kansas
Carolie Coffey, Cabrillo College, California
Paul Colomy, University of Akron, Ohio
Ed Crenshaw, University of Oklahoma
Raymonda P. Dennis, Delgado Community College, New Orleans
Lynda Dodgen, North Harris County College, Texas
David A. Edwards, San Antonio College, Texas
Laura Eells, Wichita State University, Kansas
William Egelman, Iona College, New York
Constance Elsberg, Northern Virginia Community College
Christopher Ezell, Vincennes University, Indiana
Joseph Faltmeier, South Dakota State University
Daniel E. Ferritor, University of Arkansas
Charles E. Garrison, East Carolina University, North Carolina
James R. George, Kutztown State College, Pennsylvania
Harold C. Guy, Prince George Community College, Maryland
Rose Hall, Diablo Valley College, California
Sharon E. Hogan, Blue River Community College
Michael G. Horton, Pensacola Junior College, Florida
Cornelius G. Hughes, University of Southern Colorado
Jon Ianitti, State University of New York, Morrisville
Carol Jenkins, Glendale Community College

William C. Jenné, Oregon State University
Dennis L. Kalob, Loyola University, New Orleans
Sidney J. Kaplan, University of Toledo, Ohio
Florence Karlstrom, Northern Arizona University
Diane Kayongo-Male, South Dakota State University
William Kelly, University of Texas
James A. Kithens, North Texas State University
Phillip R. Kunz, Brigham Young University, Utah
Billie J. Laney, Central Texas College
Charles Langford, Oregon State University
Mary N. Legg, Valencia Community College, Florida
John Leib, Georgia State University
Joseph J. Leon, California State Polytechnic University, Pomona
J. Robert Lilly, Northern Kentucky University
Jan Lin, University of Houston
Lisa Linares, Madison Area Technical College
James Lindberg, Montgomery College
Richard L. Loper, Seminole Community College, Florida
Ronald Matson, Wichita State University
Carol May, Illinois Central College
Rodney C. Metzger, Lane Community College, Oregon
Vera L. Milam, Northeastern Illinois University
Purna C. Mohanty, Paine College
James S. Munro, Macomb College, Michigan
Lynn D. Nelson, Virginia Commonwealth University
J. Christopher O'Brien, Northern Virginia Community College
Charles O'Connor, Bemidji State University, Minnesota
Jane Ollenberger, University of Minnesota-Duluth
Robert L. Petty, San Diego Mesa College, California
Ruth A. Pigott, Kearney State College, Nebraska
John W. Prehn, Gustavus Adolphus College, St. Peter, Minnesota
Adrian Rapp, North Harris County College, Texas
Mike Robinson, Elizabethtown Community College, Kentucky
Joe Rogers, Big Bend Community College
Will Rushton, Del Mar College, Texas
Rita P. Sakitt, Suffolk Community College, New York
Martin Scheffer, Boise State University, Idaho
Richard Scott, University of Central Arkansas
Ida Harper Simpson, Duke University, North Carolina
James B. Skellenger, Kent State Unversity, Ohio
Ricky L. Slavings, Radford University, Virginia
John M. Smith, Jr., Augusta College, Georgia
Evelyn Spiers, College of the Canyons
James Steele, James Madison University, Virginia
Michael Stein, University of Missouri–St. Louis
Barbara Stenross, University of North Carolina
Jack Stirton, San Joaquin Delta College, California
Deidre Tyler, Salt Lake Commuity College
Emil Vajda, Northern Michigan University
Henry Vandenburgh, State University of New York, Oswego
Steven L. Vassar, Mankato State University, Minnesota

Peter Venturelli, Valparaiso University
Allison Vetter, University of Central Arkansas
Leslie Wang, University of Toledo
Jane B. Wedemeyer, Sante Fe Community College, Florida
Dorether M. Welch, Penn Valley Community College
Thomas J. Yacovone, Los Angeles Valley College, California
David L. Zierath, University of Wisconsin.

About the Authors

David B. Brinkerhoff is Associate Vice Chancellor for Academic Affairs and Professor of Sociology at the University of Nebraska–Lincoln. He holds a Ph.D. in sociology from the University of Washington in Seattle, with B.S. and M.S. degrees from Brigham Young University. He has been at the University of Nebraska since 1978, having served as Associate Vice Chancellor since 1991. He has taught family and introductory sociology extensively. His research has focused on marital quality and parent-child relationships, covering such topics as children's work in the family and the effect of economic marginality on the family. He is currently concerned with issues such as introducing technology into the classroom.

Lynn K. White is Professor of Sociology at the University of Nebraska–Lincoln. She holds a Ph.D. in sociology from the University of Washington in Seattle. She has been at the University of Nebraska since 1974, having served as Chair of the Department of Sociology and Director of the Bureau of Sociological Research. She teaches social demography, family, and research methods. Her research has focused on relationships between parents and their adult children over the life course, covering such topics as the empty nest, co-residence, the link between marital quality and parenting experiences, and intergenerational exchange. Her current research on parent-child relationships in adult stepfamilies is funded by a grant from the National Institutes for Child Health and Human Development. Her work has appeared in *American Sociological Review and Social Forces* as well as in family journals.

Suzanne T. Ortega, Professor of Sociology and Vice Provost and Dean of Graduate School at the University of Washington, holds a Ph.D. in sociology from Vanderbilt University. Prior to coming to the University of Washington, she spent 5 years at the University of Missouri as Vice Provost and Graduate Dean and spent 20 years on the faculty at the University of Nebraska–Lincoln, where she taught introductory sociology, social problems, criminology, and minority-group relations. She currently serves as the chair of the GRE Board, is former chair of the Council of Graduate Schools' board, and is heavily involved in the development and implementation of a university-wide Preparing Future Faculty program.

Rose Weitz, Professor of Sociology and of Women and Gender Studies at Arizona State University, received her Ph.D. from Yale University. At Arizona State University, she has served as Director of Women's Studies, Director of Undergraduate Studies in Sociology, and Director of Graduate Studies in Sociology. She is a former president of Sociologists for Women in Society and former chair of the Medical Sociology Section of the American Sociological Association. Professor Weitz is the author of many scholarly articles, the editor of *The Politics of Women's Bodies*, and the author of the books Life with *AIDS, Rapunzel's Daughters: What Women's Hair Tells Us About Women's Lives,* and *The Sociology of Health, Illness, and Health Care: A Critical Approach.* She has won two major teaching awards at Arizona State University as well as the Pacific Sociological Association's Distinguished Contributions to Teaching Award, and has been a finalist for other teaching awards numerous times.

CHAPTER 1

The Study of Society

© Royalty-Free/Corbis

Outline

What Is Sociology?

Each of us starts the study of society with the study of individuals. We wonder why Theresa keeps getting involved with men who treat her badly, why Mike never learns to quit drinking before he gets sick, why our aunt puts up with our uncle, and why anybody ever liked the Spice Girls. We wonder why people we've known for years seem to change drastically when they get married or change jobs.

If Theresa were the only woman with bad taste in men or Mike the only man who drank too much, then we might try to understand their behavior by peering into their personalities. We know, however, that there are millions of men and women who have disappointing romances and who drink too much. We also know that women are more likely than men to sacrifice their needs to keep a romance alive, and that men are more likely than women to drown their troubles in drink. To understand Mike and Theresa, then, we must place them in a larger context and examine the forces that lead some *groups* of people to behave so differently from other groups.

Sociology is the systematic study of human society, social groups, and social interactions: It emphasizes the larger context in which Mike, Theresa, and the rest of us live.

Sociologists tend to view common human interactions as if they were plays. They might, for example, title a common human drama *Boy Meets Girl*. Just as *Hamlet* has been performed around the world for more than 400 years with different actors and different interpretations, *Boy Meets Girl* has also been performed countless times. Of course, the drama is acted out a little differently each time, depending on the scenery, the people in the lead roles, and the century, but the essentials are the same. Thus, we can read nineteenth- or even sixteenth-century love stories and still understand why those people did what they did. They were playing roles in a play that is still performed daily.

More formal definitions will be introduced later, but the metaphor of the theater can be used now to introduce two of the most basic concepts in sociology: role and social structure. By **role,** we mean the expected performance of someone who occupies a specific position. Mothers, teachers, students, and lovers all have roles. Each position has an established script that suggests appropriate lines, gestures, and relationships with others. Discovering what each society offers as a stock set of roles is one of the major themes in sociology. Sociologists try to find the common roles that appear in society and to determine why some people play one role rather than another.

The second major sociological concept is **social structure,** the larger structure of the play in which the roles appear. What is the whole set of roles that appears in this play? How are the roles interrelated? Do some actors and roles have more power than others? And how does this affect the outcome of the play? Thus, the role of student is understood in the context of the social structure we call education, a context in which teachers have more power than students, and administrators more power than teachers. Through these two major ideas of role and social structure, sociologists try to understand the human drama.

The Sociological Imagination

The ability to see the intimate realities of our lives and those of others in the context of common social structures has been called the **sociological imagination** (Mills 1959, 15). According to C. Wright Mills, the sociological imagination is what we use

---

Sociology is the systematic study of human social interaction.

A **role** is a set of norms specifying the rights and obligations associated with a status.

A **social structure** is a recurrent pattern of relationships among groups.

The **sociological imagination** is the ability to see the intimate realities of our own lives in the context of common social structures; it is the ability to see personal troubles as public issues.

when we realize that some personal troubles (such as poverty, divorce, or loss of faith) are actually common public issues that reflect a larger social context. Mills suggests that many of the things we experience as individuals are really beyond our control. They have to do with the way society as a whole is organized. For example, Mills writes:

> When, in a city of 100,000, only one man is unemployed, that is his personal trouble, and for its relief we properly look to the character of the man, his skills, and his immediate opportunities. But when in a nation of 50 million employees, 15 million men are unemployed, that is [a public] issue, and we may not hope to find its solution within the range of opportunities open to any one individual. The very structure of opportunities has collapsed. Both the correct statement of the problem and the range of possible solutions require us to consider the economic and political institutions of the society, and not merely the personal situation and character of a scatter of individuals. . . .
>
> [Similarly,] inside a marriage a man and a woman may experience personal troubles, but when the divorce rate during the first four years of marriage is 250 out of every 1000 attempts, this is an indication of a structural issue having to do with the institutions of marriage and the family and other institutions that bear on them. (Mills 1959, 9)

In everyday life, we do not define personal experiences in these terms. We rarely consider the impact of history and social structures on our own experiences. If a child becomes a drug addict, parents tend to blame themselves; if spouses divorce, their friends usually focus on their personality problems; if a student does poorly in school, most blame only the student. To develop the sociological imagination is to understand how outcomes such as these are, in part, a product of society and not fully within the control of the individual.

Some people do poorly in school, for example, not because they are stupid or lazy but because they are faced with conflicting roles and role expectations. The "this is the best time of your life" play calls for very different roles and behaviors from the "education is the key to success" play. Those who adopt the student role in the "best time of your life" play will likely earn lower grades than those in the "education is the key to success" play. Other people may do poorly because they come from a family that does not give them the financial or psychological support they need. In fact, their family may need them to earn an income to help support their younger brothers and sisters. These students may be working 25 hours a week in addition to going to school; they may be going to school despite their family's lack of understanding of why college is important, or why college students need quiet and privacy for studying. In contrast, other students may find it difficult to fail: Their parents provide tuition, living expenses, a personal computer, a car, and moral support. As we will discuss in more detail in Chapter 12, parents' social class is one of the best predictors of who will fail and who will graduate. Success or failure is thus not entirely an individual matter; it is socially structured.

The sociological imagination—the ability to see our own lives and those of others as part of a larger social structure—is central to sociology. Once we develop this imagination, we will be less likely to explain others' behavior only through their personality and will increasingly look to the roles and social structures that determine behavior. We will also recognize that the solutions to many social problems lie not in changing individuals but in changing the social structures and roles that are available to them. Although poverty, divorce, and racism are experienced as intensely personal hardships, they are unlikely to be reduced effectively through massive personal therapy. To solve these and many other social problems, we need to change social structures; we need to rewrite the play. The sociological imagination offers a new way to look at—and a new way to solve—common troubles and dilemmas that face individuals.

When thousands are out of work, it is a public issue, not simply a personal trouble. For its solution, we need to change the broader economics of society, not simply teach people how to write better resumes.

Connections

Social Policy

How can we help the unemployed? How we answer this question will depend on what we think caused the problem. Psychologists and social workers tend to focus on individual-level solutions such as offering workshops in job interviewing and short-term economic assistance to laid-off workers. Sociologists instead focus on understanding the broader social forces that cause widespread unemployment. They raise questions such as "How can we halt the movement of American jobs overseas?" and "How can we ensure that children from poorer families get the education needed to obtain good jobs in a shrinking job market?"

Sociology as a Social Science

Sociology is concerned with people and with the rules of behavior that structure the ways in which people (and groups of people) interact. Its emphasis is on relationships and patterns of interaction—how these patterns develop, how they are maintained, and how they change.

As one of the social sciences, sociology has much in common with political science, economics, psychology, and anthropology. All these fields share an interest in human social behavior and, to some extent, an interest in society. In addition, they all share an emphasis on the scientific method as the best approach to knowledge. This means that they rely on **empirical research**—research based on systematic examination of the evidence—before reaching any conclusions and expect researchers to evaluate that evidence in an unbiased, objective fashion. This empirical approach is what distinguishes the social sciences from journalism and other fields that comment on the human condition.

Empirical research is research based on systematic, unbiased examination of evidence.

Sociology differs from the other social sciences in its particular focus. Anthropologists are primarily interested in human (and nonhuman) *culture*. For example, anthropologists have studied why rape is more common in some cultures than in others and what purposes are served by cultural celebrations like Bar Mitzvahs, high school graduation parties, Mardi Gras, and *quinceañeras*. Psychologists focus on individual behavior and thought patterns, such as why some individuals experience more anxiety or gamble more than others. Political scientists study political systems and behaviors, such as how dictatorships rise and fall, and economists study how goods and services are produced, distributed, and consumed, such as why cell phones with cameras are so popular. Although sociologists are also interested in culture, individual behavior, politics, and the economy, their focus is always on how these and other issues are situated within social groups and social interactions.

The Emergence of Sociology

Sociology emerged as a field of inquiry during the political, economic, and intellectual upheavals of the eighteenth and nineteenth centuries. Rationalism and science replaced tradition and belief as methods of understanding the world, leading to changes in government, education, economic production, and even religion and family life. The clearest symbol of this turmoil is the French Revolution (1789), with its bloody uprising and rejection of the past.

Although less dramatic, the Industrial Revolution had an even greater impact. Within a few generations, traditional rural societies were replaced by industrialized urban societies. The rapidity and scope of the change resulted in substantial social disorganization. It was as if society had changed the play without bothering to tell the actors, who were still trying to read from old scripts. Although a few people prospered mightily, millions struggled desperately to make the adjustment from rural peasantry to urban working class.

This turmoil provided the inspiration for much of the intellectual effort of the nineteenth century: Charles Dickens's novels, Jane Addams's reform work, Karl Marx's revolutionary theory. It also inspired the empirical study of society. These were the years in which scientific research was a new enterprise and nothing seemed too much to hope for. After electricity, the telegraph, and the X-ray, who was to say that researchers could not discover how to turn stones into gold or how to eliminate poverty or war? Many hoped that the tools of empirical research could help in understanding and controlling a rapidly changing society.

The Founders: Comte, Spencer, Marx, Durkheim, and Weber

The upheavals of the nineteenth century in Europe and Great Britain stimulated the development of sociology as a discipline. We will look at five theorists who may be considered the founders of sociology.

August Comte (1798–1857)

The first major figure in the history of sociology was the French philosopher August Comte. He coined the term *sociology* in 1839 and is generally considered the founder of this field.

Comte was among the first to suggest that the scientific method could be applied to social events (Konig 1968). The philosophy of positivism, which he developed, suggests that the social world can be studied with the same scientific accuracy and assurance as the natural world. Once the laws of social behavior were learned, he believed, scientists could accurately predict and control events. Although thoughtful people wonder whether we will ever be able to predict human behavior as accurately as we can predict the behavior of molecules, the scientific method remains central to sociology.

Another of Comte's lasting contributions was his recognition that an understanding of society requires a concern for both the sources of order and continuity and the sources of change. These concerns remain central to sociological research, under the labels of social structure (order) and social process (change).

■ August Comte, 1798–1857

Herbert Spencer (1820–1903)

Another pioneer in sociology was the British philosopher-scientist Herbert Spencer, who advanced the thesis that evolution accounts for the development of social, as well as natural, life. Spencer viewed society as similar to a giant organism: Just as the heart and lungs work together to sustain the life of the organism, so the parts of society work together to maintain society.

These ideas led Spencer to two basic principles that still guide the study of sociology. First, Spencer concluded that each society must be understood as an adaptation to its environment. This principle of adaptation implies that to understand society, we must focus on processes of growth and change. It also implies that there is no "right" way for a society to be organized. Instead, societies will change as circumstances change.

Spencer's second major contribution was his concern with the scientific method. More than many scholars of his day, Spencer was aware of the importance of objectivity and moral neutrality in investigation. In essays on the bias of class, the bias of patriotism, and the bias of theology, he warned sociologists that they must suspend their own opinions and wishes when studying society (Turner & Beeghley 1981).

Karl Marx (1818–1883)

A philosopher, economist, and social activist, Karl Marx was born in Germany in 1818. Marx received his doctorate in philosophy at the age of 23 but never held a university appointment. Because of his radical views, he spent most of his adult life in exile and poverty (McLellan 2006).

■ Herbert Spencer, 1820–1903

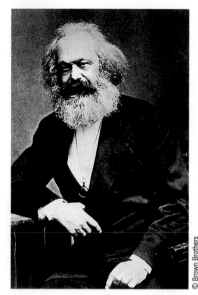

Karl Marx, 1818–1883

© Brown Brothers

Economic determinism means that economic relationships provide the foundation on which all other social and political arrangements are built.

Marx was repulsed by the poverty and inequality that characterized the nineteenth century. Unlike other scholars of his day, he was unwilling to see poverty as either a natural or a God-given condition of the human species. Instead, he viewed poverty and inequality as human-made conditions fostered by private property and capitalism. As a result, he devoted his intellectual efforts to understanding—and eliminating—capitalism. Many of Marx's ideas are of more interest to political scientists and economists than to sociologists, but he left two enduring legacies to sociology: the theories of economic determinism and the dialectic.

ECONOMIC DETERMINISM   Marx began his analysis of society by assuming that the most basic task of any human society is to provide food and shelter to sustain itself. Marx argued that the ways in which society does this—its modes of production—provide the foundations on which all other social and political arrangements are built. Thus, he believed that family, law, and religion all develop after and adapt to the economic structure; in short, they are determined by economic relationships. This idea is called **economic determinism.**

A good illustration of economic determinism is the influence of economic conditions on marriage choices. In traditional agricultural societies, young people often remain economically dependent upon their parents until well into adulthood because the only economic resource, land, is controlled by the older generation. To survive, they must remain in their parents' good graces; this means, among other things, that they cannot marry without their parents' approval. In societies where young people can earn a living without their parents' help, however, they can marry whenever and whomever they please. Marx would argue that this shift in mate selection practices is the result of changing economic relationships.

Because Marx saw all human relations as stemming ultimately from the economic systems, he suggested that the major goal of a social scientist is to understand economic relationships: Who owns what, and how does this pattern of ownership affect human relationships?

THE DIALECTIC   Marx's other major contribution was a theory of social change. Many nineteenth-century scholars applied Darwin's theories of biological evolution to society; they believed that social change was the result of a natural and more or less peaceful process of adaptation. Marx, however, argued that the basis of change was conflict between opposing economic interests, not adaptation.

Dialectic philosophy views change as a product of contradictions and conflict between the parts of society.

Marx's thinking on conflict was influenced by the German philosopher George Hegel, who suggested that for every idea (thesis), a counter idea (antithesis) develops to challenge it. As a result of conflict between the two ideas, a new idea (synthesis) is produced. This process of change is called the **dialectic** (Figure 1.1).

Marx's contribution was to apply this model of ideological change to change in economic and social systems. Within capitalism, Marx suggested, the capitalist class was the thesis and the working class was the antithesis. He predicted that conflicts between them would lead to a new synthesis. That synthesis would be a new, communistic economic system. Indeed, in his role as social activist, Marx hoped to encourage conflict and ignite the revolution that would bring about the desired change. The workers, he declared, "have nothing to lose but their chains" (Marx & Engels 1967, 258).

Although few sociologists are revolutionaries, many accept Marx's ideas on the importance of economic relationships and economic conflicts. Much more controversial is Marx's argument that the social scientist should also be a social activist, a person who not only tries to understand social relationships but also tries to change them.

Emile Durkheim (1858–1917)

Emile Durkheim's life overlapped with that of Marx. While Marx was starving as an exile in England, however, Durkheim spent most of his career occupying a prestigious professorship at the University of the Sorbonne in France. Far from rejecting society, Durkheim embraced it, and much of his outstanding scholarly energy was devoted to understanding the stability of society and the importance of social participation for individual happiness. Whereas the lasting legacy of Marx is a theory that looks for the conflict-laden and ever-changing aspects of social practices, the lasting legacy of Durkheim is a theory that examines the positive contributions and stability of social patterns. Together they allow us to see both order and change.

Durkheim's major works are still considered essential reading in sociology. These include his studies of suicide, education, divorce, crime, and social change. Two enduring contributions are his ideas about the balance between individual goals and social rules and about social science methods.

One of Durkheim's major concerns was the balance between social regulation and personal freedom. He argued that community standards of morality, which he called the collective conscience, not only confine our behavior but also give us a sense of belonging and integration. For example, many people complain about having to dress up; they complain about having to shave their faces or their legs or having to wear a tie or pantyhose. "What's wrong with jeans?" they want to know. At the same time, most of us feel a sense of satisfaction when we appear in public in our best clothes. We know that we will be considered attractive and successful. Although we may complain about having to meet what appear to be arbitrary standards, we often feel a sense of satisfaction in being able to meet those standards successfully. In Durkheim's words, "institutions may impose themselves upon us, but we cling to them; they compel us, and we love them" (Durkheim [1895] 1938, 3). This beneficial regulation, however, must not rob the individual of all freedom of choice.

In his classic study, *Suicide,* Durkheim identified two types of suicide that stem from an imbalance between social regulation and personal freedom. Fatalistic suicide occurs when society overregulates and allows too little freedom; when our behavior is so confined by social institutions that we cannot exercise our independence ([1897] 1951, 276). Durkheim gave as an example of fatalistic suicide the very young husband who feels overburdened by the demands of work, household, and family. Anomic suicide, on the other hand, occurs when there is too much freedom and too little regulation, when society's influence does not check individual passions ([1897] 1951, 258). Durkheim believed that this kind of suicide was most likely to occur in times of rapid social change. When established ways of doing things have lost their meaning, but no clear alternatives have developed, individuals feel lost. The high rate of alcohol abuse among Native Americans is often attributed to the weakening of traditional social regulation.

Durkheim was among the first to stress the importance of using reliable statistics to logically rule out incorrect theories of social life and to identify more promising theories. He strove to be an objective observer who only sought the facts. As sociology became an established discipline, this ideal of objective observation replaced Marx's social activism as the standard model for social science.

Max Weber (1864–1920)

A German economist, historian, and philosopher, Max Weber (vay-ber) provided the theoretical base for half a dozen areas of sociological inquiry. He wrote on religion, bureaucracy, method, and politics. In all these areas, his work is still valuable and

FIGURE 1.1
The Dialectic
The dialectic model of change suggests that change occurs through conflict and resolution rather than through evolution.

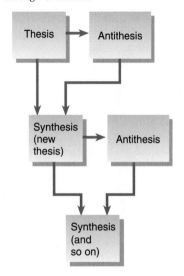

Emile Durkheim, 1858–1917

© Brown Brothers

Max Weber, 1864–1920

Value-free sociology concerns itself with establishing what is, not what ought to be.

insightful; it is covered in detail in later chapters. Three of Weber's more general contributions were an emphasis on the subjective meanings of social actions, on social as opposed to economic causes, and on the need for objectivity in studying social issues.

Weber believed that knowing patterns of behavior was less important than knowing the meanings people attach to behavior. For example, Weber would argue that it is relatively meaningless to compile statistics such as how many marriages end in divorce now compared with 100 years ago. More critical, he would argue, is understanding how the *meaning* of divorce has changed in the past hundred years. Weber's emphasis on the subjective meanings of human actions has been the foundation of scholarly work on topics as varied as religion and immigration.

Weber trained as an economist, and much of his work concerned the interplay of things economic and things social. He rejected Marx's idea that economic factors were the determinants of all other social relationships. In a classic study, *The Protestant Ethic and the Spirit of Capitalism* ([1904–05] 1958), Weber tried to show how social and religious values may be the foundation of economic systems. This argument is developed more fully in Chapter 12, but its major thesis is that the religious values of early Protestantism (self-discipline, thrift, and individualism) were the foundation for capitalism.

One of Weber's more influential ideas was his declaration that sociology must be **value-free.** Weber argued that sociology should be concerned with establishing what is and not what ought to be. Weber's dictum is at the heart of the standard scientific approach that is generally advocated by modern sociologists. Thus, although one may study poverty or racial inequality because of a sense of moral outrage, such feelings must be set aside to achieve an objective grasp of the facts. This position of neutrality is directly contradictory to the Marxist emphasis on social activism, and sociologists who adhere to Marxist principles generally reject the notion of value-free sociology. Most modern sociologists, however, try to be value-free in their scholarly work.

Although all Christians practice baptism, the typical Presbyterian baptism in which an infant's head is sprinkled with a few drops of water during a formal service is quite different in symbolic meaning from baptism by immersion, practiced by many evangelical Christians.

Sociology in the United States

Sociology developed somewhat differently in the United States than in Europe. Although U.S. sociology has the same intellectual roots as European sociology, it has some distinctive characteristics. Three features that have characterized U.S. sociology from its beginning are a concern with social problems, a reforming rather than a radical approach to these problems, and an emphasis on the scientific method.

One reason that U.S. sociology developed differently from European sociology is that our social problems differ. Slavery, the Civil War, and high immigration rates, for example, made racism and ethnic discrimination much more salient issues in the United States.

One of the first sociologists to study these issues was W. E. B. DuBois. DuBois, who received his doctorate in 1895 from Harvard University, devoted his career to developing empirical data about African Americans and to using those data to combat racism.

The work of Jane Addams, another early sociologist and recipient of the 1931 Nobel Peace Prize, illustrates the reformist approach of much U.S. sociology. Addams was the founder of Hull House, a famous center for social services and community activism located in a Chicago slum. She and her colleagues used quantitative social science data to lobby successfully for legislation mandating safer working conditions, a better juvenile justice system, better public sanitation, and services for the poor (Linn & Scott 2000).

As sociology became more established, it also became more conservative. In the years between the two World Wars, a new generation of sociology professors became convinced that social activism was incompatible with academic respectability. Beginning in the 1950s, however, sociologists such as C. Wright Mills and Ralf Dahrendorf turned renewed attention to social problems and social conflict. These days, most sociologists are interested in these concerns, although few are activists.

W. E. B. DuBois, 1868–1963

Jane Addams, 1860–1935

The first sociology course in the United States was taught at Yale University in 1876. By 1910, most colleges and universities in the United States offered sociology courses, although separate departments were slower to develop. Most of the courses were offered jointly with other departments, such as economics, history, philosophy, or general social science departments.

These days, almost all colleges and universities offer undergraduate degrees in sociology. Most universities offer master's degrees in sociology, and approximately 125 offer doctoral degree programs. Graduate sociology programs are more popular in the United States than in any other country in the world. This is partly because sociology in the United States has always been oriented toward the practical as well as the theoretical. The focus has consistently been on finding solutions to social issues and problems, with the result that U.S. sociologists not only teach sociology but also work in government and industry.

Current Perspectives in Sociology

As this brief review of the history of sociological thought has demonstrated, there are many ways of approaching the study of human social interaction. The ideas of Marx, Weber, Durkheim, and others have given rise to dozens of theories about human behavior. In this section, we summarize the three dominant theoretical perspectives in sociology today: structural functional theory, conflict theory, and symbolic interaction theory.

Structural-Functional Theory

Structural-functional theory addresses the question of social organization (structure) and how it is maintained (function).

Structural-functional theory (or *structural functionalism*) addresses the question of how social organization is maintained. This theoretical perspective has its roots in natural science and in the analogy between society and an organism. In the analysis of a living organism, the natural scientist tries to identify the various parts (structures) and to determine how they work (function). In the study of society, a sociologist with this perspective tries to identify the structures of society and how they function, hence the name *structural functionalism*.

The Assumptions Behind Structural Functionalism

All sociologists are interested in exploring how societies work. Sociologists who use the structural-functionalist perspective, however, are distinguished from others by their reliance on three major assumptions:

1. *Stability*. The chief evaluative criterion for any social pattern is whether it contributes to the maintenance of society.
2. *Harmony*. As the parts of an organism work together for the good of the whole, so the parts of society are also characterized by harmony.
3. *Evolution*. Change occurs through evolution—the mostly peaceful adaptation of social structures to new needs and demands and the elimination of unnecessary or outmoded structures.

Structural-Functional Analysis

A structural-functional analysis asks two basic questions: What is the nature of this social structure (what patterns exist)? What are the consequences of this social structure (does it promote stability and harmony)? In this analysis, positive consequences

Major Theoretical Perspectives in Sociology

	Structural Functionalism	Conflict Theory	Symbolic Interactionism
Nature of society	Interrelated social structures that fit together to form an integrated whole	Competing interests, each seeking to secure its own ends	Interacting individuals and groups
Basis of interaction	Consensus and shared values	Constraint, power, and competition	Shared symbolic meanings
Major question	What are social structures? Do they contribute to social stability?	Who benefits? How are these benefits maintained?	How do social structures relate to individual subjective experiences?
Level of analysis	Social structure	Social structure	Interpersonal interaction

are called **functions** and negative consequences are called **dysfunctions.** A distinction is also drawn between **manifest** (recognized and intended) consequences and **latent** (unrecognized and unintended) consequences.

The basic strategy of looking for structures along with their manifest and latent functions and dysfunctions is common to nearly all sociological analysis. Scholars from widely different theoretical perspectives use this framework for examining society. What sets structural-functional theorists apart from others who use this language are their assumptions about harmony and stability.

Consider, for example, new laws now under consideration in many states that would allow battered women to use the "battered women's syndrome" as a legal defense if they assault or kill their abuser. Such laws would explicitly recognize the right of women who assault or kill an abusive partner to plead not guilty by reason of temporary insanity. What would be the consequences of this new social structure? Its manifest function (intended positive outcome) is, of course, to give legal recognition

Functions are consequences of social structures that have positive effects on the stability of society.

Dysfunctions are consequences of social structures that have negative effects on the stability of society.

Manifest functions or dysfunctions are consequences of social structures that are intended or recognized.

Team sports offer a graphic metaphor of social structure. Each person on the team occupies a different status and each plays a relatively unique role. Structural functionalists focus on the benefits that these statuses and roles and the institution of sports itself provide to society.

© AP/Wide World Photos

TABLE 1.1

A Structural-Functional Analysis of the "Battered Women's Syndrome" as a Legal Defense

Structural-functional analysis examines the intended and unintended consequences of social structures. It also assesses whether the consequences are positive (functional) or negative (dysfunctional). There is no moral dimension to the assessment that an outcome is positive; it merely means that the outcome contributes to the stability of society.

	Manifest	Latent
Function	Gives legal recognition to the psychological consequences of domestic violence.	Encourages the view that women are irrational.
Dysfunction	May serve as an excuse for violence against abusers.	Makes it more difficult for victims of domestic violence to retain custody of children.

Connections

Historical Note

Until 1962, American husbands legally could beat their wives. Under laws inherited from the English system, men were assumed to be intellectually and morally superior to women. They were legally held responsible for their wives' actions and so were given the right to beat their wives if needed to control the women's behavior. The term "rule of thumb" comes from the English court ruling under which a man could beat his wife as long as the stick he used was no wider than his thumb.

Latent functions or dysfunctions are consequences of social structures that are neither intended nor recognized.

Conflict theory addresses the points of stress and conflict in society and the ways in which they contribute to social change.

to the devastating long-term psychological consequences of domestic violence. The manifest dysfunction is that some offenders might use the battered women's syndrome defense as an excuse for a malicious, premeditated assault on a significant other. A latent dysfunction may be that women who are acquitted of legal charges on the basis of a temporary insanity plea could find it difficult to retain custody of their children, given the stigma often attached to individuals with any diagnosis of mental disorder.

Another latent outcome may be the perpetuation of the view that women are irrational—that they stay with men who beat them because they are incapable of logically thinking through their options, and that they only leave when they "snap" mentally. Is this latent outcome a function or a dysfunction? Remember that structural-functional analysis typically starts from the assumption that any social action or structure that contributes to the maintenance of society and preserves the status quo is functional and that any action or structure that challenges the status quo is dysfunctional. Because perpetuating the view that women are irrational would reinforce existing gender roles, this would be judged a latent function (Table 1.1).

As this example suggests, a social pattern that contributes to the maintenance of society may benefit some groups more than others. A pattern may be functional—that is, help maintain the status quo—without being either desirable or equitable. In general, however, structural-functionalists emphasize how social structures work together to create a smooth-running society.

Conflict Theory

Whereas structural-functional theory sees the world in terms of consensus and stability, conflict theory sees the world in terms of conflict and change. Conflict theorists contend that a full understanding of society requires a critical examination of competition and conflict in society, especially of the processes by which some people become winners and others become losers. As a result, **conflict theory** addresses the points of stress and conflict in society and the ways in which they contribute to social change.

Assumptions Underlying Conflict Theory

Conflict theory is derived from Marx's ideas. The following are three primary assumptions of modern conflict theory:

1. *Competition.* Competition over scarce resources (money, leisure, sexual partners, and so on) is at the heart of all social relationships. Competition rather than consensus is characteristic of human relationships.
2. *Structural inequality.* Inequalities in power and reward are built into all social structures. Individuals and groups that benefit from any particular structure strive to see it maintained.
3. *Social change.* Change occurs as a result of conflict between competing interests rather than through adaptation. It is often abrupt and revolutionary rather than evolutionary and is often helpful rather than harmful.

Conflict Analysis

Like structural functionalists, conflict theorists are interested in social structures. The two questions they ask, however, are different. Conflict theorists ask: Who benefits from those social structures? And how do those who benefit maintain their advantages?

A conflict analysis of domestic violence, for example, would begin by noting that women are battered far more often and far more severely than are men, and that the popular term "domestic violence" hides this reality. Conflict theorists' answer to the question "Who benefits?" is that battering helps men to retain their dominance over women. These theorists go on to ask how this situation developed and how it is maintained. Their answers would focus on issues such as how some religions traditionally have taught women to submit to their husbands' wishes and to accept violence within marriage, how until recently the law did not regard woman battering as a crime, and how some police officers still consider battering merely an unimportant family matter.

Conflict theorists point out that unions exist because labor and management have different, competing interests: Workers want better pay and secure jobs, management wants to keep costs down.

Symbolic Interaction Theory

Both structural-functional and conflict theories focus on social structures and the relationships between them. What about the relationship between individuals and social structures? Sociologists who focus on the ways that individuals relate to and are affected by social structures generally use symbolic interaction theory. **Symbolic interaction theory** (or *symbolic interactionism*) addresses the subjective meanings of human acts and the processes through which we come to develop and share these subjective meanings. The theory is so named because it studies the symbolic (or subjective) meaning of human interaction.

Symbolic interaction theory addresses the subjective meanings of human acts and the processes through which people come to develop and communicate shared meanings.

Assumptions Underlying Symbolic Interaction Theory

When symbolic interactionists study human behavior, they begin with three major premises (Charon 2006):

1. Symbolic meanings are important. Any behavior, gesture, or word can have multiple interpretations (can symbolize many things). In order to understand human behavior, we must learn what it means to the participants.
2. Meanings grow out of relationships. When relationships change, so do meanings.
3. Meanings are negotiated. We do not accept others' meanings uncritically. Each of us plays an active role in negotiating the meaning that things will have for us and others.

Symbolic Interaction Analysis

These premises direct symbolic interactionists to the study of how individuals are shaped by relationships and social structures. Symbolic interactionists who study violence against women have researched how boys are socialized to consider aggression a natural part of being male through hockey games, dads who tell their sons to fight back when someone makes fun of them, older brothers who physically push them around, and the like. Symbolic interactionists also have explored how teachers unintentionally reinforce the idea that girls are inferior by allowing boys to take over schoolyards and to make fun of girls in the classroom. All these experiences, some researchers believe, set the stage for later violence against women.

Symbolic interactionists are also interested in the active role individuals play in modifying and negotiating relationships. Why do two children raised in the same family turn out differently? In part, because each child experiences subtly different relationships and situations even within the same family, and each may derive different meanings from those experiences.

Most generally, symbolic interaction is concerned with how individuals are shaped by relationships. This question leads first to a concern with childhood and the initial steps we take to learn and interpret our social worlds. It is also concerned with later relationships with lovers, friends, employers, teachers, and others. The strength of symbolic interactionism is that it focuses attention on how larger social structures affect our everyday lives, sense of self, and interpersonal relationships and encounters.

Interchangeable Lenses

Neither symbolic interaction theory, conflict theory, nor structural-functional theory is complete in itself. Together, however, they provide a valuable set of tools for understanding the relationship between the individual and society. These three sociological theories can be regarded as interchangeable lenses through which society may be viewed. Just as a telephoto lens is not always superior to a wide-angle lens, one sociological theory will not always be superior to another.

Occasionally, the same subject can be viewed through any of these perspectives. One will generally get better pictures, however, by selecting the theoretical perspective that is best suited to the particular subject. In general, structural functionalism and conflict theory are well suited to the study of social structures, or **macrosociology.** Symbolic interactionism is well suited to the study of the relationship between individual meanings and social structures, or **microsociology.** Following are three "snapshots" of female prostitution taken through the theoretical lens of structural-functional, conflict, and symbolic interaction theory.

Macrosociology focuses on social structures and organizations and the relationships between them.

Microsociology focuses on interactions among individuals.

The Functions of Prostitution

The functional analysis of female prostitution begins by examining its social structure. It identifies the recurrent patterns of relationships among pimps, prostitutes, and customers. Then it examines the consequences of this social structure. In 1961 Kingsley Davis listed the following functions of prostitution:

- It provides a sexual outlet for men who cannot compete on the marriage market— the physically or mentally handicapped or the very poor.
- It provides a sexual outlet for men who are away from home a lot, such as salesmen and sailors.
- It provides a sexual outlet for those with unusual tastes.

Some sociologists view prostitution as an outgrowth of poverty and sexism; others consider it voluntary and functional for society. Still others are interested in understanding how prostitutes maintain a positive identity in a stigmatized occupation.

Provision of these services is the manifest or intended function of prostitution. Davis goes on to note that, by providing these services, prostitution has the latent function of protecting the institution of marriage from malcontents who, for one reason or another, do not receive adequate sexual service through marriage. Prostitution is the safety valve that makes it possible to restrict respectable sexual relationships (and hence childbearing and child rearing) to marital relationships while still allowing for the variability of human sexual appetites.

Prostitution: Unequal Access to Resources

Conflict theorists analyze prostitution as part of the larger problem of unequal access to resources. Women, they argue, have not had equal access to economic opportunity. In some societies, they are forbidden to own property; in others, they suffer substantial discrimination in opportunities to work and earn. Because of this inability to support themselves, women have had to rely on economic support from men. They get this support by exchanging the one scarce resource they have to offer: sexual availability. To a conflict theorist, it makes little difference whether a woman barters her sexual availability through prostitution or through marriage. The underlying cause is the same.

Although most analyses of prostitution focus on adult women, the conflict perspective helps explain the growing problem of prostitution among runaway and homeless boys and girls. These young people have few realistic opportunities to support themselves by regular jobs: Many are not old enough to work legally and, in any case, would be unable to support themselves adequately on the minimum wage. Their young bodies are their most marketable resource.

Prostitution: Managing Self-Concepts

Symbolic interactionists who examine prostitution take an entirely different perspective. They want to know, for example, how prostitutes learn the trade and how they manage their self-concept so that they continue to think positively of themselves despite engaging in a socially disapproved profession. One such study was done by

Wendy Chapkis (1997), who interviewed more than fifty women "sex workers"—prostitutes, call girls, pornographic actresses, and others. Many of the women she interviewed felt proud of their work. They felt that the services they offered were not substantially different from those offered by day-care workers or psychotherapists, who are also expected to provide services while acting as if they like and care for their clients. Chapkis found that as long as prostitutes are able to keep a healthy distance between their emotions and their work, they can maintain their self-esteem and mental health. As one woman described: "Sex work hasn't all been a bed of roses and I've learned some painful things. But I also feel strong in what I do. I'm good at it and I know how to maintain my emotional distance. Just like if you are a fire fighter or a brain surgeon or a psychiatrist, you have to deal with some heavy stuff and that means divorcing yourself from your feelings on a certain level. You just have to be able to do that to do your job" (Chapkis, 79).

As these examples illustrate, many topics can be fruitfully studied with any of the three theoretical perspectives. Each sociologist must decide which perspective will work best for a given research project.

Researching Society

The things that sociologists study—for example, deviance, marital happiness, and poverty—have probably interested you for a long time. You may have developed your own opinions about why some people have good marriages and some have bad marriages or why some people break the law and others do not. Sociology is an academic discipline that critically examines common sense explanations of human social behavior. It aims to better understand the social world by observing and measuring what actually happens.

This is not the only means of acquiring knowledge. Some people learn what they need to know from the Bible or the Koran or the Book of Mormon. Others get their answers from their mothers, their spouses, or the television. When you ask such people, "But how do you know that that is true?" their answer is simple: "My mother told me" or "I read it in *Reader's Digest*."

Sociology differs from these other ways of knowing in that it requires empirical evidence that can be confirmed by the normal human senses. We must be able to see, hear, smell, or feel it. Before social scientists would agree that they "knew" religious intermarriage increased the likelihood of divorce, for example, they would want to see statistical evidence.

All research has two major goals: accurate description and accurate explanation. In sociology, we are concerned with accurate descriptions of human interaction (how many people marry, how many people abuse their children, how many people flunk out of school). After we know the patterns, we hope to be able to explain them, to say why people marry, abuse their children, or flunk out of school.

The Research Process

At each stage of the research process, scholars use certain conventional procedures to ensure that their findings will be accepted as scientific knowledge. The procedures used in sociological research are covered in depth in classes on research methods, statistics, and theory construction. At this point, we merely want to introduce a few ideas that you must understand if you are to be an educated consumer of research re-sults. We look at the five steps of the general research process, and in doing so review three concepts central to research: variables, operational definitions, and sampling.

Step One: Stating the Problem

The first step in the research process is carefully stating the issue to be investigated. We may select a topic because of a personal experience or out of common sense observa-tion. For example, we may have observed that African Americans appear more likely to experience unemployment and poverty than do white Americans. Alternatively, we might begin with a theory that predicts, for instance, that African Americans will have higher unemployment and poverty rates than white Americans because they have been discriminated against in schools and the workplace. In either case, we begin by review-ing the research of other scholars to help us specify exactly what it is that we want to know. If a good deal of research has already been conducted on the issue and good the-oretical explanations have been advanced for some of the patterns, then a problem may be stated in the form of a **hypothesis**—a statement about relationships that we expect to observe if our theory is correct. A hypothesis must be testable; that is, there must be some way in which data can help weed out a wrong conclusion and identify a correct one. For example, the *belief* that whites deserve better jobs than African Americans cannot be tested, but the *hypothesis* that whites receive better job offers than African Americans can be tested.

> A **hypothesis** is a statement about relationships that we expect to find if our theory is correct.

Step Two: Setting the Stage

Before we can begin to gather data, we first have to set the stage by selecting variables, defining our terms, and deciding exactly which people (or objects) we will study.

Understanding Variables

To narrow the scope of a problem to manageable size, researchers focus on variables rather than on people. **Variables** are measured characteristics that vary from one in-dividual, situation, or group to the next (Babbie 2004). If we wish to analyze differ-ences in rates of African American/white unemployment, we need information on two variables: race and unemployment. The individuals included in our study would be complex and interesting human beings, but for our purposes, we would be inter-ested only in these two aspects of each person's life.

> **Variables** are measured character-istics that vary from one individ-ual or group to the next.

When we hypothesize a cause-and-effect relationship between two variables, the cause is called the **independent variable,** and the effect is called the **dependent variable.** In our example, race is the independent variable, and unemployment is the dependent variable; that is, we hypothesize that unemployment *depends on* one's race.

> The **independent variable** is the cause in cause-and-effect relation-ships.

Defining Variables

In order to describe a pattern or test a hypothesis, each variable must be precisely de-fined. Before we can describe racial differences in unemployment rates, for instance, we need to be able to decide whether an individual is unemployed. The exact proce-dure by which a variable is measured is called an **operational definition.** Reaching general agreement about these definitions may pose a problem. For instance, people are typically considered to be unemployed if they are actively seeking work but can-not find it. This definition ignores all the people who became so discouraged in their search for work that they simply gave up. Obviously, including discouraged workers in our definition of the unemployed might lead to a different description of patterns

> The **dependent variable** is the effect in cause-and-effect relation-ships. It is dependent on the ac-tions of the independent variable.

> An **operational definition** de-scribes the exact procedure by which a variable is measured.

of unemployment. We always need to carefully check what operational definitions were used when we evaluate others' published research results.

Sampling

It would be time consuming, expensive, and probably impossible to get information on race and employment status for all adults. It is also unnecessary. The process of **sampling**—taking a systematic selection of representative cases from a larger population—allows us to get accurate empirical data at a fraction of the cost that examining all possible cases would involve.

Sampling is a systematic selection of representative cases from the larger population.

Sampling involves two processes: (1) obtaining a list of the population you want to study and (2) selecting a representative subset or sample from the list. The best samples are **random samples.** Random samples are chosen through a random procedure, such as tossing a coin, that ensures that every individual within a given population has an equal chance of being selected for the sample.

Random samples are samples chosen through a random procedure, so that each individual has an equal chance of being selected.

Once we have a list of the population, randomly selecting a sample is fairly easy. But getting such a list can be difficult or even impossible. A central principle of sampling is that a sample is only representative of the list from which it is drawn. If we draw a list of people from the telephone directory, then our sample can only be said to describe households listed in the directory; it will omit those with unlisted numbers, those with no telephones, those who use only cell phones, and those who have moved since the directory was issued. The best surveys begin with a list of all the households, individuals, or telephone numbers in a target region or group.

Step Three: Gathering Data

There are many ways of gathering sociological data, including running experiments, conducting surveys, and observing groups in action. Because this is a complex subject, we'll return to it after this discussion of the basic steps in sociological research.

Step Four: Finding Patterns

The fourth step in the research process is to analyze the data, looking for patterns. If we study unemployment, for example, we will find that African Americans are twice as likely as white Americans to experience unemployment (U.S. Bureau of the Census 2006). This finding notes a **correlation,** an empirical relationship between two variables—in this case, between race and employment.

Correlation occurs when there is an empirical relationship between two variables.

Step Five: Generating Theories

After a pattern is found, the next step in the research process is to explain it. As we will discuss in the section entitled "The Survey," finding a correlation between two variables does not necessarily mean that one variable causes the other. For example, even though there is a correlation between race and unemployment, not all African Americans are unemployed, and being African American is not the only cause of unemployment. Nevertheless, if we have good empirical evidence that being black increases the *probability* of unemployment, the next task is to explain why that should be so. Explanations are usually embodied in a **theory,** an interrelated set of assumptions that explains observed patterns. Theory always goes beyond the facts at hand; it includes untested assumptions that explain the empirical evidence.

A **theory** is an interrelated set of assumptions that explains observed patterns.

In our unemployment example, we might theorize that the reason African Americans face more unemployment than whites is because many of today's African

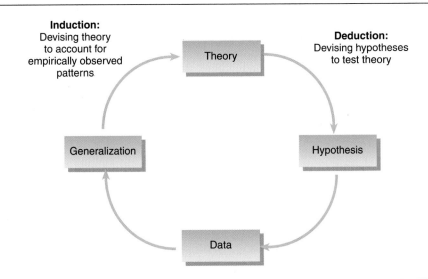

Induction:
Devising theory to account for empirically observed patterns

Theory

Deduction:
Devising hypotheses to test theory

Generalization

Hypothesis

Data

FIGURE 1.2
The Wheel of Science
The process of science can be viewed as a continuously turning wheel that moves us from data to theory and back again.
SOURCE: Adapted from Wallace, Walter. 1969. Sociological Theory. Chicago: Aldine.

American adults grew up in a time when the racial difference in educational opportunity was much greater than it is now. This simple explanation goes beyond the facts at hand to include some assumptions about how education is related to race and unemployment. Although theory rests on an empirical generalization, the theory itself is not empirical; it is . . . well, theoretical.

It should be noted that many different theories can be compatible with a given empirical generalization. We have proposed that educational differences explain the correlation between race and unemployment. An alternative theory might argue that the correlation arises because of discrimination. Because there are often many plausible explanations for any correlation, theory development is not the end of the research process. We must go on to test the theory by gathering new data.

The scientific process can be viewed as a wheel that continuously moves us from theory to data and back again (Figure 1.2). Two examples illustrate how theory leads to the need for new data and how data can lead to the development of new theory.

As we have noted, data show that unemployment rates are higher among African Americans than among white Americans. One theoretical explanation for this pattern links higher African American unemployment to educational deficits. From this theory, we can deduce the hypothesis that African Americans and whites of equal education will experience equal unemployment. To test this hypothesis, we need more data, this time about education and its relationship to race and unemployment.

A study by Lori Reid (2002) tests this hypothesis for black women. Reid asked whether educational deficits explained why African American women are more likely to lose their jobs than are whites. She found that education does play a small role. However, other factors played larger roles in explaining black unemployment. Those factors, including black women's segregation in vulnerable occupations and residence in regions where unemployment was rising, were considerably more important predictors of unemployment.

Reid's findings can be the basis for revised theories. These new theories will again be subject to empirical testing, and the process will begin anew. In the language of science, the process of moving from data to theory is called **induction,** and the process of moving from theory to data is called **deduction.** These two processes and their interrelationships are also illustrated in Figure 1.2.

Induction is the process of moving from data to theory by devising theories that account for empirically observed patterns.

Deduction is the process of moving from theory to data by testing hypotheses drawn from theory.

Research Strategies

The theories and findings reported in this book represent a variety of research strategies. Three general strategies are outlined here: experiments, survey research, and participant observation. In this section, we review each method and illustrate its advantages and disadvantages by showing how it would approach the test of a common hypothesis, namely, that alcohol use reduces grades in college.

The Experiment

The **experiment** is a method in which the researcher manipulates independent variables to test theories of cause and effect.

An **experimental group** is the group in an experiment that experiences the independent variable. Results for this group are compared with those for the control group.

A **control group** is the group in an experiment that does not receive the independent variable.

The **guinea-pig effect** occurs when subjects' knowledge that they are participating in an experiment affects their response to the independent variable.

The **experiment** is a research method in which the researcher manipulates the independent variable to test theories of cause and effect. In the classic experiment, a group that experiences the independent variable, an **experimental group,** is compared with a **control group** that does not. If the groups are equal as far as everything else, a comparison between them will show whether experience with the independent variable is associated with changes in the dependent variable.

An experiment designed to assess whether alcohol use affects grades, for example, would need to compare an experimental group that drank alcohol with a control group that did not. A hypothetical experiment might begin by observing student grades to assess students' typical performance levels. Then the class would be divided randomly into two groups. If the initial pool is large enough, we can assume that the two groups are probably similar on nearly everything. For example, we assume that both groups probably contain an equal mix of good and poor students and of lazy and ambitious students. The control group might be requested to abstain from alcohol use for 5 weeks, and the experimental group might be requested to drink three times a week during the same period. At the end of the 5 weeks, we would compare the grades of the two groups. Both groups might have experienced a drop in grades because of normal factors such as fatigue, burnout, and overwork. The existence of the control group, however, allows us to determine whether alcohol use causes a reduction in grades beyond that which normally occurs.

Experiments are excellent devices for testing hypotheses about cause and effect. They have three drawbacks, however. First, experiments are often unethical because they expose subjects to the possibility of harm. The study on alcohol use, for example, might damage student grades, introduce students to bad habits, or otherwise injure them. Because of such ethical issues, many areas of sociological interest cannot be studied with the experimental method.

A second drawback to experiments is that subjects often behave differently when they are under scientific observation than they would in their normal environment. For example, although alcohol consumption might normally lower student grades, the participants in our study might find the research so interesting that their grades would actually improve. In this case, the subjects' knowledge that they are participating in an experiment affects their response to the independent variable. This response is called the **guinea-pig effect.**

A final drawback to the experimental method is that laboratory experiments are often highly artificial. When researchers try to set up social situations in laboratories, they often must omit many of the factors that would influence the same behavior in a real-life situation. For example, the behavior of students in an experiment on drinking would likely also be affected by whether or not their friends encouraged them to drink. The result is often a very unnatural situation. Like the guinea-pig effect, this artificiality has the effect of reducing our confidence that the results that appear in the experiment can be generalized to the more complex conditions of the real world. Because of these problems, few sociologists use experiments.

The Survey

The most common research strategy in sociology is the survey. In **survey research,** the investigator asks a relatively large number of people the same set of standardized questions. These questions may be asked in a personal interview, over the telephone, online, or in a paper-and-pencil format. Because it asks the same questions of a large number of people, it is an ideal method for furnishing evidence on incidence, trends, and differentials. Thus, survey data on alcohol use may allow us to say such things as the following: 80 percent of the undergraduates of Midwestern State currently use alcohol (**incidence**); the proportion using alcohol has remained about the same over the last 10 years (**trend**); and the proportion using alcohol is higher for males than for females (**differential**). Survey research is extremely versatile; it can be used to study attitudes, behavior, ideals, and values. If you can think of some way to ask a question about such matters, then you can study it with survey research.

Most surveys use what is called a **cross-sectional design;** they take a sample (or cross section) of the population at a single point in time and look at how groups differ on the independent and the dependent variables. Thus, in our study of alcohol use we would take a sample of students, expecting to find that some of them drink and some do not. We could then compare these two groups to see which gets the best grades.

In 1997 we conducted such a survey of our undergraduate students. The results are displayed in Table 1.2. The table shows that students who drink alcohol report getting worse grades: Although 51 percent of the nondrinkers have grade point averages above 3.5, only 28 percent of those who admit drinking do. At the other end of the grade distribution, drinkers were more likely than abstainers to report a grade point average of less than 2.5.

The difficulty with the cross-sectional design is that we cannot reach any firm conclusions about cause and effect. We cannot tell whether alcohol use causes bad grades or whether bad grades lead to alcohol use. A more striking problem is that we cannot be sure there is a cause-and-effect relationship at all. Because the drinkers and the abstainers were not randomly assigned to the two categories, the two groups differ on many other variables besides drinking. For example, the drinkers may have less conventional families, come from worse neighborhoods, or simply hate reading. It could be that one of these factors is causing both the poor grades and the drinking.

Survey research is a method that involves asking a relatively large number of people the same set of standardized questions.

Incidence is the frequency with which an attitude or behavior occurs.

A **trend** is a change in a variable over time.

A **differential** is a difference in the incidence of a phenomenon across subcategories of the population.

A **cross-sectional design** uses a sample (or cross section) of the population at a single point in time.

TABLE 1.2
Cross-Tabulation of College Grades by Use of Alcohol
556 undergraduate students at a midwestern state university, 1997.

College Grades	Use Alcohol	
	No	Yes
B+ or A	51%	28%
C+ or B	42	60
Below C+	7	12
Total	100%	100%
Number	114	452

■ Survey research is an excellent way of finding the relationship between two variables, such as whether drinking affects grades among college students.

© David Young Wolff/PhotoEdit

Longitudinal research is any research in which data are collected over a long period of time.

A different strategy used in survey research is **longitudinal research,** in which a researcher collects the same sort of data over a long period of time. Researchers either interview the same group of individuals multiple times (perhaps every month, perhaps every 5 years) or interview different groups, each randomly selected from the same population but weeks, months, or years apart.

Longitudinal research has definite advantages. If we study the same individuals over time, we could tell, for example, whether students' grades began falling before or after they began drinking alcohol. If we study different (but similar) groups over time, it is harder to reach conclusions about causes, but we can easily see trends in behavior over time. The major disadvantage of a longitudinal design is, of course, that it is more expensive and time consuming than studying a group at only one point in time.

Another important drawback of survey research in general is that respondents may misrepresent the truth. Both those who drink heavily *and* those who don't drink at all may lie about their habits out of fear of what others may think of them. Such misrepresentation is known as **social-desirability bias**—the tendency for people to color the truth so that they appear to be nicer, richer, and generally more desirable than they really are. The consequences of this bias vary in seriousness depending on the research aim and topic. Obviously, it is a greater problem for sensitive topics such as alcohol use or prejudice. In the study just described, students who care strongly how others view them will be more likely both to work to get good grades and to under-report their drinking.

Social-desirability bias is the tendency of people to color the truth so that they sound more desirable and acceptable than they really are.

Survey research is designed to obtain standard answers to standard questions. It is not the best strategy for studying deviant or undesirable behaviors or for examining ideas and feelings that cannot easily be reduced to questionnaire form. An additional drawback of survey research is that it is designed to study individuals rather than contexts. Thus, it focuses on the individual alcohol user or abstainer rather than on the setting and relationships in which drinking takes place. For these kinds of answers, we must turn to participant observation.

A Global Perspective

Survey Research in Nigeria

"How do you feel about the current government?" "Do you think men and women ought to be treated equally?" "How many televisions do you own?" Without much effort, we can imagine places in the world where questions like these that are so commonplace to us would appear foolish and perhaps dangerous to ask.

In the United States and much of the developed world, we are accustomed to questions about the most intimate details of our lives. It is common for bureaucratic agencies to record data pertaining to our height, weight, IQ test scores, taxes, fertility, and credit rating. Our acceptance of intrusive questioning is based on the assumption that this information is somehow necessary to good governance and the trust that our privacy will be protected. Because we are familiar with routine bureaucratic data gathering, it is relatively easy for survey researchers to enlist our cooperation. When asked personally, most Americans will respond to survey researchers' inquiries on topics ranging from politics to religion. (The only question Americans routinely refuse to answer is, "What is your income?")

When survey research is conducted in developing nations, many of these conditions do not hold. Often people there have good reason not to trust their governments. More generally, they simply are not accustomed to opening their private lives to the probing of bureaucratic agents.

One of the most common areas in which Western survey research methods meet resistance from people in developing nations is research on fertility (birth rates) and family planning. Agencies such as the United Nations and the U.S. Agency for International Development wish to know how many children women in Kenya, Nigeria, and other high-fertility nations want so that they can learn whether the women might be interested in contraception. Agnes Riedmann's (1993) analysis of a survey research project among the Yoruba of Nigeria highlights several cultural clashes that can impede such research efforts.

- Among the Yorubans, as in many non-Western cultures, fertility is considered to be "up to God." It would be unthinkable for individuals to put their own opinions forward. A question such as, "If you had more money, would you rather have a new car or another child?" presumes that individuals have a choice about fertility and, moreover, that dollar values can be assigned to children.
- Asking a woman's opinion is often considered indecent (unless her husband is present) or simply a waste of time, because women's opinions do not count.
- Yorubans hold a profound distrust of strangers who ask personal questions. After submitting to an ill-understood interview on fertility, for example, one Yoruban respondent asked whether the police were now coming to take her away. In many cases Yoruban respondents mocked, yelled at, and ran from interviewers. The persistent inquiry of strangers into their private lives seemed to some to be just one more instance of the crazy behavior of *oyinbos,* or white people. Other Yorubans speculated that the researchers were asking these questions simply because they didn't have enough to do! If badgered into participation, Yorubans politely told interviewers whatever they thought the interviewers wanted to hear.

■ Although mothers and mothering are universal, the meanings attached to children and to childbearing are not. Survey research on fertility and family planning may give seriously misleading results in nations such as Nigeria where conception is viewed as being "up to God," not "up to the individual," and answering questions about personal topics is considered indecent, if not downright dangerous.

Participant Observation

Participant observation includes a variety of research strategies—participating, interviewing, observing—that examine the context and meanings of human behavior.

Under the label **participant observation,** we classify a variety of research strategies—participating, interviewing, observing—that examine the context and meanings of human behavior. Instead of sending forth an army of interviewers, participant observers go out into the field themselves to see firsthand what is going on. These strategies are used more often by sociologists interested in symbolic interaction theory—that is, researchers who want to understand subjective meanings, personal relationships, and the process of social life. The goals of this research method are to discover patterns of interaction and to find the meaning the patterns hold for individuals.

The three major tasks involved in participant observation are interviewing, participating, and observing. A researcher goes to the scene of the action, where she may interview people informally in the normal course of conversation, participate in whatever they are doing, and observe the activities of other participants. Not every participant observation study includes all three dimensions equally. A participant observer studying alcohol use on campus, for example, would not need to get "smashed" every night. She would, however, probably do long, informal interviews

Research Methods

Concept Summary

Controlled Experiments	
Procedure	Dividing subjects into two equivalent groups, applying the independent variable to one group only, and observing the differences between the two groups on the dependent variable
Advantages	Excellent for analysis of cause-and-effect relationships; can simulate events and behaviors that do not occur outside the laboratory in any regular way
Disadvantages	Based on small, nonrepresentative samples examined under highly artificial circumstances; unclear that people would behave the same way outside the laboratory; unethical to experiment in many areas
Survey Research	
Procedure	Asking the same set of standard questions of a relatively large, systematically selected sample
Advantages	Very versatile—can study anything that we can ask about; can be done with large, random samples so that results represent many people; good for incidence, trends, and differentials
Disadvantages	Shallow—does not get at depth and shades of meaning; affected by social-desirability bias; better for studying people than situations
Participant Observation	
Procedure	Observing people's behavior in its normal context; experiencing others' social settings as a participant; in-depth interviewing
Advantages	Seeing behavior in context; getting at meanings associated with behavior; seeing what people do rather than what they say they do
Disadvantages	Limited to small, nonrepresentative samples; dependent on interpretation of single investigator

To understand the beliefs and motives of those involved in stigmatized activities, such as this Ku Klux Klan march, researchers usually rely on participant observation as well as surveys.

© AP/Wide World Photos

with both users and nonusers, attend student parties and activities, and attempt to get a feel for how alcohol use fits in with certain student subcultures.

In some cases, participant observation is the only reasonable way to approach a subject. This is especially likely when we are examining (1) stigmatized behaviors or populations, (2) real behavior rather than attitudes, or (3) alienated populations.

When we want to study stigmatized behaviors, social-desirability bias can make it difficult to get good information on surveys. A participant-observer who takes the time to get to know a group, conduct in-depth interviews, or collect long-term observations is more likely to obtain accurate information than a researcher who simply distributes a survey. Thus, what we know about the lives of topless dancers, illegal drug users, and numbers runners rests heavily on the reports of participant observers.

Participant observation is also well suited to studying what people actually do as opposed to what they say they do. Behaviors are sometimes misrepresented in surveys simply because people are unaware of their actions or don't remember them very well. For example, individuals may believe they are not prejudiced; yet observational research might demonstrate that these same people systematically choose not to sit next to persons of another race on the bus or in public places. Sometimes, actions speak louder than words.

Finally, participant observation is often the only reliable way to obtain information about alienated populations. Groups that are alienated because they lack the skills or resources to participate in mainstream society, such as those who are illiterate, are also likely to lack the skills needed to accurately answer surveys. Groups that are alienated because they have chosen to reject mainstream society—gang members, skinheads, or rioters, for example—are not likely to agree to answer surveys.

A major disadvantage of participant observation is that it is usually based on small numbers of individuals who have not been selected according to random-sampling techniques. The data tend to be unsystematic and the samples not very representative. However, we do learn a great deal about the few individuals involved. This detail is often useful for generating ideas that can then be examined more systematically with other techniques. For this reason, participant observation may be viewed as a form of initial exploration of a research topic.

Another disadvantage of participant observation is that the observations and generalizations rely on the interpretation of one investigator. Because researchers are not robots, it seems likely that their findings reflect some of their own world view. This is a greater problem with participant observation than with survey or experimental work, but all science suffers to some extent from this phenomenon. The answer to this dilemma is **replication,** redoing the same study with another investigator or with different samples to see if the same results occur.

Replication is the repetition of empirical studies to see if the same results occur.

Alternative Strategies

The bulk of sociological research uses these three strategies. There are, however, a dozen or more other imaginative and useful methods sociologists use to do research, many of them involving the analysis of social artifacts rather than people. For example, Warren Whisenant, John Miller, and Paul M. Pedersen (2005) used job descriptions for athletic directors from 112 Texas school districts to explore how those descriptions intentionally or unintentionally suggested that women need not apply, and Terri A. Winick (2005) used articles from medical journals to illustrate how the medical profession has responded to the threat presented by the rise in alternative healing. Studies of other government records and statistics have demonstrated incidence, trends, and differentials in many areas of sociological interest.

Sociologists: What Do They Do?

A concern with social problems has been a continuing focus of U.S. sociology. This is evident both in the kinds of courses that sociology departments offer (social problems, race and ethnic relations, crime and delinquency, for example) and in the kinds of research sociologists do. That research now takes place in a wide variety of settings.

Working in Colleges and Universities

About three quarters of U.S. sociologists are employed in colleges and universities, where they are required both to teach and to do research. Much of this research is basic sociology, which has no immediate practical application and is motivated simply by a desire to describe or explain some aspect of human social behavior more fully. Even basic research, however, often has implications for social policy. In addition, many sociology professors work closely with government, businesses, or nonprofit organizations to provide immediate practical answers to social problems. For example, sociology professors who study disasters played crucial roles in helping the government, nonprofit organizations, and communities understand and respond to the environmental damage caused by Hurricane Katrina. University of New Orleans sociologist Shirley Laska, who had predicted New Orleans's vulnerability to hurricanes in a widely-cited report published a year before Katrina, is a particularly good example.

Working in Government

A long tradition of sociological work in government has to do with measuring and forecasting population trends. This work is vital for decisions about where to put roads and schools and when to stop building schools and start building nursing homes. In addition, sociologists have been employed to design and evaluate public policies in a wide variety of areas. In World War II, sociologists designed policies to increase the morale and fighting efficiency of the armed forces, and for decades, sociologists have helped plan and evaluate programs to reduce poverty.

Sociologists work in nearly every branch of government. For example, sociologists are employed by the Centers for Disease Control and Prevention (CDC), where they examine how social relationships are related to the transmission of AIDS, why intravenous drug users share needles, and why AIDS is transmitted along chains of sexual partners. While the physicians and biologists of the CDC examine the medical aspects of AIDS, sociologists work to understand the social aspects.

Working in Business

Sociologists are employed by General Motors and Pillsbury as well as by advertising and management consulting firms. Some work in internal affairs (bureaucratic structures and labor relations), but many others conduct market research. Business and industry employ sociologists so that they can use their knowledge of society to predict which way consumer demand is likely to jump. For example, the greater incidence today of single-person households has important implications for life insurance companies, food packagers, and the construction industry. To stay profitable, companies need to be able to predict and plan for such trends. Sociologists are also extensively involved in the preparation of environmental impact statements, in which they try to assess the likely impact of, say, a new coal mine on the social and economic fabric of a community.

Working in Nonprofit Organizations

Nonprofit organizations range from hospitals and clinics to social-activist organizations and private think tanks; sociologists are employed in all of them. Sociologists at the American Foundation for AIDS Research, for example, are interested in determining the causes and consequences of unsafe sexual activity, evaluating communication strategies that can be used to encourage condom use, and devising effective strategies to pursue more controversial approaches, such as distributing clean needles to addicts to prevent the transmission of HIV.

The training that sociologists receive has a strong research orientation and is very different from the therapy-oriented training received by psychologists and social workers. Nevertheless, a thorough understanding of the ways that social structures impinge on individuals can be useful in helping individuals cope with personal troubles. Consequently, some sociologists also do marriage counseling, family counseling, and rehabilitation counseling. Some of these sociologists work in organizations, others in private practice.

Working to Serve the Public

Although most sociologists are committed to a value-free approach to their work as scholars, many are equally committed to changing society for the better, whether they work in government, business, nonprofit organizations, or academia. They see sociology as a "calling"—work that is inseparable from the rest of one's life and driven by a sense of moral responsibility for people's welfare (Yamane 1994). As a result, sociologists have served on a wide variety of public commissions and in public offices to affect social change. They work for change independently, too, both as individuals and in organizations such as Sociologists Without Borders (www.sociologistswithoutborders .org), which is "committed to advancing transnational solidarities and justice."

Perhaps the clearest example of sociology in the public service is the award of the 1982 Nobel Peace Prize to Swedish sociologist Alva Myrdal for her unflagging efforts to increase awareness of the dangers of nuclear armaments. Value-free scholarship does not mean value-free citizenship.

Where This Leaves Us

Sociology is a diverse and exciting field. From its beginnings in the nineteenth century, it has grown into a core social science that plays a central role in university education. Its three major perspectives—structural functionalism, conflict theory, and symbolic interactionism—provide a complementary set of lenses for viewing the world, while its varied methodological approaches provide the tools needed to study social life in all its complexity. These lenses and tools position sociologists not only to understand the world, but to help change it for the better.

Summary

1. Sociology is the systematic study of social behavior. Sociologists use the concepts of role and social structure to analyze common human dramas. Learning to understand how individual behavior and personal troubles are affected by social structures is the process of developing the "sociological imagination."

2. The rapid social change that followed the industrial revolution was an important inspiration for the development of sociology. Problems caused by rapid social change stimulated the demand for accurate information about social processes. This social-problems orientation remains an important aspect of sociology.

3. Sociology has three major theoretical perspectives: structural-functional theory, conflict theory, and symbolic interaction theory. The three can be seen as alternative lenses through which to view society, with each having value as a tool for understanding how social structures shape human behavior.

4. Structural functionalism has its roots in evolutionary theory. It identifies social structures and analyzes their consequences for social harmony and stability. Identification of manifest and latent functions and dysfunctions is part of its analytic framework.

5. Conflict theory developed from Karl Marx's ideas about the importance of conflict and competition in structuring human behavior and social life. It analyzes social structures by asking who benefits and how these are benefits maintained. It assumes that competition is more important than consensus and that change is a positive result of conflict.

6. Symbolic interaction theory examines the subjective meanings of human interaction and the processes through which people come to develop and communicate shared symbolic meanings. Whereas structural functionalism and conflict theory emphasize macrosociology, symbolic interactionism focuses on microsociology.

7. Sociology is a social science. This means it relies on critical and systematic examination of the evidence before reaching any conclusions and that it approaches each research question from a position of neutrality. This is called value-free sociology.

8. The five steps in the research process are stating the problem, setting the stage, gathering the data, finding patterns, and generating theory. These steps form a continuous loop called the "wheel of science." The movement from data to theory is called induction, and the movement from theory to hypothesis to data is called deduction.

9. A design for gathering data depends on identifying the variables under study, agreeing on precise operational definitions of these variables, and obtaining a representative sample of cases in which to study relationships among the variables.

10. The experiment is a method designed to test cause-and-effect hypotheses. Although it is the best method for this purpose, it has three disadvantages: ethical

problems, the guinea-pig effect, and highly artificial conditions. It is most often used for small-group research and for simulation of situations not often found in real life.

11. Survey research is a method that asks a large number of people a set of standard questions. It is good for describing incidence, trends, and differentials for random samples, but it is not as good for describing the contexts of human behavior or for establishing causal relationships.

12. Participant observation is a method in which a small number of individuals who are not randomly chosen are observed or interviewed in depth. The strength of this method is the detail about the contexts of human behavior and its subjective meanings; its weaknesses are poor samples and lack of verification by independent observers.

13. Most sociologists teach and do research in academic settings. A growing minority are employed in government, nonprofit organizations, and business, where they do applied research. Regardless of the setting, sociological theory and research have implications for social policy.

Thinking Critically

1. Which of your own personal troubles might reasonably be reframed as public issues? Does such a reframing change the nature of the solutions you can see?

2. Consider how a structural-functional analysis of gender roles might differ from a conflict analysis. Would men be more or less likely than women to favor a structural-functionalist approach?

3. Can you think of situations in which a change of friends, living arrangements, or jobs has caused you to change your interpretations of the events surrounding you?

4. Consider what study design you could ethically use to determine whether drinking alcohol, living in a sorority, or growing up with a single parent reduces academic performance.

Companion Website for This Book

academic.cengage.com/sociology/brinkerhoff
Gain an even better grasp on this chapter by going to the Companion Website. This resource contains tutorial quizzes and flash cards to help you master key terms and concepts.

Suggested Readings

Babbie, Earl. 1994. *The Sociological Spirit*. Belmont, Calif.: Wadsworth. From a dedicated sociologist who believes that the world and national problems that concern us can best be addressed by sociology; a book of essays that not only introduces the sociological imagination but also motivates the reader to develop it.

Berger, Peter L. 1963. *Invitation to Sociology: A Humanistic Perspective*. Garden City, N.Y.: Doubleday Anchor. A classic introduction to what sociology is and how it differs from other social sciences. It blends a serious exploration of basic sociological understandings with scenes from everyday life.

Mills, C. Wright. 1959. *The Sociological Imagination*. New York: Oxford University Press. A penetrating account of how sociology can expand our understanding of common experiences.

Neuman, W. Lawrence. 2005. *Social Research Methods: Qualitative and Quantitative Approaches*. 6th ed. Boston: Allyn & Bacon. A textbook for undergraduates that covers the major research techniques in sociology. Up-to-date, thorough, and readable.

CHAPTER 2

Culture

© AP/World Wide Photos

Outline

Introduction to Culture

In Chapter 1 we said that sociology is concerned with analyzing the contexts of human behavior and how these contexts affect our behavior. Our neighborhood, our family, and our social class provide part of that context, but the broadest context of all is our culture. **Culture** is the total way of life shared by members of a community.

In some places, a culture cuts across national boundaries. French Canadian people and culture, for example, can be found in both Canada and the New England states. In other places, two distinct cultures may coexist within a single national boundary, as French and English culture do within Canada. For this reason, we distinguish between cultures and societies. A **society** is the population that shares the same territory and is bound together by economic and political ties. Often the members share a common culture, but not always.

Culture resides essentially in nontangible forms such as language, values, and symbolic meanings, but it also includes technology and material objects. A common image is that culture is a "tool kit" that provides us with the equipment necessary to deal with the common problems of everyday life (Swidler 1986). Consider how culture provides patterned activities of eating and drinking. People living in the United States share a common set of tools and technologies in the form of refrigerators, ovens, cell phones, computers, and coffeepots. As the advertisers suggest, we share similar feelings of psychological release and satisfaction when, after a hard day of working or playing, we take a break with a cup of coffee or a cold beer. The beverages we choose and the meanings attached to them are part of our culture. Despite many shared meanings and values, however, this example also illustrates some of the difficulties inherent in any discussion of a single common culture: Although Mormon Americans and Muslim Americans share our American culture, the former do not drink coffee and neither group drinks alcohol.

Culture can be roughly divided into two categories: material and nonmaterial. *Nonmaterial culture* consists of language, values, rules, knowledge, and meanings shared by the members of a society. *Material culture* includes the physical objects that a society produces—tools, streets, sculptures, and toys, to name but a few. These material objects depend on the nonmaterial culture for meaning. For example, Barbie dolls and figurines of fertility goddesses share some common physical features, but their meaning differs greatly and depends on nonmaterial culture.

Culture is the total way of life shared by members of a community. It includes not only language, values, and symbolic meanings but also technology and material objects.

A **society** is the population that shares the same territory and is bound together by economic and political ties.

Theoretical Perspectives on Culture

As is true in other areas of sociology, structural functionalists, conflict theorists, and symbolic interactionists each have their own approach to the study of culture.

The Structural-Functionalist Approach

The structural-functionalist approach treats culture as the underlying basis of interaction. It accepts culture as a given and is more interested in how culture shapes us than in how culture itself is shaped. Scholars taking this approach have concentrated on illustrating how norms, values, and language guide our behavior. We will return to this topic later when we discuss the carriers of culture.

The Conflict Theory Approach

In contrast, conflict theorists focus on culture as a social product. They ask why culture develops in certain ways and not others, and whose interests are served by these patterns. These scholars would be interested, for example, in how the content of television shows is affected by who owns the television stations, and how society is affected when stations are owned by the government versus corporations.

Conflict theorists are also interested in how culture can reinforce power divisions within society. They note that money—or *capital*, in economists' terms—brings with it power and status, and that cultural capital does the same (Bourdieu 1984; Lamont & Fournier 1992). **Cultural capital** refers to the attitudes and knowledge that characterize the upper social classes. If you never learned to play golf, choose a red wine, appreciate an opera, or eat a five-course meal with five different forks, you will stand out like a sore thumb at upper class events. You lack some of the cultural capital needed to marry into or work in these social circles and may be ridiculed by others if you try. In this way, culture serves as a *symbolic boundary* that keeps the social classes apart.

Finally, conflict theorists are interested in what happens when cultures come into conflict with each other. We will explore this topic further when we discuss subcultures, countercultures, and the battles over assimilation versus multiculturalism.

Cultural capital refers to having the attitudes and knowledge that characterize the upper social classes, such as knowing how to play golf.

The Symbolic Interactionist Approach

Whereas conflict theorists often focus on *what* is portrayed in the media (How many blacks are in TV shows? Is violence portrayed as fun?), symbolic interactionists are interested in how people *interpret* and *use* what they see. They are interested in the meanings people derive from culture and cultural products, and how those meanings are created in social interaction. For example, research in this tradition has documented how women find empowering messages in romance novels and horror films, how the rise of Viagra has changed the meaning of male sexuality, why people identify with pop music stars, and what "ethnic" foods (Chinese noodles, Italian pastas, southern biscuits) mean both to those who belong to ethnic groups and to outsiders (Loe 2004; Vares & Braun 2006; Vannini 2004; Bai 2003).

Bases of Human Behavior: Culture and Biology

Why do people behave as they do? What determines human behavior? To answer these questions, we must be able to explain both the varieties and the similarities in human behavior. Generally, we will argue that biological factors help explain what is common to humankind across societies, whereas culture explains why people and societies differ from one another.

Cultural Perspective

Regardless of whether they are structural functionalists, conflict theorists, or symbolic interactionists, sociologists share some common orientations toward culture: Nearly all hold that culture is *problem solving*, culture is *relative*, and culture is a *social product*.

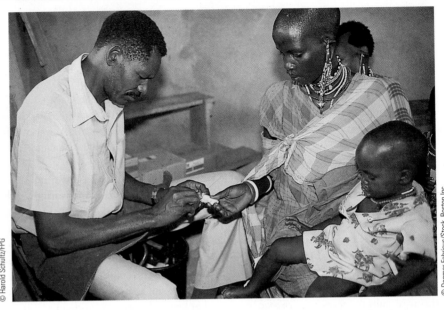

■ Despite the obvious differences between this traditional Brazilian medicine man and the modern Kenyan doctor, a close look shows many underlying similarities in their means of dealing with illness and injury.

Culture Is Problem Solving

Regardless of whether people live in tropical forests or in the crowded cities of New York, London, or Tokyo, they confront some common problems. They all must eat, they all need shelter from the elements (and often from each other), and they all need to raise children to take their place and continue their way of life. Although these problems are universal, the solutions are highly varied. For example, responsibility for child rearing may be assigned to the mother's brother, as is done in the Trobriand Islands; to the natural mother and father, as is done in the United States; or to communal nurseries, as was done in early Israeli kibbutzim.

Whenever people face a recurrent problem, cultural patterns will evolve to provide a ready-made answer. This does not mean that the answer provided is the best answer or the only answer or the fairest answer, but merely that culture provides a standard pattern for dealing with this common dilemma. One of the issues that divides conflict and functional theorists is how these answers develop. Functionalists argue that the solutions we use today have evolved over generations of trial and error and that they have survived because they work, because they help us meet basic needs. A conflict theorist would add that these solutions work better for some people than for others. Conflict theorists argue that elites manipulate culture to rationalize and maintain solutions that work to their advantage. Scholars from both perspectives agree that culture provides ready-made answers for most of the recurrent situations we face in daily life; they disagree in their answer to the question of who benefits from a particular solution.

Culture Is Relative

The solutions that each culture devises may be startlingly different. Among the Wodaabe of Niger, for example, mothers are not allowed to speak directly to their first- or second-born children and, except for nursing, are not even allowed to touch them. The babies' grandmothers and aunts, however, lavish affection and attention on them (Beckwith 1983). The effect of this pattern of child rearing is to emphasize loyalties and affections throughout the entire kinship group rather than just with one's own children or parent. This practice helps ensure that each new entrant will be loyal to the group as a whole.

Is it a good or a bad practice? That is a question we can answer only by seeing how it fits in with the rest of the Wodaabe culture and by taking the viewpoint of one or another social group. Does it help the people meet recurrent problems and maintain a stable society? If so, structural functionalists would say it works; it is functional. Conflict theorists, on the other hand, would want to know which groups are hurt and which are helped by the practice. Both sets of theorists, however, believe that each cultural trait should be evaluated in the context of its own culture. This belief is called **cultural relativity.** A corollary of cultural relativity is that no practice is universally good or universally bad; goodness and badness are relative, not absolute.

> **Cultural relativity** requires that each cultural trait be evaluated in the context of its own culture.

This type of evaluation is sometimes a difficult intellectual feat. For example, no matter how objective we try to be, most of us believe that infanticide, human sacrifice, and cannibalism are absolutely and universally wrong. Such an attitude reflects **ethnocentrism**—the tendency to use the norms and values of our own culture as standards against which to judge the practices of others. Ethnocentrism usually means that we see our way as the right way and everybody else's way as the wrong way. When American missionaries came to the South Sea Islands, they found that many things were done differently in Polynesian culture. The missionaries, however, were unable to view Polynesian practices as simply different. If those practices were not like American practices, then they must be wrong and were probably wicked. As a result, the missionaries taught the islanders that the only acceptable way (the American way) to have sexual intercourse was in a face-to-face position with the man

> **Ethnocentrism** is the tendency to judge other cultures according to the norms and values of our own culture.

© AP/Wide World Photos

■ In 1911, a British team under Robert F. Scott and a Norwegian team under Roald Amundsen raced to become the first explorers ever to reach Antarctica. The British team's ethnocentricism led to its downfall: Scott's team relied on man-hauled sleds and perished, Amundsen's team adapted Inuit dog sleds and skiing techniques and succeeded.

on top, the now-famous "missionary position." They taught the Polynesians that women and men should wear Western clothes even if the clothes don't suit the Polynesian climate, that they should have clocks and come on time to appointments, and a variety of other Americanisms that the missionaries maintained to be morally right behavior.

Ethnocentrism is often a barrier to interaction among people from different cultures, leading to much confusion and misinterpretation. It is not, however, altogether bad. In the sense that it is pride in our own culture and confidence in our own way of life, ethnocentrism is essential for social integration. In other words, we learn to follow the ways of our culture because we believe that they are the right ways; if we did not share that belief, there would be little conformity in society. Ethnocentrism, then, is a natural and even desirable product of growing up in a culture. An undesirable consequence, however, is that we simultaneously discredit or diminish the value of other ways of thinking and feeling. As a result, ethnocentrism can make it difficult for us to change our ways even if change would be in our best interests (Diamond 2005). For example, Norwegian explorers in Antarctica fared far better than did British explorers because the Norwegians adopted Inuit ("Eskimo") clothing, skis, and dog-sleds, whereas the British considered such strategies beneath them—sometimes dying as a result (Huntford 2000).

Culture Is a Social Product

A final assumption sociologists make about culture is that culture is a social, not a biological, product. The immense cultural diversity that characterizes human societies results not from unique gene pools, but from cultural evolution.

Some aspects of culture are produced deliberately. Shakespeare decided to write *Hamlet* and J.K. Rowling to write the Harry Potter books; marketing teams created the Geico gecko and the MacIntosh apple icon. Governments, bankers, and homeowners deliberately commission the designing of homes, offices, and public buildings, and people buy publishing empires so that they can spread their own version of the truth. Other aspects of culture—such as our culture's ideas about right and wrong, its dress patterns, and its language—develop gradually out of social interaction. But all these aspects of culture are human products; none of them is instinctive. People *learn* culture, and, as they use it, they modify it and change it.

Culture depends on language. Only after language emerges can bits of practical knowledge (such as "fire is good" and "don't use electricity in the bathtub") or ideas (such as "God exists") be transmitted effectively from one generation to the next. Inventions, discoveries, and forms of social organization are socially bestowed and intentionally passed on so that each new generation can elaborate on and modify the accumulated knowledge of the previous generations. In short, culture is cumulative only because of language.

Because of language, human beings are not limited to the slow process of genetic evolution in adapting to their circumstances. Cultural evolution is a uniquely human way for a species to adapt to its environment. Whereas biological evolution may require literally hundreds of generations to adapt the organism fully to new circumstances, cultural evolution allows the changes to be made within a short period of time.

Biological Perspective

As each month's *National Geographic* attests, clothing, eating habits, living arrangements, and other aspects of culture vary dramatically around the globe. It is tempting to focus on the exotic variety of human behavior and to conclude that there are no limits to what humankind can devise. A closer look, however, suggests that there are some basic similarities in cultures, such as the universal existence of the family, religion, cooperation, and warfare. When we focus on these universals, cultural explanations need to be supplemented with biological explanations.

Sociobiology is the study of the biological basis of all forms of human (and nonhuman) behavior (Alcock 2001; Wilson 1978). Sociobiologists believe that humans and all other life forms developed through evolution and natural selection. According to this perspective, change in a species occurs primarily through one mechanism: Some genes are more successful at reproduction than others. As those who carry these genes increase in number, the species comes to be characterized by the traits common among this group.

Which genes are the successful reproducers? They are genes found in people who have more children and raise more of them until they are old enough to reproduce themselves (Alcock 2001; Daly & Wilson 1983). For example, sociobiologists suggest that parents who are willing to make sacrifices for their children, occasionally even giving their life for them, are more successful reproducers; by ensuring their children's survival, they increase the likelihood that their own genes will contribute to succeeding generations. Thus, sociobiologists argue that we have evolved biological predispositions toward cultural patterns that enable our genes to continue after us.

Sociobiology provides an interesting theory about how the human species has evolved over tens of thousands of years. Most of the scholars who study the effect of biology on human behavior, however, are concerned with more contemporary questions such as "How do hormones, genes, and chromosomes affect human behavior today?" Joint work by biologists and social scientists is helping us to understand how biological and social factors work together to determine human behavior. For example, Booth and Osgood (1993) found that men were statistically more likely to engage in deviant behavior if they had *both* high levels of testosterone *and* low levels of social integration. Research such as this suggests that only by recognizing and taking into account the joint effects of culture and biology can we fully understand human behavior.

Sociobiology is the study of the biological basis of all forms of human (and nonhuman) behavior.

The Carriers of Culture

In this section, we review three vital aspects of nonmaterial culture—language, values, and norms—and show how they shape both societies and individuals. We then explore how social control pressures individuals to live within the rules of their culture.

Language

The essence of culture is the sharing of meanings among members of a society. The chief mechanism for this sharing is a common language. Language is the ability to communicate in symbols—orally, by manual sign, or in writing.

What does *communicate with symbols* mean? It means, for example, that when you see the combinations of circles and lines that appear on your textbook page as the word *orally*, you are able to understand that it means "speaking aloud." On a different level, it means that the noise we use to symbolize "dog" brings to your mind a four-legged domestic canine. Almost all communication is done through symbols. Even the meanings of physical gestures such as touching or pointing are learned as part of culture.

Scholars of sociolinguistics (the relationship between language and society) agree that language has three distinct relationships to culture: Language embodies culture, language is a symbol of culture, and language creates a framework for culture (Romaine 2000; Trudgill 2000).

Language as Embodiment of Culture

Language is the carrier of culture; it embodies the values and meanings of a society as well as its rituals, ceremonies, stories, and prayers. Until you share the language of a culture, you cannot fully participate in it (Romaine 2000; Trudgill 2000).

A corollary is that loss of language may mean loss of a culture. Of the approximately 300 to 400 Native American languages once spoken in the United States, only about 20 are expected to survive much longer (Dalby 2003, 147–148). When these languages die, important aspects of these Native American cultures will be lost. This vital link between language and culture is why many Jewish and Chinese parents in the United States send their children to special classes after school or on weekends to learn Hebrew or Chinese. This is also why U.S. law requires that people must be able to speak English before they can be naturalized as U.S. citizens. To participate fully in Jewish or Chinese culture requires some knowledge of these languages; to participate in U.S. culture requires some knowledge of English.

© Agence France Presse

When languages die, cultures often die too. This is why immigrant parents often send their children to special classes to make sure they learn their traditional language.

Language as Symbol

A common language is often the most obvious outward sign that people share a common culture. This is true of national cultures such as French and Italian and subcultures such as youth. A distinctive language symbolizes a group's separation from others while it simultaneously symbolizes unity within the group of speakers (Joseph et al. 2003; Romaine 2000; Trudgill 2000). For this reason, groups seeking to mobilize their members often insist on their own distinct language. For example, Jewish pioneers who moved in the early 1900s from the ghettos of Europe to what was then Palestine declared that everyone within their communities must speak Hebrew. Yet no one had spoken Hebrew except in prayers for hundreds of years. Nevertheless, within a few decades, Hebrew became the national language of Israel.

Similarly, in the last two decades some Americans have opposed bilingual education and pushed to declare English the official language of the United States, while French Canadians have fought to make French the official language of Quebec (Dalby 2003). Meanwhile, government bureaucracies in Mexico and France fight to keep English words from creeping into Spanish and French. All these efforts are largely symbolic; in any country, both immigrants and native-born citizens will continue to use or will quickly adopt whichever language has the most social status and social utility (Ricento & Burnaby 1998). As a result, grandchildren of Mexican immigrants to the United States rarely speak Spanish, and most French-speaking Canadians are also fluent in English (Dalby 2003).

The **linguistic relativity hypothesis** argues that the grammar, structure, and categories embodied in each language affect how its speakers see reality.

Language as Framework

According to some linguists, languages not only symbolize our culture but also help to create a framework in which culture develops. The **linguistic relativity hypothesis** associated with Whorf (1956) argues that the grammar, structure, and categories embodied in each language influence how its speakers see reality. According to this hypothesis, for example, because Hopi grammar does not have past, present, and future grammatical tenses (e.g., "I had," "I have," "I will have"), Hopi speakers think differently about time than do English speakers.

This theory has come under attack in recent years. Most linguists now believe that although differences among languages influence thought in small ways, the universal qualities of language and human thought far overshadow those differences. The difficulties of translating from one language to another illustrate the conceptual differences among languages; that translation is nonetheless possible illustrates that, despite those differences, people in all cultures have essentially the same linguistic capabilities (Trudgill 2000).

Values

After language, the most central and distinguishing aspect of culture is **values,** shared ideas about desirable goals (Hitlin & Piliavin 2004). Values are typically couched in terms of whether a thing is good or bad—desirable or undesirable. For example, many people in the United States believe that a happy marriage is desirable. In this case and many others, values may be very general. They do not, for example, specify what constitutes a happy marriage.

Some cultures value tenderness and cooperation, others value toughness and competition. Nevertheless, because all human populations face common dilemmas, certain values tend to be universal. For example, nearly every culture values stability and security, a strong family, and good health. There are, however, dramatic differences in the guidelines that cultures offer for pursuing these goals. In societies like ours, an individual may try to ensure security by putting money in the bank or investing in an education. In many traditional societies, security is maximized by having a large number of relatives. In societies such as that of the Kwakiutl of the Pacific Northwest before Western conquest, security is achieved, not by saving wealth, but by giving it away. The reasoning is that all of the people who accept your goods are now under an obligation to you. If you should ever need help, you would feel free to call on them and they would feel obliged to help. Thus, although many cultures place a value on establishing security against uncertainty and old age, the specific guidelines for reaching this goal vary. These guidelines are called norms.

Norms

Shared rules of conduct are **norms.** They specify what people *ought* or *ought not* to do. The list of things we ought to do sometimes seems endless. We begin the day with "I'm awfully tired, but I ought to get up," and may end the day with "I'd like to keep partying but I'd better go to bed." In between, we ought to brush our teeth, eat our vegetables, work hard, love our neighbors, and on and on. The list is so extensive that we may occasionally feel that we have too many obligations and too few choices. Of course, some pursuits are optional and allow us to make choices, but the whole idea of culture is that it provides a blueprint for living, a pattern to follow.

Norms vary enormously in their importance both to individuals and to society. Some, such as fashions, are short-lived. Others, such as those supporting monogamy and democracy, are powerful and long-lasting because they are central to our culture. Generally, we distinguish between two kinds of norms: folkways and mores.

Values are shared ideas about desirable goals.

Norms are shared rules of conduct that specify how people ought to think and act.

Connections

Personal Application

As you sit in your college classroom, you are following a long list of norms. Your very presence in the classroom reflects your acknowledgment that higher education is useful. No matter how bored you might be, you sit reasonably still and try not to fidget. If you are falling asleep, you pull your cap brim down to hide your droopy eyelids. You raise your hand rather than call out to demonstrate your respect for the teacher. And you write down whatever the teacher says, or at least write something down so it looks like you are taking notes.

Norms that govern daily life are usually not as explicit as in this classroom. Nevertheless, most of us figure out social norms without much trouble just from observing those around us.

Folkways

Folkways are norms that are the customary, normal, habitual ways a group does things.

The word **folkways** describes norms that are simply the customary, normal, habitual ways a group does things. Folkways is a broad concept that covers relatively permanent traditions (such as fireworks on the Fourth of July) as well as short-lived fads and fashions (such as wearing baseball caps backward).

A key feature of all folkways is that no strong feeling of right or wrong is attached to them. They are simply the way people usually do things. For example, if you choose to have hamburgers for breakfast and oatmeal for dinner, or to sleep on the floor and dye your hair purple, you will be violating U.S. folkways. Doing these things may lead others to brand you as eccentric, weird, or crazy, but you will not be regarded as immoral or criminal.

Mores

Mores are norms associated with fairly strong ideas of right or wrong; they carry a moral connotation.

Some norms are associated with strong feelings of right and wrong. These norms are called **mores** (more-ays). Whereas eating oatmeal for dinner may only cause you to be considered crazy (or lazy), other things that you can do will really offend your neighbors. If you eat your dog or spend your last dollar on liquor when your child needs shoes, you will be violating mores. At this point, your friends and neighbors may decide that they have to do something about you. They may turn you in to the police or to a child protection association; they may cut off all interaction with you or even chase you out of the neighborhood. Not all violations of mores result in legal punishment, but all result in such informal reprisals as ostracism, shunning, or reprimand. These punishments, formal and informal, reduce the likelihood that people will violate mores.

Laws

Laws are rules that are enforced and sanctioned by the authority of government. They may or may not be norms.

When mores are enforced and sanctioned by the government, they are known as **laws.** If laws cease to be supported by norms and values, they may be stricken from the record or the police may simply stop enforcing them. However, laws don't al-

Values, Norms, and Laws

Concept	Definition	Example from Marriage	Relationship to Values
Values	Shared ideas about desirable goals	It is desirable that marriage include physical love between wife and husband	
Norms	Shared rules of conduct	Have sexual intercourse regularly with each other, but not with anyone else	Generally accepted means to achieve value
Folkways	Norms that are customary or usual	Share a bedroom and a bed; kids sleep in a different room	Optional but usual means to achieve value
Mores	Norms with strong feelings of right and wrong	Thou shalt not commit adultery	Morally required means to achieve value
Laws	Formal standards of conduct, enforced by public agencies	Illegal for husband to rape wife; sexual relations must be voluntary	Legally required means; may or may not be supported by norms

Concept Summary

ways emerge from popular values. Laws requiring seat belts, for example, were created by legislators to *change* norms, not to reflect existing norms.

Social Control

From our earliest childhood, we are taught to observe norms, first within our families and later within peer groups, at school, and in the larger society. After a period of time, following the norms becomes so habitual that we can hardly imagine living any other way—they are so much a part of our lives that we may not even be aware of them as constraints. We do not think, "I ought to brush my teeth or else my friends and family will shun me"; instead we think, "It would be disgusting not to brush my teeth, and I'll hate myself if I don't brush them." For thousands of generations, no human considered it disgusting to go around with unbrushed teeth. For most people in the United States, however, brushing their teeth is so much a part of their feeling about themselves, about who they are and the kind of person they are, that they would disgust themselves by not observing the norm.

Through indoctrination, learning, and experience, many of society's norms come to seem so natural that we cannot imagine acting differently. No society relies completely on this voluntary compliance, however, and all encourage conformity by the use of **sanctions**—rewards for conformity and punishments for nonconformity. Some sanctions are formal, in the sense that the legal codes identify specific penalties, fines, and punishments that are to be meted out to individuals found guilty of violating formal laws. Formal sanctions are also built into most large organizations to control absenteeism and productivity. Some of the most effective sanctions, however, are informal. Positive sanctions such as affection, approval, and inclusion encourage normative behavior, whereas negative sanctions such as a cold shoulder, disapproval, and exclusion discourage norm violations.

Despite these sanctions, norms are not always a good guide to what people actually do, and it is important to distinguish between normative behavior (what we are supposed to do) and actual behavior. For example, our own society has powerful mores supporting marital fidelity. Yet research has shown that nearly half of all married men and women in our society have committed adultery (Laumann et al. 1994). In this instance, culture expresses expectations that differ significantly from actual

Sanctions are rewards for conformity and punishments for nonconformity.

behavior. This does not mean the norm is unimportant. Even norms that a large minority, or even a majority, fail to live up to are still important guides to behavior. The discrepancy between actual behavior and normative behavior—termed *deviance*—is a major area of sociological research and inquiry (see Chapter 6).

Cultural Diversity and Change

By definition, members of a community share a culture. But that culture is never completely homogeneous. In the following sections, we will look at two expressions of diversity within cultures—subcultures and countercultures—and at the processes by which cultures change.

Subcultures and Countercultures

No society is completely homogeneous. Instead, each society has within it a dominant culture, as well as subcultures and countercultures.

Subcultures are groups that share in the overall culture of society but also maintain a distinctive set of values, norms, and lifestyles and even a distinctive language.

Subcultures share in the overall culture of society but also maintain a distinctive set of values, norms, lifestyles, and traditions and even a distinctive language. The "Greek Life" of traditional (residential) fraternities and sororities offers an excellent example of a subculture. To enter a fraternity or sorority, prospective members must first demonstrate that their fashion style; partying or studying habits; and attitudes toward sex, drinking, service, and scholarship fit the culture of a particular house as well as of the Greek system as a whole. Those who are "tapped" must then go through the ritual of hazing, an experience that can range from humorous to dangerous and that cements ties to the fraternity or sorority and its culture. After initiation, members are taught the special traditions of the house, which can include songs, passwords, and other rituals.

Greek subculture does not have its own language, but it does have its own slang terms for members of other houses, among other things. It also has its own values, beginning with loyalty to co-members. In some fraternities and sororities, for example, members are expected to tutor fellow members when needed; in others they are expected to help others cheat on exams. Members are also expected to adopt a distinctive lifestyle: living together in sex-segregated houses and cooking, eating, and socializing primarily with other house members. Those who actively participate in this subculture are rewarded with strong, supportive bonds during college and strong social networks afterward.

Countercultures are groups whose values, interests, beliefs, and lifestyles conflict with those of the larger culture.

Subcultures differ from the dominant culture, but they are not at odds with that culture. In contrast, **countercultures** are groups whose values, interests, beliefs, and lifestyles conflict with those of the larger culture. This theme of conflict is clear among one current U.S. countercultural group—punkers. Some punkers are just part-timers who shave their heads and listen to death rock but nevertheless manage to go to school or hold a job. Hardcore punkers, however, angrily reject straight society. They refuse to work or to receive charity; they live angry and sometimes hungry lives on the streets. As one young girl explained to a (formerly punk) sociologist, "Punks are saying 'Fuck this, we're pissed off'. . . When you're really young, living with your parents in their all-white rich suburb [and you] shave your head and dye [the remainder of] your hair green, then it's saying, 'Well, fuck society, fuck the corporate world, I'm going to be against all this shit'" (Leblanc 1999, 87).

Assimilation or Multiculturalism?

Until very recently, most Americans believed it would be best if the various ethnic and religious subcultures within American society would adopt the dominant majority culture. **Assimilation** refers to the process through which individuals learn and adopt the values and social practices of the dominant group, more or less giving up their own values in the process. As discussed in more detail in Chapter 12, assimilation was, and to some extent still is, one of the major goals of our educational institutions (Spring 1997). In schools, immigrant children learn not only to read and write English but also to consider "American" foods, ideas, and social practices preferable to those of their own native culture or subculture. Children named Juan or Mei Li may be encouraged to go by John or Mary. School curricula focus on the history, art, literature, and scientific contributions of Europeans and European Americans while downplaying the contributions of U.S. minority groups and non-Western cultures.

In the last quarter century, however, more and more Americans have concluded that America has always been more of a "salad bowl" of cultures than a melting pot. Many have come to believe that this "salad bowl" is one of Americans' greatest strengths and that it should be cherished rather than eliminated. These beliefs are often referred to as *multiculturalism*. Reflecting these beliefs, many schools and universities now incorporate materials that more accurately reflect American cultural diversity.

> **Assimilation** is the process through which individuals learn and adopt the values and social practices of the dominant group, more or less giving up their own values in the process.

Case Study: Deafness as Subculture

Most of us view deafness as undesirable, even catastrophic (Dolnick 1993). At best, we see being deaf as a medical condition to be remedied. However, some deaf people maintain that deafness is not a disability but a culture (Dolnick 1993). To these individuals, the essence of deafness is not the inability to hear but a valued culture based on their shared language, American Sign Language (ASL). ASL is not just a way to "speak" English with one's hands but is a language of its own, complete with its own rules of grammar, puns, and poetry. Furthermore, it is a language that is learned and shared. Whereas babies who can hear begin to jabber nonsense syllables, deaf babies of parents who sign begin to "babble" nonsense signs with their fingers (Dolnick 1993). This shared language encourages, in turn, shared values and a positive group identity. Studies show, for instance, that many deaf people would not choose to join the "hearing" culture even if they could.

Thinking of deafness as a culture illustrates many of the points made earlier. For instance, culture is problem solving, and deaf culture embodies a way to solve the human problem of communication. Using ASL shapes deaf people's experiences, reminding them of their common values, norms, and cultural identity. For this reason, many deaf individuals have reacted with outrage to the increasing use of cochlear implants (Arana-Ward 1997). These devices, when surgically implanted in the ear, help some otherwise-deaf persons to hear sounds. Hearing sounds, however, is not the same as understanding what one hears: Many implant recipients—especially older children who were born deaf—are frustrated by a cacophony of sounds that they cannot interpret, even after months or years of training. Some deaf activists argue that most children who receive implants waste their formative years in an often futile struggle to fit into the hearing world, when they could instead have become native speakers of ASL and valued members of the deaf community. These activists, therefore, view cochlear implants not as a neutral medical technology but as an example of the ethnocentrism of hearing persons.

These deaf students believe that they share a common culture and should have rights like those given to any minority culture.

At the same time, because deaf Americans function *within* American culture (reading newspapers, holding jobs with hearing co-workers, purchasing clothes at the mall), it is most accurate to consider deafness a *sub*culture rather than a culture. Those who believe that even deaf children who receive cochlear implants should learn ASL are arguing in favor of a multicultural model in which children can feel comfortable in both the deaf and the hearing worlds. Those who argue that deaf children will only learn how to function in the modern world if they receive implants, receive constant training in speech and hearing, and never learn to sign are arguing that these children are best served by full assimilation into the hearing world.

Sources of Cultural Diversity and Change

Culture provides solutions to common and not-so-common problems. The solutions devised are immensely variable. Among the reasons for this variability are environment, isolation, cultural diffusion, technology, exposure to mass media, and dominant cultural themes.

Environment

Why are the French different from Australian aborigines, the Finns different from the Navajo? One obvious reason is the very different environmental conditions to which they must adapt. These different environmental conditions determine which kinds of economies can flourish, which kinds of clothes and foods are practical, and, to a significant extent, the degree of scarcity or abundance.

Media and Culture

Models, Magazines, and Self-Esteem

Over the last several decades, the average American has grown considerably heavier. During the same time period, however, Miss America has grown slimmer, and curvaceous actresses like Marilyn Monroe have been replaced by stick-thin ones like Kate Hudson and Lindsay Lohan. Fashion magazines, movies, and television increasingly celebrate and glamorize an extremely rare body type, far slimmer than the typical American woman's (Milkie 1999). The net result is that the gap between media images of ideal female body shape and actual female body shapes has increased substantially (Wiseman et al. 1992). What effect has this had on American culture, and on girls' self-concepts? And what is the additional impact on young African American, Asian American, Native American, and Latina women when the media still primarily showcase fair-skinned, long-haired, and European-featured girls (Collins 1991; Weitz 2004)?

Many scholars believe that unrealistic images in the media have altered cultural notions about what constitutes beauty, damaged girls' self-esteem, and contributed to epidemic levels of eating disorders (Evans et al. 1991). Others argue that both culture and girls are more resilient than this, because audiences critically evaluate what they see and read in the media rather than adopting media values automatically. Sociologist Melissa Milkie (1999) studied this issue through in-depth interviews with 60 urban and rural, African American and white high-school girls. Here is what she found.

Most of the girls in Milkie's study, both African American and white, believed that the images of female beauty shown on the pages of girls' magazines were unrealistic. As one respondent noted, even the so-called "problem" bodies shown in magazines are perfect:

> These magazines are trying to tell you "Do this and do that" . . . if you have a problem body, . . . a big butt, big chest . . . And these girls that they are showing don't have that problem. I mean you can tell they don't, and that makes me mad. . . . They say if you got a stick figure, wear a one-piece [swimsuit] . . . and I'm looking at the girl and she doesn't have a stick figure. . . . (Milkie 1999, 198–199)

Despite their critiques of unrealistic, idealized media images of beauty, Milkie found that the self-concepts of white girls in her study were harmed by those images. Even though the girls rejected these images, they believed that the values implied in those images were accepted in the culture around them. They assumed that their friends, boyfriends, and potential boyfriends judged them by how well they matched those images, and they judged themselves accordingly. Interestingly, minority girls were less affected by media. First, these girls rejected those images because they believed the images reflected only white culture. Second, they assumed that their friends, boyfriends, and potential boyfriends felt the same, and so the girls did not worry that their significant others would judge them according to media standards.

Taken together, Milkie's results suggest that (1) individuals are active consumers of media messages, (2) different audiences interpret the same media messages differently, and (3) media do shape both culture and individual beliefs and actions, at least in part because we judge ourselves through the "media-filled" eyes of others who matter to us.

■ Unrealistic images of female beauty in magazines, on TV, and in fashion advertisements contribute to poor body image and low self-esteem among young American women.

Diffusion of modern technology is particularly rapid when new tools enhance a society's ability to meet basic human needs at the same time that they are consistent with existing cultural patterns. Leaders, regardless of time, place, or the cultural bases of their authority, share a common need to communicate effectively with large numbers of followers.

Cultural diffusion is the process by which aspects of one culture or subculture are incorporated into another.

Globalization of culture is the process through which cultural elements (including musical styles, fashion trends, and cultural values) spread around the globe.

Popular culture refers to aspects of culture that are widely accessible and commonly shared by most members of a society, especially those in the middle, working, and lower classes.

Isolation

When a culture is cut off from interaction with other cultures, it is likely to develop unique norms and values. Where isolation precludes contact with others (such as in the New Guinea highlands until recently), a culture can continue on its own course, unaltered and uncontaminated by others. Since the nineteenth century, however, almost no cultures have been able to maintain their isolation from other cultures.

Cultural Diffusion

If isolation is a major reason why cultures remain both stable and different from each other, then cultural diffusion is a major reason why cultures change and become more similar over time. **Cultural diffusion** is the process by which aspects of one culture or subculture are incorporated into another culture. For example, not only have many residents of Mexico City become regular consumers of McDonald's hamburgers, but belief in the value of "fast food" is gradually replacing Mexicans' traditional belief in the value of long, family-centered meals. Meanwhile, salsa now outsells ketchup in the United States, and Heinz now offers a green ketchup specifically to compete with salsa.

At its broadest level, cultural diffusion becomes the **globalization of culture,** in which cultural elements (including fashion trends, musical styles, and cultural values) are shared by persons around the world. Nowadays, taxi drivers in Bombay, Senegal, and Peru blare U.S. popular music from their radios, while Americans relish the chance to eat in French and Chinese restaurants. The globalization of culture is likely to proceed even more rapidly in the future due to the Internet.

The globalization of culture is part of the broader topic of globalization, which we discuss further at the end of this chapter.

Technology

The tools available to a culture will affect its norms and values and its economic and social relationships. The rise of the automobile, for example, allowed young dating couples to escape the watchful eyes of concerned parents, contributing to changes in both courtship patterns and the roles of women (Scharff 1991). The automobile also encouraged individuals to live farther from their places of employment, to shop in malls rather than downtown, and to live in houses with three-car garages rather than front porches. As a result, among the unintended consequences of the automobile was a decline of inner cities, locally owned businesses, and a sense of neighborliness (Kunstler 1994).

Mass Media

The mass media are an example of **popular culture:** aspects of culture that are widely accessible and broadly shared, especially among "ordinary" folks. (In contrast, **high culture** refers to aspects of culture primarily limited to the middle and upper classes, such as opera, modern art, or modernist architecture.) The mass media includes movies; television; genre fiction such as romances, mysteries, or science fiction; and popular music styles like rock and roll or hip-hop.

An important question for researchers is whether the mass media simply reflect existing cultural values or whether the media can change values. The answer is that media probably do both. For example, for much of the twentieth century, movies and television usually portrayed African Americans as lazy or foolish and unmarried women as evil, disturbed, or unhappy (Entman & Rojecki 2000; Levy 1990). These depictions reflected American cultural beliefs of the time. Yet these days, Denzel Washington can play a romantic lead, an action hero, or a very smart lawyer in movies. Social change in American culture allowed Washington to get these roles, but seeing Washington in these roles also creates more cultural change, by suggesting to white Americans that African Americans can be attractive, ethical, smart, and professional. As this suggests, exposure to mass media can be a source of cultural change.

Dominant Cultural Themes

Cultures generally contain dominant themes that give a distinct character and direction to the culture; they also create, in part, a closed system. New ideas, values, and inventions are usually accepted only when they fit into the existing culture or represent changes that can be absorbed without too greatly distorting existing patterns. Sioux culture, for example, readily adopted rifles and horses as aids to its established cultural theme of hunting. Western types of housing and legal customs regarding land ownership, however, were rejected because they were alien to a nomadic and communal way of life.

A Case Study: American Consumer Culture

U.S. culture is a unique blend of complex elements. It is a product of the United States' environment, its immigrants, its technology, and its place in history. These days, one of the ways in which U.S. culture diverges most strongly from other cultures is in its exceptionally strong emphasis on consumerism.

Consumerism is a philosophy that says "buying is good." In turn, this philosophy is based on the belief that "we are what we buy," and that through buying certain goods we can assert or improve our social status. In American consumer culture, children attempt to improve their social status by buying certain brands of breakfast cereals, teenagers by buying certain brands of clothing, and adults by buying certain models of cars. Ironically, consumers also believe they are asserting their individuality through their purchases, rarely noticing that millions of others are buying the same goods for the same reasons.

How did this consumer culture develop? The simple answer is that more consumer goods are available and affordable than ever before. But this is only a partial answer. Research suggests that the most important cause is a change in the comparisons we use in deciding whether to make a purchase (Schor 1998). Advertising now permeates our lives more than ever before—billboards adorn public buses and sports stadiums, movie theaters show advertisements before the films, ads pop up at popular Internet sites, schools broadcast television programs laced with commercials in the classroom, and so on. All of this instills in children and adults the belief that they need certain products to be a certain sort of person (Quart 2003).

Similarly, as the number of hours Americans watch television per week has soared, so has Americans' desire for the goods they see on television. Instead of deciding what kind of shoes to wear or what kind of kitchen appliances to buy by looking at what their classmates or neighbors own, Americans now seek out consumer goods like those used by their favorite television characters. In fact, for every hour of television watched each week, individuals' annual spending on consumer goods increases by more than $200 (Schor 1998).

High culture refers to the cultural preferences associated with the upper class.

Consumerism is the philosophy that says "buying is good" because "we are what we buy."

FIGURE 2.1
Total U.S. Consumer Debt, as Percentage of Income, 1975–2005.
The rise in debts reflects both economic hard times and growing consumer desires. In total, American consumers now owe more than their yearly incomes.
SOURCE: Foster, John Bellamy. 2006. "The Household Debt Bubble." Monthly Review 58(1): 1–11.

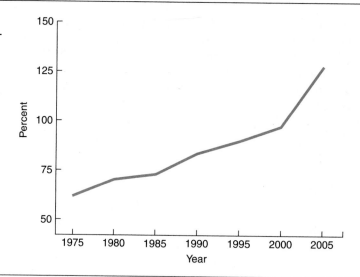

Finally, in the past, women (who do most family shopping) typically compared their belongings with those of their neighbors, whose family incomes were usually similar to their own. Now that a majority of women work outside the home, most compare their belongings with those of their fellow workers, including supervisors with much higher incomes. As a result, families now spend higher percentages of their income on consumer goods, both big and small. For example, the median house size has increased from 1500 square feet in 1973 to 2200 square feet in 2005—with prices to match (U.S. Bureau of the Census 2006).

Consumer culture affects our lives in many ways. Shopping and window-shopping have become major forms of recreation, and shopping malls have replaced parks, athletic fields, church basements, and backyards as popular gathering spots. College students put their grades at risk by working extra hours, in many cases to buy the latest gadgets or fashions. Moreover, despite these extra hours working, the average debt of graduating students rose (in constant dollars) by more than 50 percent between 1993 and 2004, and the percentage who owed more than $40,000 rose from 1 percent to 8 percent (Project on Student Debt 2006). Students who have $40,000 in debt must think twice before taking a job at a nonprofit organization, taking a year off to travel, or pursuing a graduate degree. Meanwhile, adults carry heavy debts and risk bankruptcy to buy expensive cars and houses as a way to "prove" their success and improve their social status. Figure 2.1 illustrates the rising consumer debts of Americans between 1975 and 2005. The total amount Americans owe now substantially exceeds the amount Americans earn.

Consequences of Cultural Diversity and Change

No culture remains isolated forever, and none remains forever unchanged. Although cultural diversity and change often help societies cope with existing problems, they can also create new problems. Two sociologically important types of problems caused by cultural diversity and change are cultural lag and culture shock.

Cultural Lag

Cultural change can also foster social problems when one part of a culture changes more rapidly than another part. This situation is known as **cultural lag** (Brinkman & Brinkman 1997). Most often, cultural lags occur when social practices and values do not keep up with technological changes. The recent rise in genetic testing for breast cancer illustrates some of the problems that arise when law, values, and social practices lag behind technological change (Lewin 2000).

Although most cases of breast cancer are not hereditary, women who carry the BRCA1 genetic mutation have a 50 percent lifetime risk of breast cancer. For this reason, some women who have a family history of breast cancer now opt to be tested for the mutated gene. If they do carry the gene, the only way they can reduce their chances of breast cancer is to have both breasts surgically removed. Yet half of those who test positive will never develop breast cancer and some who have their breasts removed will nevertheless die of cancer.

Because of these uncertainties, some women with family histories of breast cancer choose not to be tested. But what happens if a woman decides against getting tested and her mother (or sister or daughter) decides in favor of testing? If the mother subsequently has her breasts removed, the daughter will know immediately that her mother tested positive and will know that she, herself, is more likely to carry the mutated gene. Thus, the mother's decisions force the daughter to confront information that she may have chosen not to seek. At the same time, genetic counselors worry that they may be legally liable if their clients do not inform their relatives that they too are at risk.

All these problems reflect cultural lag. Individuals forced to decide about genetic testing must do so in the absence of agreed-upon social values, clear legal decisions, or standard social practices for how persons in their situation should act.

Culture Shock

In the long run, cultural diversity and cultural change often result in improvements in quality of life. In the short run, however, people often find both diversity and change unsettling. **Culture shock** refers to the disconcerting and unpleasant experiences that can accompany exposure to a different culture. For example, U.S. citizens who work in Greece are often taken aback by the Greek customs of hugging acquaintances and standing very close (by American standards) to anyone they are speaking with. Greeks who work in the United States are similarly confused by American customs that limit greetings to simple handshakes and dictate maintaining considerable physical distance during conversations. As a result, Americans sometimes conclude that Greeks are pushy or even sexually aggressive, while Greeks sometimes conclude that Americans are elitist or emotionally cold.

Globalization

As this discussion of cultural shock suggests, cultural change can occur not only within one society but also across societies. At its broadest, this change is referred to as the globalization of culture. More generally, **globalization** refers to the process through which ideas, resources, practices, and people increasingly operate in a worldwide rather than local framework. Because globalization is having such an impact on the world and its cultures, we devote this section to exploring its sources and effects—economic and political, as well as cultural.

Connections

Example

American education is imbued with the value of competition. Children are encouraged to best their peers in the classroom and in the schoolyard—to strive for the highest test scores or the most home runs. In many African countries, on the other hand, children are chastised by teachers as well as other students if they behave competitively. Instead, children are encouraged to help each other learn and to work together for the good of the group. These differences can create culture shock for children who emigrate from Africa to the United States or vice versa.

Cultural lag occurs when one part of culture changes more rapidly than another.

Culture shock refers to the discomfort that arises from exposure to a different culture.

Globalization is the process through which ideas, resources, practices, and people increasingly operate in a worldwide rather than local framework.

The Sources of Globalization

Globalization stems from a combination of political and technological forces. The collapse of the Soviet Union in 1991 made it possible for the nations that emerged in its wake (like the Ukraine and Estonia) as well as the nations that had been restrained by its political power (like Poland and Czechoslovakia) to move toward a more capitalistic economic system. To do so, they needed to seek out economic, political, and cultural ties to other nations that could either serve as sources of raw goods and labor or markets for their products.

The collapse of the Soviet Union also reduced political tensions that had pressed nations to adopt international trade barriers. Now that the nations of Europe are no longer fearful of Soviet might, they have united into what is in some ways a continental government, in the form of the European Union. Within this Union, goods, individuals, and services can flow more freely than ever before. Polish doctors can now seek higher paying jobs in Finland, Finnish doctors seek work in Sweden, and Swedish doctors seek work in England, with little concern about visas or immigration laws. German factories can transport and sell their products in Spain, and Spanish factories can send their products to Greece with minimal paperwork or tariffs to pay. Similarly, the North American Free Trade Agreement (NAFTA) was adopted in 1994 to reduce trade barriers between Canada, the United States, and Mexico.

All these changes were made more feasible by new technological developments. The Internet, e-mail, cellular telephone service, fax machines, and the like all made it easier, cheaper, and faster for corporations and individuals to invest and work internationally.

The Impact of Globalization

Cultural Impact

In an African urban nightclub, young people listen to American rock and roll and drink Pepsi. In New York City, young people go to Jamaican reggae concerts and watch the British television show *What Not To Wear*. In India, Hollywood films are as popular as "Bollywood" (Bombay-produced) films, and French wines are preferred by those who can afford them. All of these are examples of the global spread of culture. Increasingly, movies, television shows, music, literature, and other arts are distributed and enjoyed around the world.

These elements of popular culture carry with them not only entertainment but also cultural values. As Indian adolescents watch American films, they not only learn about the latest U.S. fashions and music but also learn to question traditional Indian practices and beliefs like arranged marriages, the subservice of women, obedience to parents, and the idea that the family is more important than the individual.

Economic Impact

Globalization has also had a striking economic impact on both the selling and producing of goods. Increasingly, economic activity takes place between people who live in different nations as goods and services are sold internationally. These days, Russians and Chinese buy Coca-Cola, and Americans buy Volvos and Toyotas. Globalization also exists when goods are *produced* internationally. Transnational corporations, such as General Motors, may buy raw goods in one country, own a factory that puts smaller pieces together in a second country, assemble the full product in a

Connections

Social Policy

France is an example of a country that has fought hard to protect itself against the cultural impact of globalization. The Académie Français is officially in charge of policing the French language against the encroachment of non-French words (like *computer* and *hamburger*) and creating new French words to fit new technologies (such as *téléphone cellulaire* for cell phone). Similarly, to combat the impact of foreign mass media, the government subsidizes the French film and television industries and requires that 40 percent of films shown on French television must be in French.

third country, contract out its data processing to a firm in a fourth country, and then sell its product worldwide.

There is considerable debate about the possible effects, good and ill, of such international economic enterprises. Some observers hope that ties of international finance will create a more interdependent (and peaceful) world, while stimulating economic growth and improving everyone's standard of living (Stiglitz 2003). Others are concerned that transnationals are exercising a thinly veiled imperialism, in which cheap labor as well as raw materials are extracted from poorer countries for the benefit of wealthier countries. In addition, these critics allege that moving labor-intensive work to less developed nations exposes workers in those countries to dangers that are banned by law in Western nations (Michalowski & Kramer 1987).

Critics have also raised questions about the impact of economic globalization even within the developed nations. In the United States, hundreds of thousands of workers have lost their jobs when corporations found it cheaper to move those jobs overseas ("NAFTA" 2003). Other workers have been forced to accept cuts in benefits or pay to keep their jobs. The question is whether this global movement of jobs raises incomes overall by shifting work from wealthier to poorer countries or instead depresses incomes overall to the level of the cheapest bidders.

© AP/Wide World Photos

As globalization spreads American products and American cultural values around the world, it can challenge the cultures of other societies. As a result, globalization can sharply increase tensions both within nations and between the United States and other nations.

Political Impact

How has globalization affected the balance of political power within nations and across nations? Some observers have noted that transnational corporations now dwarf many national governments in size and wealth. Their ability to move capital, jobs, and prosperity from one nation to another gives them power that transcends the law of any particular country (Michalowski & Kramer 1987). When a nation's economy depends on a transnational corporation, that nation can't afford to alienate the corporation. For example, Guatemala has limited ability to constrain the labor practices of United Fruit Company because the corporation could cripple the country's economy if it wanted to.

Another aspect of globalization is the sharp rise in international nongovernmental organizations. These organizations include the World Bank, the World Trade Organization, the International Monetary Fund, and the United Nation's International Criminal Court. The underlying premise of these organizations is that they will diminish the independent power of national governments and press nations to conform to international goals (such as ending torture of political prisoners, prosecuting war criminals, or reducing trade barriers). Although sometimes this benefits the individual nation, in other cases less-developed nations have suffered as a result of this loss of autonomy (Stiglitz 2003).

Where This Leaves Us

Most of the time, we think of culture simply as something that we have, in the same way that those of us who have a home or two arms take them for granted. As this chapter has shown, though, culture is dynamic: constantly changing as the world— and the balance of power within that world—changes around us. Languages, eating habits, clothing fashions, and the rest evolve, spread, or die: Ask your parents about the clothing they wore as children, the slang they spoke as teenagers, or the first time they ate a bagel or a tortilla.

Culture is also active, a force that changes us as it changes the world in which we live. The rise of American consumer culture is only one example of how culture changes and of how cultural changes affect all aspects of our lives, from how many hours we work each day to how we define ourselves as individuals.

Summary

1. Culture is a design for living that provides ready-made solutions to the basic problems of a society. It can be conceived of as a tool kit of material and nonmaterial components that help people adapt to their circumstances. Because of this, as the concept of cultural relativity emphasizes, cultural traits must be evaluated in the context of their own culture.

2. Most sociologists emphasize that culture is socially created. However, sociobiologists emphasize that human culture and behavior also have biological roots.

3. Language, or symbolic communication, is a central component of culture. Language embodies culture, serves as a framework for perceiving the world, and symbolizes common bonds among a social group.

4. Values spell out the goals that a culture finds worth pursuing, and norms specify the appropriate means to reach them.

5. The cultures of large and complex societies are not homogeneous. Subcultures and countercultures with distinct lifestyles and folkways develop to meet unique regional, class, and ethnic needs.

6. The most important factors accounting for cultural diversity and change are the physical and natural environment, isolation from other cultures, cultural diffusion, level of technological development, mass media, and dominant cultural themes.

7. Cultural diversity and change can lead to culture shock and cultural lag. Culture shock refers to the disconcerting experiences that accompany rapid cultural change or exposure to a different culture. Cultural lag occurs when changes in one part of the culture do not keep up with changes in another part.

8. Consumer culture—the philosophy that buying is good, and we are what we buy—now plays a major role in American culture.

9. Globalization refers to the process through which ideas, resources, practices, and people increasingly operate in a worldwide rather than local framework. Globalization has had political, cultural, and economic effects.

Thinking Critically

1. What features of U.S. society might explain why children are raised in small nuclear families rather than in extended kin groups?

2. Can you think of an example from U.S. culture for which values, norms, and laws are not consistent with each other? What are the consequences of these inconsistencies?

3. How do environment, isolation, technology, and dominant cultural themes contribute to the maintenance and diffusion of youth subcultures?

4. Identify three white Anglo-Saxon Protestant (WASP) American ethnic foods. (If you have trouble conceptualizing this, think about why this is difficult). If you are not WASP, also identify a favorite ethnic food from your own culture. What do these foods mean to you? What do they mean to others? When and where do you feel comfortable eating and talking about these foods? Why?

Companion Website for This Book

academic.cengage.com/sociology/brinkerhoff
Gain an even better grasp on this chapter by going to the Companion Website. This resource contains tutorial quizzes and flash cards to help you master key terms and concepts.

Suggested Readings

Fadiman, Anne. 1997. *The Spirit Catches You and You Fall Down: A Hmong Child, Her American Doctors, and the Collision of Two Cultures*. New York: Farrar, Straus, and Giroux. A heart-rending true story of the culture clash between a Hmong immigrant family and their American doctors, illustrating some of the consequences of cultural diversity, culture shock, and ethnocentrism.

Foer, Franklin. 2005. *How Soccer Explains the World: An Unlikely Theory of Globalization*. New York: Harper Collins. In this entertaining book, Foer uses the example of international soccer to explain the way in which globalization is—and is not—affecting cultures around the world.

Kephart, William M., and Zellner, William W. 2000. *Extraordinary Groups: The Sociology of Unconventional Life-Styles* (7th ed.). New York: Worth. A fascinating tour of some of the most interesting subcultures and countercultures in the United States, both past and present: the Amish, gypsies, Father Divines, and Jehovah's Witnesses.

Quart, Alissa. 2003. *Branded: The Buying and Selling of Teenagers*. Cambridge, Mass.: Perseus. Trenchant description of how advertisers aim directly at teenagers, from product placement in video games to soda machines in school cafeterias, and how teenagers use consumer goods to create their self-images.

Socialization

© David Young-Wolff/Photo Edit

What Is Socialization?

At the heart of sociology is a concern with *people*. Sociology is interesting and useful to the extent that it helps us explain why people do what they do. It should let us see ourselves, our family, and our acquaintances in a new light.

In this chapter we deal directly with individuals, focusing on the process whereby people learn the roles, statuses, and values necessary for participation in social institutions. This process is called **socialization.** Socialization is a lifelong process. It begins with learning the norms and roles of our family and our subculture and making these part of our self-concept. As we grow older and join new groups and assume new roles, we learn new norms and redefine our self-concept.

Socialization is the process of learning the roles, statuses, and values necessary for participation in social institutions.

The Self and Self-Concept

Every newborn infant has the potential to develop into a complex and fascinating human being, in some ways like every other human being and in some ways unique.

Each individual **self** may be thought of as a combination of unique attributes and normative, or socially expected, responses. Within sociology, these two parts of the self are called the *I* and the *me* (Mead 1934).

In English grammar, *I* is used when we speak of ourselves as the actor ("I disobeyed my mom."). Similarly, in sociological terminology the **I** is the spontaneous, creative part of the self. Conversely, in English we use the word *me* when we speak of ourselves as the object of others' actions ("She punished me for disobeying."). Similarly, sociologists use the term **me** to describe the self as social object, the part of the self that responds to others' expectations.

As this description of the self implies, the two parts may pull us in different directions. For example, some people naturally wake up early in the morning: their I wants to get started on the day, and their me approves. Other people face a daily conflict between their I and their me when the alarm clock goes off in the morning—their I wants to roll over and go back to sleep, but their me knows it is supposed to get up and go to class. Some of these conflicts are resolved in favor of the me and some in favor of the I. Daily behavior is the result of an ongoing internal dialogue between the I and the me.

The self is enormously complex, and we are often not fully aware of our own motives, capabilities, and characteristics. The self that we are aware of is our **self-concept.** It consists of our thoughts about our personality and social roles. For example, a young man's self-concept might include such qualities as young, male, Methodist, good athlete, decent student, shy, awkward with girls, responsible, American. His self-concept includes all the images he has of himself in the dozens of different settings in which he interacts with others.

The **self** is a complex whole that includes unique attributes and normative responses. In sociology, these two parts are called the *I* and the *me*.

The **I** is the spontaneous, creative part of the self.

The **me** represents the self as social object.

The **self-concept** is the self we are aware of. It is our thoughts about our personality and social roles.

Learning to Be Human

What is human nature? Are we born with a tendency to be cooperative and sharing or with a tendency to be selfish and aggressive? The question of the basic nature of humankind has been a staple of philosophical debate for thousands of years. It continues to be a topic of debate because it is so difficult (some would say impossible) to separate the part of human behavior that arises from our genetic heritage from the part that is developed after birth. The one thing we are sure of is that nature is never enough.

The Necessity of Nurture

Each of us begins life with a set of human potentials: the potential to walk, to communicate, to love, and to learn. By themselves, however, these natural capacities are not enough to enable us to join the human family. Without nurture—without love and attention and hugging—the human infant is unlikely to survive, much less prosper. The effects of neglect are sometimes fatal and, depending on severity and length, almost always result in retarded intellectual and social development.

How can we determine the importance of nurture? One way is to look at what happens when children are seriously neglected. Some of the clearest evidence comes from studies of children raised in low-quality orphanages, where their physical needs were met, but where staff were too overworked to give individual attention or to truly nurture each child.

Children who spend the first years of their lives in this type of institutional environment may be devastated by the experience. Because of limited personal attention, such children may withdraw from the social world; they seldom cry and often are indifferent to everything around them. Even if they are later adopted into good homes, they are significantly more likely to experience difficulties in thinking or learning. They are also more likely to experience problems in social relationships—either engaging in indiscriminate friendliness or withdrawing into autistic or near-autistic behaviors. These effects are illustrated by a study that compared children adopted by British parents from high-quality British orphanages and from low-quality Romanian institutions. The researchers found that 12 percent of the Romanian-born children exhibited autistic or quasi-autistic patterns, but none of the children from the British orphanages did (Rutter et al. 1999).

Deprivation can also occur in homes in which parents fail to provide adequate social and emotional stimulation. Children who have their physical needs met but are otherwise ignored by their parents often exhibit many of the same symptoms as institutionalized infants. The bottom line is that children need intensive interaction with others to survive and develop normally.

Much of the evidence for this conclusion, however, is derived from atypical situations in which unfortunate children have been subjected to extreme and unusual circumstances. To test the limits of these findings and to examine the reversibility of deprivation effects, researchers have turned to experiments with monkeys.

Monkeying with Isolation and Deprivation

In a classic series of experiments, psychologist Harry Harlow and his associates studied what happened when monkeys were raised in total isolation. The infants lived in individual cages with a mechanical mother figure that provided milk. Although the infant monkeys' nutritional needs were met, their social needs were not. As a result, both their physical and social growth suffered. They exhibited bizarre behavior such as biting themselves and hiding in corners. As adults, these monkeys refused to mate; if artificially impregnated, the females would not nurse or care for their babies (Harlow & Harlow 1966). These experiments provide dramatic evidence about the importance of social contact; even apparently innate behaviors such as sexuality and

Connections

Example

The true story of Genie illustrates the consequences of severe deprivation. Until the age of 13, Genie's abusive father kept her locked in a small room, tied to a chair by a sort of harness. Her mother—blind, disabled, and cowed—could do nothing to help her. Genie was never spoken to or socialized in any way. When Genie's mother finally ran away with her, Genie could not talk, walk, or even use a toilet. After years of therapy, her abilities improved, but she never reached normal intelligence levels.

maternal behavior must be developed through interaction. On the bright side, the monkey experiments affirm that some of the ill effects of isolation and deprivation are reversible: The young monkeys experienced almost total recovery when placed in a supportive social environment (Suomi, Harlow, & McKinney 1972).

Although it is dangerous to generalize from monkeys to humans, the evidence from the monkey experiments supports the observations about human infants: Physical and social development depend on interaction with others. Even being a monkey does not come naturally. Walking, talking, loving, and laughing all depend on socialization through sustained and intimate interaction with others.

Theoretical Perspectives on Socialization

Socialization refers broadly to the processes through which we come to share roughly the same norms and values as other members of our culture. But how does this learning take place? What are the processes through which, as children and adults, we translate our social experiences into part of our self-concept? In the following pages we look at some psychological and sociological theories of socialization.

Freudian Theory

The first modern theory of socialization was developed by psychoanalyst Sigmund Freud at the beginning of the twentieth century. Freud's theory of socialization links social development to biological cues. According to Freud, to become mentally healthy adults, children must develop a proper balance between their **id**—natural biological drives, such as hunger and sexual urges—and their **superego**—internalized social ideas about right and wrong. To find that balance, children must respond successfully to a series of developmental issues, each occurring at a particular stage and linked to biological changes in the body. For example, during the oral stage, infants and toddlers derive their greatest satisfaction from sucking a breast or bottle. Those who do not learn how to signal and fulfill those needs, Freud concluded, would later develop traits such as dependency and narcissism.

For Freud, the years from three to six are especially important because this is when he believed the superego developed. According to Freud, during this stage children first start noticing genitalia. When boys learn that girls lack penises, they conclude that girls must have been castrated by their fathers as punishment for some wrongdoing. To avoid this fate, boys quickly adopt their father's rules and values, thus developing a strong superego. In contrast, Freud argued that because girls need not fear castration they can never develop a strong superego (Freud 1925 [1971:241–260]).

This infant monkey and others studied by psychologist Harry Harlow grew up locked in individual cages without social contact. As a result, they had difficulty learning how to have sexual intercourse or raise their own babies. These experiments suggest that even apparently innate behaviors must be developed through interaction.

© Martin Rogers/Stock, Boston Inc.

The **id** is the natural, unsocialized, biological portion of self, including hunger and sexual urges.

The **superego** is composed of internalized social ideas about right and wrong.

Freud based his theory on his personal interpretations of his patients' lives and dreams, rather than on scientific research. Nevertheless, Freud's conception of human nature and socialization continues to permeate American culture and social science: The theory is still used (in a much revamped form) by some psychologists and even some sociologists (e.g., Chodorow 1999).

Piaget and Cognitive Development

Another influential psychological theory of socialization is cognitive development theory. This theory has its roots in the work of Swiss psychologist Jean Piaget (1954). Piaget developed his theory through intensive observations of normal young children. His goal was to identify the stages that children go through in the process of learning to think about the world.

Piaget's observations led him to conclude that there are four stages of cognitive development. In the first stage, children learn to understand things they see, touch, feel, smell, or hear, but they do not understand cause and effect. So, for example, very young children love playing peek-a-boo because it is a delightful surprise each time the person playing with them removes his or her hands to reveal his or her presence. In later stages, children may learn to use language, symbols, and numbers; to understand cause and effect; and to understand abstract concepts such as truth or justice. Not all children, however, ever reach this final stage.

Critics of Piaget's work question whether all children go through these stages, as Piaget claimed. They suggest that in addition to individual differences among children, there may also be cultural and gender differences in cognitive development. They also question whether Piaget's ideas reflect only one culture's definitions of what it means to have high cognitive development.

Structural-Functional Theory

The starting premise of all structural-functionalist analyses is that in a properly functioning society, all elements of society work together harmoniously for the good of all. The same is true for structural-functionalist analyses of socialization.

As we'll see in later chapters, structural functionalists believe that schools, religious institutions, families, and the other social frameworks in which children learn the norms and values of their society are designed to smoothly integrate the young into the broader culture, avoiding conflict or chaotic social change. In families, children learn to mind their manners, and in schools they learn to be on time and obey the rules. In places of worship children learn to practice accepted rituals (lighting candles at Mass or on the Sabbath, praying to the east or on Sundays) and to respect traditional ideas about good and bad, right and wrong.

From the perspective of structural functionalism, this socialization is all for the good: Through being properly socialized, young people learn how to become happy and productive members of society.

In the best situations, there is no question that socialization works as structural functionalists claim: Children learn to fit into the world of their elders and learn how to think for themselves sufficiently so that, when they grow up, they will be able to adapt to whatever changes might come and will be able to contribute to positive changes in the world. Critics of this theory, on the other hand, point out that in socializing children to the world as it is, we also teach them to accept existing inequalities and make it difficult for them to see how to change the world for the better. This is the perspective taken by conflict theorists.

■ Schools teach children not only reading, writing, and arithmetic, but also how to sit quietly and follow rules.

Conflict Theory

Conflict theory's approach to socialization is the inverse of structural functionalism's. Whereas structural functionalism assumes that socialization benefits everyone, conflict theory assumes it benefits only those in power and harms the rest.

Conflict theory focuses on how socialization reinforces unequal power arrangements. They look at how families socialize girls to accept that they are less important than boys (by, for example, requiring girls to clean up the kitchen after dinner while boys are allowed to go outside and play) and how teachers socialize working-class children to grow up to fill working-class jobs by punishing signs of creativity and rewarding strict obedience. They also look at how priests, ministers, rabbis, and other religious leaders can socialize congregants to believe that the privileges of the wealthy and of dominant ethnic groups have been granted by God.

Conflict theory is useful for understanding how socialization can quash dissent and social change and reproduce inequalities. It is less useful for explaining the benefits of a stable social system. In later chapters on the family, education, and religion (among others), we return to all these topics with a more detailed discussion of the strengths and weaknesses of conflict theory's perspective on socialization.

Symbolic Interaction Theory

Sociologists who use symbolic interaction theory begin with four basic premises:

1. To understand human behaviors, we must first understand what those behaviors mean to the individual actors.
2. Those meanings develop within social relationships and roles.
3. Individuals actively construct their self-concepts.
4. However, social structure and social roles limit our options for constructing our self-concepts.

According to symbolic interaction theory, the process through which individual identity is constructed centers around three concepts: the *looking-glass self*, *role taking*, and *role identity*.

The Looking-Glass Self

The **looking-glass self** is the process of learning to view ourselves as we think others view us.

Charles Horton Cooley (1902) provided a classic description of how we develop our self-concept. He proposed that we learn to view ourselves as we think others view us. He called this the **looking-glass self.** According to Cooley, there are three steps in the formation of the looking-glass self:

1. We imagine how we appear to others.
2. We imagine how others judge our appearance.
3. We actively think about, internalize, or reject these judgments.

For example, an instructor whose students doze during class is likely to realize that his students think he is a bad teacher. He may internalize their view of himself and conclude that he needs to seek another line of work. Alternatively, however, he may recall other classes that seemed to appreciate his style and may recall colleagues who have told him what a good teacher he is. As a result, he may instead conclude that this semester's students are simply not smart enough to appreciate his teaching.

As this suggests, our self-concept is not merely a mechanical reflection of those around us; rather it rests on our interpretations of and reactions to those judgments. We are actively engaged in defining our self-concept, choosing whose looking-glass we want to pay attention to and using past experiences to aid us in interpreting others' responses.

Symbolic interaction considers subjective interpretations to be extremely important determinants of the self-concept. This premise of symbolic interactionism is apparent in W. I. Thomas's classic statement: "If men define situations as real, they are real in their consequences" (Thomas & Thomas 1928, 572). People interact through the medium of symbols (words and gestures) that must be subjectively interpreted. The interpretations have real consequences—even if they are *mis*interpretations.

Role Taking

Role taking involves imagining ourselves in the role of others in order to determine the criteria others will use to judge our behavior.

The most influential contributor to symbolic interaction theory during the last century was George Herbert Mead (1934). Mead argued that we learn social norms through the process of **role taking.** This means imagining ourselves in the role of others in order to determine the criteria others will use to judge our behavior. This information is used as a guide for our own behavior.

According to Mead, role taking begins in childhood, when we learn the rights and obligations associated with being a child in our particular family. To understand what is expected of us as children, we must also learn our mother's and father's roles. We must learn to see ourselves from our parents' perspective and to evaluate our behavior from their point of view. Only when we have learned their roles as well as our own will we really understand what our own obligations are.

Mead maintained that children develop their role knowledge by playing games. When children play house, they develop their ideas of how husbands, wives, and children relate to one another. When the little boy comes in saying "I've had a hard day; I hope it's not my turn to cook dinner," or when the little girl warns her dolls not to play in the street and to wash their hands before eating, they are testing their knowledge of family role expectations.

Role playing and role taking are responsive to the expectations of **significant others**—role players with whom we have close personal relationships. Parents, siblings, and teachers, for example, are decisive in forming a child's self-concept. As children mature and participate beyond this close and familiar network, the process of role taking is expanded to a larger network that helps them understand what society in general expects of them. They learn what the bus driver, their neighbors, and their employers expect. Eventually, they come to judge their behavior not only from the perspective of significant others but also from what Mead calls the **generalized other**—the composite expectations of all the other role players with whom they interact. Being aware of the expectations of the generalized other is equivalent to having learned the norms and values of the culture. One has learned how to act like an American or a Pole or a Nigerian.

Saying that everyone learns the norms and values of the culture does not mean that everyone will behave alike or that everyone will follow the same rules. Each of us has a different set of significant others. Not only will our family experiences differ depending on the culture and subculture in which we are reared, but as we get older, we also have some freedom in choosing whose expectations will guide our behavior. Although we may know perfectly well what society in general expects of us, we may choose to march to the beat of a different drummer.

Role Identity

The concepts of a *looking-glass self* and *role taking* illustrate the ways in which individuals strive to control their own self-images. At the same time, the social structures we participate in and the roles we perform powerfully affect our identities—whether we want them to or not. This helps to explain why two close friends who go in very different directions after high school—say, one into the army and the other into a rock band—develop into such very different people despite their initial similarities. We grow into our roles, and the parts we play eventually become a part of our self-concept.

Sociologists use the term **role identity** to refer to the image we have of ourselves in a specific social role (Burke 1980, 18). For example, a woman who is a professor, a mother, and an aerobics student will have a different role identity in each setting, and her self-concept will be a composite of these multiple identities. The idea of role identities draws heavily on the analogy of life as a stage. As we move from scene to scene, we change costumes, get a new script, and come out as a different character. A young man may play the role of dutiful son at home, party animal at the dorm, and serious scholar in the classroom.

When two roles come into conflict, we typically follow the role that we believe will give us the most self-esteem. **Self-esteem** is the evaluative part of the self-concept; it is our judgment of our worth compared with others' worth. Because we would all like to think well of ourselves, we strive to sustain a self-concept that will reinforce our self-esteem (Owens, Stryker, & Goodman 2001).

Significant others are the role players with whom we have close personal relationships.

The **generalized other** is the composite expectations of all the other role players with whom we interact; it is Mead's term for our awareness of social norms.

Role identity is the image we have of ourselves in a specific social role.

Self-esteem is the evaluative component of the self-concept; it is our judgment about our worth compared with others' worth.

Agents of Socialization

Socialization is a continual process of learning. Each time we encounter new experiences, we are challenged to make new interpretations of who we are and where we fit into society. This challenge is most evident when we make major role transitions—when we leave home for the first time, join the military, change careers, or get divorced, for example. Each of these shifts requires us to expand our skills, adjust our attitudes, and accommodate ourselves to new social roles. Child psychologists have noted that these periods of transition tend to herald both intellectual and moral growth in the youngster. They constitute a crisis that challenges our old assumptions about ourselves and prompts a fundamental reappraisal of who we are.

Learning takes place in many contexts. We learn at home, in school and church, on the job, from our friends, and from television. These agents of socialization have a profound effect on the development of personality, self-concept, and the social roles we assume, especially if the messages learned in one setting are reinforced elsewhere. (See Focus on Media and Culture: Girls' Hair, Girls' Identities.) Each of these agents of socialization is discussed more fully in later chapters. They are introduced here to illustrate the importance of social structures for learning.

Family

The most important agent of socialization is the family. As the tragic cases of child neglect and the monkey experiments so clearly demonstrate, the initial warmth and nurturance we receive at home are essential to normal cognitive, emotional, and physical development. In addition, our family members—usually our parents but sometimes our grandparents, stepparents, or others—are our first teachers. From them we learn not only how to tie our shoes and hold a crayon but also beliefs and goals that may stay with us for the rest of our lives.

The activities required to meet the physical needs of a newborn provide the initial basis for social interaction. Feeding and diaper changing give opportunities for cuddling, smiling, and talking. These nurturant activities are all vital to the infant's social and physical development; without them, the child's social, emotional, and physical growth will be stunted (Rutter et al. 1999).

In addition to these basic developmental tasks, the child has a staggering amount of learning to do before becoming a full member of society. Much of this early learning occurs in the family as a result of daily interactions: The child learns to talk and communicate, to play house, and to get along with others. As the child becomes older, teaching is more direct, and parents attempt to produce conformity and obedience, impart basic skills, and prepare the child for life outside the family.

One reason the family is the most important agent of socialization is that the self-concept formed during childhood has lasting consequences. In later stages of development, we pursue experiences and activities that integrate and build on the foundations established in the primary years. Although the personality and self-concept are not rigidly fixed in childhood, we are strongly conditioned by childhood experiences.

The family is also an important agent of socialization in that the parents' religion, social class, and ethnicity influence the child's social roles and self-concept. They influence the expectations that others have for the child, and they determine the groups with which the child will interact outside the family. Thus, the family's race, class, and religion shape the child's initial experiences in the neighborhood, at school, and at work.

Media and Culture

Girls' Hair, Girls' Identities

Why does a "bad hair day" matter so much to girls and women that some will just stay home; some will go through their day cranky, unconfident, or depressed; and most will sacrifice time and money to avoid this fate? This is the question that led Rose Weitz, one of the authors of this textbook, to write the book *Rapunzel's Daughters: What Women's Hair Tells Us About Women's Lives*. A good part of the answer to this question, she found, lies in girls' socialization.

As Weitz found, parents, teachers, friends, neighbors, and even strangers passing on the street all teach girls to consider their hair central to their identity and to their position in the world. Parents praise their daughters when their hair is neatly styled, refuse to take their daughters to church or the mall when it isn't, and drag their daughters to beauty parlors even when their daughters could care less. Teachers will pull out a comb and fix girls' hair when they consider it too unruly, and strangers on the street will comment on how a girl's "beautiful blonde curls" (if she is white) or naturally long, straight hair (if she is African American) will surely garner her a rich husband.

Girls are also socialized to consider their hair central to their identity through the mass media and material culture. Through toys and other gifts, girls learn to consider hair work both fun and meaningful. Barbie dolls are an especially clear example. In addition to garden-variety Barbies, girls can get (among many others) Fashion Queen Barbie, which comes with blonde, brunette, and "titian-haired" wigs; Growin' Pretty Hair Barbie, whose hair can be pulled to make it longer; and Totally Hair Barbie, the most popular Barbie ever, which comes with hair to her toes, styling gel, a hair pick, and a styling book. The Barbie "Styling Head," which consists of nothing but a head with long hair, is also popular. Similarly, although few think it appropriate to give boys curling irons or blow dryers as gifts, many parents, aunts, and uncles give such gifts to girls, further reinforcing the importance of hair for girls.

So, too, do the mass media. Time after time, in movies like *Pretty Woman* or *America's Sweetheart*, apparently plain women get the guy once they get a new hairstyle (and ditch the glasses). If you see a girl or woman in a movie with bad hair, she is either the villain, the comic side-kick, or about to get a make-over and get the guy. Even children's cartoons follow this pattern: Smurfette, the only female on the show

The Smurfs, was created by the wicked wizard Gargamel to be an evil, conniving seductress who would cause the Smurfs' downfall. When Papa Smurf changed her into a good Smurfette, her messy, medium-length, brown hair became long, smooth, and blonde.

Magazines aimed at teenagers more directly socialize girls to focus on their hair and appearance. Weitz found that about half of advertisements and articles in recent issues of *Seventeen* (the most popular teen magazine) focused on how girls could change their hair or bodies. Similarly, makeover stories, in which ugly ducklings are transformed into swans by changing their hair, makeup, and clothes, are a regular feature on television talk shows and even news shows, as well as providing the whole focus of the daily show *The Makeover Story*.

Although some girls are more immune than others to media messages, few escape their effects fully. As discussed in the previous chapter, even when individual girls reject the idea that they should define themselves through their appearance, they still feel obligated to *act* as if they accept that idea because they believe that others accept it and judge them on that basis (Milkie 1999).

Schools

In Western societies, schooling has become accepted as a natural part of childhood. The central function of schools in industrialized societies is to impart specific skills and abilities necessary for functioning in a highly technological society.

Schools do much more than teach basic skills and technical knowledge, however; they also transmit society's central values and ideologies and are a prime area for teaching minorities and immigrants about the dominant culture (Spring 2004). Unlike the family, in which children are treated as special persons with unique needs and problems, schools expose children to situations in which authorities (teachers) have to deal with children en masse and cannot afford to pay attention to each individual.

Many teenagers change their appearance to demonstrate their membership in a specific peer group.

In schools, children first learn that levels of achievement affect status in groups and learn that they will be punished if they do not sit still, follow orders, and otherwise fit in (Parsons 1964, 133; Gatto 2002). In this sense, schools are training grounds for roles in the workplace, the military, and other bureaucracies.

Peers

In past centuries, and in some parts of the world still today, children often lived on isolated farms where their families remained almost the only important agent of socialization throughout their childhood. For the last several decades, however, compulsory education together with the late age at which most youths become full-time workers have led to the emergence of a youth subculture in modern societies. In recent years, this development has been accelerated by the tendency for both parents to work outside the home, creating a vacuum that may be filled by peer interaction (Osgood et al. 1996). Once children are exposed to others their own age, they quickly come to identify with their peers, to adopt peer culture, and to place a high value on peer acceptance (Harris 1998).

What are the consequences of peer interaction for socialization and the development of the self-concept? Because kids who hang out together tend to dress and act a lot alike, peer pressure creates conformity to the group—whether the group is cheerleaders, honors students, or gang members. As a result, conformity to peer values and lifestyles can be a source of family conflict when, for example, your friends urge you to pierce your tongue and your parents are horrified at the idea. The more time you have free to hang out with friends unsupervised by adults, the more likely it will be that your friends will affect you (Haynie and Osgood 2005).

However, the effects of peer pressure are often overestimated (Haynie & Osgood 2005). First, it may appear that kids share attitudes and behaviors because they hang out together, when in fact they chose to become friends because they *already* shared attitudes and behaviors. Mormons seek other Mormons, deadheads seek other deadheads, and heavy drinkers seek other heavy drinkers. Second, adolescents remain concerned about their parents' opinions as well as their friends'. Even if they engage in behavior with their friends that their parents disapprove of, they usually do so only if they think their parents won't find out.

Nevertheless, peer group socialization has important influences (Corsaro 2003, 2004). Because the judgments of one's peers are unclouded by love or duty, they are particularly important in helping us get an accurate picture of how we appear to others. In addition, the peer group is often a mechanism for learning social roles and values distinct from those of adults. For example, peer groups teach their own cultural norms about everything from whether one should share with another child to whether one should smoke or drink.

Mass Media

Throughout our lives we are bombarded with messages from radios, magazines, films, billboards, and other media. The most important mass medium for socialization, however, is undoubtedly television. Nearly every home has one, and the average person in the United States spends many hours a week watching it.

The effects of television viewing are vigorously debated. Many suggest that the media promote violence, sexism, racism, and other problematic ideas and behaviors, but the evidence is contradictory (Felson 1996). The most universally accepted conclusion is that the mass media can be an important means of supporting and validating what we already know. Through a process of selective perception, we tend to give special notice to material that supports our beliefs and self-concept and to ignore material that challenges us.

Television, however, may play a more active part than this. Studies suggest that characters seen regularly on television can become role models whose imagined opinions become important as we develop our own roles. For example, adolescents might look to the program *Grey's Anatomy* for ideas about how to deal with the opposite sex. Material on television may supplement the knowledge our own experience gives us about U.S. roles and norms. These findings imply that the content of television can have an important influence.

Religion

In every society, religion is an important source of individual direction. The values and moral principles in religious doctrine give guidance about appropriate roles and behaviors. Often the values we learn through religion are compatible with the ideals we learn through other agents of socialization. For example, the golden rule ("Do unto others as you would have them do unto you") taught in religious education fits easily with similar messages heard at home and at school.

The role of religion, however, cannot be reduced to a mere reinforcer of society's norms and values. As we point out in Chapter 12, religious ideals have the power to change societies and the individuals in them. Moreover, even within modern U.S. society, there are important differences in the messages delivered by, say, the Mormon, Jewish, and Baptist religions, as well as differences between the conservative and liberal wings of each of these religions. These differences account for some significant variability in socialization experiences.

Workplace

Almost all of us will spend a significant portion of our adult lives working outside the home. The environments in which we work, however, are very different. Some of us will work with machines, others with ideas; some will work with people, others on people. Work is found in cities, factories, offices, and fields. Much of it is impersonal, monotonous, and regulated by time clocks, but some is highly personal, challenging, and flexible.

Long-term research by Kohn and his associates indicates that the nature of our work affects our self-concept and behavior. The amount of autonomy, the degree of supervision and routinization, and the amount of cognitive complexity demanded by the job have important consequences. If your work demands flexibility and self-discipline, you will probably come to value these traits elsewhere—at home, in government, and in religion. If your work instead requires subordination, discipline, and routine, you will come to find these traits natural and desirable (Kohn et al. 1983).

Because of the importance of our work to our identity, studies demonstrate that losing one's job can be a major blow to the self-concept. Although some people who lose their jobs can protect their self-esteem by blaming the economy or the government, individuals more often assume they are to blame (Newman 1999a). As a result, unemployment increases the incidence of depression and physical illness (Shortt 1996).

Socialization Through the Life Course

As our discussion of agents of socialization suggested, socialization occurs throughout life, beginning in childhood and continuing throughout our adult lives, even into old age.

Childhood

Primary socialization is personality development and role learning that occurs during early childhood.

Early childhood socialization is called **primary socialization.** It is primary in two senses: It occurs first, and it is most critical for later development. During this period, children develop personality and self-concept; acquire motor abilities, reasoning, and language skills; and are exposed to a social world consisting of roles, values, and norms.

During the period of primary socialization, children are expected to learn and embrace the norms and values of society. Most learn that conforming to social rules is an important key to gaining acceptance and love, first from their family and then from others. Because young children are so dependent on the love and acceptance of their family, they are under especially strong pressure to conform to their family's expectations. This is a critical step in turning them into conforming members of society. If this learning does not take place in childhood, conformity is unlikely to develop in later life.

Adolescence

Anticipatory socialization is role learning that prepares us for roles we are likely to assume in the future.

Adolescence serves as a bridge between childhood and adulthood. As such, the central task of adolescence is to begin to establish one's independence from one's parents.

During adolescence, we often engage in **anticipatory socialization**—role learning that prepares us for roles we are likely to assume in the future. Until about 1980, for instance, all American girls were required to take "home economics" courses to learn how to sew and cook. Boys, in turn, were required to take "shop" courses to learn

A Global Perspective

Preschools and Socialization in Japan

focus on

Because each culture holds different values and traditions, each culture socializes its children differently. This is easily illustrated by comparing Japanese and American kindergartens (Small 2001, 129–132).

A central value of Japanese culture is the sense of belonging to a unified, homogeneous nation with common goals. Reflecting this, kindergartens across Japan are state-regulated, have similar facilities, and use similar curricula, which have changed little over the years. In this way, the country ensures that all children—across generations and regions—are socialized into the same values, and that all have more or less equal access to the resources they

need to be successfully socialized and to succeed in their later studies. In contrast, American culture values individual rights *vis á vis* the state, and state's rights *vis á vis* the national government. As a result, we expect and accept great differences in how and what young children are taught.

Another central value of Japanese culture is an emphasis on cooperation and group accomplishment over individual achievement. Japanese adults are expected to take pride in the successes of their work groups and to humbly downplay their own successes. Similarly, national preschool curricula in Japan stress cooperation over individual achievement. Teachers speak to their students in ways that emphasize the students' "groupness;" use games, songs, and other activities designed to teach students to work together and to think of themselves as a group; and continually urge children to consider

how their actions affect others. When children misbehave, teachers integrate the children back into the group and into acceptable behavior, rather than highlighting the misbehaviors. In contrast, in the United States kindergarteners are taught from the start to interpret their successes as resulting from their individual achievements, rather than from group support or activity, and are taught to take pride in those successes. Teachers goad children to perform better by praising individuals who are succeeding and correcting or chastising those who are not. Finally, teachers quickly come to conclude that certain students are troublemakers, best dealt with by isolating them from the group rather than trying to integrate them into it.

In sum, socialization in both Japanese and U.S. kindergartens both reflects and reinforces the different cultural values of these two countries.

Whereas American kindergartens emphasize individual achievement and pride, Japanese kindergartens socialize children to value cooperation and group accomplishment.

Anticipatory socialization prepares us for the roles we will take in the future. Children everywhere play out their visions of how mommies and daddies ought to behave.

Professional socialization is role learning that provides individuals with both the knowledge and a cultural understanding of their profession.

Role exits are the processes by which individuals leave important social roles.

how to fix cars and use wood-working tools. These days, boys can take cooking and girls can take wood-working. Nevertheless, teenagers' household chores, part-time jobs, and volunteer work still tend to divide along traditional lines. While boys sometimes help around the house, girls more often are expected to care for their younger siblings, cook, and clean house (Lee, Schneider, & Waite 2003). If a boy does help at home, he's likely to take on such "masculine" tasks as mowing the lawn or washing the car. Similarly, the volunteer work and part-time jobs that girls hold more often involve emotional nurturance (such as working as a hospital volunteer), whereas boys more often work in outdoor or mechanical tasks. In all these ways, girls and boys prepare for the family and work roles they anticipate holding as adults.

Adulthood

Because of anticipatory socialization, most of us are more or less prepared for the responsibilities we will face as spouses, parents, and workers. Goals have been established, skills acquired, and attitudes developed that prepare us to accept and even embrace adult roles. Because anticipatory socialization is never complete, however, anyone who wants to enter a professional field must first undergo **professional socialization.** The purpose of professional socialization is to learn not only the knowledge but also the *culture* of a profession. Medical training provides an example of this process.

Most commonly, people choose to become doctors out of a desire to help others. Yet one of the primary tenets of medical culture is that doctors should be emotionally detached—distancing themselves from their patients and avoiding any show of emotion (Weitz 2007). According to sociologist and medical school professor Frederic Hafferty (1991), this cultural norm is taught from the beginning of medical education. Through his observations, Hafferty discovered that when new students come to a medical school for the first time, second-year students almost invariably take them to the school's anatomy laboratory. There the second-year students proudly display the most grotesque partially dissected human cadaver available. Although officially they do so to display the school's laboratory facilities, their true purpose seems to be to elicit emotional reactions from the new students. The laughter and snickers these reactions evoke in the second-year students demonstrate to the new students that such behavior is shameful while demonstrating to the second-year students how "tough" they have become. This is a particularly vivid example of professional socialization, but every job change we make as adults requires some socialization to new responsibilities and demands.

Age 65 and Beyond

More and more Americans now live for many years past age 65. During these years, most will give up their former careers, and many will lose spouses and experience declining physical abilities. As a result, **role exits,** which can exist at any life stage, are especially common in later life.

The concept of role exits was first defined and studied by Helen Rose Fuchs Ebaugh (1988), a former nun who became a sociologist. Through comparing her experiences and those of other ex-nuns to the experiences of ex-alcoholics, ex-spouses, and others, she identified a common process through which people leave important social roles. Individuals first go through a stage of questioning their original role.

Subsequently, they begin to consider alternative roles and to imagine what their lives would be like in those roles. During this time, they begin disengaging themselves emotionally from their original role. Eventually, the pull of the imagined new role becomes stronger than the ties to the former role.

Individuals who leave important social roles never do so completely, however; an individual can cease being a nun or a convict, but there is no way to cease being an ex-nun or ex-convict. Many ex-nuns never become comfortable wearing skimpy blouses, and many ex-alcoholics decide never to return to social drinking. In addition, other people who know of their past may continue to interact with them on the basis of their former statuses. Employers may refuse to hire ex-convicts and friends may refrain from swearing around ex-nuns.

As these examples suggest, the process of role exit is followed by the process of socialization to new roles. Even during later life, many individuals who leave earlier roles develop new careers, engage in volunteer work, or return to school. Others must adapt to new roles within intimate relationships: learning to live alone or with new spouses or learning to allow their children or others to take care of them.

Resocialization

Most of the time, socialization and role change are gradual processes. Sometimes, though, changes are abrupt and extreme. The most extreme example of role change comes about when we abandon our self-concept and way of life for a radically different one. This is called **resocialization.** Changing the social behavior, values, and self-concept acquired over a lifetime of experience is difficult, and few people undertake the change voluntarily. In this section, we look at the process of resocialization and at certain locations, known as *total institutions,* where resocialization often occurs.

A drastic example of resocialization occurs when people become permanently disabled. Those who become paralyzed experience intense resocialization to adjust to their handicap. All of a sudden, their social roles and capacities are changed. Their old self-concept no longer covers the situation. They may be unable to control bladder and bowels, be severely limited in their ability to get around, or be unable to function sexually in the ways they had previously. If they are single, they must face the fact that they may never marry or have children; if they are older, they may have to reevaluate their adequacy as spouses or parents. These changes require a radical redefinition of self. If self-esteem is to remain high, priorities will have to be rearranged and new, less physically active roles given prominence.

Resocialization may also be deliberately imposed by society. When an individual's behavior leads to social problems—as is the case with habitual criminals, problem alcoholics, and mentally disturbed individuals—society may decree that the individual must abandon the old identity and accept a more conventional one.

Total Institutions

Generally speaking, a radical change in self-concept requires a radical change in environment. Drug counseling one night a week is not likely to drastically alter the self-concept of a teenager who spends the rest of the week among kids who are constantly "wasted." Thus, the first step in the resocialization process is to isolate the individual from the past environment.

This is most efficiently done in **total institutions**—facilities in which all aspects of life are strictly controlled for the purpose of radical resocialization (Goffman 1961a).

Resocialization occurs when we abandon our self-concept and way of life for a radically different one.

Total institutions are facilities in which all aspects of life are strictly controlled for the purpose of radical resocialization.

Historical Note

Goffman developed the idea of *total institutions* after working as an orderly in a public mental hospital in the 1950s. In that hospital, as in most others during that time period, all clothing and other belongings were taken from patients at admission. Patients were given shapeless hospital gowns to wear, their hair was cut short, and they were forced to live regimented lives in which virtually all everyday decisions were taken away from them. Contact with the outside world was severely limited. In these ways, patients' individuality was diminished and their susceptibility to resocialization increased.

Monasteries, prisons, boot camps, and mental hospitals are good examples. Within them, past statuses are wiped away. Social roles and relationships that formed the basis of the previous self-concept are systematically eliminated. New statuses are symbolized by regulation clothing, rigidly scheduled activity, and new relationships. Inmates are encouraged to engage in self-analysis and self-criticism, a process intended to reveal the inferiority of past perspectives, attachments, and statuses.

A Case Study: Prison Boot Camps and Resocialization

Polls repeatedly show that most people and politicians in the United States want to get tough with criminals. At the same time, most of us want to believe that criminals can be rehabilitated. One approach that combines both these goals are prison boot camps or "shock incarceration programs" for young offenders. Such camps, which first appeared in the 1980s, have become common across the country.

Like all resocialization programs, prison boot camps begin with the premise that inmates must radically alter their lifestyles, values, and beliefs. To become productive members of society, inmates must come to reject their old identity and often their old relationships. How are such goals accomplished with groups of young lawbreakers, who often have a long history of prior delinquent acts?

First, new prisoners are segregated or isolated from competing social environments. Shaved heads and derogatory names remind prisoners that they are leaving one self-concept behind and taking on another. Second, all aspects of daily life are strictly controlled, and interaction with other inmates is closely regulated. Days typically begin with an hour of strenuous calisthenics in the early morning darkness and proceed through long hours of military drilling, hard physical labor, drug counseling, and study. Meals last for eight minutes and absolutely no talking is allowed. Prisoners

© R. Maiman/CORBIS Sygma

Uniforms, shaved heads, harsh discipline, and rigorous physical demands are all designed to encourage young prison boot-camp inmates to cast off their old deviant identity and adopt a new conformist self-concept.

never walk; they run. They are taught to stand at attention before approaching a prison official and never to look him or her in the eye (Anderson 1998).

To many, boot camps sound like a great idea. They promise to teach young offenders self-discipline, respect for the law, and the value of work by placing them in a rigorous military-style setting while simultaneously satisfying the public's apparent need for "seeing some civility pounded into thugs who terrorize their neighborhoods" (Katel, Liu, & Cohn 1994). As a result, such camps have also become popular options for parents who feel they have lost control over their children, even if the children have not (yet) been arrested for anything.

Unfortunately, although a few studies have suggested that boot camps work, when researchers have looked at all the available evidence from multiple studies, they find that the camps have no effect (MacKenzie, Wilson, & Kider 2001). Camp graduates are just as likely as prison parolees to end up back in trouble with the law.

Given the disparity between the norms espoused in the prison camps and those in the peer groups to which most inmates return, it is little wonder that so many return to crime. Both proponents of boot camps and their critics agree that without educational programs, drug and vocational counseling, and a well-designed program of postincarceration supervision, the changes in self-concept and values that prison boot camps can generate are insufficient to enable offenders to overcome the difficulties they face when they return home (MacKenzie, Wilson, & Kider 2001).

Where This Leaves Us

Each of us is unique, a product of our individual biology, abilities, personality, experiences, and choices. But each of us is also a social creation. Through socialization we come to learn the roles and values expected of us and, more often than not, to take on those roles and values as our own. Sometimes that process is obvious: a parent slapping a child's hand for grabbing a cookie without permission, a minister preaching a sermon on the wages of sin, one girl giving another girl pointers on how to flirt. Other times, we are no more aware of the socialization process than a fish is aware of water; it is simply a part of the life around us. The typical American, for example, now spends several hours each day watching television. During those hours we not only learn who murdered this week's victim on *Law & Order*, who is this season's *American Idol*, and who married Joe Millionaire, but we also learn ways of looking at the world: to fear random violence and trust the police; to value success, talent, and fame; to honor wealth; and so on.

Summary

1. Understanding the effects of socialization does not mean accepting that we are victims of that socialization and have no choices in our lives. But unless we understand the ways in which we have been socialized, we will be unable to see our choices clearly and to turn those choices into realities.

2. The self is a combination of unique qualities and shared norms. The two key components of the self are the I and the me.

3. Although biological capacities enter into human development, our identities are socially bestowed and socially sustained. Without human relationships, even our natural capacities would not develop.

4. Freudian theory links social development to biological cues. Freud believed that to become a healthy adult, children must develop a reasonable balance between id and superego.

5. Piaget theorized that cognition develops through a series of stages. Only in the last stage do children develop the capacity to understand and think abstractly,
and some children may never reach that stage.

6. Structural functionalists theorize that socialization—
in schools, religious institutions, families, and elsewhere—is designed to smoothly integrate the young
into the broader culture, avoiding conflict or chaotic
social change. It is most useful for explaining the benefits of a stable social system.

7. Conflict theory focuses on how socialization reinforces
unequal power arrangements. It is most useful for understanding how socialization can quash dissent and
social change and reproduce inequalities.

8. Symbolic interaction theory emphasizes that self-
concept develops through actively interpreting our interactions with others and the images of ourselves that
we glean from others. Three important concepts in this
theory are the looking-glass self, role taking, and role
identity.

9. Socialization, the process of learning the norms and
values necessary for participation in social institutions,
occurs throughout life at four basic levels: primary socialization, anticipatory socialization, professional socialization, and resocialization.

10. The family is responsible for the early nurturance that
helps the infant develop into a human being, and is
central in laying the foundation of socialization and of
the self-concept. In addition, the family provides a
background (social class, religion, and so on) that determines much of the child's other interactions. Other
important agents of socialization include peers,
schools, mass media, the workplace, and religion.

Thinking Critically

1. What role identity is most important to you now? Do
you think that might change over the course of your life?
Why or why not?

2. Think about the student role in your own life. To what
extent did you *take* it? To what extent did you *make* it?

3. List some specific ways that a family's social class might
influence the way a child is socialized. Can you think of
any ways that living in the city versus living in the country might matter?

4. Thinking back to your childhood, what values might
you have learned from the four or five television shows
that you watched the most? Did you learn from them?
Why or why not?

Companion Website for This Book

academic.cengage.com/sociology/brinkerhoff
Gain an even better grasp on this chapter by going to
the Companion Website. This resource contains tutorial
quizzes and flash cards to help you master key terms and
concepts.

Suggested Readings

Croteau, David, and Hoynes, William. 2000. *Media/Society:
Industries, Images, and Audiences.* Thousand Oaks, Calif:
Pine Forge Press. Includes excellent sections on how
the media represent different social problems and social issues and how audiences interpret and respond to
media portrayals.

Eberstadt, Fernanda. 2006. *Little Money Street: In Search of Gypsies and Their Music in the South of France.* New York:

Knopf. A journalist's witty and fascinating account of her encounters with gypsies. Along the way, she learns a great deal about how gypsy children are socialized by their families—and how schools fail to make any impact on them at all.

Goffman, Erving. 1961. *Asylums*. Garden City, N.Y.: Anchor/Doubleday. A classic and penetrating account of how total institutions affect self-concept. Focuses on mental hospitals and mental patients, but the analysis is applicable to other total institutions.

Thorne, Barrie. 1993. *Gender Play: Girls and Boys at School*. New Brunswick, N.J.: Rutgers University Press. Detailed observations of young girls and boys in classrooms and schoolyards identify how both teachers and children themselves teach and create ideas about what it means to be male or female.

CHAPTER 4

Social Structure and Social Interaction

© Robert Caputo/Stock, Boston Inc.

Intertwining Forces: Social Structure and Social Interaction

Most people who become sociologists do so because they are interested in studying particular social problems, such as homelessness, mental illness, or racial inequality. Each of these problems has roots in and consequences for both broad social structures and everyday social interactions. For example, racial inequality in the United States in part stems from the nature of our national economy and political institutions: There simply aren't enough good-paying jobs near nonwhite communities, and these communities rarely have enough political power to entice corporations to bring in good jobs. But racial inequality is also reinforced on a day-to-day basis whenever teachers spend less time with nonwhite than with white students or police officers assume that nonwhites are more likely than whites to be criminals. As this example suggests, to fully understand society and social problems, sociologists must look at both social structure and social interaction. This chapter describes these two basic features of society. As we will see, research on social structures often draws on structural-functionalist or conflict theories, whereas research on social interaction typically draws on symbolic interaction theory.

Social Structures

Many of our daily encounters occur in patterns. Every day we interact with the same people (our family or best friends) or with the same kinds of people (salesclerks or teachers). These patterned relationships are called social structures. Each of these dramas has a set of actors (mother/child or buyer/seller) and a set of norms that defines appropriate behavior for each actor.

Formally, a **social structure** is a recurrent pattern of relationships. Social structures can be found at all levels in society. Baseball games, friendship networks, families, and large corporations all have patterns of relationships that are repeated day after day. Some of these patterns are reinforced by formal rules or laws, but many more are maintained by force of custom.

> A **social structure** is a recurrent pattern of relationships.

The patterns in our lives are both constraining and enabling (Giddens 1984). If you would like to be free to set your own schedule, you will find the nine-to-five, Monday-to-Friday work pattern a constraint. On the other hand, preset patterns provide convenient and comfortable ways of handling many aspects of life. They help us to get through cross-town traffic, find dates and spouses, and raise our children.

Whether we are talking about a Saturday afternoon ball game, families, or the workplace, social structures can be analyzed in terms of three concepts: status, role, and institution.

Status

The basic building block of society is **status**—a person's position in a group, relative to other group members. Sociologists who want to study the status structure of a society examine two types of statuses: achieved and ascribed. An **achieved status** is a position (good or bad) that a person can attain in a lifetime. Being a father is an

> A **status** is a specialized position within a group.

> An **achieved status** is optional, one that a person can obtain in a lifetime.

An **ascribed status** is fixed by birth and inheritance and is unalterable in a person's lifetime.

achieved status; so is being a convict. An **ascribed status** is a position generally assumed to be fixed by birth or inheritance and unalterable in a person's lifetime. For example, although some people have gender reassignment surgery and some people "pass" as members of a different race, we assume that sex and race are unchangeable. Hence sociologists generally consider sex and race to be ascribed statuses.

Sociologists who analyze the status structure of a society are typically concerned with four related issues (Blau 1987): (1) identifying the number and types of statuses that are available in a society; (2) assessing the distribution of people among these statuses; (3) determining how the consequences—the rewards, resources, and opportunities—differ for people who occupy one status rather than another; and (4) ascertaining what combinations of statuses are likely or even possible.

A Case Study: Race as a Status

To illustrate how our lives are structured by status membership, we apply this approach to one particular ascribed status and ask how being African American affects relationships and experiences in the United States.

To begin: How many racial statuses are there in the United States? The 2000 census asked us to identify ourselves as belonging to one of six racial categories: white, African American, Native American, Asian, mixed race, or other. This same question with more or less the same list of possible answers appears on almost every social survey. The apparent consensus on which racial statuses are significant and the nearly universal concern about them should alert us to the possible consequences that racial status has in our daily lives.

It is not just the number of statuses that has consequences. The numerical distribution of the population among racial statuses also encourages or discourages certain patterns of behavior. In 2000, for example, 2.1 million African Americans lived in New York City, but only 3 lived in Worland, Wyoming. Consequently, white New Yorkers have a far greater chance, statistically, of marrying an African American than do white residents of Worland.

Of course, numbers alone do not tell the whole story. By nearly every measure that one might choose, there is substantial inequality in the rewards, resources, and opportunities available to African American and white people in the United States. African American unemployment is twice that of whites; the infant mortality rate is more than twice as high; the likelihood of being murdered is six times higher (U.S. Bureau of the Census 2006). Similarly, when Hurricane Katrina hit New Orleans in 2005, African Americans were far more likely than white residents to stay in the city. The cause was poverty: African Americans were far less likely than others to have transportation out of the city, money to rent hotel rooms elsewhere, and well-off relatives with large homes who could take them in for an extended stay. Obviously, racial status has enormous consequences on the structures of daily experiences.

Although racial inequality persists, racial status does not correspond as directly with occupational and educational statuses as it once did, and different combinations of statuses are possible. Forty years ago, being African American meant probably having much less education and a much lower status occupation than whites. Today, knowing a person's ascribed status (race) is not such an accurate guide to his or her achieved statuses (education or occupation). Nevertheless, 33 percent of all nurse's aides, attendants, and orderlies but only 5 percent of all physicians in the United States are African American (U.S. Bureau of the Census 2005). The processes through which these overlapping racial, political, and economic statuses are maintained are discussed further in Chapter 8.

Roles

The status structure of a society provides the broad outlines for interaction. These broad outlines are filled in by **roles,** sets of norms that specify the rights and obligations of each status. To use a theatrical metaphor, the status structure is equivalent to the cast of characters ("a young girl, her father, and their maid," for example) whereas roles are equivalent to the scripts that define how the characters ought to act, feel, and relate to one another. This language of the theater helps to make a vital point about the relationship between status and role: People occupy statuses, but they play roles. This distinction is helpful when we analyze how structures work in practice— and why they sometimes don't work. A man may occupy the status of father, but he may play the role associated with it very poorly.

Sometimes people fail to fulfill role requirements despite their best intentions. It is hard to be a good provider, for instance, when there are no jobs available. Failure is also particularly likely when people are faced with incompatible demands because of multiple or complex roles. Sociologists distinguish between two types of incompatible role demands: When incompatible role demands develop within a single status, we refer to **role strain;** when they develop because of multiple statuses, we refer to **role conflict.** For example, role strain occurs when parents don't have enough time to wash their children's clothes, cook their dinner, help them with homework, and play a game together all in the same evening. Role conflict occurs when a parent's need to take time off to care for a sick child conflicts with an employer's expectation that the parent put work obligations ahead of family obligations.

As this suggests, social roles are always changing and flexible. We do not simply play the parts we are assigned with machinelike conformity. Instead, each individual plays a given role differently, depending on their other social statuses and roles, their resources, and the social rewards or punishments they will face as a result of how they play their role.

Institutions

Social structures vary in scope and importance. Some, such as those that pattern a Friday night poker game, have limited application. The players could change the game to Saturday night or up the ante, and it would not have a major effect on the lives of anyone other than members of the group. If a major corporation changed seniority or family leave policies, it would have somewhat broader consequences, not only affecting employees of that firm but also setting a precedent for other firms. Still, the impact of change in this one corporation (or social structure) would likely be limited to certain sorts of businesses. In contrast, changes in other social structures have the power to shape the basic fabric of all our lives. We call these structures social institutions.

An **institution** is an enduring and complex social structure that meets basic human needs. Its primary features are that it endures for generations; includes a complex set of values, norms, statuses, and roles; and addresses a basic human need. Embedded in the statuses and roles of the family institution, for example, are enduring patterns for dating and courtship, child rearing, and care of the elderly. Because the institution of family is composed of millions of separate families, however, the exact rules and behaviors surrounding dating or elder care will vary.

Despite these variations, institutions provide routine patterns for dealing with predictable problems of social life. Because these problems tend to be similar across societies, we find that every society tends to have the same types of institutions.

Connections

Personal Application

Many college students experience role conflict due to the multiple roles they play. Your teachers expect you to read assigned books and your boss expects you to show up on time and work cheerily and efficiently. If you are on a team, your coach expects you to get enough sleep and come to practices prepared to work hard. Your girlfriend or boyfriend expects you to go out with him or her at least once a week. If you sometimes feel there aren't enough hours in the day, you are probably experiencing role conflict.

Roles are sets of norms specifying the rights and obligations associated with status.

Role strain is when incompatible role demands develop within a single status.

Role conflict is when incompatible role demands develop because of multiple statuses.

An **institution** is an enduring social structure that meets basic human needs.

Religion is one of the five basic institutions. Although doctrines and rituals vary enormously, all cultures and societies include a structured pattern of behavior and belief that provides individuals with explanations for events and experiences that are beyond their own personal control.

Basic Institutions

There are five basic social institutions:

1. The family, to care for dependents and rear children.
2. The economy, to produce and distribute goods.
3. Government, to provide community coordination and defense.
4. Education, to train new generations.
5. Religion, to supply answers about the unknown or unknowable.

These institutions are basic in the sense that every society provides some set of enduring social arrangements designed to meet these important social needs. These arrangements may vary from one society to the next, sometimes dramatically. Government institutions may be monarchies, democracies, dictatorships, or tribal councils. However, a stable social structure that is responsible for meeting these needs is common to all healthy societies.

In simple societies, all of these important social needs—political, economic, education, and religious—are met through one major social institution, the family or kinship group. Social relationships based on kinship obligations serve as a basis for organizing production, reproduction, education, and defense.

As societies grow larger and more complex, the kinship structure is less able to furnish solutions to all the recurrent problems. As a result, some activities are gradually transferred to more specialized social structures outside the family. The economy, education, religion, and government become fully developed institutionalized structures that exist separately from the family. (The institutions of the contemporary United States are the subjects of Chapters 10 to 13.)

As the social and physical environments of a society change and the technology for dealing with that environment expands or contracts, the problems that individuals have to face change. Thus, institutional structures are not static; new structures emerge to cope with new problems—or a society will collapse into chaos (Diamond 2005).

Institutional Interdependence

Each institution of society can be analyzed as an independent social structure, but none really stands alone. Instead, institutions are interdependent; each affects the others and is affected by them.

In a stable society, the norms and values embodied in the roles of one institution will usually be compatible with those in other institutions. For example, a society that stresses male dominance and rule by seniority in the family will also stress the same norms in its religious, economic, and political systems. In this case, interdependence reinforces norms and values and adds to social stability.

Sometimes, however, interdependence is an important mechanism for social change. Because each institution affects and is affected by the others, a change in one tends to lead to change in the others. Changes in the economy lead to changes in the family; changes in religion lead to changes in government. For example, when years of schooling become more important than hereditary position in determining occupation, hereditary position will also be endangered in government, the family, and religion.

Institutions as Agents of Stability or Inequality

Sociologists use two major theoretical frameworks to approach the study of social structures: structural functionalism and conflict theory. The first focuses on the part that institutions play in creating social and personal stability; the second focuses on the role of institutions in legitimizing inequality. Because each framework places a different value judgment on stability and order, each prompts us to ask different questions about social structures.

Structural-functional theorists begin with the question "How do institutions help to stabilize a society? To answer this question, they focus on the ready-made, shared patterns for responding to everyday problems that are offered by institutions. By keeping us from having to reinvent the social equivalent of the wheel with each new encounter and each new generation, structural functionalists argue, these patterns and the institutions that underlie them allow social life to run smoothly in stable and predictable ways. Moreover, because these patterns have been sanctified by tradition, we tend to experience them as morally right. As a result, we find satisfaction and security in social institutions.

In contrast, although conflict theorists acknowledge that institutions meet basic human needs, they raise the questions "Why this social pattern rather than another?" The answers they provide typically focus on who benefits from existing institutions and typically illustrate how institutions support the interests of those already in power. Because institutions have existed for a long time, we tend to think of our familial, religious, and political systems as not merely one way of fulfilling a particular need but as the only acceptable way. Just as an eleventh-century Christian might have thought, "Of course, witches should be burned at the stake," so we tend to think, "Of course women should sacrifice their careers for their children." In both cases, the cloak of tradition obscures our ability to recognize inequalities, making inequality seem normal and even desirable. As a result, conflict theorists argue that institutions stifle social change and help maintain inequality.

Types of Societies

Institutions give a society a distinctive character. In some societies, the church is the dominant institution; in others, it is the family or the economy. Whatever the circumstance, recognizing the institutional framework of a society is critical to understanding how it works.

Societies range greatly in complexity. In simple societies, we often find only one major social institution—the family or kinship group. Complex, modern societies, however, have as many as a dozen institutions. What causes this expansion of institutions? The triggering event appears to be economic change. When changes in technology, physical environment, access to resources, or social arrangements increase the level of economic surplus, the possibilities of institutional expansion arise (Lenski 1966; Diamond 1997). In this section we sketch a broad outline of the institutional evolution that accompanied three revolutions in production.

Hunting, Fishing, and Gathering Societies

The chief characteristic of hunting, fishing, and gathering societies is that they have subsistence economies. This means that they rarely produce a surplus. In some years, game and fruit are plentiful, but in many years scarcity is a constant companion.

In hunting, fishing, and gathering societies like that of the Kung Bushmen, tasks tend to be divided along gender lines. Individuals accumulate few personal possessions because there is little surplus and because possessions would be difficult to move.

The basic units of social organization in these societies are the household and the local clan, both of which are based primarily on kinship. Most of the activities of hunting and gathering are organized around these units. A band rarely exceeds 50 people in size and tends to be nomadic or semi-nomadic. Because of their frequent wanderings, members of these societies accumulate few personal possessions.

The division of labor is simple, based on age and sex. The common pattern is for older boys and men (other than the elderly) to participate in hunting and deep-sea fishing and for older girls and women to participate in gathering, shore fishing, and preserving. Aside from inequalities of status by age and sex, few structured inequalities exist in subsistence economies. Members possess little wealth; they have few, if any, hereditary privileges; and the societies are almost always too small to develop class distinctions. In fact, a major characteristic of subsistence societies is that individuals are homogeneous, or alike. Apart from differences occasioned by age and sex, members generally have the same everyday experiences.

Horticultural Societies

The first major breakthrough from subsistence economy to economic surplus was the development of agriculture. When people began to plant and cultivate crops, rather than just harvesting whatever nature provided, stable horticultural societies developed. The technology was often primitive—a digging stick, occasionally a rudimentary hoe—but it produced a surplus.

The regular production of more than the bare necessities revolutionized society. It meant that some people could take time off from basic production and turn to other pursuits: art, religion, writing, and frequently warfare. Of course, the people who participate in these alternate activities are not picked at random; instead, a class hierarchy develops between the peasants, who must devote themselves full time to food production, and those who live off the surplus produced by the peasants.

Because of relative abundance and a settled way of life, horticultural societies tend to develop complex and stable institutions outside the family. Some economic activity is carried on outside the family, a religious structure with full-time priests may develop, and a stable system of government—complete with bureaucrats, tax collectors, and a hereditary ruler—often develops. Such societies are sometimes very large. The Inca Empire, for example, had an estimated population of more than 4 million.

Agricultural Societies

Approximately 5,000 to 6,000 years ago, a second agricultural revolution occurred, and the efficiency of food production was doubled and redoubled through better technology. The advances included the harnessing of animals, the development, in time, of metal tools, the use of the wheel, and improved knowledge of irrigation and fertilization. These changes dramatically altered social institutions.

The major advances in technology meant that even more people could be freed from direct production. The people not tied directly to the land congregated in large urban centers and developed a complex division of labor. Technology, trade, reading and writing, science, and art grew rapidly as larger and larger numbers of people were able to devote full time to these pursuits. Along with greater specialization and occupational diversity came greater inequality. In the place of the rather simple class structure of horticultural societies, a complex class system developed, with merchants, soldiers, scholars, officials, and kings—and, of course, the poor peasants who comprised the bulk of the population and on whose labor the rest all ultimately depended.

One of the common uses to which societies put their new leisure time and other new technology was warfare. With the domestication of the horse (cavalry) and the invention of the wheel (chariot warfare), military technology became more advanced and efficient. Military might was used as a means to gain even greater surplus through conquering other peoples. The Romans were so successful at this that they managed to turn the peoples of the entire Mediterranean basin into a peasant class that supported a ruling elite in Italy.

Industrial Societies

The third major revolution in production was the advent of industrialization about 200 years ago in Western Europe. The substitution of mechanical, electrical, and fossil-fuel energy for human and animal labor caused an explosive growth in productivity, not only of goods but also of knowledge and technology. In the space of a few decades, agricultural societies were transformed. The enormous increases in energy, technology, and knowledge freed the bulk of the workforce from agricultural production and increasingly also from industrial production. The overall effect on society has been to transform its political, social, and economic character. Old institutions such as education have expanded dramatically, and new institutions such as science, medicine, and law have emerged.

A Case Study: When Institutions Die

Throughout most of history, changes in production, reproduction, education, and social control occurred slowly. When these changes occurred gradually and harmoniously, institutions could continue to support one another and to provide stable patterns that met ongoing human needs. On other occasions, however, old institutions—along with old roles and statuses—are destroyed before new ones can evolve. When this happens, societies and the individuals within them are traumatized, and societies and people fall apart.

In 1985, Anastasia Shkilnyk chronicled just such a human tragedy when she described the plight of the Ojibwa Indians of Northwestern Ontario in her book *A Poison Stronger Than Love*. Although the details are specific to the Ojibwa, her story is helpful in understanding what happened to Native Americans in general (including Ojibwas in the United States) and to other traditional societies when rapid social change altered social institutions.

A Broken Society

In 1976, Shkilnyk was sent by the Canadian Department of Indian Affairs to Grassy Narrows, an Ojibwa community of 520 people, to advise the band on how to alleviate economic disruption caused by mercury poisoning in nearby lakes and rivers. Grassy Narrows was a destroyed community. Drunken 6-year-olds roamed winter streets when the temperatures were 40 degrees below zero. The death rate for both children and adults was very high compared with that for the rest of Canada. Nearly three quarters of all deaths were linked directly to alcohol and drug abuse. A quote from Shkilnyk's journal evokes the tragedy of life in Grassy Narrows:

> *Friday.* My neighbor comes over to tell me that last night, just before midnight, she found 4-year-old Dolores wandering alone around the reserve, about 2 miles from her home. She called the police and they went to the house to investigate. They found Dolores's

3-year-old sister, Diane, huddled in a corner crying. The house was empty, bare of food, and all the windows were broken. The police discovered that the parents had gone to Kenora the day before and were drinking in town. Both of them were sober when they deserted their children (Shkilnyk 1985, 41).

Like Dolores and Diane's parents, most of the adults in Grassy Narrows were binge drinkers. When wages were paid or the welfare checks came, many drank until they were unconscious and the money was gone. Often children waited until their parents had drunk themselves unconscious and then drank the liquor that was left. If they could not get liquor, they sniffed glue or gasoline.

Yet 20 years before, the Ojibwa had been a thriving people. How was a society so thoroughly destroyed?

Ojibwa Society before 1963

The Ojibwa have been in contact with whites for two centuries. In 1873, they signed the treaty that defines their relationship with the government of Canada and that established the borders of their reservation.

In the decades that followed, the Ojibwa continued their traditional lives as hunters and gatherers. The family was their primary social institution. A family group consisted of a man and his wife, their sons and their wives and children, or of several brothers and their wives and children. The houses or tents of this family group would all be clustered together, perhaps as far as a half mile from the next family group.

Economic activities were all carried out by family groups. These activities varied with the season. In the late summer and fall, there was blueberry picking and harvesting of wild rice. In the winter, there was hunting and trapping. In all of these endeavors, the entire family participated, with everybody packing up and going to where the work was. The men would trap and hunt, the women would skin and prepare the meat, and the old people would come along to take care of the children and teach them. They used their reserve only as a summer encampment. From late summer until late spring, the family was on the move.

Besides being the chief economic and educational unit, the family was also the major agent of social control. Family elders enforced the rules and punished those who violated them. In addition, most religious ceremonials were performed by family elders. Although a loose band of families formed the Ojibwa society, each family group was largely self-sufficient, interacting with other family groups only to exchange marriage partners and for other ceremonial activities.

The earliest changes brought by white culture did not disrupt this way of life particularly. Even the development of boarding schools, which removed many Indian children from their homes for the winter months, had only a limited effect on Ojibway life: The boarding schools took the children away but did not disrupt the major social institutions of the society they left behind. When the children returned home each summer, their families could still educate them into Ojibway culture and social structure.

The Change

In 1963, however, the government decided that the Ojibwa should be brought into modern society and given the benefits thereof: modern plumbing, better health care, roads, and the like. To this end, they moved the entire Ojibwa community from the old reserve to a government-built new community about 4 miles from their traditional encampment. The new community had houses, roads, schools, and easy access to "civilization." The differences between the new and the old were sufficient to destroy the fragile interdependence of Ojibwa institutions.

First, all the houses were close together in neat rows, assigned randomly without regard for family group. As a result, the kinship group ceased to exist as a physical unit. Second, the replacement of boarding schools with a local community school meant that mothers had to stay home with the children instead of going out on the trap line. As a result, adult women overnight became consumers rather than producers, shattering their traditional relationships with their husbands and community. As a consequence of the women's and children's immobility, men had to go out alone on the trap line. Because they were by themselves rather than with their family, the trapping trips were reduced from several weeks to a few days, and trapping ceased to be a way of life for the whole family. The productivity of the Ojibwa reached bottom with the government order in May 1970 to halt all fishing because of severe mercury contamination of the water caused by a white-owned paper mill. Then the economic contributions of men as well as women were sharply curtailed; the people became heavily dependent on the government rather than on themselves or on each other.

What happened was the total destruction of old patterns of doing things—that is, of social roles, statuses, and institutions. The relationships between husbands and wives were no longer clear. What were their rights and obligations to each other now that their joint economic productivity was at an end? What were their rights and obligations to their children when no one cared about tomorrow?

The Future of the Ojibwa

In 1985, the Ojibwa finally reached a $16.7 million out-of-court settlement with the government and the paper mill to compensate for damages to their way of life arising from both government policies and mercury pollution. However, environmental pollution remains a serious health and economic problem. In addition, mining and clear-cutting of the land by outside corporations now pose new threats to the tribe and its environment. Nevertheless, Ojibwa society has begun the process of healing and recovery. It is using money from the settlement to develop local industries that will provide an ongoing basis for a productive and thriving society, and it is organizing politically against these new threats to its environment, health, and culture (Envirowatch 2006). In the process, it is rebuilding old social institutions and creating new ones.

A Sociological Response

Unfortunately, the Ojibwa are not an exceptional case. Their tragedy has been played out in tribe after tribe, band after band, all over North America. In some tribes alcoholism touches nearly every family. Compared with other Americans, Native American youths and adults are about twice as likely to report abuse of alcohol or illicit drugs (NHSDA Report 2003). As a result, they are significantly more likely to die from chronic liver disease, cirrhosis, accidents, homicide, and suicide (National Center for Health Statistics 2005). In addition, experts estimate that methamphetamine abuse is now twice as common on Indian reservations as elsewhere in the country (Wagner 2006).

High levels of alcohol and drug use are health problems, economic problems, and social problems. Among the related issues are fetal alcohol syndrome, child and spouse abuse, unemployment, teenage pregnancy, nonmarital births, and divorce. How can these interrelated problems be addressed? To paraphrase C. Wright Mills

Connections

Social Policy

Like the Ojibwa, many Native Americans in the United States have lost contact with their culture and suffered greatly as a result. One response to this has been the development of the Native American Church, which combines Christian beliefs with the cultural and religious traditions of various tribes. The Church has been particularly active in U.S. prisons, where it has given many young Native Americans both pride in their culture and hope for the future and helped individuals to exchange abuse of alcohol for controlled, ritual use of peyote.

(see Chapter 1), when one or two individuals abuse alcohol or drugs, this is an individual problem, and for its relief we rightfully look to clinicians and counselors. When large segments of a population have alcohol or drug problems, this is a public issue and must be addressed at the level of social structure.

A sociological response to reducing alcohol and drug problems among Native Americans begins by asking what social structures encourage substance abuse. Conversely, why don't social structures reward those who avoid substance abuse?

The answer depends on one's theoretical framework. Structural functionalists would be likely to focus on the destruction of Native American institutions and the absence of harmony between their remaining institutions and those of white society. Conflict theorists would be likely to focus on how whites used their power to damage or destroy Native American societies by systematically and violently stripping Native Americans of their means of economic production.

Regardless of theoretical position, it is obvious that Native Americans are severely economically disadvantaged. Unemployment is often a way of life; on some reservations, up to 85 percent of the adults are unemployed. Lack of employment is a critical factor in substance abuse in all populations. Having a steady, rewarding job is an incentive to avoid substance abuse; it also reduces the time available for drinking and drug use, which are essentially leisure-time activities. From this perspective, the solution to high levels of substance abuse among Native Americans must include changing economic institutions to provide full employment and bolstering Native American culture and pride, as well as hiring more doctors, counselors, and others to help individuals fight addiction.

In many ways, fighting substance abuse is like fighting measles. We cannot eradicate the problem by treating people after they have it; we have to *prevent* it in the first place. When substance abuse is epidemic in a community, it requires community-wide efforts for prevention. Statuses, roles, and institutions must be re-formed so that people have a reason to avoid abusing drugs or alcohol.

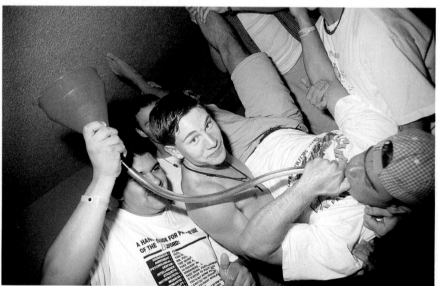

Native Americans are hardly the only subculture in American society to have problems with alcohol abuse. Heavy drinking and drinking games are built into many collegiate activities, such as parties and spring break.

Social Interaction and Everyday Life

Why do people do what they do? The answer depends in part on which social roles they are playing, but it also depends on individuals' social status and resources, on the situation, and on the individual role player. Two people playing the role of physician will do so differently, and the same individual will play the role differently with different patients and in different circumstances. Social structure explains the broad outlines of why we do what we do, but it doesn't deal with specific concrete situations. This is where the sociology of everyday life comes in. Researchers who study the **sociology of everyday life** focus on the social processes that structure our experience in ordinary face-to-face situations.

The **sociology of everyday life** focuses on the social processes that structure our experience in ordinary face-to-face situations.

Managing Everyday Life

Much of our daily life is covered by routines. The most important routines we use for interaction with others are carried out through talk. We all learn dozens of these verbal routines and can usually pull out an appropriate one to suit each occasion. Small rituals such as "Hello. How are you?" "Fine. How are you?" will carry us through dozens of encounters every day. If we supplement this ritual with half a dozen others, such as "thanks/ you're welcome" and "excuse me/no problem," we will be equipped to meet most of the repetitive situations of everyday life.

Nevertheless, each encounter is potentially problematic. What do you do when you say "How are you?" to someone purely as a social gesture, and they then regale you with their troubles for the next twenty minutes? What do you do when your father asks where his car keys are, and you know your brother took them without permission? Although, as Chapter 2 discussed, our culture provides a "tool kit" of routines, each of us must constantly decide which routine to employ, how, when, and why.

At the beginning of any encounter, then, individuals must resolve two issues: (1) What is going on here—what is the nature of the action? and (2) What identities will be granted—who are the actors? All action depends on our answers to these questions. Even the decision to ignore a stranger in the hallway presupposes that we have asked and answered these questions to our satisfaction. How do we do this?

A **frame** is an answer to the question, what is going on here? It is roughly identical to a definition of the situation.

What is going on here? In order to plan an effective course of action, one of the first issues to resolve is defining just what kind of a situation it is. Our response would certainly be different if we believed that this Los Angeles scene had resulted from an earthquake rather than from a riot.

Frames

The first step in any encounter is to develop an answer to the question, What is going on here? The answer forms a frame, or framework, for the encounter. A **frame** is roughly identical to a definition of the situation—a set of expectations about the nature of the interaction episode that is taking place.

All face-to-face encounters are preceded by a framework of expectations—how people will act, what they will mean by their actions, and so on. Even the most simple encounter, say, approaching a salesclerk to buy a pack of gum, is covered by dozens of expectations: In most parts of the United States we expect that the salesclerk will speak English, will wait on the person who got to the counter first, will not try to barter with us over the price, and will not put us down if we are overweight. These expectations—the frame—give us guidance on how we should act and allow us to evaluate the encounter as normal or as deviant.

Our frames will be shared with other actors in most of our routine encounters, but this is not always the case. We may simply be wrong in our assessment of what is going on, or other actors in the encounter may have an entirely different frame. The final frame that we use to define the situation will be the result of a negotiation between the actors.

Identity Negotiation

After we have put a frame on an encounter, we need to answer the second question: Which identities will be acknowledged? This question is far more complex than simply attaching names to the actors. Because each of us has a repertoire of roles and identities from which to choose, we are frequently uncertain about which identity an actor is presenting *in this specific situation.*

To some extent, identities will be determined by the frame being used. If a student's visit to a professor's office is framed as an academic tutorial, then the professor's academic identity is the relevant one. If the professor is a friend of the family, then their interaction might be framed as a social visit, and other aspects of the professor's identity (hobbies, family life, and so on) become relevant.

Typically, identities are not problematic in encounters. Although confusion about identities is a frequent device in comedy films, in real life, a few minutes chatting will usually resolve any confusion about actors' identities. In some cases, however, identity definitions are a matter of serious conflict. For example, Jennifer may want Mike to regard her as an equal, but Mike may prefer to treat her as an inferior.

Resolving the identity issue involves negotiations about both your own and the other's identity. How do we negotiate another's identity? We do so by trying to manipulate others into playing the roles we have assigned them. Mostly we handle this through talk. For example, "Let me introduce Mary, the computer whiz," sets up a different encounter than "Let me introduce Mary, the party animal." Of course, others may reject your casting decisions. Mary may prefer to present a different identity than you have suggested. In that case, she will try to renegotiate her identity.

Identity issues can become a major hidden agenda in interactions. Imagine a newly-minted male lawyer talking to an established female lawyer. If the man finds this situation uncomfortable, he may try to define it as a man/woman encounter rather than a junior lawyer/senior lawyer encounter. He may start with techniques such as "How do you, as a woman, feel about this?" To reinforce this simple device, he might follow up with remarks such as "You're so small, you make me feel like a giant." He may interrupt her by remarking on her perfume. He may also use a variety of nonverbal strategies such as stretching his arm across the back of her chair to assert dominance. Through such strategies, actors try to negotiate both their own and others' identities.

A Case Study: The Ex-Wife at the Funeral

How we act in any encounter depends in large part on the frame we have developed and the identities we and the other participants have worked out. One of the many processes of negotiation that goes on in social encounters is determining whose definition of the situation will be accepted. The person who enters the situation with the most power and resources is most likely to win, but either way, whoever wins this negotiation gains a powerful advantage. A paper by Riedmann (1987) on the role of the ex-wife at the funeral is an insightful illustration of this process.

In this case, a 44-year-old man dies suddenly and accidentally at the home he shares with his new wife of six months. Immediately after the accident, his 20-year-old daughter, who is staying with them, calls her mother to break the news. The ex-wife is terribly upset: She went steady with Bill from age 16 to 22 and was married to him for 20 years; even after the divorce and his remarriage, they remained friends and had lunch together frequently. Although they no longer lived together, they shared a concern for their two children and half a lifetime of memories. His parents were almost as close to her as her own; she had spent many summer afternoons with his brothers and their children.

When informed of his death, therefore, she felt bereaved. Her first response was "What? Our dad is dead?" Her definition of the situation was that she had had a death in her family. Acting on this definition, she went over to the home of her former parents-in-law. Here she came up against a different definition. From the in-laws' position, it was a death in their family but not in her family. Although they did not say so, they wanted her to leave quickly before the current wife came over.

Over the next few days, the ex-wife again and again ran into the dilemmas posed by contrasting definitions of the situation. The obituary did not mention her among the relatives left behind. When she went to view the body at the mortuary, she was denied admittance on the grounds that only members of the immediate family were eligible. When the minister asked everybody but the immediate family to leave the graveyard, she was supposed to leave. Although the ex-wife defined Bill's death as a death in her family, her definition was not commonly shared.

The ex-wife and the new wife each formed teams, with each trying to pursue her own definition of the situation. The "old family" team (made up of the ex-wife, one of her children, her relatives, and one brother-in-law) and the "new family" team competed to be recognized as the "real family." At the funeral service itself, each team held down one corner of the lobby. When family and friends entered the lobby, they were faced with the predicament of which team to console first. Because many of Bill's old friends knew the ex-wife well and the new wife not at all, the "old family" team may be considered to have won this round.

In the end, though, the ex-wife was unable to impose her definition of the situation. The mortician and the priest were paid agents of the "new family" team. The ex-wife was ultimately forced to reframe the situation and to recognize her powerlessness to impose her definition of the situation on others.

Dramaturgy

The management of everyday life is the focus of a sociological perspective called dramaturgy. **Dramaturgy** is a version of symbolic interaction that views social situations as scenes (such as those just described at the funeral home and cemetery) manipulated by actors to convey their desired impression to the audience.

The chief architect of the dramaturgical perspective is Erving Goffman (1959, 1963). To Goffman, all the world was a theater. Like actors, each of us uses our appearance to establish our character—something we do each morning as we choose which clothes to wear, how to style our hair, and whether this would be a good day to show off our tattoos or piercings (if we have any). And like actors, we can use facial expressions, eye contact, posture, and other body language to enhance, reinforce, or even contradict the things we say. For example, telling a worried friend that "Your dress looks fine" doesn't mean as much if you say it without looking up from your cell phone.

Sociologists who use dramaturgy also point out that life, like the theater, has both a front region (the stage) where the performance is given, and a back region where rehearsals take place and different behavioral norms apply. For example, waiters at expensive restaurants are acutely aware of being on stage and act in a dignified

Dramaturgy is a version of symbolic interaction that views social situations as scenes manipulated by the actors to convey the desired impression to the audience.

◼ This boy's body language radiates his dissatisfaction.

and formal manner. Once in the kitchen, however, they may be transformed back into rowdy college kids.

The ultimate back region for most of us, the place where we can be our real selves, is at home. Nevertheless, even here front-region behavior is called for when company comes. ("Oh yes, we always keep our house this clean.") On such occasions, a married couple functions as a team in a performance designed to manage their guests' impressions. People who were screaming at each other before the doorbell rang suddenly start calling each other "dear" and "honey." The guests are the audience, and they too play a role. By seeming to believe the team's act, they contribute to a successful visit/performance.

Identity Work

So far, we've mostly focused on *what* people do in everyday encounters. But it's also important to ask *why* people do what they do. The answer most often supplied by scholars studying everyday behavior is that people are trying to enhance their self-esteem (Owens, Stryker, & Goodman 2001). Social approval is one of the most important rewards that human interaction has to offer, and we try to manage our identities so that this approval is maximized.

Managing identities to support and sustain our self-esteem is called **identity work** (Snow & Anderson 1987). It consists of two general strategies: avoiding blame and gaining credit (Tedeschi & Riess 1981).

Avoiding Blame

There are many potential sources of damage to our social identity and self-esteem. We may have lost our job, flunked a class, been unintentionally rude, or said something that we immediately feared made us look stupid. When we behave in ways that make us look bad, or when we fear we are on the verge of doing so, we need to find ways to protect our identities.

Most of this work is done through talk. C. Wright Mills (1940, 909) noted that we learn what kinds of accounts will justify our norm violations more or less at the same time that we learn the norms themselves. If we can successfully explain away our rule-breaking, we can present ourselves as people who normally obey norms and who deserve to be thought well of by ourselves and others. The two basic strategies we use to avoid blame are accounts and disclaimers.

Accounts

Much of the rule-breaking that occurs in everyday life is of a minor sort that can be explained away. We do this by giving **accounts,** explanations of unexpected or untoward behavior. Accounts fall into two categories: excuses and justifications (Scott & Lyman 1968). **Excuses** are accounts in which an individual admits that the act in question is bad, wrong, or inappropriate but claims he or she couldn't help it. **Justifications** are accounts that explain the good reasons the violator had for breaking the rule; often this is done by appealing to some higher rule (Scott & Lyman, 47).

One of the most fertile fields for both excuses and justifications occurs in the student role. This is highlighted in a research project conducted by sociology pro-

Identity work is the process of managing identities to support and sustain our self-esteem.

Accounts are explanations of unexpected or untoward behavior. They are of two sorts: excuses and justifications.

Excuses are accounts in which one admits that the act in question is wrong or inappropriate but claims one couldn't help it.

Justifications are accounts that explain the good reasons the violator had for choosing to break the rule; often they are appeals to some alternate rule.

A Global Perspective

Identity Work Among Australian Aborigines

I want to owe you five dollars," one Australian Aborigine says to another, requesting a loan of that amount. According to anthropologist Nicolas Peterson (1993), "bumming" like this is an expected part of everyday life among Aborigines living in northeastern Australia. Unlike Native Americans, Australian Aborigines retain sufficient control over enough land that some of them can continue to be nomadic hunters and gatherers. But like most hunters and gatherers, they have virtually no surpluses of food or other needed goods. In a society where food must be found anew each day, sharing makes good sense. Consequently, sharing is an important part of Australian Aboriginal culture. Very little spontaneous sharing occurs outside of the household, however. Instead, food and other things are typically shared because they are demanded.

Unspoken demands for food and other items are common among Aborigines. Simply presenting yourself when food is being prepared and eaten means you have to be included. (We in the United States have this norm, too.)

Adults usually present themselves for something to eat only when large quantities of food have been brought into camp by one household. Children, on the other hand, commonly "bum" food. When children hear someone talking about food, they often go to have a look. This behavior has given rise to the mild rebuke, "Your ears have fingers." Another interesting practice is known as *warmarrkane*. At a public event, A may give B a compliment, such as "You are a fine dancer." A then has the right to make a substantial demand of B.

Demand sharing puts individuals in difficult positions when they don't have enough for themselves or want to accumulate resources for hard times. Since a demand should not be directly refused, what do Australian Aborigines do when they don't want to

share? Aboriginal culture provides individuals with a number of acceptable strategies for avoiding excessive demands. Men and women can hide their resources to avoid requests. By custom, men (but not women) also can stop a demand simply by saying the name of the demander's mother-in-law. Men also can officially give their spears and guns to their mothers or other elderly women so that they can legitimately refuse requests from other men who want to use their weapons.

Demanding, relinquishing, and concealing are all routine rituals of everyday Australian Aboriginal life. An individual's prestige depends largely on being able and willing to share. Demanding, in turn, is one way of reaffirming and making others recognize the demander's rights. When thought of in this way, demand sharing is part of identity work. By demanding, receiving, sharing, and skillfully avoiding the demands of others, Australian Aborigines reinforce both their status and their self-esteem.

---

fessor Kathleen Kalab, who asked her students to explain in writing why they missed class. The answers she received show how students try to explain away their norm violations so that they can preserve a positive identity. For example, one student offered the following *excuse*:

> I am so sorry I missed your class. . . . The reason I missed is a simple yet probably unacceptable reason, I slept until 1:45 p.m. Not only did I sleep through Sociology, I also slept through these things: Geography, trash pick-up at Poland Hall, maintenance workers sawing down a tree outside my window, [and] my alarm clock. (Kalab 1987, 79)

Other students *justified* missing class by explaining that they had to fulfill other, more important obligations. For example, one wrote, "Sorry I wasn't here but Friday my parents called and our entire herd of cattle broke out and were all over the county and then I got to castrate two 500-lb bull calves. So needless to say I was extremely busy" (Kalab 1987, 75).

Accounts such as these are verbal efforts to resolve the discrepancy between what happened and what others legitimately expected to happen. If accounts are accepted, self-identity and social status are preserved, and interaction can proceed normally.

Disclaimers

A person who recognizes that he or she is likely to violate expectations may preface that action with a **disclaimer,** a verbal device used, in advance, to defeat any doubts and negative reaction that might result from conduct (Hewitt & Stokes 1975, 3). Students commonly begin a query with "I know this is a stupid question, but . . ." The disclaimer lets the hearer know that the speaker knows the rules, even though he or she doesn't know the answer.

Disclaimers occur before the act; accounts occur after the act. Nevertheless, both are verbal devices we use to try to maintain a good image of ourselves, both in our own eyes and in the eyes of others. They help us to avoid self-blame for rule-breaking, and they try to reduce the blame that others might attribute to us. If we are successful in this identity work, we can retain fairly good reputations and our social status despite occasional failures in meeting our social responsibilities.

Gaining Credit

To maintain our self-esteem, we need not only to avoid blame but also to get credit for anything good we do. To do so, we employ a variety of verbal devices to associate ourselves with positive outcomes. Just as there are a variety of ways to avoid blame, there are many ways we can claim credit. One way is to link ourselves to situations or individuals with high status. This ranges from dropping the names of popular students we happen to know, to wearing a baseball cap from a winning team, to making a $1,000 donation at a political fund-raiser so we can get a signed photograph of the President to hang on our wall.

Claiming credit is a strategy that requires considerable tact. Bragging is generally considered inappropriate, and if you pat yourself too hard on the back, you are likely to find that others will refuse to do so. The trick is to find the delicate balance where others are subtly reminded of your admirable qualities without your actually having to ask for or demand praise.

Again the negotiation of student identity provides a good example of how we do identify work. Daniel and Cheryl Albas (1988) studied the ways their students at the University of Manitoba managed identity when their papers were returned. "Aces" are students who have done very well on examinations, and their task is to claim all the credit they can without bragging or being condescending to classmates who have done poorly, the "bombers." The Albases note that aces differ in sophistication: While all leave their examinations and grades prominently displayed on their desks, some, in addition, announce their score to others. In the most successful strategy, the ace begins by complaining about how hard the test was and then asks how other students did. Only when asked does the ace reveal his or her own score. Another strategy that enhances the ace's credit while protecting the bomber's identity is to blame both success and failure on luck. When talking to a bomber, an ace may say "It was only luck that I studied the stuff that was on the exam." This strategy gives the ace double credit: for being a good student and for being a nice person.

A Case Study: Identity Work and Homeless Kids

One of the best ways to understand identity work is to look at individuals who have what Goffman (1961b) called *spoiled identities*—identities that are extremely low in status. Examples include sex offenders, traitors, and people with disfiguring facial scars. How do people with spoiled identities sustain their self-esteem?

Like this mouse breeder showing off his awards, most of us seek ways to enhance our credit with others.

© Patrick Ward/Stock, Boston Inc.

A **disclaimer** is a verbal device employed in advance to ward off doubts and negative reactions that might result from one's conduct.

A study among homeless kids in transitional settings (such as shelters and motels) in San Francisco investigated just this question. Anne Roschelle spent four years volunteering at drop-in centers for homeless kids, observing their activities and conversations, and talking with them formally and informally (Roschelle & Kaufman 2004).

The kids Roschelle met were keenly aware of their spoiled identities. They knew that local newspapers often ran stories on the "homeless problem," and that local politicians gained votes by vowing to remove the homeless from the city. As one kid explained, "Everyone hates the homeless because we represent what sucks in society. If this country was really so great there wouldn't be kids like us" (Roschelle & Kaufman 2004, 30). How, then, did these kids maintain their self-esteem and develop a positive identity?

Roschelle and her co-author, Peter Kaufman, found that the kids used two sets of strategies to protect themselves: fitting in and fighting back. *Fitting in* could take various forms. Kids formed friendships with volunteers and with other homeless kids so they would feel they were valued as individuals. They also tried to fit in by dressing, talking, and acting as much like nonhomeless kids as they could: selecting the most stylish coats from the donations box rather than the warmest ones, for example. Kids also chose their words carefully to hide their homelessness. At school, they called caseworkers their "aunts," called homeless shelter staff their "friends,"

Even homeless teens can protect their identities and self-esteem. Some attempt to hide their homelessness and fit in, others, like these teens adopt a tough veneer as a way of fighting back.

© Royalty-Free/CORBIS

and referred to friends who slept three cots away as friends who lived three houses away.

Homeless kids also protected their identities by *fighting back*. First, homeless kids fought back through using "gangsta" clothes, gestures, and actions to intimidate nonhomeless kids. Second, homeless kids adopted sexual behaviors and attitudes far beyond their years and took pride in their sexual "conquests." Finally, kids bolstered their social position by loudly criticizing homeless street people who were more stigmatized than themselves:

> Rosita: Man, look at those smelly street people, they are so disgusting, why don't they take a shower?
> Jalesa: Yeah, I'm glad they don't let them into Hamilton [shelter] with us.
> Rosita: Really, they would steal our stuff and stink up the place!
> Jalesa: Probably be drunk all the time too. (Roschelle & Kaufman 2004, 37)

By contrasting themselves with more stigmatized others, Rosita, Jalesa, and other kids could feel better about themselves.

The homeless kids that Roschelle and Kaufman studied had all the ingredients for the formation of a negative identity: They were hungry, poor, ragged, and homeless in a society that values wealth and blames poverty on the poor. Yet many nevertheless managed to feel good about themselves and construct positive identities. Their experiences reaffirm the assumption made by the interaction school: Even in the face of a spoiled identity, we can negotiate a positive self-concept. But their experiences also illustrate that tactics used to maintain a positive self-identity can also be harmful: Thirteen year olds who take pride in "seducing" 33 year olds or in threatening others with knives and guns are likely to suffer in the long run.

Where This Leaves Us

In the 1950s, structural-functional theory dominated sociology, and a great deal of emphasis was placed on the power of institutionalized norms to determine behavior. Beginning in the 1960s, however, sociologists grew increasingly concerned that this view of human behavior reflected an "oversocialized view of man" (Wrong 1961). In 1967, Garfinkel signaled rebellion against this perspective when he argued that the deterministic model presented people as "judgmental dopes" who couldn't do their own thinking.

Since then, scholars have increasingly tended to view social behavior as more negotiable and less rule bound and have increasingly focused on how people resist rather than accommodate to social pressures (Weitz 2001). This change is most obvious in the sociology of everyday life, but it is also evident in most other areas of sociology, including studies of mental hospitals, businesses, and complex organizations (Crozier & Friedberg 1980; Fine 1996).

This does not mean that the rules don't make a difference. Indeed, they make a great deal of difference, and there are obvious limits to the extent to which we can negotiate given situations. Each actor's ability to negotiate depends on his or her access to resources and power, both of which are strongly determined by social structure.

The perspective of life as problematic and negotiable is a useful balance to the role of social structure in determining behavior. Our behavior is neither entirely negotiable nor entirely determined.

Summary

1. The analysis of social structure—recurrent patterns of relationships—revolves around three concepts: status, role, and institution. Statuses are specialized positions within a group and may be of two types: achieved or ascribed. Roles define how status occupants ought to act and feel.

2. Because societies share common human needs, they also share common institutions: enduring and complex social structures that meet basic human needs. The common institutions are family, economy, government, education, and religion.

3. Institutions are interdependent; none stands alone, and a change in one results in changes in others. Structural functionalists point out that institutions regulate behavior and maintain the stability of social life across generations. Conflict theorists point out that these patterns often benefit one group more than others.

4. An important determinant of institutional development is the ability of a society to produce an economic surplus. Each major improvement in production has led to an expansion in social institutions.

5. The sociology of everyday life is a perspective that analyzes the patterns of human social behavior in concrete encounters in daily life.

6. Deciding how to act in a given encounter requires answering two questions: What is going on here? and Which identities will be granted? These issues of framing and identity resolution may involve competition and negotiation between actors or teams of actors.

7. Dramaturgy is a symbolic interactionist perspective pioneered by Erving Goffman. It views the self as a strategist who is choosing roles and setting scenes to maximize self-interest.

8. The desire for approval is an important factor guiding human behavior. To maximize this approval, people engage in active identity work to sustain and support their self-esteem. This work takes two forms: avoiding blame and gaining credit.

Thinking Critically

1. Is social class an achieved or ascribed status? What would a structural functionalist say? A conflict theorist? A symbolic interactionist?

2. Consider religion as an institution. How would a conflict theorist view it? What might a structural functionalist say? Which position is closest to your own view and why?

3. Pick a social problem that affects you personally; for example, alcoholism, unemployment, racism, sexism, illegal immigration. Describe a social structural solution—one that focuses on changing the underlying social structural causes of the problem, rather than on improving individuals' situations one by one.

4. Describe a time when you disagreed with someone about his or her identity. What kind of situation was it, and why was identity problematic? In the end, whose definition of identity was accepted? Why?

Companion Website for This Book

academic.cengage.com/sociology/brinkerhoff
Gain an even better grasp on this chapter by going to the Companion Website. This resource contains tutorial quizzes and flash cards to help you master key terms and concepts.

Suggested Readings

Barnes, J. A. 1994. *A Pack of Lies: Toward a Sociology of Lying.* New York: Cambridge University Press. An interesting book on lying in everyday life. Barnes argues that almost all of us lie once in a while, especially in socially ambiguous situations. Barnes also looks at lying among politicians, bureaucrats, and medical practitioners.

Diamond, Jared M. 1997. *Guns, Germs, and Steel: The Fates of Human Societies.* New York: W.W. Norton. A fascinating,

Pulitzer-Prize winning account of the factors that allow societies to thrive, based on research data across history and around the world.

Goffman, Erving. 1959. *The Presentation of Self in Everyday Life.* New York: Doubleday. The book in which Goffman laid out the basic ideas behind dramaturgy. Each of Goffman's books is enjoyable reading and easily accessible to the average undergraduate.

Kennedy, Michelle. 2006. *Without a Net: Middle Class and Homeless (with Kids) in America.* New York: Penguin. Kennedy explains how she fell from being a middle-class college student to living in her car with her kids, and how she strove to maintain her sense of self despite her circumstances.

Culture, Social Structure, and Social Change

A major theoretical debate in sociology has revolved around the question of which comes first, culture or social structure? As noted in Chapter 1, Karl Marx believed that the structure of economic institutions determines every other facet of social life, including culture. In contrast, Max Weber believed that cultural values were important causes of social structure. Most sociologists agree with both points of view. On Marx's side, by regulating access to social rewards, social structures (including but not limited to economic institutions) clearly reinforce certain values and norms and discourage and inhibit others. But social institutions exist only to the extent that they are expressed in the roles that individuals play. Thus, as Weber would have it, changes in cultural values and norms eventually lead to changes in individuals, which in turn lead to changes in the statuses and roles that make up the larger social structure.

To illustrate this debate, consider current concerns over the environment. These days, the Amazon forest is rapidly losing all its trees to logging, Glacier National Park will soon have no glaciers due to global warming, and 40 percent of the world's population now suffers from serious water shortages. In northern China, entire villages have collapsed due to underground coal mining, and mountains of coal waste pollute the air, while in many poor U.S. communities, exposure to toxic waste has led to a surge in major birth defects. How did things get this bad? Did culture create social structures that facilitated this environmental destruction? Or did social structures create cultures that did so?

Writing in 1967, historian Lynn White, Jr., argued that the roots of the modern environmental crisis lie in Western cultural values. White believed that a technology capable of efficiently exploiting natural resources (and other human beings, for that matter) began its unparalleled growth in medieval Europe at least partly because of the Christian belief that God granted humans dominion over nature. In contrast, White and others argued, traditional Hindu, Native American, and African cultures believed that human beings are spiritually connected to all elements of nature and are obligated to consider the impact on the environment before taking any actions. From this perspective, culture created social structures that protected the environment (White 1967; Moshoeshoe II 1993; Singh 1993).

More recent historical, anthropological, and geographic research suggests that White's benign view of nonwestern culture is based more on myth than reality (Krech 1999; Cronon 2003; Diamond 1997, 2005). In addition, these critics argue, *regardless* of a society's traditional culture, as its social structure (especially its access to economic and environmental resources) changes, its culture will change as well, in ways that can either protect or harm the environment.

Research has shown how successful societies around the world, with wildly different cultures, eventually collapsed when their environments decayed beyond a certain point: Peasants lost faith in their kings and revolted, tribes decided they needed to take over the land of other tribes to survive, and individuals came to believe that abandoning weak relatives and children made sense.

The most influential statement of this position appears in the work of geographer Jared Diamond. In his book, *Collapse: How Societies Choose to Fail or Succeed,* Diamond (2005) shows how numerous successful societies around the world, with wildly different cultures (such as Rwandan culture, Mayan culture in ancient Mexico, and the Polynesian-based culture of Easter Island) eventually collapsed when their environments decayed beyond a certain point. In these cases, first social structures (like economic institutions and the government) collapsed and then cultural changes followed: Peasants lost faith in their kings and revolted, tribes decided they needed to take over the land of other tribes to survive, and individuals came to believe that abandoning weak relatives and children made sense.) As social structures and culture changed, protecting the environment from further decay seemed less important than maintaining individual survival.

In contrast, in other societies Diamond studied, cultures adapted once they realized the dangers posed by fragile environments, leading to new social structures (such as sustainable irrigation systems) that protected the environment and warded off societal collapse. In these cases, culture eventually affected social structure.

Our world is now at a turning point, and further environmental destruction could cause modern society to vanish as completely as did the Mayan empire. But we still have the potential to change both our culture and our social structure so as to avoid this fate. Recycling, for example, is now far more common because individuals who believed in the need for environmental protection banded together to push for both cultural change (the *idea* that protecting the environment was important) and social structural change (the provision of community recycling facilities). If access to environmental resources becomes more expensive (structural change) and protecting those resources begins to seem cost-effective (cultural change), businesses and politicians may also come to support environmental protection and even more cultural and structural change will be possible.

CHAPTER 5

Groups, Networks, and Organizations

© Tony Freeman/ PhotoEdit

Outline

Human Relationships

Sociology is the study of relationships. We are concerned with how relationships are formed and the consequences these relationships have for the individual and the community. In this chapter we review the basic types of human relationships, from small and intimate groups to large and formal organizations, and discuss some of the consequences of these relationships.

Social Processes

Some relationships are characterized by harmony and stability; others are made stressful by conflict and competition. We use the term **social processes** to describe the types of interaction that go on in relationships. This section looks closely at four social processes that regularly occur in human relationships: exchange, cooperation, competition, and conflict.

Exchange

Exchange is voluntary interaction in which the parties trade tangible or intangible benefits with the expectation that all parties will benefit (Stolte, Fine, & Cook 2001). A wide variety of social relationships include elements of exchange. In friendships and marriages, exchanges usually include intangibles such as companionship, moral support, and a willingness to listen to the other's problems. In business or politics, an exchange may be more direct; politicians, for example, openly acknowledge exchanging votes on legislative bills—I'll vote for yours if you'll vote for mine.

Exchange relationships are based on the expectation that people will return favors and strive to maintain a balance of obligation in social relationships (Molm & Cook 1995). This expectation is called the **norm of reciprocity** (Gouldner 1960; Uehara 1995). If you help your sister-in-law move, then she is obligated to you. Somehow she must pay you back. If she fails to do so, the social relationship is likely to be strained. A corollary of the norm of reciprocity is that you avoid accepting favors from people with whom you do not wish to enter into a relationship. For example, if someone you do not know very well volunteers to type your term paper, you will probably be suspicious. Your first thought is likely to be, "What does this guy want from me?" If you do not want to owe this person a favor, you will say that you prefer to type your own paper. Nonsociologists might sum up the norm of reciprocity by concluding that there's no such thing as a free lunch.

Exchange is one of the most basic processes of social interaction. Almost all voluntary relationships are entered into with the expectation of exchange. In traditional U.S. families, these exchanges were clearly spelled out in social norms. Although not always followed in practice, the husband was expected to support the family, which obligated the wife to keep house and look after the children; or, conversely, she bore the children and was expected to keep house, which obligated him to support her.

An exchange relationship persists only if each party to the interaction is getting something out of it. This does not mean that the rewards must be equal; in fact, rewards are frequently very unequal. Nor does this mean that each party to the exchange

---

Social processes are the forms of interaction through which people relate to one another; they are the dynamic aspects of society.

Exchange is the voluntary interaction from which all parties expect some reward.

The **norm of reciprocity** is the expectation that people will return favors and strive to maintain a balance of obligation in social relationships.

Connections

Personal Application

The norm of reciprocity also applies in dating relationships. If you are a man and buy your date dinner or a movie, you may feel that she now owes you something in return—gratitude, a good night kiss, or more. If you are the woman, you also may believe that you are now indebted to your date, and so may do things you really don't want to do in exchange. When couples disagree on who owes what to whom, situations like these can escalate to sexual assault or even date rape.

relationship has equal power; rather, the actor with greater control over a more valu-able resource always has more power. In children's play groups, for example, one child may be treated badly by the other children and be allowed to play with them only if he agrees to give them his lunch or allows them to use his bicycle. If this boy has no one else to play with, he may find this relationship more rewarding than the alternative of playing alone. The continuation of very unequal exchange relationships usually rests on a lack of desirable alternatives (Molm 2003; Stolte, Fine, & Cook 2001).

Cooperation

Cooperation occurs when people work together to achieve shared goals. Exchange is a trade: I give you something and you give me something else in return. Coopera-tion is teamwork. It is characteristic of relationships in which people work together to achieve goals that they cannot achieve alone. Consider, for example, a four-way stop sign. Although it may entail some waiting, in the long run we will all get through more safely and more quickly if we cooperate and take turns. Most continu-ing relationships have some element of cooperation. Spouses cooperate in raising their children; children cooperate in tricking their substitute teachers.

Cooperation is also important at a much broader social level. Neighborhood residents may work together to fight against a proposed high-rise apartment build-ing, and a nation's citizens may support higher taxes to provide health care for the needy. Individuals are most likely to cooperate when faced with a common threat, when cooperation seems in their economic self-interest, when they share a sense of community identity, and when they value belonging to a community (Van Vugt & Snyder 2002).

> **Cooperation** is interaction that occurs when people work to-gether to achieve shared goals.

Competition

It is not always possible for people to reach their goals by exchange or cooperation. If your goal and my goal are mutually exclusive (for example, I want to sleep and you want to play loud music on your stereo), we cannot both achieve our goals. Similarly, in situations of scarcity, there may not be enough of a desired good to go around. In these situations, social processes are likely to take the form of either competition or conflict.

A struggle over scarce resources that is regulated by shared rules is **competition.** The rules usually specify the conditions under which winning will be considered fair and losing will be considered tolerable. When the norms are violated and rule-breaking is uncovered, competition may erupt into conflict.

Competition is a common form of interaction in individualistic societies like the United States. Jobs, grades, athletic honors, sexual attention, marriage partners, and parental affection are only a few of the scarce resources for which individuals or groups compete. One positive consequence of competition is that it stimulates achievement and heightens people's aspirations. It also, however, often results in per-sonal stress, reduced cooperation, and social inequalities (elaborated on in Chapters 7 through 9).

Because competition often results in change, groups that seek to maximize sta-bility often devise elaborate rules to avoid the appearance of competition. Competi-tion is particularly problematic in informal groups such as friendships and marriages. Friends who want to stay friends will not compete for anything of high value; they might compete over computer game scores, but they won't compete for each other's

> **Competition** is a struggle over scarce resources that is regulated by shared rules.

■ When the struggle for scarce resources (including children's toys) is not regulated by norms that specify the rules of fair play, conflict often results.

spouses. Similarly, couples who value their marriage will not compete for their children's affection or loyalty. To do so might destroy their marriage.

Conflict

Conflict is a struggle over scarce resources that is not regulated by shared rules; it may include attempts to destroy, injure, or neutralize one's rivals.

When a struggle over scarce resources is not regulated by shared rules, **conflict** occurs (Coser 1956). Because no tactics are forbidden and anything goes, conflict may include attempts to neutralize, injure, or destroy one's rivals. Conflict creates divisiveness rather than solidarity.

Conflict with outsiders, however, may enhance the solidarity of the group. Whether the conflict is between warring superpowers or warring street gangs, the us-against-them feeling that emerges from conflict with outsiders causes group members to put aside their jealousies and differences to work together. Groups from nations to schools have found that starting conflicts with outsiders is a useful device for restraining the more disgruntled members of their own group. Critics of U.S. foreign policy suggest that our foreign adventures (such as invading Iraq in 2003) often seem to be timed to improve politicians' approval ratings by diverting the public's attention away from economic recessions and other divisive domestic issues.

Social Processes in Everyday Life

Exchange, cooperation, competition, and even conflict are important aspects of our relationships with others. Few of our relationships involve just one type of group process. Even friendships usually involve some competition as well as cooperation and exchange; similarly, relationships among competitors often involve cooperation.

We interact with people in a variety of relationships. Some of these relationships are temporary and others permanent; some are formal and others informal. In the rest of this chapter, we discuss the wide variety of relationships we have under three general categories: groups, social networks, and organizations.

Groups

A **group** is a collection of two or more people that has two special characteristics: (1) Its members interact within a shared social structure of statuses, roles, and norms, and (2) its members recognize that they are mutually dependent. Groups may be large or small, formal or informal; they range from a pair of lovers to the residents of a local fraternity house to the employees of IBM.

A **group** is two or more people who interact on the basis of shared social structure and recognize mutual dependency.

The distinctive nature of groups stands out when we compare groups to other collections of people. *Categories* of people who share a characteristic, such as all dorm residents, bald-headed men, or Hungarians, are not groups because most members of this category never meet, let alone interact. Similarly, *crowds* who temporarily cluster together on a city bus or in a movie theater are not groups because they are not mutually dependent. Although they share certain norms, many of those norms (such as not staring) are designed to *reduce* their interactions with each other.

The distinguishing characteristics of groups hint at the rewards of group life. Groups are the people we take into account and the people who take us into account. They are the people with whom we share many norms and values. Thus, groups can be a major source of solidarity and cohesion, reinforcing and strengthening our integration into society. When groups function well, they offer benefits ranging from sharing basic survival and problem-solving techniques to satisfying personal and emotional needs. Conversely, when groups function poorly, they can create anxiety, conflict, and social stress.

Types of Groups

Some groups affect our lives far more than others. Almost all of you, for example, belong to a family group as well as to the student body of your college or university. But your membership in the family is probably far more important to you than is your membership in the student body. Sociologists call small, informal, and lasting groups *primary groups;* they call larger, impersonal, less permanent groups *secondary groups*.

Primary Groups

Primary groups are typically informal, small, and personal, characterized by face-to-face interaction (Cooley [1909] 1967). The family is a primary group, as are friendship networks, coworkers, and gangs. The relationships formed in these groups are relatively permanent, generate a strong sense of loyalty and belongingness, constitute a basic source of identity, and strengthen our sense of social integration into society.

Primary groups are groups characterized by intimate, face-to-face interaction.

The major purpose of primary groups is to serve *expressive needs*: to provide individuals with emotional support and a sense of social integration. Your family, for example, probably provides an informal support group that is bound to help you, come rain or shine. You should be able to call on your family and friends to bring you some soup when you have the flu, to pick you up in the dead of night when your car breaks down, and to listen to your troubles when you are blue.

Because we need primary groups so much, they have tremendous power to bring us into line. From the society's point of view, this is the major function of primary groups: They are the major agents of social control. The reason most of us don't shoplift is because we would be mortified if our parents, friends, or coworkers found out. The reason most soldiers go into combat is because their buddies are going. We tend to dress, act, vote, and believe in ways that will keep the support of our primary

■ One of the most powerful mechanisms of social control is the threat of exclusion from valued groups. To avoid the threat of exclusion, we conform: We dress as other members do, think as they think, and do as they do.

© Left Lane Productions/CORBIS

groups. In short, we conform. The law would be relatively helpless at keeping us in line if we weren't already restrained by the desire to stay in the good graces of our primary groups. One corollary of this, however, which Chapter 6 addresses, is that if our primary groups consider shoplifting or tax evasion acceptable, then our primary-group associations may lead us into deviance rather than conformity.

Secondary Groups

Secondary groups are groups that are formal, large, and impersonal.

By contrast, **secondary groups** are formal, large, and impersonal. Whereas the major purpose of many primary groups is to serve expressive needs, secondary groups usually form to serve *instrumental needs*—that is, to accomplish some specific task. The quintessential secondary group is entirely rational and contractual in nature; the participants interact solely to accomplish some purpose (earn credit hours, buy a pair of shoes, get a paycheck). Their interest in each other does not extend past this contract. The Concept Summary shows the important differences between primary and secondary groups.

The major purpose of secondary groups is accomplishing specific tasks. If you want to build an airplane, raise money for a community project, or teach introductory sociology to 2,000 students a year, then secondary groups are your best bet. They are responsible for building our houses, growing and shipping our vegetables, educating our children, and curing our ills. In short, we could not do without them.

The Shift to Secondary Groups

In preindustrial society, there were few secondary groups. Vegetables and houses were produced by families, not by Del Monte or Del Webb. Parents taught their own children, and neighbors nursed one another's ills. Under these conditions, primary groups served both expressive and instrumental functions. As society has become more industrialized, more and more of our instrumental needs are met by secondary rather than primary groups.

In addition to losing their instrumental functions to secondary groups, primary groups have suffered other threats in industrialized societies. In the United

Differences between Primary and Secondary Groups

Concept Summary

	Primary Groups	**Secondary Groups**
Size	Small	Large
Relationships	Personal, intimate	Impersonal, aloof
Communication	Face to face	Indirect—memos, telephone, etc.
Duration	Permanent	Temporary
Cohesion	Strong sense of loyalty, we-feeling	Weak, based on self-interest
Decisions	Based on tradition and personal feelings	Based on rationality and rules
Social structure	Informal	Formal—titles, officers, charters, regular meeting times, etc.
Purpose	Meet expressive needs—provide emotional support and social integration	Meet instrumental goals—Accomplish specific tasks

States, for example, approximately 13 percent of all households move each year (U.S. Bureau of the Census 2006). This fact alone means that our ties to friends, neighborhoods, and coworkers are seldom really permanent. People change jobs, spouses, and neighborhoods. One consequence of this breakdown of traditional primary groups is that many people rely on secondary groups even for expressive needs; if they have marriage problems, they may join a support group rather than talk to a parent, for example.

Many scholars have suggested that these inroads on the primary group represent a weakening of social control; that is, the weaker ties to neighbors and kin mean that peo-

Many of the groups we participate in combine characteristics of primary and secondary groups. The elementary school classroom is a secondary group, yet many of the friendships developed there will last for 6, 12, or even 40 years.

© Bob Daemmrich/Stock, Boston Inc.

FIGURE 5.1

Patterns of Communication

Patterns of communication can affect individual participation and influence. In each figure the circles represent individuals and the lines are flows of communication. The all-channel network provides the greatest opportunity for participation and is more often found in groups in which status differences are not present or are minimal. The wheel, by contrast, is associated with important status differences within the group.

All-channel

Circle

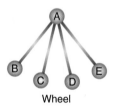

Wheel

ple feel less pressure to conform. They don't have to worry about what the neighbors will say because they haven't met them; they don't have to worry about what mother will say because she lives 2,000 miles away, and what she doesn't know won't hurt her. There is some truth in this suggestion, and it may be one of the reasons that small towns with stable populations are more conventional and have lower crime rates than do big cities with more fluid populations (an issue addressed more fully in Chapter 14).

Interaction in Groups

We spend much of our lives in groups. We have work groups, family groups, and peer groups. In class we have discussion groups, and everywhere we have committees. Regardless of the type of group, its operation depends on the quality of interaction among members. This section reviews some of the more important factors that affect interaction in small groups. As we will see, interaction is affected by group size, physical proximity, and communication patterns.

Size

The smallest possible group is two people. As the group grows to three, four, and more, its characteristics change. With each increase in size, each member has fewer opportunities to share opinions and contribute to decision making or problem solving: Think of the difference between being in a class of 15 students versus a class of 1,500. In many instances, the larger group will be better equipped for solving problems and finding answers, but this practical utility may be gained at the expense of individual satisfaction. Although there will be more ideas to consider, the likelihood that each person's ideas will be influential diminishes. As the group gets larger, interaction becomes more impersonal, more structured, and less personally satisfying.

Physical Proximity

Dozens of laboratory studies demonstrate that interaction is more likely to occur among group members who are physically close to one another. This effect is not limited to the laboratory. You are more likely to become friends with the student who sits next to you in class or who rooms next to you than with the student who sits at the end of your row or who rooms at the end of your hall.

Communication Patterns

Interaction of group members can be either facilitated or hindered by patterns of communication. Figure 5.1 shows some common communication patterns for five-person groups. The communication pattern allowing the greatest equality of participation is the *all-channel network*. In this pattern, each person can interact with every other person with approximately the same ease. Each participant has equal access to the others and an equal ability to become the focus of attention.

The other two common communication patterns allow for less interaction. In the *circle pattern*, people can speak only to their neighbors on either side. This pattern reduces interaction, but it does not give one person more power than others have. In the *wheel pattern*, on the other hand, not only is interaction reduced but a single, pivotal individual gains greater power in the group. The wheel pattern is characteristic of the traditional classroom. Students do not interact with one another; instead they interact directly only with the teacher, thereby giving that person the power to direct the flow of interaction.

Communication patterns are often created, either accidentally or purposefully, by the physical distribution of group members. The seating of committee members at a round table tends to facilitate either an all-channel or a circle pattern, depending on the size of the table. A rectangular table gives people at the ends and in the middle of the long sides an advantage. They find it easier to attract attention and are apt to be more active in interactions and more influential in group discussions. Consider the way communication is structured in the classes and groups you participate in. How do seating structures encourage or discourage communication?

Cohesion

One of the important dimensions along which groups vary is their degree of **cohesion,** or solidarity. A cohesive group is characterized by higher levels of interaction and by strong feelings of attachment and dependency. Because its members feel that their happiness or welfare depends on the group, the group may make extensive claims on the individual members (Hechter 1987). Cohesive adolescent friendship groups can enforce unofficial dress codes on their members; cohesive youth gangs can convince new male members to commit random murders and can convince new female members to submit to gang rapes.

Marriage, church, and friendship groups differ in their cohesiveness. What makes one marriage or church more cohesive than another? Among the factors that contribute to cohesion are small size, similarity, frequent interaction, long duration, a clear distinction between insiders and outsiders, and few ties to outsiders (McPherson, Popielarz, & Drobnic 1992; McPherson & Smith-Lovin 2002). Although all legal marriages in our society are the same size (two members), a marriage in which the partners are more similar, spend more time together, and so on will generally be more cohesive than one in which the partners are dissimilar and see each other for only a short time each day.

Group Conformity

When a man opens a door for a woman, do you see traditional courtesy or sexist condescension? When you listen to Kanye West, LeAnn Rimes, or the Foo Fighters, do you hear good music or irritating noise? Like taste in music, many of the things we deal with and believe in are not true or correct in any absolute sense; they are simply what our groups have agreed to accept as right. Researchers who look at individual decision making in groups find that group interaction increases conformity.

The tremendous impact of group definitions on our own attitudes and perceptions was cleverly documented in a classic experiment by Asch (1955). In this experiment, the group consists of nine college students, all supposedly unknown to each other. The experimenter explains that the task at hand is an experiment in visual judgment. The subjects are shown two cards similar to those in Figure 5.2 and are asked to judge which line on Card B is most similar to the line on Card A. This is not a difficult task; unless you have really poor vision, you can tell that Line 2 most closely matches the line on the first card.

The experimental part of this research consists of changing the conditions under which the subjects make their judgments. Each group must make 15 decisions, and, in the first few trials, all of the students agree on the obviously correct answer. In subsequent trials, however, the first eight students—in reality, all paid stooges of the experimenter—all give an obviously wrong answer. The real test comes in seeing what the last student—the real subject of the experiment—will do. Will he go along with everybody else, or will he publicly set himself apart? Photographs of the experiment show that the real subjects wrinkled their brows, squirmed in their seats, gaped at their neighbors, and, in 37 percent of the trials, agreed with the wrong answer.

Cohesion in a group is characterized by high levels of interaction and by strong feelings of attachment and dependency.

FIGURE 5.2
The Cards Used in Asch's Experiment
In Asch's experiment, subjects were instructed to select the line on Card B that was equal in length to the line on Card A. The results showed that many people will give an obviously wrong answer in order to conform to the group.
SOURCE: From "Opinions and Social Pressure," by Solomon E. Asch. Copyright © November 1955 by Scientific American, Inc. All rights reserved.

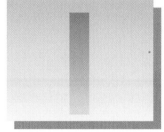
Card A

Card B

■ In this sequence of pictures, a subject (in the middle) shows the strain that comes from disagreeing with the judgments of other members of the group. Some subjects in the Asch experiment disagreed on all 12 trials of the experiment. However, 75 percent of the experimental subjects agreed with the majority on at least one trial. Subjects who initially yield to the majority find it increasingly difficult to make independent judgments as the experiment progresses.

In the case of this experiment, it is clear what the right answer should be. Many of the students who agreed with the wrong answer probably were not persuaded by group opinion that their own judgment was wrong, but they decided not to make waves. When the object being judged is less objective, however—whether Eminem is better than Britney Spears, or football more interesting than basketball—the group is likely to influence not only public responses but also private views. Whether we go along because we are really convinced or because we are avoiding the hassles of being different, we all have a strong tendency to conform to the norms and expectations of our groups.

Yet small groups rarely have access to legal or formal sanctions—they usually can't throw those who disagree with them in jail or the like—so why do individuals so often go along with groups' opinions? The answer is fear—fear of not being accepted by the group (Douglas 1983). The major weapons that groups use to punish nonconformity are ridicule and contempt, but their ultimate sanction is exclusion from the group. From "you're fired" to "you can't sit at our lunch table anymore," exclusion is one of the most powerful threats we can make against others. This form of social control is most effective in cohesive groups, but Asch's experiment shows that fear of rejection can induce conformity even in groups of strangers (Asch 1955).

Group Decision Making

One of the primary research interests in the sociology of small groups is how group characteristics (size, cohesion, and so on) affect group decision making. This research has focused on a wide variety of actual groups: flight crews, submarine crews, business meetings, and juries, to name a few (Davis & Stasson 1988; Robert & Carnevale 1997).

Generally, groups strive to reach consensus; they would like all their decisions to be agreeable to every member. As the size of the group grows, consensus requires lengthy and time-consuming interaction so that everybody's objections can be clearly understood and incorporated. Thus, as groups grow in size, they often adopt the more expedient policy of majority rule. This policy results in quicker decisions,

© Joel Gordon

but often at the expense of individual satisfaction. It therefore reduces the cohesiveness of the group.

Choice Shifts

One of the most consistent findings of research is that it is seldom necessary to resort to majority rule in small groups. Both in the laboratory and in the real world, there is a strong tendency for opinions to converge. One of the classic experiments on convergence was done by Sherif in 1936. In this experiment, strangers were put into a totally dark room. A dot of light was flashed onto the wall, and each participant was asked to estimate how far the light moved during the experimental period. (The dot of light was, in fact, stationary.) After the first session, the participants recorded their own answers and then shared them with the other participants. There was quite a bit of variation in the estimates. Then they did the experiment again. This time there was less difference. After four trials, all participants agreed on an estimate that was close to the average of the initial estimates.

This tendency toward convergence has been demonstrated in dozens of studies since. Convergence, however, is not always to a middle position. Sometimes, the group reaches consensus on an extreme position. This is called the *risky shift* when the group converges on a risky option and the *tame shift* when the choice is extremely conservative. Sometimes these choice shifts depend on persuasive arguments put forward by one or more members, but often they result from general norms in the group that favor either conservatism or risk (Davis & Stasson 1988). For example, one might expect a church steering committee to choose the safest option and a terrorist group to choose the riskiest option.

A special case of choice shift is *groupthink* (Janis 1982; Street 1997). Groupthink occurs when pressures to agree are so strong that they stifle critical thinking. For example, sociologist Diane Vaughan (1996) showed how groupthink contributed to the tragic 1986 explosion of the space shuttle *Challenger*. The engineers working on the *Challenger* all knew before the launch that the shuttle's O-rings probably would suffer some damage. But political pressures to launch the shuttle, coupled with a culture within NASA that rewarded risk taking, created a situation in which the engineers essentially convinced each other and themselves that the risk of O-ring failure was within acceptable limits. As this example illustrates, groupthink often results in bad decisions.

Social Networks

Each of us belongs to a variety of primary and secondary groups. Through these group ties we develop a **social network.** This social network is the total set of relationships we have. It includes our family, our insurance agent, our neighbors, some of our classmates and coworkers, and the people who belong to our clubs. Through our social network, we are linked to hundreds of people in our communities and perhaps across the country.

Your social network does not include everybody with whom you have ever interacted. Many interactions, such as those with some classmates and neighbors, are so superficial that they cannot truly be said to be part of a relationship at all. Unless contacts develop into personal relationships that extend beyond a brief hello or a passing nod, they would not be included in your social network.

Research suggests that social networks are vital for integration into society; they reduce the odds of suicide, increase participation in civic issues, encourage conformity,

Connections

Example

In 2003, sociologists asked 24,613 volunteers to send an e-mail message to one of 18 people in 13 countries (Dodds, Muhamad, & Watts 2003). Each volunteer was asked to forward the message to someone he or she knew, with the goal of eventually delivering the message to the target person. Only 384 e-mail chains reached their target, with most chains breaking down quickly because people didn't forward the message. Successful chains were more likely than unsuccessful chains to rely on weak ties: Individuals forwarded the message to anyone they knew in the target's occupation or country, rather than sending it to individuals with whom they had strong ties.

A **social network** is an individual's total set of relationships.

■ A critical part of our social network is our strong ties—the handful of people to whom we feel intense loyalty and intimacy. Like these two sisters, women are somewhat more apt than men to choose their strong ties from among family members.

and build a firm sense of self-identity (Bearman and Moody 2004; Putnam 2000; Wellman 1999; Wellman & Berkowitz, 1988). Because of their importance for the individual and society, documenting the trends in social networks is an important part of sociological study.

Strong and Weak Ties

Although our insurance agent and our mother are both part of our social network, there is a qualitative difference between them. We can divide our social networks into two general categories of intimacy: strong ties and weak ties. **Strong ties** are relationships characterized by intimacy, emotional intensity, and sharing. **Weak ties** are relationships that are characterized by low intensity and emotional distance (Granovetter 1973). Coworkers, neighbors, fellow club members, distant cousins, and in-laws generally fall in this category. If you and the person you sit next to in class often chat about how you spent the weekend, and occasionally trade notes, but never get together outside of class, you have a weak tie. If the two of you often hang out together, and you'd feel comfortable asking him or her for advice on your romantic relationships, you have a strong tie.

Strong ties are relationships characterized by intimacy, emotional intensity, and sharing.

Weak ties are relationships characterized by low intensity and intimacy.

Strong Ties

Strong ties are crucial for social life. If you are sick, or broke, or your car breaks down just when you need to get to campus for a final exam, it is your strong ties you will call on for help. These are the people who care the most about you, and whom you are most likely to care deeply about. Strong ties give us emotional support, financial help, and all sorts of practical help when needed. However, strong ties can't always be relied on: When those you turn to are also financially or emotionally stressed to the limit, they may not be able to give you the help you need (Menjívar 2000). Not surprisingly, this problem is most severe among poor people, who need the most assistance but whose strong ties are least able to afford to help.

Across socioeconomic groups, Americans' strong ties decreased dramatically between 1985 and 2004 (McPherson, Smith-Lovin, & Brashears 2006). When, in 1985, a national random sample of Americans were asked to name the people with whom they could discuss personally important matters, the most common response (the mode) was to give three names. When the question was repeated with a similar sample in 2004, the most common response was "No one." This is a dramatic shift, in only 19 years. In both surveys, the most common confidants were friends and spouses, but reliance on friends declined while reliance on spouses increased. Most telling, respondents were far less likely in 2004 to report that they turned to parents, children, siblings, co-workers, neighbors, or co-members of groups.

Several factors affect the number and composition of strong ties (McPherson, Smith-Lovin, & Brashears 2006). The most important of these factors is education. People with more education have more strong ties, have a greater diversity of strong ties, and rely less on kinship ties. People with more education are also more likely to have strong ties to influential people—lawyers and doctors rather than plumbers and mechanics. Nonwhites have fewer strong ties than do whites, especially with regard to kinship ties. Neither age nor gender affect the average number or type of strong ties (McPherson, Smith-Lovin, & Brashears 2006), but having children in the home does make a difference: Children reduce women's strong ties by pulling them away from friends and work, whereas children increase men's strong ties to relatives (Munch, McPherson, & Smith-Lovin 1997).

Weak Ties

Weak ties are also important to social life. For example, research indicates that many people first hear about jobs and career opportunities through weak ties (Granovetter 1974; Newman 1999b). In this and other instances, the more people you know, the better off you are.

As this suggests, weak ties are crucial whenever you need to learn or obtain something that requires a broad network. If you have a question about Microsoft Word, for example, you can easily find someone to answer it. But unless you are a computer jock, you'll need a large network of weak ties if you want to get the answer to a question about Linux software.

One of the best things about the Internet is the ease with which you can use it to create weak ties: If you have a rare disease, enjoy an unusual hobby, or are a fan of an obscure band, you can easily create weak ties with others who share your needs or interests.

Ties Versus Groups

The distinction between strong and weak ties obviously parallels the distinction between primary and secondary groups. The difference between these two sets of concepts is that strong and weak apply to one-to-one relationships, whereas primary and secondary apply to the group as a whole. We can have both strong and weak ties within the primary as well as the secondary group. (See Figure 5.3 for an illustration.)

For example, the family is obviously a primary group; it is relatively permanent, with strong feelings of loyalty and attachment. We are not equally intimate with every family member, however. We may be very close to our mother but estranged from our brother. Similarly, although the school as a whole is classified as a secondary group, we may have developed an intimate relationship, a strong tie, with one of our schoolmates. *Strong* and *weak* are terms used to describe the relationship between two individuals; *primary* and *secondary* are characteristics of the group as a whole.

A Global Perspective

Finding Jobs, Changing Jobs in Singapore and China

How will you go about looking for a job after you graduate? After you have read over all of the classified ads and gone to the Career Planning and Placement Office, what will you do next? For most of us, the answer seems obvious: ask friends and family if they are aware of any challenging, well-paying, secure job opportunities for persons with just our credentials. Research conducted in the United States suggests that this might be the wrong strategy (Granovetter 1973, 1974). The people to whom we are most strongly tied are a lot like us; our friends and family tend to be interested in the same things we are, share similar educational and economic backgrounds, and know the same people we do. Consequently, they are un-

likely to offer much information about job possibilities that we didn't already know or couldn't find out. In contrast, our weak ties—people whom we know less well and interact with less frequently—are more likely to have useful new information about jobs. Furthermore, because our weak ties are most likely to bridge us to people with higher status than our own, weak ties give us access to information about more prestigious jobs and to people who may have some influence in hiring decisions (Lin 1990). But are weak ties equally important in finding work in other countries with different economic and social structures?

In the late 1980s and early 1990s, Yanjie Bian and Soon Ang (1997) went to Tianjin, China, and to Singapore to study this question. Tianjin is the third largest city in China (Map 5.1) and in the late 1980s had still not begun the transition to a more Westernized market economy. Jobs in Tianjin and elsewhere in China were assigned by the

government, and although new workers could express their job preferences, these preferences did not necessarily affect their final assignments. In Singapore, the situation was somewhat different. Because of labor shortages, Singaporean workers had a very broad range of jobs from which to choose. And because advertising for new employees was costly, Singaporean employers had a strong stake in trying to identify potential employees who would be loyal to the company and stay with them for a long time.

Compared with the United States, in both Tianjin and Singapore, individuals rely much more directly and explicitly on their social networks of strong ties for favors of all kinds. The guiding principle for individual social networks is *guanxi* (pronounced gwanshe). *Guanxi* literally means "relationship" or "relation" and refers to a set of interpersonal connections that are based on the reciprocal exchange of favors. *Guanxi* relations are strong ties, and

Voluntary Associations

Voluntary associations are nonprofit organizations designed to allow individuals an opportunity to pursue their shared interests collectively.

In addition to relationships formed with individuals, many of us voluntarily choose to join groups and associations. We may join a Bible study group, a soccer team, the Elks, or the Sierra Club. These groups, called **voluntary associations,** are nonprofit organizations designed to allow individuals an opportunity to pursue their shared interests collectively. They vary considerably in size and formality. Some—for example, the Elks and the Sierra Club—are very large and have national headquarters, elected officers, formal titles, charters, membership dues, regular meeting times, and national conventions. Others—for example, soccer teams and knitting groups—are small, informal groups that draw their membership from a local community or neighborhood.

Voluntary associations are an important mechanism for enlarging our social networks. Most of the relationships we form in voluntary associations will be weak ties. But voluntary associations also can introduce us to people with whom we will develop strong ties as close friends and intimates.

Voluntary associations perform an important function for individuals. Studies document that people who participate in them generally report greater satisfaction and personal happiness, longer life, greater self-esteem, more political effectiveness, and a greater sense of community (Burman 1988; Ellison 1991; Moen, Dempster-McLain, & Williams 1989; Prestby et al. 1990; Rietschlin 1998).

those who fail to reciprocate exchanges lose face (Bian 1997; Bian & Soon 1997).

So what did Bian and Ang find? In contrast to the United States, finding work and changing jobs in Singapore and China depends more on strong ties than on weak ones. In China, the trust embedded in *guanxi* allows people to "bend government rules" in helping job seekers find more desirable employment; in Singapore, employers have more faith in the future loyalty of potential employees who are personally recommended to them through *guanxi* networks. Nevertheless, in both China and Singapore, the best strategy of all appears to be one that involves both weak ties and strong *guanxi* ties. Getting help in the job search process depends first and foremost on ties of trust and reciprocity. The highest level job placements occur, however, when a job seeker uses strong ties to approach a higher-

ranked person with whom the job seeker has only a weak tie. In China and Singapore, as in the United States, looking for work is a lot easier when friends have well-placed friends.

MAP 5.1
Tianjin, China, and Singapore.

The correlation between high participation and greater satisfaction does not necessarily mean that joining a voluntary association is the road to happiness. At least part of the relationship between participation and happiness is undoubtedly due to the fact that happy persons who feel politically effective and attached to their communities are more likely than others to join voluntary associations. It also appears to be true, however, that greater participation can be an avenue for achievement and can lead to feelings of integration and satisfaction.

The Mediation Hypothesis

An important characteristic of voluntary associations is that they combine some of the features of primary and secondary groups—for example, the companionship of a small group and the rational efficiency of a secondary group. Some scholars have therefore suggested that voluntary associations mediate (provide a bridge) between primary and secondary groups (Pollock 1982). They allow us to pursue instrumental goals without completely sacrificing the satisfactions that come from participation in a primary group. Through participation in voluntary associations, we meet some of our needs for intimacy and association while we achieve greater control over our immediate environment. Take, for example, a recreational hunter who wishes to protect the right to have guns. This individual can write letters to his representatives in Congress, but he

FIGURE 5.3
Jill's Ties and Groups
Everyone belongs to both primary and secondary groups. Within these groups we each have both weak and strong ties. Jill has strong ties to some of her sorority sisters, one of her classmates, her boyfriend, her mother, and her sister. She has only weak ties to the rest of her family, classmates, and sorority sisters. She also has a strong tie to her boyfriend, who does not belong to her primary or secondary groups. The orange boxed figures show Jill's strong ties.

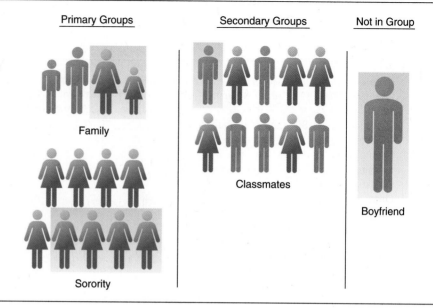

Primary Groups

Family

Sorority

Secondary Groups

Classmates

Not in Group

Boyfriend

will believe, rightly, that as an individual he is unlikely to have much clout. If this same individual joins with others in the National Rifle Association, he will have the enjoyment of associating with other like-minded individuals and the satisfaction of knowing that a paid lobbyist is representing his opinions in Washington. In this way, voluntary associations provide a bridge linking the benefits of primary and secondary groups.

Participation in Voluntary Associations

Although some social critics have argued that membership in U.S. voluntary associations has declined—a thesis popularized in the book *Bowling Alone*, by Robert Putnam (2000)—most observers believe that, if anything, participation has increased (Rich 1999). It is true that some large voluntary associations, such as the Elks and bowling leagues, have seen declines in membership. Other groups, however, are burgeoning, especially small local associations, groups focused on ethnicity or gender issues, alternative religious organizations, and Internet-based groups (Rich 1999).

Americans belong to an average of two voluntary associations, considerably above the average for industrialized nations (Curtis, Baer, and Grabb 2001). Among those who report membership, a large proportion are passive participants who belong in name only. They buy a membership in the Parent-Teacher Association (PTA) when pressured to do so, but they don't go to meetings. Similarly, anyone who subscribes to *Audubon* magazine is automatically enrolled in the local Audubon Club, but few subscribers become active members. Because so many of our memberships are superficial, they are also temporary. Nevertheless, most people in the United States maintain continuous membership in at least one association.

Membership in voluntary associations is highest among middle-aged, married, well-educated, and middle-class individuals (Curtis et al. 1992). In addition, having school-age children draws both men and women into youth-related groups and so increases voluntary association membership (Rotow 2000). Interestingly, marriage increases men's participation in associations but not women's, primarily by drawing

Joining an amateur baseball league or other voluntary association is guaranteed to increase our network of weak ties. If we become close friends with any fellow members or teammates, we also increase our network of strong ties.

men into church-related groups. Conversely, full-time employment increases women's participation but not men's, primarily by drawing women into job-related groups. Taken together, these findings suggest that individuals are more likely to participate in voluntary associations when their neighborhood, work, children, or some other aspect of their life provides them with opportunities to do so.

Community

In everyday life, we often hear about the benefits of having "community." Yet we rarely hear a clear definition of what this means. According to sociologists, a **community** is a collection of individuals characterized by dense, cross-cutting social networks (Wellman & Berkowitz 1988). A community is strongest when all members are linked to one another through complex overlapping ties.

A **community** is a collection of individuals characterized by dense, cross-cutting social networks.

Yet network ties need not be strong to have important consequences for individuals and the community. For example, research shows that even when neighbors share only weak ties, they often help each other in many ways—loaning tools, picking up the mail when a family is out of town, and the like (Wellman & Wortley 1990). Similarly, neighborhoods experience substantially less crime and delinquency when neighbors enjoy weak ties and so believe they have both the right and the obligation to sanction teenagers who throw trash, shout profanity, or otherwise misbehave (Sampson & Raudenbush 1999; Sampson, Morenoff, & Earls 1999; Sampson, Morenoff, & Gannon-Rowley 2002).

Computer Networks and Communities

With the exponential rise in use of the Internet, many individuals now seek and find online networks and communities (DiMiaggio et al. 2001; Wellman 1999; Wellman et al. 1996). The Internet's potential for promoting both strong and weak ties has been most impressively demonstrated by the spectacular rise of "social networking"

Participating in computer games can increase social networks by helping each individual make new friends and by cementing existing friendships.

sites such as Friendster, Facebook, and Myspace. Similarly, online discussion groups and chat rooms allow anyone to quickly send out a comment or request to a large and diverse audience. Although the quality of information and relationships obtained via the Internet can vary widely, the Internet does provide a wide network of weak ties to many people who might otherwise be isolated. Furthermore, because individuals often forward the comments or requests they receive to others, this network of weak ties can grow both broadly and quickly.

Although less common, online networks also can provide strong ties and a true sense of community. Even when individuals initially enter online groups simply to obtain information, those who stay typically do so because they enjoy the social support, companionship, and sense of community the group offers. Relatively strong online communities can form over anything from organizing political efforts to writing and sharing personal journals, each group fulfilling a different combination of instrumental and expressive functions. Many of the most popular online groups link people who share a health problem. Within these groups, individuals share not only suggestions regarding medical treatment but also their fears, sorrow, and triumphs as they grapple with their injuries or illnesses.

Interestingly, even participating in computer games—seemingly a highly individual activity—can *increase* individuals' social networks. A national survey (Jones 2003) conducted in 2002 found that 65 percent of college students regularly or occasionally play computer games (including Internet games and computerized video games). Surprisingly, two thirds of computer gamers believe that playing computer games does not reduce the time they spend with family or friends, and 20 percent believe that gaming helped them make new friends and cement existing friendships. In observations on college campuses, the researchers found that students often sit together in computer labs and trade tips on new games or new strategies for older games. Other students participate in chat options on interactive, multi-player games. Even computer games, it seems, can increase both strong and weak ties.

Complex Organizations

Few people in our society escape involvement in large-scale organizations. Unless we are willing to retreat from society altogether, a major part of our lives is organization-bound. Even in birth and death, large, complex organizations (such as hospitals and vital statistics bureaus) make demands on us. Throughout the in-between years, we are constantly adjusting to organizational demands.

Sociologists use the term **complex organizations** to refer to large, formal organizations with complex status networks (Handel 2002). Examples include universities, governments, corporations, churches, and voluntary associations such as fraternities or the Kiwanis Club.

These complex organizations make a major contribution to the overall quality of life within society. Because of their size and complexity, however, they don't supply the cohesion and personal satisfaction that smaller groups do. In fact, members often feel as if they are simply cogs in the machine rather than important people in their own right. This is nowhere more true than in a bureaucracy.

Bureaucracy is a special type of complex organization characterized by explicit rules and a hierarchical authority structure, all designed to maximize efficiency. In popular usage, bureaucracy often has a negative connotation: red tape, silly rules, and unyielding rigidity. In social science, however, it is simply an organization in which the roles of each actor have been carefully planned to maximize efficiency.

Complex organizations are large formal organizations with complex status networks.

Bureaucracy is a special type of complex organization characterized by explicit rules and hierarchical authority structure, all designed to maximize efficiency.

The "Ideal" Bureaucracy: Weber's Theory

Most large, complex organizations are also bureaucratic: IBM, General Motors, U.S. Steel, the Catholic Church, colleges, and hospitals. The major characteristics of an "ideal" bureaucracy were outlined almost 100 years ago by Max Weber ([1910] 1970a):

1. *Division of labor.* Bureaucratic organizations employ specialists in each position and make them responsible for specific duties. Job titles and job descriptions specify who is to do what and who is responsible for each activity.

2. *Hierarchical authority.* Positions are arranged in a hierarchy so that each position is under the control and supervision of a higher position. Frequently referred to as chains of command, these lines of authority and responsibility are easily drawn on an organization chart, often in the shape of a pyramid.

3. *Rules and regulations.* All activities and operations of a bureaucracy are governed by abstract rules or procedures. These rules are designed to cover almost every possible situation that might arise: hiring, firing, and the everyday operations of the office. The object is to standardize all activities.

4. *Impersonal relationships.* Interactions in a bureaucracy are supposed to be guided by the rules rather than by personal feelings. Consistent application of impersonal rules is intended to eliminate favoritism and bias.

5. *Careers, tenure, and technical qualifications.* Candidates for bureaucratic positions are supposed to be selected on the basis of technical qualifications such as education, experience, or high scores on civil service examinations. Once selected for a position, persons should advance in the hierarchy by means of achievement and seniority, and should be able to keep their job as long as their performance holds up.

6. *Efficiency.* Bureaucratic organizations are intended to maximize efficiency by coordinating the activities of a large number of people in the pursuit of organizational goals. From the practice of hiring on the basis of credentials rather than personal contacts to the rigid specification of duties and authority, the whole system is

Bureaucracies are characterized by alienating, rigid job conditions, such as those experienced by these customer service representatives.

constructed to keep individuality, whim, and favoritism out of the operation of the organization.

When Weber described these characteristics as the "ideal model" of a bureaucracy, he didn't mean to imply that this was the *best* form of bureaucracy. An *ideal model* refers merely to what something is *expected* to be like. Weber realized that few if any bureaucracies totally meet this description: workers often must do tasks beyond those they are assigned, lines of authority are sometimes unclear, environments often change before new rules evolve to deal with those changes, biases like sexism and racism certainly can lead to the hiring of unqualified or less qualified persons, and organizations can, at times, be wildly inefficient. In addition, over the last quarter century American corporations, a major form of bureaucracy, have downsized and now contract out many services. In this new environment, workers have less guarantee of tenure and corporations can't be as hierarchical, since they can't exert as much control over contracted workers—especially if the workers live half a world away (Scott 2004). Still, Weber's list of bureaucratic characteristics helps us understand at least the intended role and nature of bureaucracies.

Real Bureaucracies: Organizational Culture

Weber's classic theory of bureaucracy almost demands not individuals, but robots who will follow every rule to the letter. Yet when workers really do follow every rule, no matter how nonsensical or unnecessary, work quickly grinds to a halt. In fact, in cities where police cannot legally strike, police unions sometimes instead protest through "slowdowns," in which officers follow every rule for the purpose of throwing the system into chaos. Not surprisingly, therefore, few organizations try to be totally bureaucratic. Instead, organizations strive to create an atmosphere of goodwill and common purpose among their members so that they all will apply their ingenuity and best efforts to meeting organizational goals (Kunda 1993). This goodwill is as essential to efficiency as are the rules.

Organizational culture refers to the pattern of norms and values that structures how business is actually carried out in an organization.

Sociologists use the term **organizational culture** to refer to the pattern of norms and values that structures how business is actually carried out in an organization (Kunda 1993). The key to a successful organizational culture is cohesion, and most organizations strive to build cohesion among their members. They do this by encouraging interaction and loyalty among employees (providing lunchrooms; sponsoring after-hours sports leagues; having company picnics and newsletters; and developing unifying symbols, such as mascots, company colors, or uniforms). This is most apparent in wealthy organizations that require creative employees, such as Google, which is famous for such on-site "perks" as free massages, gourmet meals, and volleyball courts. But it is also characteristic of some transnational corporations, especially in Japan and Scandinavia. When organizational managers succeed at motivating loyalty, workers may be willing to skip vacations, work extremely long hours, and sacrifice time with family and friends; it is not uncommon for workers at "fun" companies like Google to work 12, 15, or even 24 hours a day when facing a deadline.

Compared with a business like Ford Motor Company, businesses like Google also stand out for their emphasis on flexibility and informal decision making. Why would this be so? Companies that develop software require creativity from their employees and must change strategies rapidly in response to changes in the broader environment and changes made by their competitors. In contrast, changes come more slowly to the factory line at Ford. The degree of bureaucratization in an organization is related to the degree of uncertainty in the organization's activities. When activities tend to be routine and predictable, the organization is likely to emphasize rules, central planning, and hierarchical chains of command. This explains why, for example, classrooms tend to be less bureaucratic and factories more bureaucratic.

Critiques of Bureaucracies

Bureaucracy is the standard organizational form in the modern world. Organizations from churches to governments are run along bureaucratic lines. Yet despite the widespread adoption of this organizational form, it has several major drawbacks. Three of the most widely acknowledged are as follows:

1. *Ritualism*. Rigid adherence to rules may mean that a rule is followed regardless of whether it helps accomplish the purpose for which it was designed. The rule becomes an end in itself rather than a means to an end. For example, individuals may struggle to arrive at 8 A.M. and leave at 4 P.M. when they could work more effectively from 10 to 6. Although the existence of a bureaucracy per se doesn't always breed rote adherence to rules (Foster 1990), an overemphasis on bureaucratic rules can stifle initiative and prevent the development of more efficient procedures.

2. *Alienation*. The emphasis on rules, hierarchies, and impersonal relationships can sharply reduce the cohesion of the organization. This has several drawbacks: It reduces social control, reduces member satisfaction and commitment, and increases staff turnover. All of these may interfere with the organization's ability to reach its goals.

3. *Structured inequality*. Critics charge that the modern bureaucracy with its multiple layers of authority is a profoundly antidemocratic organization. The purpose of the bureaucratic form is to concentrate power in one or two decision makers whose decisions are then passed down as orders to subordinates.

In addition to these three criticisms, more recent criticism has focused on the dangers of McDonaldization (Ritzer 1996). **McDonaldization** refers to incorporation of business principles derived from the fast-food restaurant industry into other business sectors. These four principles are efficiency, calculability, predictability, and control. For example, McDonald's streamlined its procedures to serve customers extremely rapidly (efficiency) and shifted from advertising how good their burgers taste (something that can't be measured) to advertising how many ounces are in them (calculability). McDonald's guarantees that a Big Mac in New York tastes exactly like a Big Mac in Des Moines (predictability). And McDonald's requires its employees to follow strict guidelines for work procedures (control) and pressures customers to order and leave quickly by offering limited menus and uncomfortable seats (control and efficiency). These principles have now been adopted by all kinds of organizations around the world, from drop-off laundries to "dial-a-porn" sex businesses.

Ironically, the attempt to rationalize bureaucratic structures through McDonaldization often produces *irrational* consequences for the society as a whole. The

McDonaldization is the process by which the principles of the fast-food restaurant—efficiency, calculability, predictability, and control—are coming to dominate more sectors of American society.

disadvantages of McDonaldization stem directly from each supposed advantage. For instance, although it is more efficient for businesses to use voice-mail systems instead of operators, it is less efficient for the customers who must listen to a series of menus, hoping that they will reach the department they seek before getting disconnected. Moreover, businesses lose customers when customers hang up in frustration after getting lost in voice-mail mazes. Similarly, businesses like McDonald's make decisions based on calculations of how they can best generate a profit, but they do not calculate the impact of their business decisions on the environment or on the quality of life of their customers or workers. The predictability that chain stores and restaurants offer makes the world a less interesting place, as large national businesses drive out unique local businesses. And the control that McDonaldized organizations offer is frequently dehumanizing—something you have probably experienced every time your name and identity are replaced by an institutional identification number.

Where This Leaves Us

Humans are social beings. Our lives are lived within relationships, groups, networks, and—whether we like it or not—complex organizations. Without these human connections we cannot survive, let alone thrive. Groups, networks, and organizations help us obtain the very basics of life—food, clothing, work, shelter, companionship, love. They also enable us to make our mark on the world, as we raise children within families, create better communities through voluntary associations, strive for success within complex organizations from schools to corporations, and so on. Yet working with others also carries risks, for exchange and cooperation can turn into competition and conflict, and groups can affect our ideas and behaviors in ways we may not even recognize. A sociological understanding of groups, networks, and organizations can help us understand, prevent, and, where necessary, counteract these effects.

Summary

1. Relationships are characterized by four basic social processes: exchange, cooperation, competition, and conflict.
2. Groups are distinguished from crowds and categories in that group members take one another into account, and their interactions are shaped by shared expectations and interdependency.
3. Primary groups are characterized by intimate, face-to-face interaction. They are essential to individual satisfaction and integration, and they are also primary agents of social control in society. Secondary groups are large, formal, and impersonal. They are generally task-oriented and perform instrumental functions for societies and individuals.
4. Group interaction is affected by group size, proximity, and communication patterns. Group interaction can lead to conformity and consensus among group members,

sometimes around obviously incorrect decisions. The amount of interaction in turn affects group cohesion.
5. Each person has a social network that consists of both strong and weak ties. The number of strong ties is generally greater for individuals who are white and educated.
6. Strong ties are the people we can count on when we really need help of some sort. Weak ties, however, are more useful when we need to pull on a broad social network, such as in searching for work in the United States.
7. Voluntary associations are nonprofit groups that bring together people with shared interests. They combine some of the expressive functions of primary groups with the instrumental functions of secondary groups.
8. When individuals are linked by dense, cross-cutting networks, they form a community. Communities have important influences on members, even when social ties within the community are relatively weak.

9. Complex organizations are large, formal organizations with complex status networks. Bureaucracies are complex organizations whose goal is to maximize efficiency. The chief characteristics of a bureaucracy are division of labor; hierarchical authority; rules and regulations; impersonal social relations; an emphasis on careers, tenure, and technical qualifications; and an emphasis on efficiency.

10. Although most contemporary organizations are built on a bureaucratic model, many are far less rational than the classic model suggests. Critics of McDonaldization suggest that the bureaucratic emphasis on rationality can have irrational consequences. In addition, all effective bureaucracies must rely on organizational culture to inspire employees to give their best efforts and to help meet organizational goals.

Thinking Critically

1. Do social networking sites like Facebook serve as primary groups or secondary groups? Do they enforce group conformity? Explain, with examples.
2. Can you think of a situation in your life in which the primary source of social control was a secondary rather than a primary group?
3. Suppose you were trying to get help for a family member's substance-abuse problem. Who would you turn to, your strong or your weak ties?
4. From your experience, what are some of the functions of bureaucracy? What are some of the dysfunctions?

Companion Website for This Book

academic.cengage.com/sociology/brinkerhoff
Gain an even better grasp on this chapter by going to the Companion Website. This resource contains tutorial quizzes and flash cards to help you master key terms and concepts.

Suggested Readings

Menjívar, Cecilia. 2000. *Fragmented Ties*. Berkeley, Calif.: University of California Press. Extensive interviews and observations of Salvadoran immigrants to California show how social networks actually function in individuals' lives.

Rheingold, Howard. 2002. *Smart Mobs: The Next Social Revolution*. New York: Perseus Publishing. "Smart mobs" ("mobs" rhymes with "robes") are individuals who use mobile phones and Internet devices to work and play together. Smart mobs include everyone from teenagers looking for others to "hang out" with to prostitutes seeking customers and terrorists coordinating attacks.

Ritzer, George. 1996. *The McDonaldization of Society*, revised edition. Thousand Oaks, Calif.: Pine Forge Press. A fascinating account by a scholar and social activist on how McDonaldization is changing everything from how we are born to how we die.

Putnam, Robert and Lewis Feldstein. 2003. *Better Together: Restoring the American Community*. New York: Simon & Schuster. Illustrates how groups across America, from the online bulletin board Craigslist to a New Hampshire public art program, are fighting the decline of voluntary associations, neighborhoods, and community ties.

CHAPTER 6

Deviance, Crime, and Social Control

© AP/Wide World Photos

Conformity and Deviance

In providing a blueprint for living, our culture supplies norms and values that structure our behavior. These norms and values tell us what we ought to believe in and what we ought to do. Because we are brought up to accept them, for the most part we do what we are expected to do and think as we are expected to think. Only "for the most part," however, because none of us follows all the rules all the time.

Previous chapters concentrated on how norms and values structure our lives and how we learn them through socialization. This chapter considers some of the ways individuals break out of these patterns—from merely eccentric behaviors to serious violations of others' rights.

Understanding Conformity

To understand why people *break* social norms, we first must understand why most people, most of the time, conform. The forces and processes that encourage conformity are known as **social control.** Social control takes place at three levels:

1. Through internalized self-control, we police ourselves.
2. Through informal controls, our friends and intimates reward us for conformity and punish us for nonconformity.
3. Through formal controls, the state or other authorities discourage nonconformity.

Self-control occurs because individuals **internalize** the norms of their group, making them part of their basic belief system and their very identity. Most of us do not murder, rape, or rob, not because we fear arrest but because it would never occur to us to do these things; they would violate our sense of self-identity.

This self-control is reinforced by **informal social control:** all the small and not-so-small ways that friends, co-workers, and others around us informally keep us from behaving improperly. Thus, even if your own values do not prevent you from breaking into your professor's office to steal the answers to your midterm test, you might be deterred by the thought of how others will respond if they find out. Your friends might consider you a cheat, your family would be disappointed in you, your professor might publicly embarrass you by denouncing you to the class.

If none of these considerations is a deterrent, you might be scared into conformity by the thought of **formal social controls:** administrative sanctions such as fines, expulsion, or imprisonment. Those who steal test answers, for example, face formal sanctions such as automatic failing grades, loss of scholarships, and dismissal from school.

Whether we are talking about cheating on examinations or murder, social control rests largely on self-control and informal social controls. Few formal agencies have the ability to force compliance to rules that are not supported by individual or group values. Sex is a good example. In many states, sex between unmarried persons is illegal, and theoretically you could be fined or imprisoned for it. Even if the police devoted a substantial part of their energies to stamping out illegal sex, however, they would probably not succeed. In contemporary United States, a substantial proportion of unmarried people are not embarrassed about having sexual relations; they do not care if their friends know about it, and their friends might even be cheering them on. In such conditions, formal sanctions cannot enforce conformity. Prostitution,

Social control consists of the forces and processes that encourage conformity, including self-control, informal control, and formal control.

Internalization occurs when individuals accept the norms and values of their group and make conformity to these norms part of their self-concept.

Informal social control is self-restraint exercised because of fear of what others will think.

Formal social controls are administrative sanctions such as fines, expulsion, or imprisonment.

Although your parents are not likely to be pleased if you pierce your nose, doing so does not violate any major norms or arouse too much public disapproval. It is an example of nonconformity but not of deviance.

marijuana use, underage drinking—all are examples of situations in which laws unsupported by public consensus have not produced conformity.

Defining Deviance

People may break out of cultural patterns for a variety of reasons and in a variety of ways. Whether your nonconformity is regarded as deviant or merely eccentric depends, among other things, on the seriousness of the rule you violate. If you wear bib overalls to church or carry a potted palm with you everywhere, you will be challenging the rules of conventional behavior. Probably nobody will care too much, however; these are minor kinds of nonconformity. Norm violations only become **deviance** when they exceed the tolerance level of the community and bring negative sanctions. Deviance is behavior of which others disapprove to such an extent that they believe something significant ought to be done about it.

Defining deviance as behavior of which others disapprove has an interesting implication: It is not the *act* that is important but the *audience*. The same act may be deviant in front of one audience but not another, deviant in one place but not another.

Few acts are intrinsically deviant. Even taking another's life may be acceptable in war, police work, or self-defense. Whether an act is regarded as deviant often depends on the time, the place, the individual, and the audience. For this reason, sociologists stress that *deviance is relative*. For example, alcohol use is deviant for adolescents but not for adults, having two wives is deviant in the United States but not in Nigeria, wearing a gun in town (if you are a civilian) is deviant now but wasn't 150 years ago, and wearing a skirt is deviant for American men but not for American women.

The sociology of deviance has two concerns: why people break the rules of their time and place and how those rules become established. In the following sections, we review several major theories of deviance before looking at crime—a type of deviance—in the United States.

Deviance refers to norm violations that exceed the tolerance level of the community and result in negative sanctions.

Theoretical Perspectives on Deviance

Biological and psychological explanations for deviant behavior typically focus on how processes within the individual lead to deviance. Such theories often look for the causes of deviance in genetics, neurochemical imbalances, or childhood failures to internalize appropriate behavior or attitudes. Most sociologists agree that biology and psychology play a role in causing deviance but consider social forces even more important. Sociological theories, therefore, search for the causes of deviance within the social structure rather than within the individual (see the Concept Summary "Theories of Deviance").

Structural-Functional Theories

In Chapter 1, we said that the basic premise of structural-functional theory is that the parts of society work together like the parts of an organism. From this point of view, deviance can be useful for a society—at least up to a point. Consider spring break: It's easier to settle down and do your work for the rest of the semester if you had a

Theories of Deviance

	Major Question	Major Assumption	Cause of Deviance	Most Useful for Explaining Deviance of
Structural-Functional Theory				
Strain theory	Why do people break rules?	Deviance is an abnormal characteristic of the social structure.	A dislocation between the goals of society and the means to achieve them	The working and lower classes who cannot achieve desired goals by prescribed means
Conflict Theory	How does unequal access to scarce resources lead to deviance?	Deviance is a normal response to competition and conflict over scarce resources.	Inequality and competition	All classes: Lower class is driven to deviance to meet basic needs and to act out frustration; upper class uses deviant means to maintain their privilege
Symbolic Interaction Theories				
Differential association theory	Why is deviance more characteristic of some groups than others?	Deviance is learned like other social behavior.	Subcultural values differ in complex societies; some subcultures hold values that favor deviance; these are learned through socialization	Delinquent gangs and those integrated into deviant subcultures and neighborhoods
Deterrence theories	When is conformity not the best choice?	Deviance is a choice based on cost/benefit assessments.	Failure of sanctioning system (benefits of deviance exceed the costs)	All groups, but especially those lacking a "stake in conformity"
Labeling theory	How do acts and people become labeled *deviant*?	Deviance is relative and depends on how others label acts and actors.	People whose acts are labeled *deviant* and who accept that label become career deviants	The powerless who are labeled *deviant* by more powerful individuals

few days to carouse in Florida or the Caribbean. In addition, according to structural-functionalists, deviance can help nudge a society toward needed, incremental social changes. But when deviance becomes extreme, they argue, it is *dysfunctional* (disruptive) to the society.

This perspective was first applied to the explanation of deviance by Emile Durkheim. Durkheim recognized the potential benefits of minor deviance. In his classic study of suicide ([1897] 1951), however, he focused on the causes of dysfunctional, extreme deviance. To explore this issue, Durkheim raised the question of why people in industrialized societies are more likely to commit suicide than are people in agricultural societies. He suggested that in traditional societies the rules tend to be well known and widely supported. As a society grows larger, becomes more diverse, and experiences rapid social change, the norms of society may become unclear or no longer apply. Durkheim called this situation **anomie** and believed it was a major cause of suicide in industrializing nations.

Anomie is a situation in which the norms of society are unclear or no longer applicable to current conditions.

Importantly, Durkheim and later structural-functional theorists define deviance as a social problem rather than a personal trouble; it is a property of the social structure, not of the individual (Passos & Agnew 1997). As a consequence, the solution to deviance lies not in reforming the individual deviant but in changing the dysfunctional aspects of the society.

Explaining Individual Deviance: Strain Theory

Strain theory suggests that deviance occurs when culturally approved goals cannot be reached by culturally approved means.

The classic structural-functionalist theory of crime is Robert Merton's (1957) **strain theory.** Strain theory begins by noting that most of us are conformists, who (as Merton defined the term) accept both our society's culturally approved *goals* and its culturally approved *means* for reaching these goals. Strain theory argues that deviance results when individuals cannot reach those goals by using culturally approved means. This theory is most commonly used to explain lower-class crime.

American culture places strong emphasis on economic success. Although this goal is widely shared by Americans, the means to obtain this goal are not. Few lower class Americans are able to achieve success through culturally approved means, such as attending school to become a lawyer or computer programmer. According to Merton, lower-class persons turn to crime not because they *reject* American values but because they *accept* them: They believe that only through crime can they achieve our shared cultural goal of economic success.

Of course, few people who find society's norms inapplicable to their situation respond by turning to a life of crime. Merton identifies four ways in which people adapt to anomie without becoming criminals (see the Concept Summary "Merton's Types of Deviance"): innovation, ritualism, retreatism, and rebellion. In Merton's terms, *innovation* refers to people who accept society's goals but reject accepted means, instead using illegitimate means to achieve their goals. Innovators include poor teenagers who steal flashy cars, students who cheat on tests, and athletes who use steroids to boost their performances. *Ritualism* refers to people who continue to use culturally approved means for achieving socially desired goals even though they have rejected—or at least given up on—those goals. The primary example of the ritualist is the worker who follows all bureaucratic procedure but just wants to avoid notice and keep his or her job, not get ahead.

Retreatism refers to those who have given up on both society's goals and its accepted means. They are society's dropouts: the vagabonds, drifters, and street people.

Merton's Types of Deviance

Merton's strain theory of deviance suggests that deviance results whenever there is a disparity between goals and the institutionalized means available to reach them. Individuals caught in this dilemma may reject the goals or the means or both. In doing so, they become deviants.

Modes of Adaptation	Cultural Goals	Institutional Means
Innovation	Accepted	Rejected
Ritualism	Rejected	Accepted
Retreatism	Rejected	Rejected
Rebellion	Rejected/replaced	Rejected/replaced

Concept Summary

Like retreatism, *rebellion* also refers to those who abandon society's goals and means, but rebels additionally adopt alternative values. These are people like revolutionaries, Rastafarians, or the Rainbow Tribe who hope to create an alternative society.

Explaining Neighborhood Crime Rates: Collective Efficacy Theory

Whereas strain theory hopes to explain why some *individuals* are more likely to engage in crime than are others, collective efficacy theory hopes to explain why some *neighborhoods* have higher rates of crime than others (Sampson & Raudenbush 1999; Sampson, Morenoff, & Earls 1999; Sampson, Morenoff, & Gannon-Rowley 2002). Collective efficacy theory is also a structural-functionalist theory because it, too, assumes that crime or deviance occurs when the parts of a society no longer work together smoothly.

■ Body builders who use steroids to increase their chances of winning contests fit Merton's definition of innovators: They accept society's goals of success, fame, and wealth, but use illegitimate means (steroids) to achieve those goals.

Collective efficacy refers to the extent to which individuals in a neighborhood share the expectation that neighbors will intervene and work together to maintain social order. If your neighbors believe it is important to work together to control neighborhood crime and delinquency and are likely to call the police when teenagers race cars down the block or scrawl graffiti on a wall, then you live in a neighborhood with high collective efficacy. Collective efficacy is most common in neighborhoods that experience few structural disadvantages: They have high rates of employment and home ownership, many residents whose work and incomes give them a sense of control over their lives, and police and municipal services that they can count on for help when needed. According to collective efficacy theory, crime is most likely in neighborhoods that suffer extreme structural disadvantage and, as a result, experience low collective efficacy. This theory has strong empirical support and is growing in influence.

Collective efficacy refers to the extent to which individuals in a neighborhood share the expectation that neighbors will intervene to stop social disorder and deviance and will work together to maintain social order.

Conflict Theory

Structural-functional theory suggests that deviance results from a lack of integration among the parts of a social structure (norms, goals, and resources); it is viewed as an abnormal state produced by extraordinary circumstances. Conflict theorists, however, see deviance as a natural and inevitable product of competition in a society in which groups have different access to scarce resources. They suggest that the ongoing processes of competition should be the real focus of deviance studies (Lemert 1981).

Conflict theory proposes that deviance results from competition and class conflict. Class conflict affects deviance in two ways (Reiman 2005): (1) Class interests determine how the criminal justice system defines and responds to crime and (2) economic pressures can lead to crime, particularly property crimes, among the poor.

Defining and Responding to Crime

Conflict theorists argue that the law is a weapon used by the ruling class to maintain the political and economic status quo (Arrigo 1998; Liska, Chamlin, & Reed 1985; Reiman 2005). Supporters of this position argue that the very definitions of crime sometimes reflect the interests of the wealthy. Corporations can kill or injure thousands when they sell cars, contact lenses, or other goods that they know are harmful. They can endanger workers when they cut corners on factory safety and endanger whole communities when they dump dangerous chemicals into the water or soil. And they can impoverish workers and investors through shady business practices, even while their executives earn multimillion dollar salaries. Yet these actions are often defined by the courts as ordinary and necessary business practices, rather than as crimes.

Similarly, conflict theorists argue that the criminal justice system's response to behaviors labeled criminal also reflects the interests of the wealthy. Our system spends more money deterring muggers than embezzlers and more money arresting prostitutes than arresting their clients. Courts impose much more severe sentences for street crimes than for corporate crimes and impose much heavier sentences against those who use drugs favored by the poor (such as "crack" cocaine) than against those who use drugs favored by the more affluent (such as other forms of cocaine). Police are more likely to arrest those who assault members of the ruling class (well-off whites) than those who assault the powerless (nonwhites and the poor) (Reiman 2005). Finally, even when people from the upper and lower classes commit similar crimes, those from the lower class are more likely to be arrested, prosecuted, and sentenced (Reiman 2005).

As this suggests, most conflict theorists reject structural functionalism's assumption that poor people are unusually likely to commit crimes. Instead, and as research suggests, most poorer people adjust their goals downward sufficiently so that they can meet their goals through respectable means (Simons & Gray 1989). Meanwhile, many highly successful individuals adjust their goals so far upward that they cannot reach them by legitimate means. Recent court cases that reveal Microsoft's illegal attempts to gain a monopoly over Internet services and tobacco manufacturers' attempts to make cigarettes more addictive provide clear evidence that the means-versus-goals discrepancy is not limited to the lower class. Conflict theorists argue that it only appears that rich people commit fewer crimes because rich people control the state, schools, and courts, and so are often able to avoid criminal labels (Reiman 2005).

Lower-Class Crime

Although the preceding view of the way crime is defined would be accepted by all conflict theorists, some believe that individuals in the lower class really are more likely to commit criminal acts. One critical criminologist has declared that crime is a rational response for the lower class (Quinney 1980). These criminologists generally seem to agree with Merton that a means/ends discrepancy is particularly acute among the poor and that it may lead to crime (Reiman 2005). They believe, however, that this is a natural condition of an unequal society rather than an unnatural condition.

Symbolic Interaction Theories

Symbolic interaction theories of deviance suggest that deviance is learned through interaction with others and involves the development of a deviant self-concept. Deviance is believed to result not from broad social structure but from specific face-to-face interactions. This argument takes three forms: differential association theory, deterrence theories, and labeling theory.

Differential Association Theory

Differential association theory argues that people learn to be deviant when more of their associates favor deviance than favor conformity.

Not surprisingly, researchers have found that those who have more delinquent friends are more likely to become delinquent themselves (Haynie and Osgood 2006). **Differential association theory,** first proposed by Edwin Sutherland, explains this finding by arguing that people *learn* to be deviant through their associations with others.

How does differential association encourage deviance? There are two primary mechanisms. First, if our interactions are mostly with deviants, we may develop a biased image of the generalized other. We may learn that, "of course, everybody steals" or, "of course, you should beat up anyone who insults you." The norms that we internalize may be very different from those of conventional society. Second, if we interact mostly within a deviant subculture, that subculture will reward us not for *following* conventional norms but for *violating* those norms. Through these mechanisms, we can learn that deviance is acceptable and rewarded.

Deterrence Theory

Differential association theory can only explain deviance that occurs in settings and groups that encourage it. Deterrence theory provides a broader explanation of deviance. This theory suggests that individuals will engage in deviance when they believe it will offer more rewards than will conformity *and* when they believe the potential risks and costs of deviance are low. **Deterrence theory** combines elements of structural-functional and symbolic interaction theories. Although they place the primary blame for deviance on an inadequate (dysfunctional) system of rewards and punishments, they also believe that individuals actively make a cost/benefit decision about whether to engage in deviance (McCarthy 2002; Paternoster 1989; Piliavin et al. 1986). When social structures do not provide adequate rewards for conformity, more people will choose deviance.

For example, people with no jobs or only dead-end jobs are more likely than others to believe they have little to lose and much to gain from crime or other forms of deviance, especially if they believe the risk of arrest is low (Crutchfield 1989; Devine, Shaley, & Smith 1988; McCarthy 2002). Conversely, those who have strong bonds with their parents, do well in school, feel a part of their school, and are employed are more likely to avoid deviance because they feel they have too much to lose (Haynie & Osgood 2005).

Labeling Theory

A third theory of deviance, which combines symbolic interaction and conflict theories, is labeling theory. **Labeling theory** is concerned with how the label *deviant* comes to be attached to specific people and behaviors. This theory takes to heart the maxim that deviance is relative. As the chief proponent of labeling theory puts it, "Deviant behavior is behavior that people so label" (Becker 1963, 90).

EXPLAINING INDIVIDUAL DEVIANCE   The process through which a person becomes labeled as deviant depends on the reactions of others toward nonconforming behavior. The first time a child acts up in class, it may be owing to high spirits or a bad mood. This impulsive act is *primary deviance*. What happens in the future depends on how others interpret the act. If teachers, counselors, and other children label the child a troublemaker and if she accepts this definition as part of her self-concept, then she may take on the role of a troublemaker. Continued rule violation because of a deviant self-concept is called *secondary deviance*.

The major limitation of labeling theory is that (1) it doesn't explain why primary deviance occurs and (2) it cannot explain repeated deviance by those who haven't been caught—that is, labeled. For this reason, it is less popular today as an explanation of why individuals become deviants.

© Alon Reininger/Contact Press Images

Differential association theory points out that people who grow up in crime-ridden neighborhoods are more likely to grow up to be criminals themselves. It is easy to see how differential association theory might apply to gang members like these, but can you think of ways it might also help to explain white-collar crime?

Deterrence theories suggest that deviance results when social sanctions, formal and informal, provide insufficient rewards for conformity.

Labeling theory is concerned with the processes by which labels such as *deviant* come to be attached to specific people and specific behaviors.

■ Young women like these, who have received significant rewards for conventional behavior, are highly unlikely to be tempted by opportunities for deviance

© Bob Daemmrich/Stock, Boston Inc.

Moral entrepreneurs are people who attempt to create and enforce new definitions of morality.

Connections

Example

In the last decade, a growing anti-smoking movement has fought to outlaw tobacco smoking in public buildings, private businesses such as restaurants and bars, and even on the street. Anti-smoking activists have also battled to stigmatize smokers by, for example, developing advertising campaigns that accuse pregnant women who smoke of endangering their fetuses and parents who smoke of endangering their children. Sociologists would describe these activists as moral entrepreneurs, who are working to create and enforce new definitions of morality and deviance.

EXPLAINING DEVIANCE LABELING Labeling theory is more useful as an explanation of how behaviors become labeled as deviant. Many labeling theorists take a conflict perspective in exploring this topic. They assume that one of the strategies groups use in competing with one another is to get the other groups' behavior labeled as deviant. Because one group is trying to "sell" their moral ideas about who should be labeled deviant, in the same way that entrepreneurs sell their ideas for new businesses, sociologists refer to those who attempt to create new definitions of deviance as **moral entrepreneurs.** Naturally, the more power a group has, the more likely it is to be able to brand others as deviant. This, labeling theorists allege, explains why lower-class deviance is more likely to be subject to criminal sanctions than is upper-class deviance.

But groups can fight back against those who would label them deviant. For example, the Parents Television Council (PTC) is a nonprofit organization that campaigns against television shows that offend its conservative morality. One of its targets is the World Wrestling Federation, which PTC has lambasted for its violent and sexually explicit shows (Lowney 2003). The Federation responded in two ways. First, it attacked with humor, by forming a wrestling team called the Right to Censor that pretended to preach the PTC's moral values while brazenly cheating during fights. Second, it successfully sued the PTC for libel and slander. Through both these strategies, the Federation protected its public image and fended off the PTC's efforts to label the Federation's shows as deviant in the public's eyes.

Case Study: Mental Illness as Deviance

In recent years, more and more behaviors once labeled as deviant have become labeled as mental illnesses. Labeling theory's emphasis on subjective meanings and conflict theory's emphasis on the power to define the situation give us a framework for understanding this shift.

Five hundred years ago, the most powerful social institution in Western society was the church. At that time, those who routinely became drunk in public were regarded as sinners and publicly castigated by ministers (the moral entrepreneurs of the time). But by the 1800s, the state and the criminal justice system had become more powerful than the church. Although ministers still railed against those who drank alcohol, public drunks were now treated as criminals and thrown into jails.

These days, churches and judges vie for power with doctors and the health care establishment. Individuals whose drinking gets publicly out of control will still be regarded as criminals by some and as sinners by others. Still others, however, argue that these individuals suffer from the disease of alcoholism. The behavior hasn't changed, and it's still considered deviant. But a different group (doctors) now define what is deviance. As with heavy drinking, other criminal behaviors like child abuse, gambling, murder, and rape are also regarded by some as signs of mental illness, better treated by doctors than by sheriffs (Conrad & Schneider 1992). This process of redefining "badness" or oddness into illness is referred to as the **medicalization of deviance** (a topic we will return to in Chapter 10).

What happens when a behavior is medicalized? Individuals who acquire the *ill* label rather than the *bad* label are entitled to treatment rather than punishment and may be excused from blame for their behavior (Conrad & Schneider 1992). As you might expect—and as labeling and conflict theory would both predict—people in positions of power are more apt to be successful in claiming the sick label. The upper-class woman who shoplifts is treated for obsessive-compulsive disorder, whereas the lower-class woman who does so is arrested for theft. The middle-class boy who acts up in school is medicated for hyperactivity, whereas the lower-class boy is jailed for juvenile delinquency.

At present we have reached few firm decisions regarding when and whether deviant behaviors should be medicalized. The public seems to believe that some alcohol abusers, murderers, rapists, and so on are mentally ill and should be treated by physicians, but that others are just bad and should be put in jail.

> The **medicalization of deviance** is the process through which "badness" or oddness is redefined as illness, such as the historic change from thinking of problem drinking as a crime to thinking of it as a mental illness known as alcoholism.

Crime

Most deviant behavior is subject only to informal social controls. When deviance becomes labeled crime, it becomes subject to legal penalties. This is, in fact, the definition of **crime**: behavior that is considered so unacceptable that it is subject to legal penalties. Most, though not all, crimes violate social norms and are subject to informal as well as legal sanctions. In this section, we briefly discuss the different types of crimes, look at crime rates in the United States, and describe who is most likely to commit these crimes.

> **Crime** is the behavior that is subject to legal or civil penalties.

Violent and Property Crimes

Each year the federal government publishes the Uniform Crime Report (UCR), which summarizes the number of criminal incidents known to the police for five major crimes (Federal Bureau of Investigation 2006):

- *Murder and nonnegligent manslaughter.* Overall, murder is a rare crime; yet some segments of society are touched by it much more than others. More than 47 percent of all murder victims in 2004 were African American and 78 percent were male.
- *Rape.* Rape accounted for 7 percent of all reported violent crimes in 2004 (Federal Bureau of Investigation 2006). Even though most rapes go unreported, more than 90,000 women reported being raped that year. A large national sample surveyed in 1995/96 found that 15 percent of all American women and 2 percent of all men had been raped at some point in their lives (Tjaden & Thoennes 1998).

- *Robbery.* Robbery is defined as taking or attempting to take anything of value from another person by force or threat of force. Unlike simple theft or larceny, robbery involves a personal confrontation between the victim and the robber. Nationally, the rate of robbery has fallen almost 50 percent since 1990.
- *Assault.* Aggravated assault is an unlawful attack for the purpose of inflicting severe bodily injury. Seventy-two percent of assaults during 2004 involved a weapon.
- *Property crimes (burglary, larceny-theft, motor-vehicle theft, and arson).* Property crimes are much more common than are crimes of violence and account for 88 percent of the crimes covered in the UCR (Federal Bureau of Investigation 2006).

Figures 6.1 and 6.2 depict trends in seven crimes since 1980. They show that crime rates have been declining more or less steadily for the last decade or so and are now lower than they were 30 years ago. The causes of this decline are hotly debated. However, most observers agree that a major reason is that young people commit most crimes, and there are now fewer young people than in earlier generations.

FIGURE 6.1

Percentage Change in Violent Crime Rates, 1980–2004

Violent crime rose during the late 1980s and early 1990s but has been declining sharply since then. Murder, rape, and robbery are now less common than they were in 1980.

SOURCE: Federal Bureau of Investigation (2006); U.S. Department of Justice (1995).

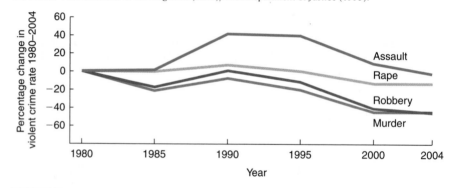

FIGURE 6.2

Percentage Change in Property Crime Rates, 1980–2001

Burglary, larceny, and motor vehicle theft have all declined steadily since 1990 and are now less common than they were in 1980.

SOURCE: Federal Bureau of Investigation (2006).

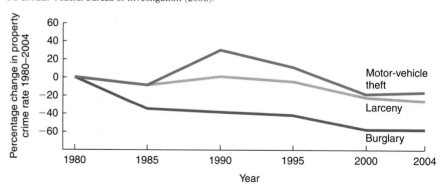

Media and Culture

focus on

The Culture of Fear

Since 1990, rates of crime and of drug use have dropped dramatically. Yet Americans' *fear* of crime and drug use is higher than ever (Glassner 2000, 2004). We live in constant expectation of danger, not only from ordinary criminals but also from Satanists, flesh-eating bacteria, alien kidnapping, and all sorts of other things. Where does this "culture of fear" come from?

Many groups and individuals benefit from promoting fear. Politicians use fear-mongering to get votes, businesses "sell" fear to sell products (from guns to antibacterial soap), and advocacy groups promote fear to gain support for their causes (whether fighting alcohol abuse or raising money to research a rare disease). But when asked why their fears have grown, most Americans point to the media (Blendon & Young 1998). Television reporters and producers, especially, seek stories that can be told in a 3-minute spot: emotionally gripping, visually exciting, and with clear villains and victims (Altheide 2002). As a result, they often choose their headline stories according to the principle "If it bleeds, it leads."

How do the media teach people to overestimate the dangers of crime? Sociologist Barry Glassner offers three answers. First, the media *misidentify* isolated events as trends, such as describing the Columbine school massacre as part of a broader pattern of school homicides, even though no such pattern existed. Second, the media *misdirect* us, making crime seem important by ignoring more serious problems. For example, television news shows spend far more time discussing the very rare school homicides than discussing, for example, the millions of American children who leave home hungry each morning to go to dilapidated, segregated schools. Third, the media *repeat* exaggerated claims of dangers so often that we believe them. True, the media do sometimes try to debunk myths of dangers, but such stories appear infrequently and are often buried far back in newspapers or later in newscasts.

"Valid fears," Glassner writes (2000, 2) "have their place; they cue us to danger. False and overdrawn fears only cause hardship." Our fears of crime have led us to imprison millions of our citizens and to spend more on criminal justice than on education—even though spending on education is more likely to reduce crime rates. Moreover, because increased spending on crime control heightens awareness of crime, it actually *increases* public fears. This is especially a problem for the elderly. According to Glassner, many elderly Americans are so upset by all the murder and mayhem they see on their television screens that they are terrified to leave their homes. Some become so isolated, studies found, that they do not get enough exercise and their physical and mental health deteriorates. In the worst cases they actually suffer malnutrition as a consequence of media-induced fear of crime. Afraid to go out and buy groceries, they literally waste away in their homes. The pattern becomes self-perpetuating; the more time elderly people spend at home, the more TV they tend to watch, and the more fearful they grow.

Until we sort out the real dangers from the false ones, we will spend our tax dollars fighting the wrong battles—while the real problems are left untouched.

Victimless Crimes

The so-called **victimless crimes**—such as drug use, prostitution, gambling, and pornography—are voluntary exchanges between persons who desire illegal goods or services from one another (Schur 1979). They are called victimless crimes because participants in the exchange typically do not see themselves as being victimized or as suffering from the transaction: There are no complaining victims.

There is substantial debate about whether these crimes are truly victimless. Some argue that prostitutes, drug abusers, and pornography models *are* victims (Chapman & Gates 1978; Dworkin 1981) because individuals usually enter these situations only if they feel they have no reasonable alternatives. Others believe that such activities are legitimate areas of free enterprise and free choice (Jenness 1990). These observers argue that although prostitutes and drug users might benefit from laws against pimping or selling contaminated drugs, they are only further victimized by laws against prostitution or drug use *per se*.

Victimless crimes such as drug use, prostitution, gambling, and pornography are voluntary exchanges between persons who desire illegal goods or services from each other.

Because there are no complaining victims, these crimes are difficult to control. The drug user is generally not going to complain about the drug pusher, and the illegal gambler is unlikely to bring charges against a bookie. In the absence of a complaining victim, the police must find not only the criminal but also the crime. Efforts to do so are costly and divert attention from other criminal acts. As a result, laws relating to victimless crimes are irregularly and inconsistently enforced, most often in the form of periodic crackdowns and routine harassment.

White-Collar Crimes

White-collar crime refers to crimes committed by respectable people of high status in the course of their occupation.

Crime committed by respectable people of high social status in the course of their work is called **white-collar crime** (Sutherland 1961; Shover & Wright 2000). White-collar crime occurs at several levels. Embezzlement, for example, is committed by employees against companies. But companies also commit white-collar crime when they engage in price fixing, sell defective products, evade taxes, or pollute the environment.

When companies are the perpetrators, white-collar crime is often referred to as *corporate crime.* Sometimes, corporate crime more closely parallels organized crime than it does anything else. For example, accountants, auditors, and executives working for Enron Corporation worked together to hide the company's debts, exaggerate its profits, and pull in money from investors whom they tricked into buying their stock for much more than it was worth (Eichenwald 2005). Meanwhile, corporate executives took home multimillion-dollar salaries. When its false bookkeeping became known and the company was forced into bankruptcy, Enron retirees lost their pensions, 4,000 Enron employees lost their jobs, and thousands of small investors lost their life savings. The only thing distinguishing this type of corporate crime from organized crime is that the Enron embezzlers wore Armani suits and white shirts while "mobsters" (at least on the *Sopranos*) wear black shirts and white ties.

White-collar crime brings heavy costs to society. Most scholars and law enforcement officials believe that the dollar loss due to corporate crime dwarfs that lost through street crime (Hagan 2002). In addition to the economic cost, there are social costs as well. Exposure to repeated tales of corruption breeds distrust and cynicism and, ultimately, undermines the integrity of social institutions. If you think that all members of Congress are crooks, then you quit voting. If you think that police officers can be bought, then you cease to respect the law. Finally, white-collar crime can cost lives when tainted medicines or dangerous cars are sold, safety precautions are flouted on factory lines, or toxic chemicals are dumped by manufacturers into rivers and streams. Thus, the costs of white-collar crime go beyond the actual dollars involved in the crime itself.

The reasons for white-collar crime are similar to those for street crimes: People want more than they can legitimately get and think the benefits of a crime outrun its potential costs (Shover & Wright 2000). Differential association also plays a role. In some corporations, organizational culture winks at or actively encourages illegal behavior. Speaking of the insider trading scandals that rocked Wall Street during the 1980s, one participant said:

> You gotta do it. . . . Everybody else is. [It] is part of the business. . . . You work at a deli, you take home pastrami every night for free. It's the same thing as information on Wall Street. . . . I know you want to help your mother and provide for your family. This is the way to do it. Don't be a shmuck. Nobody gets hurt. (As quoted in Reichman 1989, 198)

The magnitude of white-collar crime in our society challenges the popular image of crime as a lower-class phenomenon. Instead, it appears that people of different statuses simply have different opportunities to commit crime. Those in lower statuses are hardly in the position to engage in price fixing, stock manipulation, or tax evasion. They are in a position, however, to engage in high-risk, low-yield crimes such as robbery and larceny. In contrast, higher-status individuals are in the position to engage in low-risk, high-yield crimes (Reiman 2005; Schur 1979).

Similarly, the lenient treatment received by most convicted white-collar criminals mocks the idea of equal justice. For example, one study found that health-care professionals guilty of Medicaid fraud were much less likely to be incarcerated than persons charged with grand theft, despite the fact that the dollar losses from the Medicaid crimes, on the average, were much greater (Tillman & Pontell 1992). The 2006 convictions of Ken Lay, Jeff Skilling, and numerous other Enron executives, however, signal an increased awareness (at least among government prosecutors) of the importance of white-collar crime.

Correlates of Crime: Age, Sex, Class, and Race

In 2004, only 46 percent of violent crimes reported in the UCR and 17 percent of property crimes were cleared by an arrest (U.S. Department of Justice 2004). Murder was the crime most likely to be cleared, and burglary was least likely. This means that the people arrested for the criminal acts summarized in the UCR represent only a sample of those who commit these crimes; they are undoubtedly not a random sample. Nor do they represent at all those who commit white-collar crimes, which are not included in the UCR. As a result, we must be cautious in generalizing from arrestees to the larger population of criminals.

With this caution in mind, we note that the persons arrested for criminal acts are disproportionately male, young, and from minority groups. Figure 6.3 shows the pattern of arrest rates by sex and age. As you can see, crime rates, especially for men, peak sharply during ages 15 to 24; during these peak crime years, young men are about three to four times more likely to be arrested than women of the same age. Minority data are not available by age and sex, but the overall rates show that African Americans and Hispanics are more than three times as likely as whites to be arrested.

What accounts for these differentials? Can the theories reviewed earlier help explain these patterns?

Age Differences

The age differences in arrest rates noted in Figure 6.3 are both longstanding and characteristic of nearly every nation in the world that gathers crime statistics (Cook & Laub 1998; Hirschi & Gottfredson 1983). Researchers disagree over the reasons for the high arrest rates of young adults, but deterrence theories have the most promise for explaining this age pattern.

In many ways, adolescents and young adults have less to lose than other people. They don't have a "stake in conformity"—a career, a mortgage, or a credit rating (Steffensmeier et al. 1989). When young people do have jobs and especially when they have good jobs, their chances of getting into trouble are much less (Allan & Steffensmeier 1989).

Delinquency is basically a leisure-time activity. It is strongly associated with spending large blocks of unsupervised time with peers (Osgood & Haynie 2005). When young people have "nothing better to do," a substantial portion will get their fun by causing trouble. Conversely, deviance is deterred by having a close attachment to parents or school.

FIGURE 6.3

Arrest Rates by Age and Sex, 2001
Arrest rates in the United States and most other nations show strong and consistent age and sex patterns. Arrest rates peak sharply for young people ages 15 to 24; at all ages, men are considerably more likely than women to be arrested.
SOURCE: U.S. Department of Justice (2001a).

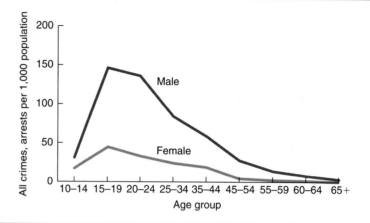

Sex Differences

The sex differential in arrest rates has both social and biological roots. Women's smaller size and lesser strength make them less able to engage in the types of crimes emphasized in the UCR; they have learned that, for them, these are ineffective strategies. Evidence linking male hormones to aggressiveness indicates that biology also may be a factor in women's lower inclination to engage in violent behavior.

Among social theories of deviance, deterrence theory seems to be the most effective in explaining these differences. Generally, girls are supervised more closely than boys, and they are subject to more social control, especially in less affluent families (Chesney-Lind & Shelden 2004; Hagan, Gillis, & Simpson 1985; Heidensohn 1985; Thompson 1989). Whereas parents may let their boys wander about at night unsupervised, they are much more likely to insist on knowing where their daughters are and with whom they are associating. The greater supervision that girls receive increases their bonds to parents and other conventional institutions; it also reduces their opportunity to join gangs or other deviant groups.

These explanations raise questions about whether changing roles for women will affect women's participation in crime. Will increased equality in education and labor-force participation and increased smoking and drinking also carry over to greater equality of criminal behavior? So far, the answer appears to be no (Chesney-Lind & Shelden 2004; Steffensmeier & Allan 1996). Although the crime rate for women has increased, most of this increase is in minor property crimes and drug possession (as opposed to drug dealing) (Chesney-Lind & Shelden; Maher & Daly 1996). Meanwhile, the gender gap in rates of violent and major property crime has actually increased.

This pattern of change lends support to feminist theories of crime. Whereas deterrence theory argues that men's higher crime rates reflect their relatively weaker bonds to conventional *authority*, feminist sociologists instead argue that those rates reflect men's strong bonds to conventional *gender roles* (Bourgois 1995; Katz 1988; Messerschmidt 1993). According to these theories, to be considered "masculine," boys and men must challenge authority and act aggressively or even violently, at least in certain times and places. This theory is particularly useful for explaining crimes against women by groups of men, such as gang rapes (Lefkowitz 1997; Sanday 1990).

Feminist sociologists also have noted that *victimization* of females by males explains a significant proportion of crime among females (Chesney-Lind & Shelden 2004). Girls and women who have been sexually or physically abused by men (including male relatives) are more likely to run away from home, turn to drugs, enter prostitution, and respond violently to their abusers and others.

Social-Class Differences

The effect of social class on crime rates is complex. Braithwaite's (1981) review of more than 100 studies leads to the conclusion that lower-class people commit more of the direct interpersonal types of crimes normally handled by the police than do people from the middle class. These are the types of crimes reported in the UCR. Middle-class people, on the other hand, commit more of the crimes that involve the use of power, particularly in the context of their occupational roles: fraud, embezzlement, price fixing, and other forms of white-collar crime. There is also evidence that the social-class differential may be greater for adult crime than for juvenile delinquency (Thornberry & Farnworth 1982).

Nearly all the deviance theories we have examined offer some explanation of the social-class differential. Strain theorists and some conflict theorists suggest that the lower class is more likely to engage in crime because of blocked avenues to achievement, which explains why crime rises along with unemployment (Grant & Martinez-Ramiro 1997). Deterrence theorists argue that the lower classes commit more crimes because they receive fewer rewards from conventional institutions such as school and the labor market. All these theories accept and seek to explain the social-class pattern found in the UCR, where the lower class is overrepresented.

Labeling and conflict theories, on the other hand, argue that this overrepresentation is not a reflection of underlying social-class patterns of deviance but of bias in the law and within social control agencies (Williams & Drake 1980). Evidence suggests, for instance, that the disproportionately high lower-class homicide rates found in most modern societies result from governmental failure to provide the least privileged with the same legal means of conflict resolution as is typically provided to the social elite (Cooney 1997). Overrepresentation of the lower class also reflects the particular mix of crimes included in the UCR; if embezzlement, price fixing, and stock manipulations were included in the UCR, we would see a very different social-class distribution of criminals.

Race Differences

Although African Americans compose less than 13 percent of the population, they make up 32 percent of those arrested for rape, 33 percent of those arrested for assault, and 47 percent of those arrested for murder (Federal Bureau of Investigation 2006). Hispanics, who compose about 14 percent of the total population, represent about 28 percent of those imprisoned for violent crimes. These strong differences in arrest and imprisonment rates are explained in part by social-class differences between minority and white populations. Even after this effect is taken into account, however, African Americans and Hispanics are still much more likely to be arrested for committing crimes.

The explanation for this is complex. As we will document in Chapter 8, race and ethnicity continue to represent a fundamental cleavage in U.S. society. The continued and even growing correlation of minority status with poverty, unemployment, inner-city residence, and female-headed households reinforces the barriers between nonwhites and whites in U.S. society. An international study confirms that the larger the number of overlapping dimensions of inequality, the higher the "pent-up aggression which manifests itself in diffuse hostility and violence" (Messner 1989). The root cause of higher minority crime rates, from this perspective, is the low quality of minority employment—which leads directly to unstable families and neighborhoods (Newman 1999b; Sampson & Raudenbush 1999; Sampson, Morenoff, & Earls 1999).

Poverty and segregation combine to put African American children in the worst neighborhoods in the country, where getting into trouble is a way of life and where lack of resources makes conventional achievement almost impossible (Newman 1999b). Differential association theory thus explains a great deal of the racial difference in arrest rates. Deterrence theory is also important. Compared to non-Hispanic whites, African American children are much more likely to live in a fatherless home and Hispanic children are somewhat more likely to do so, leaving them without an important social bond that might deter deviant behavior.

But these differences in crime rates between minorities and nonminorities are to some extent more apparent than real. It is true that on average minorities commit more crimes than do whites. But when we compare minorities and whites who engage in the *same* behavior—from causing trouble in school to committing murder—minorities are more likely than whites to be cited, arrested, prosecuted, and convicted (Austin & Allen 2000; Cureton 2000). As a result, UCR rates overestimate the percentage of crime actually committed by minorities.

The Criminal Justice System

The responsibility for dealing with crime rests with the criminal justice system, the subject of this section. Any assessment of this system must begin with the question Why punish?

Why Punish?

Traditionally, there have been four major rationalizations for punishment (Conrad 1983):

1. *Retribution.* Society punishes offenders to avenge the victim and society as a whole.
2. *Prevention.* By imprisoning, executing, or otherwise controlling offenders, society keeps them from committing further crimes.
3. *Deterrence.* Punishment is intended to scare both previous offenders and non-offenders away from a life of crime.
4. *Reform.* By building character and improving skills, former criminals are enabled and encouraged to become law-abiding members of society.

Today, social control agencies in the United States represent a mixture of these different philosophies and practices. However, the increasing emphasis since the 1970s has been on long—even life-long—sentences. For example, under "three strikes and you're out" laws passed around the country, individuals convicted of three felonies, regardless of the circumstances, must serve at least 25 years in prison without probation. These laws don't differentiate between a serial killer and someone who breaks into a store to steal food. The shift toward mandatory, long sentences, combined with the dearth of educational programs and psychological counseling in jails and prisons, suggests that reformation is only a minor goal of our criminal justice system.

In the United States, this system consists of a vast network of agencies set up to deal with persons who deviate from the law: police departments, probation and parole agencies, rehabilitation agencies, criminal courts, jails, and prisons.

The Police

Police officers occupy a unique and powerful position in the criminal justice system. They can make arrests even if no one has filed a complaint against an individual, and even if no one is there to oversee their actions. Although they are supposed to enforce the law fully and uniformly, everyone realizes that this is neither practical nor possible. In 2004, there were 3 full-time police officers for every 1,000 persons in the nation (Federal Bureau of Investigation 2006). This means that the police ordinarily must give greater attention to more serious crimes. Minor offenses and ambiguous situations are likely to be ignored.

In most situations, the police officer is out on her or his own, away from supervision and direction, and must make snap decisions about whether to pursue violations or disregard them.

Police officers have a considerable amount of discretionary power in determining the extent to which the policy of full enforcement is carried out. Should a drunk and disorderly person be charged or sent home? Should a juvenile offender be charged or only reported to his or her parents? Should a strong odor of marijuana in an otherwise orderly group be overlooked or investigated? Unlike decisions meted out in courts of law, decisions made by police officers on the street are relatively invisible and thus hard to evaluate.

The Courts

Once arrested, an individual starts a complex journey through the criminal justice system. This trip can best be thought of as a series of decision stages. A significant proportion of those who are arrested are never prosecuted. Of those who are prosecuted, however, almost 90 percent are eventually convicted, with almost all convictions resulting from pretrial negotiations, rather than public trials (U.S. Department of Justice 2001b). Thus, the pretrial phases of prosecution are far more crucial to arriving at judicial decisions of guilt or innocence than are court trials themselves. Like the police, prosecutors have considerable discretion in deciding whom to prosecute and what charges to file.

Throughout the entire process, the prosecution, the defense, and the judges participate in negotiated plea bargaining. The accused is encouraged to plead guilty in the interest of getting a lighter sentence, a reduced charge, or, in the case of multiple offenses, the dropping of some charges. In return, the prosecution is saved the trouble and cost of a trial. As a result, court decisions reflect much more—and much less—than simple guilt and innocence.

Prisons

For most people, getting tough on crime means locking criminals up and throwing away the key. Indeed, almost three quarters (74 percent) of the U.S. public believes that courts do not deal harshly enough with criminals (U.S. Department of Justice 2001b). Presidential politics, the strength of the Republican Party, the rise of conservative

■ Overcrowded prisons in which inmates are depersonalized by assigned numbers, identical uniforms, and unvarying routines breed anger, violence, boredom, and further deviance.

religious denominations, and overall public opinion have contributed to rapid expansion in the law enforcement sector and a rapid rise in imprisonment (Curry 1996; Jacobs & Helms 1997; Kraska & Kappeler 1997). By the end of 2004, there were more than 1.4 million people in U.S. federal and state prisons—almost four times the number in 1977 (Crary 2006). Rates of imprisonment are now higher in the United States than anywhere else in the world. This is primarily due to harsher sentencing policies, especially for drug-related crimes, such as "mandatory minimums" and "three strikes and you're out" laws (Figure 6.4).

Prison residents are disproportionately young men who are uneducated, poor, and African American. As of 2004, about 41 percent of all prisoners are African American males (Bureau of Justice Statistics 2005). Even more shockingly, 8.4 percent of *all* African American males ages 25 to 29 are in prison, compared with 2.5 percent of Hispanics and 1.4 percent of whites.

The sharp increase in the use of imprisonment has resulted in a crisis in prison (and jail) conditions. Many facilities are housing twice as many inmates as they were designed to hold, in inhumane conditions; in Arizona, jail inmates—most of whom have not even been tried yet, let alone convicted—are housed in tents in the desert with temperatures rising up to 125 degrees. When inmates consider these conditions unjust, they become a major cause of prison riots (Useem & Goldstone 2002). As a result, prisons in more than 30 states are under court order to reduce crowding and improve conditions. This is an enormously expensive undertaking.

Other Options

Do we really need to spend billions and billions of dollars to build more prisons to warehouse a growing proportion of those accused or convicted of crime? Maybe not.

A growing number of empirical studies demonstrate that the certainty of getting caught deters crime more effectively than does long sentences (McCarthy 2002).

FIGURE 6.4

Number of People in Prison During 2006 (per 100,000 Population)

The United States leads the world in imprisoning its own population. Not only do we imprison more people than do similar countries like Canada, we even imprison more people than do dictatorships like Libya and Cuba.

SOURCE: www.prisonstudies.org.

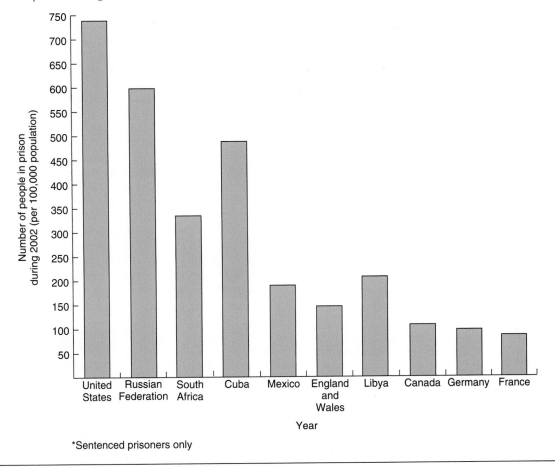

*Sentenced prisoners only

These findings suggest that we are pursuing the wrong strategy. Rather than building more prisons to warehouse criminals for longer periods of time, we need to put more money into law enforcement.

Another approach to solving the prison crisis is to change the way we deal with convicted criminals. As the cost of imprisoning larger numbers of people balloons to crisis proportions, community-based corrections has emerged as an alternative to long prison sentences. New intensive supervision probation programs are being used across the country to safely release convicts from prison earlier. They include curfews, mandatory drug testing, supervised halfway houses, mandatory community service, frequent reporting and unannounced home visits, restitution, electronic surveillance, and split sentences (incarceration followed by supervised probation). A review conducted for the U.S. National Institute of Justice found that when these programs included treatment for drug addiction and other supportive services, they increased the chances of rehabilitation and reduced overall costs to the system (Petersilia 1999).

American Diversity

focus on

Capital Punishment and Racism

In 1972, in the case of *Furman v. Georgia*, three African American defendants appealed their death sentences to the U.S. Supreme Court on the grounds that capital punishment constituted cruel and unusual punishment (Ogletree & Sarat 2006). Their argument was that other defendants, many of whom were white, committed equally or more serious crimes but were not sentenced to death. After reviewing the data, the Supreme Court agreed with the defendants, holding that the uncontrolled discretion of judges and juries reflected racist biases and denied defendants constitutionally guaranteed rights to due process.

The *Furman* decision put a temporary stop to capital punishment, but states attempted to solve the problem of disparity in sentencing by giving judges and juries less discretion in these cases. Have these changes eliminated earlier racial biases in capital punishment?

Unfortunately, no. Studies conducted in the post-*Furman* era continue to show that race strongly predicts who is sentenced to death. Between 1977 and 1990, African Americans comprised about 12 percent of the U.S. population but 39 percent of executed convicts (Culver 1992). Nonwhites now account for 55 percent of all Americans awaiting execution (Death Penalty Information Center 2006). This is much higher than the percentage of non-whites in the general population and also much higher than the percentage of nonwhites among convicted murderers. Moreover, black defendants who look stereotypically black are twice as likely as other black defendants to be convicted of murders (Eberhardt et al. 2006).

Race of the victim affects the likelihood of receiving a death sentence even more than does race of the defendant: Those convicted of killing whites are significantly more likely to receive the death penalty than are those convicted of killing African Americans (Reiman 1998; Schaefer et al. 1999; U.S. General Accounting Office 1996; Williams & Holcomb 2001).

The importance of eliminating racism (and other biases) from death penalty cases is highlighted by the growing realization that innocent people can and do get convicted, if infrequently. Between 2000 and 2006, more than 100 people were exonerated based on DNA testing (Innocence Project, www.innocence-project.org, accessed May 2006). The most common cause of false convictions is mistaken identifications by witnesses, which is most common when the witness is white and the accused is nonwhite. In addition, most defendants in these cases were poor, many lacked proper legal representation, and many were pressed by the police into making false confessions (Innocence Project 2006; Ogletree & Sarat 2006).

Execution is forever. If we are to continue to impose the death penalty, we must eliminate racial bias from the process.

■ Among those accused of murder, dark-skinned African Americans are twice as likely as lighter-skinned African Americans to be convicted. In addition, among those *convicted* of murder, African Americans are significantly more likely than whites to receive the death penalty. This raises serious questions about the fairness of capital punishment.

© Joel Gordon

Where This Leaves Us

The conservative approach to confronting deviance and crime has generally been to make deviance illegal and to increase penalties for convicted criminals. This approach has dominated since the 1970s, which is why prison populations have soared. An alternative approach is, first, to develop greater tolerance for victimless crimes and other forms of deviance that are relatively inconsequential. For more serious forms of deviance and crime, we can address the social problems that give rise to these activities. A leading criminologist (Currie 1998) advocates five major strategies for doing so:

1. Reduce inequality and social impoverishment.
2. Replace unstable, low-wage, dead-end jobs with decent jobs.
3. Prevent child abuse and neglect.
4. Increase the economic and social stability of communities.
5. Improve the quality of education in all communities.

These strategies would require a massive commitment of energy and money. They are not only expensive but also politically risky. Whereas law-and-order advocates want to get tough on crime by sending more criminals to jail, a policy incorporating these five strategies would channel dollars and beneficial programs into high-crime neighborhoods. Such a policy calls for more teachers and good jobs rather than for more police officers and prisons.

Observers from all sociological perspectives and all political parties recognize that social control is necessary. They recognize that rape, assault, and drug-related crimes are serious problems that must be addressed. The issue is how to do so. The sociological perspective suggests that crime can be addressed most effectively by changing the social institutions that breed crime rather than by focusing on changing individual criminals after the fact.

Summary

1. Most of us conform most of the time. We are encouraged to conform through three types of social control: (1) self-restraint through the internalization of norms and values, (2) informal social controls, and (3) formal social controls.
2. Nonconformity occurs when people violate expected norms of behavior. Acts that go beyond eccentricity, challenge important norms, and result in social sanctions are called deviance. Crimes are deviant acts that are also illegal.
3. Deviance is relative. It depends on society's definitions, the circumstances surrounding an act, and the particular groups or subcultures one belongs to.
4. Structural functionalists use strain theory to explain how individual deviance is linked to social disorganization and use collective efficacy to explain why some neighborhoods have higher rates of crime than others. Symbolic interactionists propose differential association, deterrence, and labeling theories, which link deviance to interaction patterns that encourage deviant behaviors and a deviant self-concept. Conflict theorists locate the cause of deviance, and of laws defining what is criminally deviant, in inequality and class conflict.

5. Rates of violent and property crimes rose from 1960 to 1990 but have fallen steadily since then.
6. Many arrests are for victimless crimes—acts for which there is no complainant. Laws relating to such crimes are the most difficult and costly to enforce.
7. The high incidence of white-collar crimes—those committed in the course of one's occupation—indicates that crime is not merely a lower-class behavior.
8. Males, minority-group members, lower-class people, and young people are disproportionately likely to be arrested for crimes. Some of this disparity is due to their greater likelihood of committing a crime, but it is also explained partly by their differential treatment within the criminal justice system.
9. The criminal justice system includes the police, the courts, and the correctional system. Considerable discretion in the execution of justice is available to authorities at each of these levels.
10. The United States faces a "crisis of penalty," as our "get-tough" approach to crime is populating prisons far beyond capacity. Evidence suggests that longer sentences may not be necessary. Alternatives to imprisonment include community-based corrections and social change to reduce the causes of crime.

Thinking Critically

1. Explain how differential association theory can or cannot explain why some children who grow up in bad neighborhoods do not become delinquent.
2. Why do you think most Americans view street crime as more serious than corporate crime? What would a conflict theorist say? A structural functionalist?
3. Describe a deviant whom you have known well—someone who got in trouble with the law or should have. Evaluate the theories of deviance in light of this one person. Which theory best explains why your acquaintance deviated rather than conformed? Which theory best explains whether or not your acquaintance was arrested and imprisoned for his or her behavior?
4. Devise a strategy for deterring white-collar or corporate crime, keeping in mind what you have read in this chapter.
5. From a sociological perspective, why would the race of the *victim* be as important as the race of the *defendant* in predicting whether a convicted killer will be sentenced to death? What does this tell us about racial and ethnic relations in our society? If racial discrimination exists in death sentencing, is that a good reason to stop capital punishment altogether? Why or why not?

Companion Website for This Book

academic.cengage.com/sociology/brinkerhoff
Gain an even better grasp on this chapter by going to the Companion Website. This resource contains tutorial quizzes and flash cards to help you master key terms and concepts.

Suggested Readings

Gonnerman, Jennifer. 2004. *Life on the Outside: The Prison Odyssey of Elaine Bartlett.* New York: Farrar, Straus and Giroux. A moving and disturbing account of a woman who served 16 years in prison for her first criminal offense, and the difficulties that awaited her in readjusting to life outside of prison.

Conover, Ted. 2000. *Newjack: Guarding Sing Sing.* New York: Alfred Knopf. Investigative journalist Ted Conover spent a year working as a prison guard at Sing Sing. His book details the impact of prison life on guards as well as prisoners, and the impact of "prison culture" on all of us.

Prejean, Helen. 1993. *Dead Man Walking: An Eye Witness Account of the Death Penalty in the United States.* New York: Vintage. A deeply moving, thoughtful book written by a Roman Catholic nun who has worked for many years both with convicted murderers on death row and with the families of their victims.

Reiman, Jeffrey. 2005. *The Rich Get Richer and the Poor Get Prison: Ideology, Class, and Criminal Justice* (7th ed.). Boston: Allyn & Bacon. A conflict perspective on how the police and legal system ignore the crimes of the wealthy while cracking down on both unconventional and criminal behavior among the poor.

Stratification

© AP/Wide World Photos

Outline

Structures of Inequality

Inequality exists all around us. For example, perhaps your parents gave your brother more help with college tuition than they gave you because they think he is nicer. If so, this is an example of personal inequality and is not a subject for sociological inquiry.

Sociologists study a particular kind of inequality called stratification. **Stratification** is an institutionalized pattern of inequality in which those who hold some social statuses get more access to scarce resources than do others. If you are female and your parents gave your brother more economic help than they gave you simply because he was male, that was stratification.

Inequality becomes stratification when two conditions exist:

1. The inequality is *institutionalized*, backed up both by social structures and by long-standing social norms.
2. The inequality is based on membership in a status (such as oldest son or blue-collar worker) rather than on personal attributes.

The scarce resources that we focus on when we talk about inequality are generally of three types: material wealth, prestige, and power. When inequality in one of these dimensions is supported by social structures and long-standing social norms, and when it is based on status membership, then we speak of stratification.

Types of Stratification Structures

Stratification is present in every society. All societies have norms specifying that some categories of people ought to receive more wealth, power, or prestige than others. There is, however, wide variety in how inequality is structured.

A key difference among structures of inequality is whether the categories used to distribute unequal rewards are based on ascribed or achieved statuses. As noted in Chapter 4, *ascribed statuses* are those that are fixed by birth and inheritance and are unalterable during a person's lifetime. *Achieved statuses* are optional ones that a person can obtain in a lifetime. Being African American or male, for example, is an ascribed status; being an ex-convict or a physician is an achieved status.

Every society uses some ascribed and some achieved statuses in distributing scarce resources, but the balance between them varies greatly. Stratification structures that rely largely on ascribed statuses as the basis for distributing scarce resources are called **caste systems**; structures that rely largely on achieved statuses are called **class systems.**

Caste Systems

In a caste system, whether you are rich or poor, powerful or powerless depends almost entirely on who your parents are (Smaje 2000). Whether you are lazy and stupid or hardworking and clever makes little difference. Your parents' position determines your own, and you are expected to follow the same occupation as they did. Moreover, in a caste system you are only allowed to marry someone whose social position matches yours, and so your children are guaranteed to share the same status as you and your spouse.

India provides the best-known example of a caste system. Under its caste system, all Hindus (the majority religion) are divided into castes, roughly comparable

---

Stratification is an institutionalized pattern of inequality in which social statuses are ranked on the basis of their access to scarce resources.

Caste systems rely largely on ascribed statuses as the basis for distributing scarce resources.

Class systems rely largely on achieved statuses as the basis for distributing scarce resources.

Connections

Personal Application

Everyone has both ascribed and achieved statuses. You now have the *achieved* statuses of high school graduate and college student, and hope to have the achieved status of college graduate. If your parents graduated college, you also have the *ascribed* status of coming from an educated family, which you will keep whether or not you graduate college. How others view you will depend on both your achieved and your ascribed statuses.

to occupational groups, which differ substantially in prestige, power, and wealth. Caste membership is unalterable; it marks one's children and one's children's children. Although the caste system was officially outlawed more than 50 years ago, it continues to determine individuals' life chances—especially for the 25 percent of the population who are Dalit, or "untouchables."

Class Systems

In a class system, achieved statuses are the major basis of unequal resource distribution. Occupation remains the major determinant of rewards, but it is not fixed at birth. Instead, you can achieve an occupation far better or far worse than those of your parents. The amount of rewards you receive is influenced by your own talent, ambition, and work—or lack thereof.

The primary difference between caste and class systems is not the level of inequality but the opportunity for achievement. The distinctive characteristic of a class system is that it permits **social mobility**—a change in social class, either upward or downward. Mobility can occur between one generation and another; if you graduate college, and your parents didn't, you will experience upward social mobility. It can also occur within one's lifetime; a middle-aged engineer who is forced to retire early by "downsizing" and ends up working as a Wal-Mart greeter experiences downward social mobility.

Social mobility is the process of changing one's social class.

Even in a class system, ascribed characteristics matter. Whether you are male or female, Hispanic or non-Hispanic, Jewish or Protestant is likely to influence which doors are thrown open and which barriers have to be surmounted. Nevertheless, these factors are much less important in a class society than in a caste society. Because class systems predominate in the modern world, the rest of this chapter is devoted to them.

Classes—How Many?

A class system is an ordered set of statuses. Which statuses are included? And how are they divided? Two theoretical answers and two practical answers to these questions are presented in this section.

Marx: The Bourgeoisie and the Proletariat

Karl Marx (1818–1883) believed that there were only two classes. We could call them the haves and the have-nots; Marx called them the bourgeoisie (boor-zhwah-zee) and the proletariat. The **bourgeoisie** are those who own the tools and materials necessary for their work—the means of production. The **proletariat** are those who do not. The latter must therefore support themselves by selling their labor to the former. In Marx's view, **class** is determined entirely by one's relationship to the means of production.

The **bourgeoisie** is the class that owns the tools and materials for their work—the means of production.

Relationship to the means of production obviously has something to do with occupation, but it is not the same thing. According to Marx, your college instructor, the manager of the Sears store, and the janitor are all proletarians because they work for someone else. If your garbage collector works for the city, he is also a proletarian; if he owns his own truck, however, he is a member of the bourgeoisie. The key factor is not income or occupation but whether individuals control their own tools and their own work.

The **proletariat** is the class that does not own the means of production. They must support themselves by selling their labor to those who own the means of production.

Marx, of course, was not blind to the fact that in the eyes of the world, store managers are regarded as more successful than truck-owning garbage collectors. Probably managers think of themselves as being superior to garbage collectors. In Marx's eyes, this is **false consciousness** —a lack of awareness of one's real position in the class structure. Marx, a social activist as well as a social theorist, hoped that

In Marxist theory, **class** refers to a person's relationship to the means of production.

False consciousness is a lack of awareness of one's real position in the class structure.

FIGURE 7.1

Weber's Model of Social Class
Weber identifies three independent dimensions of stratification. This multidimensional concept is sometimes called social class.

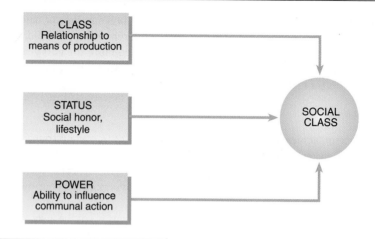

managers and janitors would develop **class consciousness**—an awareness of their true class identity. If they did, he believed, a revolutionary movement to eliminate class differences would be likely to occur.

Weber: Class, Status, and Power

Several decades after Marx wrote, Max Weber developed a more complex system for analyzing classes. Instead of Marx's one-dimensional ranking system, which provided only two classes, Weber proposed three independent dimensions on which people are ranked in a stratification system (Figure 7.1). One of them, as Marx suggested, is class. The second is **status,** the amount of social honor or prestige afforded one individual or group relative to another. Individuals who share similar status typically form a community of sorts. They invite one another to dinner, marry one another, engage in the same kinds of recreation, and generally do the same things in the same places. The third dimension is **power,** the ability to influence or force others to do what you want them to do, regardless of their own wishes.

Weber argued that although status and power often follow economic position, they may also stand on their own and have an independent effect on social inequality. In particular, Weber noted that status often stands in opposition to economic power, depressing the pretensions of those who "just" have money. Thus, for example, a member of the Mafia may have a lot of money. He may in fact own the means of production (a brothel, a heroin manufacturing plant, or a casino). But he will not have social honor.

Measuring Social Class

Marx and Weber provide us with theoretical concepts we can use in understanding class systems. Modern researchers, however, need practical ways of measuring class, not just theoretical definitions. These days, most researchers focus not on class (as Marx defined it) but on *social* class. **Social class** is a category of people who (as Weber suggested) share roughly the same class, status, and power and who have a sense of identification with one another. When we speak of the upper class or of the working class, we are speaking of social class in this sense.

The most direct way of measuring social class is simply to ask people what social class they belong to. The results of a 2004 survey are presented in Figure 7.2. As you can see, only tiny minorities see themselves as belonging to the upper and

Class consciousness occurs when people understand their relationship to the means of production and recognize their true class identity.

Status is social honor, expressed in lifestyle.

Power is the ability to direct others' behavior even against their wishes.

Social class is a category of people who share roughly the same class, status, and power and who have a sense of identification with each other.

lower classes, and the bulk of the population is split nearly evenly between working- and middle-class identification. Studies show that the difference between working- and middle-class identification has important consequences, affecting what church you go to, how you vote, and how you raise your children.

Another common way to measure social class is by **socioeconomic status** (SES), which ranks individuals on income, education, occupation, or some combination of these. SES measures do not produce self-aware social-class groupings, but result in a ranking of the population from high to low on criteria such as years of school completed, family income, or the prestige of one's occupation (as ranked by surveys of the population). Occupational prestige is quite stable over time: Surveys consistently find that doctors, lawyers, and scientists have high prestige and that maids, plumbers, and file clerks have low prestige.

Inequality in the United States

Stratification exists in all societies. In Britain, India, and China, social structures ensure that some social classes routinely receive more rewards than do others. This section considers how stratification works in the United States.

Economic Inequality

Income inequality is very high in all class systems but is especially high in the United States. Of the 29 industrialized nations that participate in the long-term Luxembourg Income Study (2000), only two, Mexico and Russia, have more income inequality than the United States.

Income inequality in the United States has increased steadily since 1970, even for white men with full-time, year-round jobs (DeNavas-Walt & Cleveland 2002). The income gap, however, is most pronounced at the two ends of the income spectrum: The poorest 10 percent of the population has become significantly poorer, while the wealthiest 10 percent has become significantly wealthier. When we divide the U.S. population into five equal-sized groups (quintiles), we find that the poorest 20 percent of American households now receive only 3.4 percent of all personal income, whereas the richest 20 percent receive 50 percent—more than 14 times as much (Figure 7.3). In contrast, in Sweden, for example, doctors and lawyers earn on average only about twice what waitresses and gas station mechanics earn.

The rise in income inequality stems primarily from changes in the U.S. economic structure (Morris & Western 1999). As we will discuss in more detail in Chapter 13, 80 percent of all Americans now work in service or retail jobs. These jobs typically pay far less than the manufacturing jobs that once dominated the U.S. economy. Meanwhile, across all economic sectors, employers are laying off permanent employees and replacing them with lower-paid temporary or part-time workers. Other employers are replacing well-paid American workers with cheaper workers either in Southern states or, increasingly, in foreign countries.

FIGURE 7.2
Social-Class Identification in the United States
Social class is a very real concept to most Americans. They are aware of their own social-class membership: They feel that, in a variety of important respects, they are similar to others in their own social class and different from those in other social classes. The great majority of Americans place themselves in either the working or the middle class.
General Social Survey 2004.
http://sda.berkeley.edu/cgi-bin/hsda3.

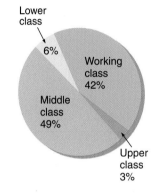

Socioeconomic status (SES) is a measure of social class that ranks individuals on income, education, occupation, or some combination of these.

Income inequality is higher in the United States than in any other industrialized nation and is continuing to increase.

FIGURE 7.3

Income Inequality in the United States, 2001

Imagine dividing all U.S. citizens into five equal-size groups (quintiles). If all income in the country was also divided equally, each quintile (20 percent) of Americans would receive 20 percent of all income. In reality, the richest 20 percent (quintile) of Americans receives half of all the income, and the poorest 20 percent of the population receives less than 4 percent.

SOURCE: U.S. Census Bureau, Current Population Survey, 2004 and 2005 Annual Social and Economic Supplements. http://www.census.gov/hhes/www/poverty/effect2004/table_2_020106.pdf.

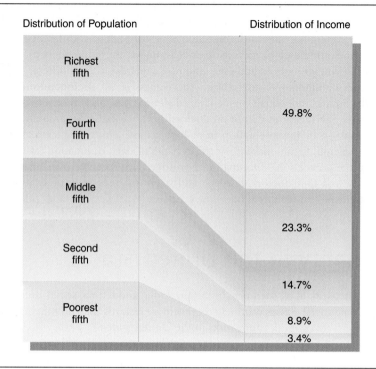

Distribution of Population / Distribution of Income

Richest fifth — 49.8%
Fourth fifth — 23.3%
Middle fifth — 14.7%
Second fifth — 8.9%
Poorest fifth — 3.4%

Combined with declining union membership and a stagnant minimum wage, these changes have kept down the incomes of poor and working-class Americans.

As bad as income inequality is, looking only at that measure actually understates the levels of economic inequality in the United States. If we measure inequality by the distribution of *wealth*—all that a given household accumulates over the years (savings, investments, homes, land, cars, and other possessions)—we find that the richest 20 percent of households by income own *69 percent* of all wealth (McClain 2005). Historical research suggests that this unequal distribution of wealth is a long-standing pattern in the United States, dating back at least to 1810. However, wealth inequality has increased over the last two decades and is now higher in the United States than in any other industrialized nation (Keister & Moller 2000).

The Consequences of Social Class

The following chapters point out the influence of social class in a number of areas—among them religious participation, divorce, prejudice, and work satisfaction. Here it suffices to say that almost every behavior and attitude we have is related to our social class. Do you prefer bowling or tennis? foreign films or American? beer or sherry? These choices and nearly all the others you make are influenced by your social class. Knowing a person's social class will often tell us more about an individual than any other single piece of information. This is why "Glad to meet you" is often followed by "What do you do for a living?"

But social class differences go beyond mere preferences to real consequences. Consider the following examples:

• People with incomes of less than $7,500 a year are two and a half times as likely to have been the victim of a violent crime as those with incomes over $75,000 (U.S. Bureau of the Census 2006).

When children grow up in very unequal backgrounds, they are likely to grow up to lead very different and unequal lives.

- Infants whose mothers fail to graduate from high school are 60 percent more likely to die before their first birthday than infants whose mothers attend college (U.S. Department of Health and Human Services 2005).
- Compared to those from more affluent homes, students from poor and working-class homes are much more likely to attend community colleges rather than four-year colleges and to drop out regardless of which type of college they attend (Correspondents of the New York Times 2005).

On these and many other indicators, individuals who have more money enjoy a higher quality of life.

Theoretical Perspectives on Inequality

According to *Forbes* magazine, Steven Spielberg is now worth $2.7 *billion* and earns many millions each year. Meanwhile, the average police officer or teacher earns about $39,000, and 20 percent of American families have annual incomes below $17,970 (DeNavas-Walt & Cleveland 2002). How can we account for such vast differences in income? Why isn't anybody doing anything about it? We begin our answers to these questions by examining the social structure of stratification—that is, instead of asking about Steven Spielberg or Officer Malloy, we ask why some *statuses* routinely get more scarce resources than others. After we review these general theories of stratification, we will turn to explanations about how individuals are sorted into these various statuses.

Structural-Functional Theory

The structural-functional theory of stratification begins (as do all structural-functional theories) with the question, Does this social structure contribute to the maintenance of society? The classic statement of this position was given by Kingsley Davis and

Wilbert Moore (1945), who argued that stratification is necessary and justifiable because it contributes to the maintenance of society. Their argument begins with the premise that each society has essential tasks (functional prerequisites) that must be performed if it is to survive. The tasks associated with shelter, food, and reproduction are some of the most obvious examples. Davis and Moore argue that we need to offer high rewards as an incentive to make sure that people are willing to perform these tasks. The size of the rewards must be proportional to three factors:

1. *The importance of the task.* When a task is very important, very high rewards may be necessary to guarantee that it is done.
2. *The pleasantness of the task.* When the task is relatively enjoyable, there will be no shortage of volunteers and high rewards need not be offered.
3. *The scarcity of the talent and ability necessary to perform the task.* When relatively few have the ability to perform an important task, high rewards are necessary to motivate this small minority to perform the necessary task.

From this perspective, it makes sense to pay doctors more than childcare workers: Although both fields are necessary, far fewer people have the intelligence, skills, and talent needed to enter medicine, especially since it requires long years of training and long hours of work in sometimes unpleasant and stressful circumstances. To motivate people who have this relatively scarce talent to undertake such a demanding and important task, Davis and Moore would argue that we must hold out the incentive of very high rewards in prestige and income. Society is likely to decide, however, that little reward is necessary to motivate women to fill the even more vital task of having and raising children. Although the function is essential, the potential to fill the position is widespread (most women between the ages of 15 and 40 can do it), and the job has sufficient noncash attractions that no shortage of volunteers has arisen. To structural functionalists, then, the fact that doctors are paid more than childcare workers is a rational response to a social need.

Criticisms

This theory has generated a great deal of controversy. Among the major criticisms are these: (1) High demand (scarcity) can be artificially created by limiting access to good jobs. For example, keeping medical schools small and making admissions criteria unnecessarily stiff reduce supply and increase demand for physicians. (2) Social-class background, sex, and race or ethnicity probably have more to do with who gets highly rewarded statuses than do scarce talents and ability. (3) Many highly rewarded statuses (rock stars and professional athletes, but also plastic surgeons and speechwriters) are hardly necessary to the maintenance of society.

The Conflict Perspective

Conflict theorists take a very different approach to inequality. These theorists argue that inequality results not from consensus over how to meet social needs but from class conflict.

Karl Marx provided the classic conflict theory of inequality. He argued that inequality was rooted in private ownership of the means of production. Those who own the means of production seek to maximize their own profit by minimizing the amount of return they must give to the proletarians, who have no choice but to sell

Removing garbage is both unpleasant and absolutely essential to modern life, yet most garbage collectors are paid low wages. Structural-functional theory attributes their low wages to their lack of skill, whereas conflict theory attributes it to their lack of power.

their labor to the highest bidder. In this view, stratification is neither necessary nor justifiable. Inequality does not benefit society; it benefits only the rich.

Like classic Marxist theory, modern conflict theory recognizes that the powerful can oppress those who work for them by claiming the profits from their labor (Wright 1985). It goes beyond Marx's focus on ownership, however, by considering how control also may affect the struggle over scarce resources and how class battles are played out in governmental politics (Grimes 1989). In addition, modern conflict theory looks at noneconomic sources of power, especially gender and race. These theorists argue, for example, that in the same way that capitalists benefit from the productive labor of workers, men gain benefit from the "reproductive" labor of women. The term **reproductive labor** is usually used to describe traditionally female tasks such as cooking, cleaning, and nurturing—those tasks that often make it possible for others to work and play. Modern conflict theorists point out that in most families, those with the least power do the most reproductive labor; as a result, these individuals end up having fewer opportunities to earn the good incomes that might otherwise increase their power within the family (Cancian & Oliker 2000).

Reproductive labor refers to traditionally female tasks such as cooking, cleaning, and nurturing that make it possible for a society to continue and for others to work and play.

Criticisms

There is little doubt that people who have control (through ownership or management) systematically use their power to extend and enhance their own advantage. Critics, however, question the conclusion that this means that inequality is necessarily undesirable and unfair. First, people *are* unequal. Some people are harder working, smarter, and more talented than others. Unless forcibly held back, these people will pull ahead of the others—even without force, fraud, and trickery. Second, coordination and authority *are*

A Comparison of Two Models of Stratification

Basis of Comparison	Structural-Functional Theory	Conflict Theory
1. Society can best be understood as…	Groups *cooperating* to meet common needs	Groups *competing* for scarce resources
2. Social structures…	Solve problems and help society adapt	Maintain current patterns of inequality
3. Causes of stratification are…	Importance of vital tasks, unequal ability, pleasantness of tasks	Unequal control of means of production maintained by force, fraud, and trickery
4. Conclusion about stratification…	Necessary and desirable	Unnecessary and undesirable, but difficult to eliminate
5. Strengths…	Consideration of unequal skills and talents and necessity of motivating people to work	Consideration of conflict of interests and how those with control use the system to their advantage
6. Weaknesses…	Ignores importance of power and inheritance in allocated rewards; functional importance overstated	Ignores the functions of inequality and importance of individual differences

functional. Organizations work better when those trying to do the coordinating have the power or authority to do so.

Symbolic Interaction Theory

Unlike structural-functional theory and conflict theory, symbolic interaction theory does not attempt to explain why some statuses are so much better rewarded than others. Instead, it asks *how* these inequalities are perpetuated in everyday life.

One of the major contributions of symbolic interaction theory is its identification of the importance of **self-fulfilling prophecies.** Self-fulfilling prophecies occur when something is *defined* as real and therefore *becomes* real in its consequences. This social dynamic is one of the ways that social class statuses are reinforced. For example, when teachers assume that lower-class students are less intelligent and less able to do intellectual work, the teachers are less likely to spend time helping them learn. Instead, teachers may shuffle lower-class students off to vocational classes that emphasize discipline and mechanical skills rather than intellectual skills. After several years of such "schooling," lower-class students may, in fact, have fewer intellectual skills than do others.

Symbolic interaction theory also helps us understand how everyday interactions reinforce inequality by constantly reminding us of our place in the social order. For example, in most restaurants, waiters and waitresses must enter through the back doors (Paules 1991). They often must use separate bathrooms that are far less pleasant than those used by customers, take their breaks in windowless rooms that lack air conditioning, and wear clothes that make them look like maids and butlers. Customers often speak rudely (or crudely) to serving staff, who are expected to smile in response. And, at the end of the evening, the customer decides whether the waiter or waitress deserves a tip. In all these ways, normal restaurant interactions reinforce customers' sense of social superiority and servers' sense of social inferiority.

Self-fulfilling prophecies occur when something is *defined* as real and therefore *becomes* real in its consequences.

The Determinants of Social-Class Position

With each generation, the social statuses in a given society must be allocated anew. Some people will get the good positions and some will get the bad ones; some will receive many scarce resources and some will not. In a class system, this allocation process depends on two things: the opportunities available to specific individuals and the overall opportunities available in a society's labor market. We refer to these, respectively, as micro- and macro-level factors that affect achievement.

Microstructure: Individual Opportunities

Unlike in a caste system, in the United States social status is not directly or completely inherited. Yet people tend to hold social class status similar to that of their parents. How does this come about? The best way to describe the system is as an **indirect inheritance model.** Parents cannot fully determine their children's social status, but they strongly affect whether their children will have the opportunities needed to obtain or maintain a higher social status.

The best predictor of your eventual social class is your parents' income (Corcoran 1995). Your parents' income affects your life chances in many ways (Corcoran 1995; Harris 1996; Bettie 2003). If your parents are middle or upper class, you are more likely to be born healthy and more likely to get good nutrition and health care during childhood. As a result, you are less likely to have mental or physical disabilities that might reduce your potential income. Your parents will have the time and money to give you a stimulating environment in which your intellectual capacities can thrive and will most likely live in neighborhoods with good schools, where teachers will assume you are "college material." Similarly, as we discussed in Chapter 2, your parents will have endowed you with cultural capital: values, interests, knowledge, and social behavior patterns that mark you as middle or upper class.

Class differences in home environment and in parents' support for school also have important effects on children's success. Bright and ambitious lower-class children often find it hard to do well in school when they have to study at a noisy kitchen table, have no funds for SAT tutoring or extracurricular activities, have to work part-time job to help support their family, and know their parents need them to get full-time employment as soon as possible. In contrast, middle-class children who grow up in supportive environments often find it hard to fail even if their ambitions and talents are modest.

In addition, if your parents went to college or have middle-class jobs, then you have probably always assumed that you, too, would go to college, and automatically signed up for algebra and chemistry in high school. Your parents may have given you money to take an SAT prep course, to visit colleges around the country, and to pay for as many applications as you chose to submit. If your parents didn't attend college, they may have encouraged you to start earning an income right away rather than seeking further education. Your high school advisor, too, is more likely to have encouraged you to register for shop or sewing rather than algebra or other courses needed for college entrance (Bettie 2003). If you later decided you wanted to go to college, you first had to overcome all these barriers.

If your parents graduated college, the benefits to you will continue even after you graduate college yourself. Your parents are likely to have both the income and

The **indirect inheritance model** argues that children have occupations of a status similar to that of their parents because the family's status and income determine children's aspirations and opportunities.

the contacts that will help you get into a good school. After you graduate, they are likely to know people who can help you get good jobs. They may also help you buy clothes for your job interviews, purchase your first home, or pay for family vacations, allowing you to invest your earnings in a new business. They might even invest in the business themselves. All these factors make parents' income a powerful predictor of their children's eventual income (Corcoran 1995).

Macrostructure: The Labor Market

The indirect inheritance model explains how some people come to be well prepared to step into good jobs, whereas others lack the necessary skills or credentials. By themselves, however, skills and credentials do not necessarily lead to class, status, or power. The other variable in the equation is the labor market: If there is a major economic depression, you will not be able to get a good job no matter what your education, motivation, or aspirations. Indeed, most observers believe that changes in the nation's economic structure and labor market will offer fewer opportunities for upward mobility over the next generation.

As Figure 7.4 shows, the proportion of positions at the top of the U.S. occupational structure has increased dramatically over the last century. Not everyone, however, has benefited equally from these new opportunities for upward mobility. Although women and minorities now have an easier time entering high-earning occupations, they tend to find themselves in the lower-earning positions within those occupations. They are more likely to be public defenders than corporate lawyers, more likely to be pediatricians than surgeons.

Labor market theorists suggest that the United States has a *segmented labor market*: one labor market for good jobs (usually in the big companies) and one labor market for poor jobs (usually in small companies). Women and minorities are disproportion-

© AP/Wide World Photos

■ When the number of jobs available in a geographic area or occupation shrink, finding employment becomes difficult regardless of one's education, achievement, motivation, or aspirations. This is the situation facing these men, who were all seeking jobs at Microsoft.

FIGURE 7.4

The Changing Occupational Structure, 1900–2000

Since the beginning of the twentieth century, the occupational structure of the United States has shifted away from farming, fishing, and forestry. Today there are many more white-collar, professional, and managerial jobs.

SOURCE: U.S. Bureau of the Census, 1999a, 424 and U.S. Census Bureau, Census 2000 summary file 3, matrices P49, P50, and P51.

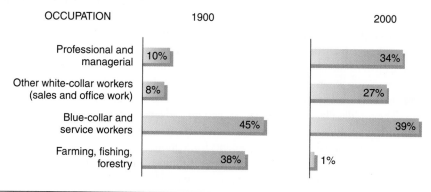

ately directed into companies with low wages, low benefits, low security, and short career ladders. (Segmented labor markets and the contemporary economic structure are discussed in more detail in Chapter 13.)

The American Dream: Ideology and Reality

In any stratification system, there are winners and losers. Why do the losers put up with it?

The answer lies in **ideology.** Ideology refers to any set of beliefs that strengthen and support a social, political, economic, or cultural system. Each stratification system has an ideology that rationalizes the existing social structure and motivates people to accept it. In India, for example, the Hindu religion teaches that a low caste is a punishment for poor behavior in a previous life, but that if you live morally in this life, you can expect to be born into a higher caste in the next life. This ideology offers individuals an incentive to accept their lot in life.

In the United States, the major ideology that justifies inequality is the *American Dream*. This ideology proposes that equality of opportunity exists in the United States and that anyone who works hard enough will get ahead. Conversely, anyone who does *not* succeed must be responsible for his or her own failure. Belief in this ideology is considerably stronger in the United States than elsewhere in the world (Kohut & Stokes 2006).

Although upper class Americans are the most likely to believe in the American Dream, this belief is widespread (Bettie 2003). For example, according to a Gallup Poll conducted January 23, 2003, almost one third of Americans believe it very or somewhat likely that they will someday be rich. Consequently, most Americans are less interested in changing the rules of the game than in being dealt into it.

The American Dream is, in part, a reality: most Americans who were poor as children do not remain poor as adults. But many do: About half of African Americans and one-quarter of whites from persistently poor families also experience poverty as adults—much higher than the percentage of non-poor children who experience poverty as adults (Corcoran 1995).

An **ideology** is a set of norms and values that rationalizes the existing social structure.

Explaining Upward Mobility

A major reason that the American Dream ideology can survive is because there is, indeed, some upward social mobility. Given all the social forces that hinder mobility, how can we explain why some people do indeed rise above their parents' social class?

It would be easy to assume that the reason some rise and others don't is because of intelligence and hard work, and certainly these factors matter. But many highly intelligent people nonetheless earn low incomes, and some of the hardest working people have the lowest status jobs.

Sociologist Julie Bettie's ethnographic research is particularly useful for understanding upward social mobility. Bettie (2003) spent nine months intensively observing and interviewing at a predominantly working-class high school. Overwhelmingly, she found students were treated by the school in ways that reinforced their existing class status: The middle-class "preps" were tracked into advanced classes and celebrated for their academic achievements, students from stable working-class homes were encouraged to take vocational classes, and students from poorer homes were ignored, marginalized, and expected to fail. In addition, minority students also suffered discrimination and low teacher expectations, whether they were middle-class or poorer.

Nonetheless, some working-class students seemed destined for upward social mobility. All of these upwardly mobile students were smart and hard-working. But they also benefited from resources not available to other working- and lower-class students. Some had become part of middle-class peer groups and received "middle-class treatment" from teachers and advisors because they belonged to mostly middle-class athletic teams or had attended middle-class elementary schools. Some had older siblings who had gone to college and could help them both financially and culturally (by, for example, explaining the importance of earning a four-year degree). All benefited from attending a high school that included college-track, middle-class students rather than a school that was uniformly working- or lower-class. Finally, some students were the children of immigrants who had belonged to the middle class before coming to this country. Although these students lacked the financial resources available to middle-class students, they still had the cultural resources that come with college-educated parents.

Similarly, Darlon Conley (2004) found that differential access to resources explains differences in social mobility *within* families. A son who is already in college when his parents divorce or his father loses his job is more likely to graduate college than is his younger brother who was still in high school when these events occurred. Conversely, when parents' incomes rise over time, they are better able to support their last child through school than their first child. By the same token, when parents lack the money to invest in all their children's education, they may pay for their sons' education but not their daughters' education, pay for their first child but run out of money for the rest, or invest only in the child who seems most likely to succeed. Those who receive the most help from their parents are the ones most likely to experience upward social mobility. As with the students studied by Julie Bettie, social mobility depends on access to resources.

Social Class and Social Life

To a large extent, your social class determines how you live your life. This section briefly reviews the special conditions of each of the classes in the United States.

The Upper Class

In 2004, a family living in the United States required an income of $157,000 to be in the richest 5 percent of Americans (U.S. Census Bureau 2005). Thus, a variety of more-or-less ordinary salespersons, doctors, lawyers, and managers in towns and cities across the nation qualify as very rich compared with the majority. Although their incomes are nothing to sneeze at, most of this upper 5 percent is still middle class. Like members of the working class, they would have a hard time making their mortgage payments if they—or their spouses—lost their jobs and were out of work for a few months. This is because although their current income is quite high, their wealth—their investments, savings, and assets they could easily sell—may not add up to much more than their debts.

The true upper class, on the other hand, is made up of two overlapping groups: those whose families have had high incomes and statuses for more than a generation and those who themselves earn incomes in the millions of dollars. The central institution that cements the first group, whose upper-class status is inherited from their parents, is the private preparatory school, especially New England boarding schools such as Philips Academy, St. Paul, and Choate (Higley 1995). Many graduates of these schools attend Ivy League colleges, such as Harvard, Yale, and Princeton. After graduation, they are likely to join selective country clubs and high-status Episcopalian or perhaps Presbyterian churches, and to serve on the boards of high-culture organizations such as art museums, symphonies, opera companies, and the like.

Unlike those who inherit their millions, other members of the true upper class earned at least part of their wealth. There are about a half million millionaires in the United States. Few went from rags to riches, however. Most had middle- or upper-class parents who sent them to excellent schools and helped them financially in many ways (Table 7.1).

© Gerd Ludwig/Woodfin Camp & Associates

■ Only a small elite ever have the opportunity to drink champagne and eat hors d'ouevres in a skybox at a football game.

TABLE 7.1
The Ten Richest People in the United States, 2005
Half of these fabulously wealthy individuals inherited their fortunes. Another four grew up in affluent families with many advantages, but also played a role in generating their own vast wealth. Only Larry Ellison is truly a self-made man.

Rank	Name	Net Worth ($ million)	Age	Residence	Source
1	Gates, William Henry III	51,000	49	Medina, WA	Microsoft
2	Buffett, Warren Edward	40,000	75	Omaha, NE	Berkshire Hathaway
3	Allen, Paul Gardner	22,500	52	Seattle, WA	Microsoft, investments
4	Dell, Michael	18,000	40	Austin, TX	Dell
5	Ellison, Lawrence Joseph	17,000	61	Silicon Valley, CA	Oracle
6	Walton, Christy	15,700	50	Jackson, WY	Wal-Mart inheritance
6	Walton, Jim C	15,700	57	Bentonville, AR	Wal-Mart inheritance
8	Walton, S Robson	15,600	61	Bentonville, AR	Wal-Mart inheritance
9	Walton, Alice L.	15,500	56	Fort Worth, TX	Wal-Mart inheritance
10	Walton, Helen R	15,400	86	Bentonville, AR	Wal-Mart inheritance

SOURCE: The Forbes Four Hundred, 2005.

The Middle Class

The middle class is a large and diverse group. Ranging from professionals with graduate degrees to salespersons and secretaries, middle-class workers have widely varying incomes, with some earning less than the typical working-class individual. Compared with those in the working class, however, middle-class workers tend to have more job security and more opportunities for promotions and advancement. Middle-class workers also can expect to have important benefits such as health insurance and sick leave and to have incomes that will continue to increase over most of their working lives. The middle class is also united by having at least a high school education and, in most cases, at least some college.

Middle-class culture differs from both elite upper-class culture and working-class culture. Compared with working-class individuals, middle-class individuals are less likely to decorate their homes with religious icons or to belong to bowling leagues and are more likely to value education and equality between the sexes; compared with upper-class individuals, members of the middle class are less likely to decorate their homes with modern art or to belong to golf leagues (Halle 1993). Middle-class parents spend time explaining to their children why they need to follow rules and consider themselves responsible for shepherding their children to activities and providing them with entertainment (Lareau 2003). In contrast, working-class parents more often expect children to entertain themselves and to obey orders.

Media and Culture

Karaoke Class Wars

Prior to the 1990s, sociologists consistently found that the cultural taste of the middle and working classes not only differed significantly, but that middle-class Americans used their cultural taste to distinguish themselves from the working class. Having an original oil painting on the wall, for example, or listening to classical music, not only *showed* that someone was middle-class but was *intended* to have that effect (if only subconsciously).

By the 1990s, however, analysts noticed that middle-class Americans seemed increasingly to be adopting aspects of working-class culture (Peterson & Simkus 1992; Brooks 2000). Listening to rap or country music and wearing "bohemian" clothes signaled that a person might be middle-class, but was still "hip."

Research by Rob Drew (2005), however, questions the extent to which these two cultures are actually blurring. Based on participant-observation at 30 karaoke bars around the country, as well as questionnaires and many informal conversations with participants, Drew found that when middle-class individuals adopt working-class culture (in this case, performing karaoke), they do so in ways that identify them as really middle class.

Karaoke first took root in the United States in working-class neighborhoods. Middle-class commentators reacted with scorn, lambasting the "no-talent" singers and the "death" of true music (Drew 2005). More recently, however, karaoke has become increasingly popular among middle-class Americans. As Drew notes, though, whereas working-class karaoke singers and audiences regard karaoke as a skill deserving of respect, middle-class par-

ticipants regarded karaoke as acceptable only if treated as an object of humor. Middle-class participants

> would sing with exaggerated and obviously insincere feeling, or they would sing in inappropriate styles (for instance, singing a ballad in a hard rock voice), or adopt comic voices, or sing [parodies of] lyrics, or (the simplest and commonest option) they would simply flood out with laughter throughout their performances. (Drew, 376).

Furthermore, Drew observed, middle-class performers not only made fun of the songs (most of which were chosen from "working-class" genres like country music and heavy metal), but also made fun of the very idea of a karaoke singer.

In sum, far from suggesting the blurring of class boundaries in cultural taste, middle-class adoption of karaoke reinforced those boundaries.

The Working Class

Who are the members of the working class? The answer is determined partly by income, but mostly by occupation, education, self-definition, and lifestyle. Generally, the working class includes those who work in blue-collar industries and their families. They are the men and women who work in factories, on loading docks, and in beauty parlors; they drive trucks, work as secretaries, build houses, and work for maid services. Although they sometimes receive excellent wages and benefits, it is the working class that has 10 to 30 percent unemployment during economic recessions and slumps. And although a majority are high school graduates, an eleventh-grade education is more common than a year of college.

Quite a few members of the working class have incomes as good as or better than those at the lower end of the middle-class spectrum. Truck drivers, for example, often make more than do nurses and public school teachers. As a result, working-class families may live in the same neighborhoods as middle-class families. Their economic prospects differ, however, in three ways. First, working-class people have little or no chance of promotion, and their incomes rarely rise much over their lifetimes. Second, their jobs are rarely secure, especially now that the American economy is shifting away from manufacturing to service industries (Newman 1999a). Third, they are much less likely than members of the middle class to receive pensions, health insurance, and other benefits. For all these reasons, working class people are much less likely to have savings or other assets. As a result, layoffs, illnesses, or injuries can quickly drive working class families into poverty (Newman).

As a result of low prospects and economic uncertainty, members of the working class tend to place a higher value on security than do others. Whereas middle- and upper-class people typically associate having choices with having freedom and control, working-class people associate having choices with insecurity, doubts, and fear (Schwartz, Markus, & Snibbe 2006). So, for example, middle-class Americans more often enjoy rock music and its celebration of individual freedom, whereas working-class Americans more often enjoy country music, which frequently warns about the dangers of choices (such as when George Jones sang "Now I'm living and dying with the choices I've made.")

The Poor

Each year, the U.S. government sets an official *poverty level*, or poverty line: the minimum amount of money a family needs to have a decent standard of living. The poverty level adjusts for family size, and in 2004 (the latest data available as of 2006), the poverty level for a family of two adults and two children was $19,157. Most observers agree that in reality, a family of four really needs about $38,000—twice the poverty level—to escape poverty. Hence, statistics based on the official level drastically underestimate the percentage of Americans who are poor.

The proportion of Americans in or near poverty increased dramatically during the 1990s, a situation that shows no signs of improving (Rank 2004; Eckholm 2006). In 2004, 37 million people fell below the poverty level and were classified by the government as poor, and another 54 million were classified as near-poor—above the poverty level but still poor by any reasonable measure (Eckholm 2006).

Who Are the Poor?

Poverty cuts across several dimensions of society. It is found among white Americans as well as among nonwhites, in small towns and big cities, among those with and without full-time jobs, and in traditional nuclear families as well as in female-headed households. But poverty does not affect all groups equally. As Table 7.2 indicates, African Americans and Hispanics are far more likely to be poor than are whites or

■ Although we often think of poverty as a problem of urban, racial minorities, most poor Americans are white and about half live in nonurban areas.

© Nathan Benn/CORBIS

TABLE 7.2
Americans Living Below the Poverty Level, 2004

	Millions of People	Percentage of Group in Poverty
Total	36.9	12.7
Ethnicity		
White non-Hispanic	16.9	8.6
African American	9.0	24.7
Hispanic	9.1	21.9
Asian/Pacific	1.2	9.8
Age		
Under 18	13.0	17.8
18–64	20.5	11.3
65 and older	3.5	9.8
Citizenship/nativity		
Native-born	31.0	12.1
Naturalized citizen	1.3	9.8
Noncitizen	4.7	21.6
Household composition		
Married couple	3.2	5.5
Female-headed, no husband	4.0	28.4
Male-headed, no wife	0.7	13.5

SOURCE: U.S. Census Bureau 2005

Asians; children are more likely to be poor than are middle-aged or elderly persons; noncitizens are more likely to be poor than are citizens (whether native-born or not); and households run by single mothers are more likely to be poor than are households run by single fathers or by two parents (U.S. Census Bureau 2005).

Those who live below the poverty level face crises every day: Parents go hungry so their children can eat, finding clothing for growing children is a nightmare, and a simple cold can easily turn into pneumonia because everyone is under stress and undernourished and no one can afford a doctor's visit. The worst off of the poor have nowhere to call home: In any given year, about 3.5 million Americans—more than one third of whom are children—experience homelessness (National Coalition for the Homeless 2006).

Those who live at or just above the poverty level typically have a roof over their heads and food to eat most of the time. On the other hand, their car is broken down and so is the television, the landlord is threatening to evict them, they are eating too much macaroni, they have to think long and hard before going to a doctor, and they have probably lost a few teeth because they couldn't afford dental care. Although they may not be absolutely poor, they are deprived in terms of what is regarded as a decent standard of living.

Causes of Poverty

Earlier in this chapter, we said that both micro- and macro-level processes determine social-class position. The causes of poverty are simply a special case of these larger processes. At the micro level, some believe poverty can be explained by

Connections
Historical Note

Homelessness has been a major problem for the United States since the early 1980s, when the federal government slashed funds for low-income housing while increasing subsidies for "gentrifying" good-quality older buildings in inner-city neighborhoods. Although the latter policy was intended to improve quality of life in these neighborhoods, its unintended consequence was to raise rents. Meanwhile, the value of the minimum wage (adjusted for inflation) declined, and public assistance became harder to get and lower in value. As a result, Americans now must earn *twice* the minimum wage to afford a modest, two-bedroom apartment. The result has been an epidemic of homelessness.

various "cultures of poverty"; at the macro level, some believe poverty is better explained by the lack of adequate opportunities.

CULTURES OF POVERTY   The idea that poverty is caused (or perpetuated) by a **culture of poverty** was first promoted by anthropologist Oscar Lewis (1969). Lewis argued that poor people hold a set of values—the culture of poverty—that emphasizes living for the moment rather than thrift, investment in the future, or hard work. Recognizing that success is not within their reach and that no matter how hard they work or how thrifty they are, they will not make it, the poor come to value living for the moment.

The **culture of poverty** is a set of values that emphasizes living for the moment rather than thrift, investment in the future, or hard work.

More recently, other scholars have argued that families remain in poverty over generations because a lack of "family values" promotes teen pregnancy and single motherhood or because children raised on welfare conclude that it's smarter to have babies and stay on welfare than to seek employment (Mead 1986, 1992; Murray 1984). Others argue that poor youths (especially nonwhites) grow up to be poor adults because they actively reject work, education, and marriage as symbols of a middle-class culture that they despise.

Comprehensive reviews of 30 years of research on poverty provide little support for any of these culture of poverty theories (Corcoran 1995; Small & Newman 2001; Newman & Massengill 2006). Researchers have found that poor people overwhelmingly share the same attitudes toward welfare, work, education, and marriage as do middle-class people. In fact, far from devaluing work, most of America's poor work as many hours as they can find employment, with many working far more than 40 hours a week (Newman 1999b). This research suggests that teen pregnancy, a "live for the moment" culture, and so on are the *result* of poverty, not the cause (Edin & Kefalas 2006; Newman & Massengill 2006).

THE CHANGING LABOR MARKET   The culture of poverty theories implicitly blame the poor for perpetuating their condition. Critics of these theories suggest that we cannot explain poverty by looking at micro-level processes. To understand poverty, they argue, we need to look at the changing labor market. If there are no good-paying jobs available, then we don't need to psychoanalyze people in order to figure out why they are poor.

The changing labor market is particularly critical for understanding contemporary poverty. As we documented in Figure 7.4, the shift from an agricultural to an industrial society produced major structural pressure for upward mobility earlier in this century. Now, at the end of the century, the deindustrialization of the United States is squeezing the traditional working class out of jobs that once paid living wages (Newman 1999a, 1999b). Instead of the good union jobs that their parents held, today's high school graduates often find themselves working at dead-end jobs, with no benefits, for the minimum wage. A little arithmetic shows that the minimum wage means poverty. In sum, a major cause of poverty is the absence of good jobs.

Social Class and Public Policy

If the competition is fair, inequality is acceptable to most people in the United States. The question is how to ensure that no one has an unfair advantage. Politicians, activists, and social scientists have promoted various approaches to fostering equality. Two of these are fair wage movements and increasing educational opportunities.

Fair Wage Movements

One obvious way to foster income equality is to add income to those on the lower end of the social scale. Since the nineteenth century, labor unions have worked to increase wages for American workers, especially in working-class occupations (Lichtenstein 2003). Unions have used such tactics as boycotts, strikes, and collective bargaining to pressure employers to meet what the unions consider fair demands for fair wages. Unions played a major role in improving the working and living conditions of workers during the first half of the twentieth century. Since the 1970s, however, manufacturing industries have declined, taking many union jobs with them.

Currently, many who are interested in income equality are focusing on raising the minimum wage (Waltman 2000). After adjusting for the effects of inflation, the value of the minimum wage ($5.15 per hour as of 2006) is now worth about one-quarter *less* than in 1979 (Economic Policy Institute 2006). Individuals who work full-time, year-round at minimum wage jobs earn far less than is needed to move themselves out of poverty, let alone to support even a small family. Raising the minimum wage would at least lighten their burdens.

Increasing Educational Opportunities

Education is widely believed to be the key to reducing unfair disadvantages associated with poverty. Pre-kindergarten classes designed to provide intellectual stimulation for children from deprived backgrounds, special education courses for those who don't speak standard English, and loan and grant programs to enable the poor to go as far in school as their ability permits—all these are designed to increase the chances of students from lower-class backgrounds getting an education.

These programs have had some success: Colleges and universities have many more students from disadvantaged backgrounds than they used to. Because students spend only 35 hours a week at school, however, and another 130 hours a week with their families and neighbors, the school cannot overcome the entire deficit that hinders disadvantaged children. For example, researchers have found that during the school year, poor children and better-off children perform at almost the same level in first- and second-grade mathematics. For poor children, however, every summer means a loss in learning, whereas every summer means a gain for wealthier children (Entwisle & Alexander 1992). The home environments of less-advantaged children do not often include trips to the library and other activities that encourage them to use and remember their schoolwork. As a result, many now argue for year-round schools or summer enrichment programs for students from poorer families.

Inequality Internationally

In the same way that inequality can exist *within* a nation, inequality also exists *between* nations. Indeed, a central fact in our world today is the vast international inequality. For example, in 2003 (the latest data available as of 2006), gross domestic product per capita was $37,562 in the United States but only $548 in Sierra Leone (United Nations Development Programme 2005). Average life expectancy in the United States is 77 years; the average in Sierra Leone is 41. The massive disparities not only in wealth and health but also in security and justice are the driving mechanism of current international relationships.

◼ In the least-developed nations, most jobs pay only a few dollars per day, and many can find no jobs at all.

Development refers to the process of increasing the productivity and standard of living of a society—longer life expectancies, more adequate diets, better education, better housing, and more consumer goods.

Most-developed countries are those rich nations that have relatively high degrees of economic and political autonomy.

Less-developed countries are those nations that have lower living standards than the most-developed countries but are substantially better off than the least-developed nations.

Least-developed countries are those nations that are characterized by poverty and political weakness and that are considerably behind on every measure of development.

Because massive inequality leads to political instability and to unjustifiable disparities in health and happiness, nearly every nation—whether more or less developed—supports reducing international inequality. The most accepted way to do this is through development—that is, by raising the standard of living of the less-developed nations.

What is development? First, development is *not* the same as Westernization. It does not necessarily entail monogamy, three-piece suits, or any other cultural practices associated with the Western world. **Development** refers to the process of increasing the productivity and the standard of living of a society, leading to longer life expectancies, better diets, more education, better housing, and more consumer goods.

Importantly, development is not a predictable, unidirectional process. Some countries, such as South Korea, have developed faster than others. Other countries, such as Russia and Argentina, have become *less* developed over time or have fluctuated over the years.

Three Worlds: Most- to Least-Developed Countries

Almost all societies in the world have development as a major goal: They want more education, higher standards of living, better health, and more productivity. Just as social scientists often think of three social classes in the U.S. stratification system—upper, middle, and lower—nations of the world can also be stratified into roughly three levels.

The **most-developed countries** are those rich nations that have relatively high degrees of economic and political autonomy. Examples include the United States, the Western European nations, Japan, Canada, Australia, and New Zealand. Taken together, these nations make up roughly 20 percent of the world's population, produce 82 percent of the gross world product, and own 79 percent of the world's telephone mainlines (United Nations Development Programme 2005). Politically, economically, scientifically, and technologically, they dominate the international political economy.

Less-developed countries include the former Soviet Union and the former Communist bloc nations of Eastern Europe, plus several nations in Southeast Asia and Central and South America. These nations hold an intermediate position in the world political economy. They have far lower living standards than the most-developed nations but are substantially better off than the least-developed nations.

The remaining 75 percent or so of the world's population lives in the **least-developed countries.** These countries are characterized by poverty and political weakness. Although they vary in population, political ideologies, and resources, they are considerably behind on every measure of development.

The Human Development Index

The differences among the world's nations are obvious: In the most-developed countries, people are healthier, more educated, and richer. But how important are these differences to the average person's life?

One approach to answering this question is to develop an index that measures the average achievements of a country along the basic dimensions of human experi-

MAP 7.1
Human Development Around the World
The human development index measures the average achievements in a country in three basic dimensions of human development—life expectancy, educational attainment, and a decent standard of living.
SOURCE: United Nations Human Development Report 2003.

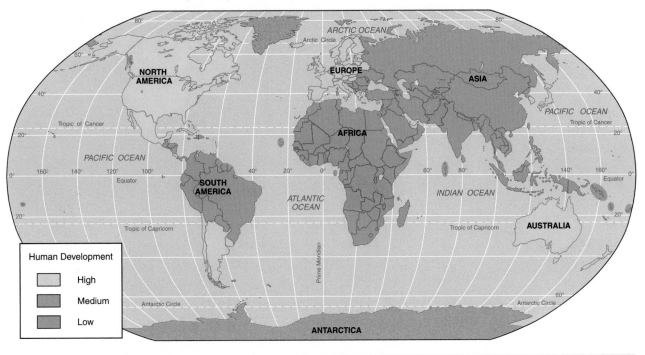

ence: life expectancy, educational attainment, and a decent standard of living. Another approach not only focuses on these three aspects of development but also takes into account the unequal opportunities of men and women. Map 7.1 shows the location of the most- to least-developed countries worldwide. Table 7.3 compares several basic quality-of-life indicators for 15 nations, representing most-developed, less-developed, and least-developed countries. In addition to information about longevity and economic productivity, Table 7.3 also includes each country's overall ranking on the composite Human Development Index and the Gender-Related Development Index. These indexes are based on information about adult literacy rates and educational attainment, life expectancy, and per capita gross domestic product; the greater the disparity between men's and women's quality of life, the lower a country's Gender-Related Development ranking will be compared with its overall Human Development ranking.

In general, the more productive a nation is, the better its quality of life. Norway, with a per capita gross domestic product of $37,670, also has one of the lowest infant mortality rates in the world: Each year, for every 1,000 live births in Norway, only 4 children die before their first birthday. In contrast, in one of the world's poorest nations—Niger—per capita gross national product is only $835, and 154 out of every 1,000 children die before their first birthday (United Nations Development Programme 2005). Note, however, that the relationship between economic productivity and quality of life is not perfect. For example, per capita income is 15 percent

TABLE 7.3
The Extent of International Inequality

Type of Country	GDP Per Capita (U.S. dollars)	Life Expectancy at Birth	Infant mortality Rate/1,000 live births)	Human Development Ranking	Gender-related Development Ranking
Most-developed countries					
Norway	$37,670	79.4	3	1	1
United States	37,562	77.4	7	10	8
Canada	30,677	80	5	5	5
Japan	27,967	82	3	11	14
Rep. of Korea	17,971	77.0	5	28	27
Less-developed countries					
Saudi Arabia	13,226	71.8	22	77	65
Russian Federation	9,230	65.3	16	62	56**
Brazil	7,790	70.5	33	63	52
China	5,003	71.6	30	85	64
El Salvador	4,761	70.9	32	104	80
Least-developed countries					
Haiti	1,742	51.6	76	153	122**
Rwanda	1,268	43.9	118	159	122
Congo, Dem. Rep.	697	43.1	129	167	131
Ethiopia	711	47.6	112	170	134
Sierra Leone	548	40.8	166	176	139

SOURCE: Human Development Report (2005).
*Out of 176 countries
**2003 data

higher in the United States than in Norway, but, according to the United Nations Development Programme (2003, 2005), Norwegians enjoy a higher quality of life than do Americans. In large part, Norway's rankings on human development reflect the fact that access to health care, education, and adequate nutrition is more universally available there than in the more affluent United States.

The gap between quality of life for the world's wealthiest and poorest is enormous. The world's richest 10 percent of people earn as much income each year as do the world's poorest 54 percent. Put another way, the ratio of the income of the top 10 percent of the world's population to that of the poorest 10 percent is 103 to 1. Women and children are particularly at risk in poor nations. A half-million women in the developing nations die each year during pregnancy or childbirth, at rates up to 100 times those found in the most-developed nations. Worldwide, one out of every three preschool children suffer from malnutrition (United Nations Development Programme 2005). International inequality is indeed dramatic.

No nation wants to be poor and underdeveloped. Why are some nations poor, and what can be done about this? We examine two general theories of development—modernization and world-systems theory—and their implications for reducing global inequality.

Structural-Functional Analysis: Modernization Theory

Modernization theory sees development as the natural unfolding of an evolutionary process in which societies go from simple to complex economies and institutional structures. This is a structural-functional theory based on the premise that adaptation is the chief determinant of social structures. According to this perspective, developed nations are merely ahead of the developing nations in a natural evolutionary process. Given time, the developing nations will catch up.

Modernization theory emerged in the 1950s and 1960s, when many believed that developing nations would follow pretty much the same path as the developed nations. Greater productivity through industrialization would lead to greater surpluses, which could be used to improve health and education and technology. Initial expansion of industrialization would lead to a spiral of ever-increasing productivity and a higher standard of living. These theorists believed this process would occur more rapidly in the least-developed nations than it had in Europe because of the direct introduction of Western-style education, health care, and technology (Chodak 1973).

Events have shown, however, that development is far from a certain process. Some "developed" nations such as those in the former Soviet Union have regressed over time. Thailand and the Republic of Korea have modernized quickly, Haiti has modernized hardly at all, and Mexico has gone through wild economic upswings and downturns. The least-developed countries have not caught up with the developed world, and, in many cases, the poor have simply become poorer, while the rich have become richer.

Why haven't the less-developed nations followed in the footsteps of developed nations? The primary reason is that they encounter many obstacles not faced by nations that developed earlier: population pressures of much greater magnitude, environments ravaged by the developed nations since they were colonial powers, and the disadvantage of being latecomers to a world market that is already carved up. These formidable obstacles have given rise to an alternative view of world modernization—world-systems theory.

> **Modernization theory** sees development as the natural unfolding of an evolutionary process in which societies go from simple to complex economies and institutional structures.

Conflict Analysis: World-Systems Theory

Conflict theorists' interpretations of modernization begin by arguing that the entire world is a single economic system, dominated by capitalism for the past 200 years. Nation-states and large **transnational corporations** (i.e., corporations that produce and distribute goods in more than one country) are the chief actors in a free-market system in which goods, services, and labor are organized to maximize profits (Chirot 1986; Turner & Musick 1985). This system includes an international division of labor in which some nations extract raw materials and others fabricate raw materials into finished products.

Nation-states can pursue a variety of strategies to maximize their profits on the world market. They can capture markets forcibly through invasion, they can manipulate markets through treaties or other special arrangements (such as NAFTA), or they can simply do the international equivalent of building a better mousetrap. The Japanese auto industry (indeed, all of Japanese industry) is a successful example of the last strategy.

World-systems theory is a conflict analysis of the economic relationships between developed and developing countries. It looks at this economic system with a

> **Transnational corporations** are large corporations that produce and distribute goods internationally.

> **World-systems theory** is a conflict perspective of the economic relationships between developed and developing countries, the core and peripheral societies.

Because of their economic and political power, transnational corporations based in the most-developed nations are able to capture markets in less-developed nations, as FedEx is trying to do in Vietnam.

distinctly Marxist eye. Developed countries are the bourgeoisie of the world capitalist system, and underdeveloped and developing countries are the proletariat. The division of labor between them is supported by a prevailing ideology (capitalism) and kept in place by an exploitive ruling class (rich countries and transnational corporations) that seeks to maximize its benefits at the expense of the working class (underdeveloped and developing countries).

World-systems theory distinguishes two classes of nations: core societies and peripheral societies. **Core societies** are rich, powerful nations that are economically diversified and relatively free from outside control. They arrive at their position of dominance, in part, through exploiting other (peripheral) societies.

Peripheral societies, by contrast, are poor and weak, with highly specialized economies over which they have relatively little control (Chirot 1977). Some of the poorest countries rely heavily on a single cash crop for their export revenue. For example, 70 percent of Guinea-Bissau's export earnings come from cashews, and more than half of Uganda's earnings come from coffee (Central Intelligence Agency 2002). The economies of these and many other developing nations are vulnerable to conditions beyond their control: world demand, crop damage from infestation, flooding, drought, and so on.

A key element of world-systems theory is the connectedness between core society prosperity and peripheral society poverty. According to this theory, our prosperity is their poverty. In other words, our inexpensive shoes, transistors, bananas, and the rest depend on someone in a least-developed nation receiving low wages, often while working for a company based in one of the developed nations. Were their wages to rise, our prices would rise, and our standard of living would drop.

Core societies are rich, powerful nations that are economically diversified and relatively free from outside control.

Peripheral societies are poor and weak, with highly specialized economies over which they have relatively little control.

Challenges to the Status Quo

Awareness of vast inequalities within a nation can lead to organized nationalist movements, resulting in violent, revolutionary challenges to the status quo (Kerbo 1991, 513). Since 1970, there have been nationalist revolutions or violent class or ethnic struggles in Angola, Ethiopia, Nicaragua, Iraq, Sri Lanka, and Afghanistan, to name just some.

Inequality also can cause conflict *between* nations. Information technology, the media, and increasing international travel all contribute to a rising sense of deprivation and frustration as citizens of peripheral nations are exposed to the lifestyles of more affluent countries. They also can lead citizens of the less-developed nations to feel that their culture is under attack by Western culture, or to recognize how their countries' economies and governments are controlled by foreign corporations and governments. Now that the Cold War is over and the Soviet Union no longer poses a military threat, many observers believe that the greatest threats to the United States will come from the less-developed nations—as the events of September 11 demonstrated. Nearly a half-dozen peripheral and semi-peripheral nations—Argentina, Brazil, India, Pakistan, and South Africa—now have nuclear weapons. Even more important than the threat of nuclear attack is the threat of terrorism, since many individuals in developing nations, like the terrorists who destroyed the World Trade Center, can easily obtain the resources needed to attack the United States and other developed countries.

A Case Study: Global Inequality and 9/11

The terrorist attacks of September 11, 2001, which destroyed the World Trade Center and damaged the Pentagon, partly resulted from global inequality (Amanat 2001; Barber 2001; Jacquard 2002; Stern 2003).

One underlying cause of the 9/11 attacks is the deepening belief among many Muslims that their nations and religion are under political attack. This belief has roots in the Russian invasion of Afghanistan and the civil war in Bosnia that pitted Muslims against Christians. Actions taken by the United States have also played a large role in creating this sense of victimization among many Muslims. The United States consistently has supported Israel against the Palestinians. It has also used an economic blockade to try to starve out the Iranian and Iraqi governments, invaded Iraq, and based military troops in Saudi Arabia, where Islam's holiest sites are located. These actions on the part of the United States and other non-Muslim nations have left many Muslims feeling that not only are individual Muslim governments under attack, but their religion is as well. It also has contributed to a sense of wounded pride among Muslims who feel that they no longer control their own national destinies.

To these external problems are added internal economic and political problems. The Middle East and the Muslim countries of Asia have been wracked by war for the last 50 years. Poverty is very high, inequality is extreme, governments by and large are corrupt, and access to education is often limited. As a result, millions of individuals are suffering, and thousands are willing to become the foot-soldiers of Al Qaeda and other similar groups. Even members of the middle and upper classes, who are protected from the worst impacts of these forces, still live in a culture of alienation, despair, and wounded pride. These are the individuals, like Osama bin Laden and the 19 terrorists who attacked the United States on 9/11, who become the leaders and lieutenants in global terrorism. It is probably no coincidence that the terrorists of 9/11 attacked the World Trade Center and the Pentagon—icons of American economic and military power.

Connections

Personal Application

If you have ever traveled to a less-developed country, you have seen the consequences of economic dependence. You probably were warned not to drink water from the faucets, because these nations lack the economic resources to provide safe drinking water. Because wages are so low in these nations, you could buy meals, clothes, and souvenirs very cheaply and could afford to purchase services like taxis that you could never use at home. And because even at these low wages, many can find no jobs at all, you might have seen beggars or prostitutes and been warned to watch out for thieves.

■ Pro-Taliban supporters in Pakistan one week after the 9/11 attacks on the United States. Such demonstrations reflected fear not only of American cultural and economic power but also of American political and military power.

Finally, because of the mass media and information technology, people throughout the Muslim world are now inundated with American culture. In countries where women are expected to cover themselves from head to toe, American television shows display nearly naked women. Rap music boasts of sexual conquests, and Hollywood romances feature independent women and men whose lifestyles are the antithesis of traditional Muslim values. Around the world, America has become the symbol of the good life, but also a symbol of materialism, violence, promiscuity, and the attack on traditionalism. The terrorist actions of 9/11 were designed as blows against all these forces.

Where This Leaves Us

Research on stratification leads to one basic conclusion: As long as some people are born in tenements to poorly educated parents who lack the time, money, and cultural resources needed to provide their children with intellectual stimulation, while others are born to wealthy, educated parents with excellent connections and "cultural capital," there can never be true equality of opportunity. Similarly, as long as some nations have a greater share of resources—money, oil, media outlets, good schools—other nations can never flourish. Those nations that lack power will have lower life expectancies, many homeless people, and many who experience malnutrition or even starvation. Thus the only way to create equal opportunity, either within or across nations, is to attack these underlying problems. The question for Americans is, should we care?

To answer this question from a moral perspective, we might point out that quality of life among poor and affluent Americans and between poor and wealthy nations are directly related to each other. Wealthy Americans enjoy a good life because they can cheaply hire maids and taxi drivers and can buy houses and other goods produced by poorly paid American workers. Similarly, citizens of the United States and other developed nations enjoy raw goods and products obtained cheaply from countries where people work for pennies an hour, and enjoy the security of knowing that other nations cannot challenge our military and economic power. By the same token, American culture is spreading around the world in part because of our economic and political power and because other cultures lack the power to oppose it.

But this is a sociology textbook, not an ethics textbook. From a sociological perspective, perhaps the most important thing to keep in mind is the ways in which both individuals and societies benefit when inequality is reduced. Of course, no wealthy individual or wealthy nation wants to give resources away. On the other hand, inequality carries costs for everyone. In nations where inequality is high, everyone—including the wealthy—experiences more stress, more crime, worse health, and lower life expectancies (Marmot 2004; Wilkinson 1996, 2005). It's just not as much fun being wealthy or even middle class when you have to lock your doors all the time, worry about crime, fear that you might lose your job to a cheaper worker, and fear that your standard of living might plummet if you were ill or injured. Similarly, wealthy nations can never relax their guard when other nations envy their economic and cultural position. The events of 9/11 demonstrate what can happen when resentment of wealthy nations rises.

Summary

1. Stratification is distinguished from simple inequality in that (a) it is based on membership in social categories rather than on personal characteristics, and (b) it is supported by norms and values that justify unequal rewards.

2. There are two types of stratification systems. In a caste system, your social position depends entirely on your parents' position. In a class system, social position is based on educational and occupational attainment so individuals can have higher or lower positions than their parents.

3. Marx believed that there was only one important dimension of stratification: class. Weber added two further dimensions, and most sociologists now rely on his three-dimensional view of stratification: class, status, and power.

4. Inequality in income and wealth is substantial in the United States and has increased steadily since 1970. Income inequality is higher in the United States than in any other industrialized nation.

5. Structural-functional theorists argue that inequality is a necessary and justifiable way of sorting people into positions. Conflict theorists believe that inequality arises from conflict over scarce resources, in which those with the most power manipulate the system to enhance and maintain their advantage. Symbolic interaction theory focuses on how social status is reinforced through self-fulfilling prophecies.

6. Allocation of people into statuses includes macro and micro processes. At the macro level, the labor market sets the stage by creating demands for certain statuses. At the micro level, the status attainment process is largely governed by indirect inheritance.

7. Despite high levels of inequality, most people in any society accept the structure of inequality as natural or just. In the United States, the ideology that teaches people to accept inequality is the *American Dream*, which suggests that success or failure is the individual's choice.

8. Upward mobility is most common among those who have better access to economic, educational, and cultural resources, whether compared with their siblings or with children from other families.

9. Approximately 13 percent of the U.S. population falls below the poverty level. Although some have argued that "cultures of poverty" explain why people stay poor, research suggests that the shrinking options provided by the current labor market is a more likely explanation.

10. Among the approaches proposed for reducing poverty are fair wage movements and increasing educational opportunities.

11. International inequality is a key factor in today's world. Reducing this disparity through the development of less-developed and least-developed countries is a common international goal. Development is not the same

as Westernization; it means increasing productivity and raising the standard of living.

12. The world's nations can be divided into the rich, diversified, independent, most-developed core nations; the least-developed nations of the periphery, and the less-developed nations, which fall in between these two extremes.

13. The Human Development Index and the Gender-Related Development Index use literacy and educational attainment, life expectancy, and economic productivity to assess quality of life overall, as well as quality of life adjusted for the effects of gender inequality.

14. Modernization theory, a functionalist perspective of social change, argues that less-developed countries will evolve toward industrialization by adopting the technologies and social institutions used by the developed countries.

15. World-systems theory, a conflict perspective, views the world as a single economic system in which the industrialized countries, known as core societies, control world resources at the expense of the less-developed, peripheral societies.

16. Inequality within nations can lead to nationalist revolutions or violent class or ethnic struggles. Inequality *between* nations can lead to resentment of outside economic or political influence and is an underlying source of warfare and terrorism.

Thinking Critically

1. Can you think of any ways in which the U.S. system of stratification resembles a caste system?

2. To what social class do you belong? How do you know? How are you affected by your social class?

3. What *wealth* does your family own? What cultural capital does your family have? How has your family's wealth and cultural capital, or lack of wealth and cultural capital, affected you?

4. *You* are a hundred times better off than the average person in Haiti. Is this a necessary and just reflection of your greater contribution to society? How do you benefit from this inequality? How are you harmed by it?

5. Critically evaluate the components of the Human Development Index. Which seems most important to you? Could this index be used to understand group differences in quality of life within the United States?

Companion Website for This Book

academic.cengage.com/sociology/brinkerhoff
Gain an even better grasp on this chapter by going to the Companion Website. This resource contains tutorial quizzes and flash cards to help you master key terms and concepts.

Suggested Readings

Correspondents of the New York Times. 2005. *Class Matters.* New York: Times Books. 2005. The acclaimed *New York Times* series on social class in America, from top to bottom, and its implications for the way we live our lives.

Ehrenreich, Barbara. 2001. *Nickel and Dimed: On (Not) Getting By in America.* New York: Metropolitan. Investigative journalist Barbara Ehrenreich spent a year trying to support herself at minimum wage jobs: waitress, maid, nursing home aide, and Wal-Mart salesclerk. Her book details the extraordinary physical and emotional toll exacted by working- and lower-class jobs. A gripping read.

Lappé, Frances Moore, Collins, Joseph, and Rosset, Peter. 1998. *World Hunger: Twelve Myths*. New York: Grove Press. An excellent explanation of the critical role inequality within and between nations plays in world hunger.

LeBlanc, Adrian Nicole. 2003. *Random Family: Love, Drugs, Trouble, and Coming of Age in the Bronx*. New York: Scribner. Journalist LeBlanc spent ten years following an extended family that lives in one of the country's worst slums. Her richly detailed book paints a vivid portrait of life at the bottom.

Shipler, David K. 2005. *The Working Poor: Invisible in America*. New York: Vintage. A blistering account by an award-winning journalist of how the working poor stay poor and of the consequences of poverty.

CHAPTER 8

Racial and Ethnic Inequality

© Cleo Photography/Photoedit

Outline

Race and Ethnicity

Race and ethnicity are ascribed characteristics that define categories of people. Each has been used in various times and places as bases of stratification; that is, cultures have thought it right and proper that some people receive more scarce resources than others simply because they belong to one category rather than another. In the following section, we provide a basic framework for looking at racial and ethnic inequality before focusing on the patterns of inequality that exist in the United States.

Understanding Racial and Ethnic Inequality

How is it possible for groups to interact on a daily basis within the same society and yet remain separate and unequal? In this section, we begin by introducing some basic concepts needed to understand racial and ethnic inequality: the social construction of race and ethnicity, how disadvantages multiply, and the concepts of majority and minority groups.

The Social Construction of Race and Ethnicity

A **race** is a category of people treated as distinct because of *physical* characteristics to which *social* importance has been assigned. An **ethnic group** is a category whose members are thought to share a common origin and to share important elements of a common culture—for example, a common language or religion (Marger 2003). Both race and ethnicity are handed down to us from our parents, but the first refers to the presumed genetic transmission of physical characteristics, whereas the second refers to the transmission of cultural characteristics through socialization.

In actuality, racial differences among humans account for only a tiny fraction of all genes. Both race and ethnicity are based loosely if at all on physiological characteristics such as skin color. For this reason, sociologists talk of the **social construction of race and ethnicity:** the process through which a culture defines what constitutes a race or an ethnic group. As this suggests, this process is based more on social ideas than biological facts; indeed, biologists are almost unanimous in believing that race has no biological reality.

How are racial and ethnic identities socially constructed? Consider the changes in racial definitions that emerged during the 1930s. Before this, the modern concept of a "white race" really didn't exist. Instead, people talked of multiple races, including an Anglo-Saxon race, a Mediterranean race (Italians and Greeks), a Hebrew race (Jews), and Slavic races (Jacobson 1998). Around 1930, doctors, politicians, lawyers, legislators, anti-immigrant activists, journalists, and others dropped these distinctions and instead began describing whites as a unitary racial group, in professional journals, newspapers, law courts, and the mass media. Sociologists would say that these professionals and activists, whether or not they realized it, were engaging in the social construction of whiteness as a racial category. At the same time, the U.S. Bureau of the Census

A **race** is a category of people treated as distinct on account of *physical* characteristics to which *social* importance has been assigned.

An **ethnic group** is a category whose members are thought to share a common origin and to share important elements of a common culture.

The **social construction of race and ethnicity** is the process through which a culture (based more on social ideas than on biological facts) defines what constitutes a race or an ethnic group.

declared that Mexican Americans would be classified in the census as nonwhite. The Mexican government complained, and the Bureau reversed itself. Currently the Census Bureau defines Hispanic Americans as an ethnic group, whose members can belong to any race. Similarly, the shift from using the term "black" to using the term "African American" reflected changing social ideas about race and ethnicity, not any new information about the biological origins of that group. Each of these examples illustrates the social construction of race and ethnicity: a political process in which groups compete over how racial and ethnic categories should be defined.

The growing number of multiracial births in the United States is beginning to blur the very concept of race (Kalish 1995; Morganthau 1995). Yet many Americans continue to feel uncomfortable when they cannot wedge an individual into a predetermined racial slot. Golf superstar Eldrick "Tiger" Woods has had to fight constantly against journalists and others who want to describe him simply as African American, even though two of his eight great-grandparents were Native American, four were Asian, and one was European American. The 2000 census was the first U.S. census that allowed individuals to describe themselves as multiracial, rather than forcing them to choose a single racial category.

Elsewhere in the world, new ethnic identities are forged by changing national borders. Only during the twentieth century did Sicilians, Napolitanos, Milanese, and others begin developing a common Italian language, culture, and ethnic identity. Conversely, after the break-up of the former Soviet Union, Lithuanians, Latvians, Kazakhs, and others began rebuilding ethnic, linguistic, and cultural traditions and identities that had been suppressed or even abandoned during the Soviet years (Nagel 1994). As these examples illustrate, racial and ethnic statuses are not fixed. Over time, individuals may change their racial and ethnic identification, and society, too, may change the statuses it recognizes and uses.

Multiplying Disadvantages

Most contemporary scholars use some form of conflict theory to explain how racial and ethnic inequalities are developed and maintained. This theory suggests that in the conflict over scarce resources, historical circumstances such as access to technology and slavery gave some groups advantages while holding other groups back. To maintain their power, those who have advantages work to keep others from getting access to them (Tilly 1998). These inherited advantages have left us with two stratification systems, one based on class and one based on race and ethnicity.

These two stratification systems work together to multiply disadvantages and inequality. We can see how this works in Table 8.1. As the table shows, in 2000 (the latest data available as of 2006), the races displayed very similar patterns of *internal* inequality: In each of these three populations, the wealthiest 20 percent of families receive almost half of all income for that population. On the other hand, the median income of white families is more than one and one-half times that of Hispanic and African American families.

The differences become even more extreme when we look not at income but at *wealth*. Wealth refers to all assets owned by a household, including stocks, bonds, and savings; houses, cars, and land; and anything else that can be sold for money or that essentially is money. As Table 8.1 shows, the median net worth of white non-Hispanic families was $79,400. This is *ten times* higher than the median for African American families and eight times higher than the median for Hispanic families. These racial differences in wealth, not racial differences in income, are primarily responsible for the continuing U.S. racial divide (Oliver & Shapiro 1997; Shapiro 2004).

TABLE 8.1

Income and Wealth of Families by Race and Ethnicity, 2000

The United States is stratified by both race and class, and each form of stratification reinforces the other. Within each racial or ethnic group, the richest 20 percent receive almost 50 percent of all income for that group, indicating real social class differences. At the same time, whites as a group have considerably more income and wealth than do African Americans or Hispanics, indicating real race and ethnic differences.

Income Quintile	Percentage of Income Received		
	African American	Hispanic	White
Poorest fifth	3%	4%	5%
Second fifth	9	9	10
Third fifth	16	15	16
Fourth fifth	25	23	23
Richest fifth	47	49	47
Median income	$34,192	$35,054	$53,256
Median wealth	$7,500	$9,750	$79,400

SOURCE: U.S. Bureau of the Census 2002b, 2003a.

Case Study: Environmental Racism

One example of how poverty and racism combine to multiply inequality is environmental racism. The term **environmental racism** is used to describe the disproportionately large number of health and environmental risks that minorities, especially if they are poor, face daily in their neighborhoods and workplaces (Bullard, Warren, & Johnson 2001; Camacho 1998). For example, landfills for hazardous waste are disproportionately located in African American and Hispanic communities. Farm workers and their children, most of whom are Hispanic and very poor, are exposed to poisons whenever the crops they pick are sprayed with pesticides. On poor Native American reservations where uranium mining is often the only well-paid job, mining has poisoned thousands of workers, as well as their spouses and children, when mine waste seeps into the water or is blown into the air. This unequal environmental burden exists because manufacturers, mining companies, and the like find it easiest to locate polluting industries in poor minority communities that lack the political power to enforce environmental restrictions and that are desperate for jobs, no matter the environmental cost.

The best predictor of exposure to environmental pollution is race; the second best predictor of exposure is poverty (Bullard, 1993; Stretesky & Hogan, 1998). These environmental hazards reinforce as well as reflect ethnic and class inequality: Children who are exposed to toxic chemicals or air pollution, for example, risk mental retardation, developmental delays, and physical illnesses such as asthma that can lead to their missing school days. As a result, they are less likely to succeed in school and more likely to continue to live in poverty as adults.

Environmental racism refers to the disproportionately large number of health and environmental risks that minorities face daily in their neighborhoods and workplaces.

Majority and Minority Groups

In addition to talking specifically about whites and African Americans or Jews and Arabs, sociologists interested in race and ethnicity also talk more broadly of majority and minority groups. A **majority group** is one that is culturally, economically, and

A **majority group** is a group that is culturally, economically, and politically dominant.

Connections

Historical Note

In 1492, all Jews in Spain were ordered by royal decree to either convert to Christianity or leave the country. Those who refused were hanged or burned at the stake as heretics. Over the centuries, descendants of Jews who stayed in Spain and converted came to blend more and more into Spanish culture. Today, they are so fully acculturated that they barely exist as an ethnic group anymore. The only signs of their ancestors' religion are a few remnant rituals, like lighting candles on Friday nights. These remnants have become simply family traditions, separate from the religious meanings and prayers that once accompanied them.

A **minority group** is a group that is culturally, economically, and politically subordinate.

Accommodation occurs when two groups coexist as separate cultures in the same society.

Social integration exists when a minority group is fully a part of the institutions and social life of a society and belonging to a minority group no longer affects individuals' social position.

politically dominant. A **minority group** is a group that, because of physical differences, is regarded as inferior and is kept culturally, economically, and politically subordinate. Although minority groups are often smaller than majority groups, that is not always the case. In the Republic of South Africa, for example, whites made up only 15 percent of the population but until recently were, sociologically, the majority group, controlling all major political and social institutions. Similarly, some scholars regard women as a minority group because, based on physical sex differences, they have been economically, politically, and culturally subordinate to men.

Patterns of Interaction

Relations between majority and minority groups generally take one of four forms: conflict, accommodation, acculturation, or social integration.

CONFLICT   As we discussed in Chapter 5, conflict is a struggle over scarce resources that is not regulated by shared rules; it may include attempts to neutralize, injure, or destroy one's rivals. Although some intergroup conflicts are expressed in violence (for example, genocide), conflict may also be expressed in laws forbidding social, political, or economic participation by the minority group. When African Americans or Hispanics are systematically excluded from voting or from some forms of political or social participation, this is a form of conflict. It excludes them from the competition for scarce resources.

ACCOMMODATION   When two groups coexist as separate cultures in the same society, we speak of **accommodation.** They are essentially parallel cultures, each with its own institutions. The term was originally developed to apply to situations such as Canada's French and English provinces. As any French Canadian will tell you, separate is seldom truly equal. Nevertheless, accommodation gives at least outward support to the norm of equality. Sometimes referred to as pluralism, these systems are difficult to maintain over the long run, as is illustrated by the current conflict between Kurds, Shiite Moslems, and Sunni Moslems in Iraq.

ACCULTURATION   Another possible outcome of intergroup contact is for the minority group to adopt parts of the culture of the majority group while still holding onto valued aspects of its original culture. This process is called acculturation. It includes learning the language, history, and manners of the majority group; it also involves accepting some of the loyalties and values of the majority group as one's own. As middle-class African Americans have discovered, however, acculturation does not necessarily mean full acceptance or an end to discrimination (Feagin & Sikes 1994).

SOCIAL INTEGRATION   When full acceptance comes—when the minority group is fully a part of the institutions and social life of society and ceases to be a subordinate group—we speak of **social integration.** Social integration means that belonging to a particular race, ethnic group, or religion does not affect individuals' social position or social relationships or how they are viewed by members of the dominant group. The term "assimilation" is also sometimes used for this concept, but has fallen somewhat out of favor because it can imply that people achieve social integration only by abandoning their ethnic culture completely.

Full social integration rarely occurs in human societies, but we can tell that social integration is increasing when minority group members go to the same schools as the dominant group, live in the same neighborhoods, belong to the same social groups, and are willing to marry one another. Despite all the remaining problems, compared with other societies the United States has done a remarkable job of integrating its minorities (Alba & Nee 2003).

Maintaining Racial and Ethnic Inequality

To understand the persistence of racial and ethnic inequality, we need to look at the processes that promote what sociologists call **social distance.** Social distance between ethnic groups is operationally defined by questions such as, "Would you be willing to have a member of this group as a neighbor? A close friend? A work colleague?" It is a measure of the degree of intimacy and equality in the relationship between two groups.

Prejudice and discrimination are processes that allow social distance to be maintained even when physical distance is absent. Most societies also use segregation, or physical distance, as an aid to maintaining social distance.

Prejudice

The foundation of prejudice is stereotyping, a belief that people who belong to a given category share common characteristics—for example, that athletes are dumb or that African Americans are naturally good dancers. Stereotyping does have its uses. It's probably a good idea to assume you should stay away from someone who is waving a gun in the air and mumbling to himself, and it's probably a safe bet that a very fashionably dressed woman can give you directions to a high-end shopping mall. Life would be very difficult if we had to start absolutely from scratch in every social interaction, with no idea of how this individual might be similar to or different from others we've met (or heard about) in the past.

On the other hand, stereotypes also *hinder* social interactions when they lead us to make false assumptions about others. The man waving the gun around might be an actor, and the fashionably dressed woman might be wearing clothes her sister chose for her. Some Asians are good at math, and some aren't. Some men are good at sports, and some are utterly uninterested. Some computer jocks are also punk rockers, and some punk rockers also enjoy knitting.

Prejudice moves beyond stereotyping in that it is always a negative image and always irrational. It exists despite the facts rather than because of them. A person who believes that all Italian Americans are associated with the Mafia will ignore all instances of the law-abiding behavior of Italian Americans. If confronted with an exceptionally honest man of Italian descent, the bigot will rationalize him as the exception that proves the rule. **Racism** is a form of prejudice. It is the belief that inherited physical characteristics associated with racial groups determine individuals' abilities and are a legitimate basis for unequal treatment.

A startling example of racist prejudice was the decision by the United States to put Japanese American citizens in concentration camps during World War II (Smith 1995). This decision, which occurred in the absence of any evidence suggesting that Japanese Americans were disloyal to the United States, demonstrates the irrationality of prejudice. That irrationality was epitomized in the words of then-General John DeWitt (1943), who claimed that "The very fact that no sabotage [by Japanese

Connections

Example

Prejudice and stereotypes are not limited to ethnic group relationships. For example, many younger Americans stereotype older people as hags, crones, dirty old men, and the like. Even when we admire elderly persons as wonderful grandparents, we tend not to see them as whole and unique individuals (who, for example, may still be sexually active or have interests far beyond their grandchildren). When these stereotypes coalesce into prejudice, we refer to that prejudice as ageism: the belief that older persons are less capable and worthy than younger people and so deserve unequal treatment.

Social distance is the degree of intimacy and equality in relationships between two groups.

Prejudice is an irrational, negative attitude toward a category of people.

Racism is the belief that inherited physical characteristics associated with racial groups determine individuals' abilities and characteristics and provide a legitimate basis for unequal treatment.

■ Across nations, the causes and consequences of prejudice and discrimination are very similar. Here, Turkish victims of neo-Nazi violence mourn the victims of an arson attack that left three children injured and five adults dead.

Americans] has taken place to date is a disturbing and confirming indication that such action will be taken."

Learning Prejudice

What causes prejudice? We learn to hate and fear in the same way we learn to love and admire. Prejudice is a shared meaning that we develop through our interactions with others. Most prejudiced people learn prejudice when they are very young, along with other social norms. This prejudice may then grow or diminish, depending on whether groups and institutions encountered during adulthood reinforce these early teachings (Wilson 1986).

Prejudice arises from and is reinforced by institutionalized patterns of inequality. In a stratified society, we tend to rate ourselves and others in terms of economic worth. If we observe that no one pays highly for a group's labor, we are likely to conclude that the members of the group are not worth much. Through this learning process, members of the minority as well as the majority group learn to devalue the minority group (Wilson 1992).

Prejudice is also powerfully reinforced by patterns of segregation and discrimination. A child growing up in a society where separation by race or sex is well established is very likely to learn prejudice. "They are not like us" can lead quickly to "They are not as good as us."

Competition over scarce resources (such as good jobs, nice homes, and admission to prestigious universities) also increases prejudice. For all racial and ethnic groups, prejudicial attitudes are closely associated with the belief that gains for other racial and ethnic groups will spell losses for one's own group (Bobo & Hutchings 1996).

Maintaining Prejudice: The Self-Fulfilling Prophecy

In Chapter 7, we introduced the concept of the self-fulfilling prophecy—where acting on the belief that a situation exists causes the situation to become real. The self-fulfilling prophecy is one very important mechanism for maintaining prejudice. A classic example is the situation of American women until the last few decades. Because women were considered to be inferior and capable of only a narrow range of social roles, they were given limited education and barred from participation in the institutions of the larger society. That they subsequently knew little of science, government, or economics was then taken as proof that they were indeed inferior and suited only for a role at home. In fact, many women were unsuited for any other role: Being treated as inferiors had made them ignorant and unworldly. The same process reinforces boundaries between racial and ethnic groups. For example, if we believe that Jews think they are better than others, then we don't invite them to our homes. When we subsequently observe that they associate only with one another, we take this as confirmation of our belief.

Personal Factors and Prejudice

Even though prejudice is learned, all individuals are not equally susceptible to it. Three factors that dispose people to prejudice are authoritarianism, frustration, and beliefs about stratification.

Authoritarianism is a tendency to be submissive to those in authority and aggressive and negative to those lower in status (Pettigrew 1982). Authoritarians in the United States tend to be strongly anti-African American and anti-Semitic (as well as homophobic and sexist).

Frustration is another characteristic associated with prejudice. People or groups who are blocked in their own goal attainment are likely to blame others for their problems. This practice, called **scapegoating,** has appeared time and again. For example, anti-Semitism exploded in Nazi Germany during the Great Depression of the 1930s, when the German economy collapsed and many Germans were left jobless and impoverished.

Finally, prejudice is more likely to exist among individuals who believe strongly in the American Dream. People who subscribe to the view that we can all get ahead if we work hard and that poor people have only themselves to blame are substantially more likely to attribute poverty or disadvantage to personal deficiencies. In the case of disadvantaged minorities, the American Dream ideology supports the belief that the disadvantage is the fault of undesirable traits within the minority group (Kluegel & Smith 1983; Pettigrew 1982).

> **Authoritarianism** is the tendency to be submissive to those in authority, coupled with an aggressive and negative attitude toward those lower in status.

> **Scapegoating** occurs when people or groups who are blocked in their own goal attainment blame others for their failures.

Discrimination

Treating people unequally because of the categories they belong to is **discrimination.** Prejudice is an attitude; discrimination is behavior. Most of the time the two go together: If your boss thinks that African Americans are less intelligent than whites (prejudice), he will likely pay his African American workers less (discrimination). Some people, however, are inconsistent, usually because their own values differ from others around them. They may be prejudiced, but they nonetheless avoid discriminating because they don't want to be sued for unfair treatment. Or they might *not* be prejudiced but nonetheless discriminate because it is expected of them—perhaps by a boss who opposes hiring minorities, or by a parent who opposes interracial romance.

Public policy directed at racism is aimed mostly at reducing discrimination rather than at trying to change prejudices. As Martin Luther King, Jr., remarked, "The law may not make a man love me, but it can restrain him from lynching me, and I think that's pretty important" (as quoted in Rose 1981, 90).

> **Discrimination** is the unequal treatment of individuals on the basis of their membership in categories.

■ Racial segregation remains a fact of life in the United States. Even among the middle class, African Americans are more likely than European Americans to live in a poor neighborhood.

© Ted Spiegel/CORBIS

Segregation

Segregation refers to the physical separation of minority- and majority-group members.

Prejudice and discrimination may occur between groups in close, even intimate, contact; they create social distance between groups. Differences between groups are easier to maintain, however, if social distance is accompanied by **segregation**—the physical separation of minority- and majority-group members. Thus, most societies with strong divisions between racial or ethnic groups have ghettos, barrios, and Chinatowns where, by law or custom, members of the minority group live apart.

Historical studies suggest that high levels of residential segregation of Hispanic, Asian, and African Americans have existed since at least 1940. Such segregation is no longer established in law, but it is no historical accident. Segregation continues for two reasons: economic differences across racial/ethnic groups and continuing prejudice and discrimination (Iceland & Wilkes 2006).

Economic differences certainly matter. Lower-income Hispanics, African Americans, and Asians are all more likely to live in ethnically segregated neighborhoods than are wealthier members of the same groups. This suggests that if minorities' social class increases, segregation will decline.

But economic differences alone can't explain segregation. Whereas Hispanics and Asians are significantly less likely to live in segregated neighborhoods if they are at least middle class, this is not true for African Americans. Similarly, even when African Americans are educated, affluent, and move to the suburbs, they remain substantially less likely than whites to escape "distressed" neighborhoods (Crowder, South, & Chavez 2006; Iceland & Wilkes 2006; Alba, Logan, & Stults 2000). This suggests that prejudice and discrimination continue to foster segregation of African Americans. Studies find that, compared with others with similar incomes, African Americans are less likely to be shown homes in "nicer" areas by real estate agents and are more likely to be turned down for mortgages or to face hostility from potential neighbors (Iceland & Wilkes 2006; Ross & Turner 2005).

Current data suggest that we should be guardedly optimistic. Real estate agents and mortgage brokers are less likely to discriminate against African Americans than in the recent past, and African American segregation has declined somewhat since the 1970s. Segregation of other groups shows little decline, but this is mostly explained by recent immigration from Asia and Latin America (Iceland & Wilkes 2006).

Racial and Ethnic Inequality in the United States

Racial and ethnic inequality is not new. In this section, we discuss the past, present, and future social positions of selected racial and ethnic groups in the United States.

White Americans

The earliest voluntary immigrants to North America were English, Dutch, French, and Spanish. By 1700, however, English culture was dominant on the entire Eastern seaboard. The English became the majority group, and everybody else became a minority group. In the 1840s, employers posted signs saying "No Irish need apply." In the 1860s Chinese and Japanese were targeted, in the 1890s Jews and Italians. This pattern of prejudice and discrimination continues to the present day (if more rarely) against groups as diverse as French Canadians and Arab Americans.

White Ethnicity Today

Despite and in part because of this history of prejudice and discrimination, many white Americans continue to have a strong sense of connection to their ethnic roots. They are proud to be Italians, Greeks, Norwegians, or Poles, and enjoy eating the foods, celebrating the holidays, and singing the songs of their ethnic group. By the third or fourth generation after immigration, however, ethnic identity is largely symbolic and a matter of choice (McDermott & Samson 2005). This

Although being Irish has little impact on most Irish Americans' lives these days, many still enjoy celebrating their cultural heritage, like these boys at a St. Patrick's Day parade.

© Sandy Felsenthal/CORBIS

choice carries few risks because white ethnicity rarely presents a barrier to social integration or personal advancement.

Other white Americans no longer can claim an ethnic identity. Some come from families that emigrated to this country generations ago, and others come from families of such mixed heritage that they can no long identify with a single ethnic group, or even a couple of ethnic groups. These individuals' only ethnic identity is as white Americans.

This shift from Italian-American, Polish-American, and other ethnic identities to "unhyphenated American" identities led some past observers to suggest that America had become a "melting pot," in which (white) ethnic groups had blended together into a new American identity. The reality is more complex. Certainly, our language is peppered with words borrowed from other languages (*frankfurter, ombudsman, hors d'oeuvre, chutzpah*), and some of us are such mixtures of nationalities that we would be hard pressed to identify our national heritage. Instead of a blending of all cultures, however, what has occurred is acculturation to the dominant language and culture of the United States. To gain admission into U.S. society and to be eligible for social mobility, one has to learn "correct" English with the "correct" accent, speak without using your hands too much, work on Saturday and worship on Sunday, and, in general, adopt the culture of the northern and western Europeans who dominated the United States for generations.

White Racial Identity Today

As white Americans' connections to their different ethnic identities have declined, sociologists have begun to focus on whiteness as a *racial* identity (McDermott & Samson 2005). The most important things to understand about white racial identity are (1) it is typically invisible and (2) it carries considerable if unacknowledged privileges. Except in unusual circumstances, such as when they live surrounded by non-whites, white Americans rarely think of themselves as even having a race. Because of this, they often assume that the life they enjoy—living in relatively safe neighborhoods, having relatively good jobs, going to relatively good schools—was fully earned, and in no way resulted from structured racial inequalities built into the system long before they were born. Yet these inequalities continue to haunt the lives of African Americans, Asian Americans, Hispanic Americans, and Native Americans, as we will see in the sections that follow.

African Americans

African Americans now comprise 12.8 percent of the U.S. population. Until very recently, they were the largest minority group in the country, but they were recently passed by Hispanics. Still, their importance goes beyond their numbers: They have made innumerable contributions to U.S. history and culture, and their circumstances have long challenged the United States's view of itself as a moral and principled nation.

The social position of African Americans has its roots in one central fact: Most African Americans are descended from people who were brought here as slaves. After slavery ended, both legal barriers (such as patently unfair "literacy tests" that barred African Americans from voting) and illegal barriers (such as lynching any who challenged white authority) prevented most African Americans from rising in the American social and economic structure. Real change did not take place until the Civil Rights Act of 1964 and the Civil Rights activism of the late 1960s.

These days, more African Americans are middle class than ever before. At the same time, however, a troubling fissure has emerged within the African American population: on the one hand, a working- and middle-class population that is increasingly integrated into U.S. society; on the other, a population of poor African Americans living in segregated neighborhoods with few employment opportunities. In fact, for this second group, the situation has deteriorated (Wilson 1996).

Current Concerns

Since World War II, white attitudes toward African Americans have improved dramatically; most whites now support integration in principle, are comfortable living in neighborhoods where African Americans form a small minority, and no longer disapprove of interracial marriage (Krysan 2000). Similarly, important improvements have been made in many areas of African American life.

Nevertheless, neighborhood segregation remains so high and has such damaging consequences that one scholar has referred to it as "American apartheid" (Massey 1990). African American infants are still twice as likely as white infants to die before their first birthday (U.S. Bureau of the Census 2006), and African American men's life expectancy is still six years less than white men's.

Similarly, although African Americans are rapidly catching up with whites in their educational attainment, they still lag behind (Kao & Thompson 2003). Moreover, even when whites and African Americans have the same levels of education, whites have higher incomes. Almost one quarter of African American families live below the poverty line, and the median income for African American families is only 62 percent that of white families (U.S. Bureau of the Census 2006). This striking economic disadvantage is due to two factors: African American workers earn less than white workers, and African American families are less likely to have two earners.

LOW EARNINGS   One of the reasons individual African Americans earn less than whites is that they are twice as likely to be unemployed. Even when they are employed full time and year round, however, median income for African Americans is only 72 percent that of whites.

African Americans (especially older persons) are less well educated than whites, and a relatively high proportion live in the South, where wages are low; the average African American worker is also somewhat less experienced than the average white worker. African Americans also find themselves disproportionately in very low-paying jobs. For example, African Americans continue to make up 36 percent of the nation's home health and nursing aides (U.S. Bureau of the Census 2006).

These differences account for only part of the earnings gap between African American and white people in the United States (Cancio, Evans, & Maume 1996). The other part is the result of a pervasive pattern of discrimination that produces a very different occupational distribution, pattern of mobility, and earnings picture for African Americans and whites in the United States. Thanks largely to government employment opportunities, there is a growing African American middle class (Hout 1986). Yet, African American professionals often are kept outside the true corporate power structure (Collins 1993, 1997): They are given less authority at work and receive lower economic returns as compared with whites employed in comparable positions (Smith 1997; Wilson 1997).

FEMALE-HEADED FAMILIES   About half of the gap between African American and white family incomes is due to the fact that African American families are less likely to include an adult male. Because women earn less than men and because a one-earner family is obviously disadvantaged relative to a two-earner family, these female-headed households have income far below those of husband–wife families.

The fact that so many more African American than white families are headed by females—45 percent compared with 14 percent—has led some commentators to conclude that poverty is the result of bad decisions by African American men and women. This type of argument is an example of "blaming the victim," and empirical evidence suggests that it simply isn't true. Rather than *causing* poverty, research indicates that female headship *results* from poverty: African American women are less likely to marry because relatively few men in their community can support a family (Lichter, LeClere, & McLaughlin 1991; Luker 1996; Newman 1999b).

Hispanic Americans

Hispanics or Latinos are an ethnic group rather than a racial category, and a Hispanic may be white, black, or some other race. This ethnic group includes immigrants and their descendants from Puerto Rico, Mexico, Cuba, and other Central or South American countries. Hispanics constitute 14.4 percent of the U.S. population, making them the largest minority group in the country. About two thirds of Hispanics in the United States are of Mexican origin, with the rest originating in Central and South America and the Caribbean.

It is almost impossible to speak of Hispanics as if they were a single group. The experiences of different Hispanic groups in the United States have been and continue to be very different. For example, Cubans who emigrated in the 1960s shortly after the Cuban Revolution typically came from wealthier backgrounds, were lighter skinned, and so were accepted more readily by white Americans, and acculturated more rapidly than either later waves of Cuban immigrants or immigrants from Mexico or Guatemala.

Figure 8.1 compares the various Hispanic groups to one another and to the non-Hispanic white, Asian, and African American populations on three measures: education, poverty, and family structure. On two of these measures, a Hispanic group comes out at the very bottom: Mexican Americans are the most poorly educated racial or ethnic group, and Puerto Ricans (many of whom are considered black by other Americans) are the most likely to live in poverty. In addition, Puerto Ricans are second most likely, after African Americans, to live in female-headed households.

Despite the difficulties many Hispanics now face, they remain optimistic about their current and future prospects in the United States (Romero & Elder 2003). In a national poll conducted in 2003 by the *New York Times* and CBS, 81 percent of Hispanics believed that Hispanics had the same or better chances of getting ahead in America as do other Americans.

Perhaps because 57 percent of Hispanics who participated in the survey were immigrants, 75 percent believed that they had a better chance to succeed than did their parents. (In contrast, only 56 percent of surveyed non-Hispanics believed their chances were better than were their parents' chances.) Finally, only 28 percent of surveyed Hispanics believed that the police treat Hispanics more harshly than others.

Current Concerns

Government researchers now estimate that Hispanics will continue to increase as a proportion of the U.S. population until at least 2050 (U.S. Bureau of the Census 2006). This rapid growth has raised two concerns among many Americans. First, because most of the new immigrants are young, poorly educated, and (especially if they are undocumented immigrants) willing to accept very low wages, some fear they are bringing down the wage scale for American citizens. Second, some fear that Hispanic culture and the Spanish language will "take over" the country. These fears are height-

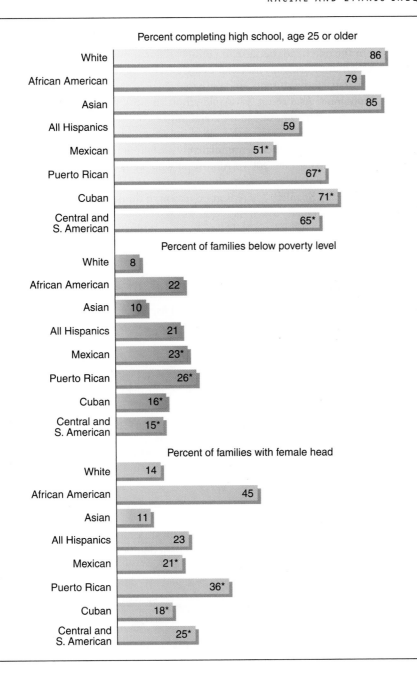

Percent completing high school, age 25 or older

White — 86
African American — 79
Asian — 85
All Hispanics — 59
Mexican — 51*
Puerto Rican — 67*
Cuban — 71*
Central and S. American — 65*

Percent of families below poverty level

White — 8
African American — 22
Asian — 10
All Hispanics — 21
Mexican — 23*
Puerto Rican — 26*
Cuban — 16*
Central and S. American — 15*

Percent of families with female head

White — 14
African American — 45
Asian — 11
All Hispanics — 23
Mexican — 21*
Puerto Rican — 36*
Cuban — 18*
Central and S. American — 25*

FIGURE 8.1

Education, Poverty, and Family Structure by Race and Hispanic Origin, 2004

Compared with other groups, Hispanics—especially Mexicans—are the most likely to lack a high school education, partly because many are recent immigrants. Hispanics and African Americans are more likely than whites and Asians to live below the poverty level, and African Americans are the most likely to live in female-headed households, followed by Puerto Ricans (many of whom are also of African descent). SOURCE: U.S. Bureau of the Census (2006); Ramirez & de la Cruz (2003).

ened by the (slightly) increasing residential segregation of Hispanics in the United States (Iceland & Wilkes 2006), which has led some to question whether these new immigrants will ever become socially integrated into U.S. society.

Are new Hispanic immigrants in fact driving down wages? Research on this topic is mixed. Some have concluded that immigrants stimulate the economy overall and thus benefit all Americans (Card 2005). Others argue that immigration improves the quality of life of affluent Americans (by making cheap labor available), but depresses the wages of Americans who lack high school degrees by as much as 5 percent (Porter 2006; Borjas & Katz 2006).

Hispanic Americans are now the largest and fastest growing ethnic minority in the United States. Because racism against Hispanic Americans is relatively mild, by the second and third generation, most Hispanics can translate their educational attainment into well-paying jobs

Will Hispanics become socially integrated into the United States? On the one hand, because of continued immigration from Latin America, U.S.-born Hispanics now can easily enjoy salsa dancing, Mexican fiestas, Guatemalan restaurants, and perhaps romance and marriage with a recent immigrant. As a result, Hispanic ethnic identity is being reinforced even among those whose families emigrated here much earlier (Waters & Jimenez 2005). On the other hand, these earlier generations of Hispanic immigrants nevertheless are relatively socially integrated into the United States (Alba & Nee 2003). Almost all who were born in this country are fluent in English, and those whose parents were also born here often speak little if any Spanish. There is good reason to think the same will be true of new immigrants.

In addition, the caste-like barrier separating races operates much less dramatically for Hispanics. White prejudice against Hispanics is far less strong than against African Americans, and although Hispanic segregation has increased, it remains modest. As a result, the main barrier Hispanics face is class rather than ethnicity—at least if they are white. As a result, by the second and third generation, most Hispanics can translate educational attainment into well-paying jobs and leave the segregated barrios (Iceland & Wilkes 2006).

In sum, there is good reason to be optimistic about the effect of Hispanic immigration on the United States. Nevertheless, concern about rising immigration has fueled recent demands for stricter border. In turn, these demands have led to a surge of political activism among Hispanic immigrants and their supporters, calling for more humane treatment of immigrants, and perhaps guest worker programs or "amnesty" programs for undocumented immigrants. The result of these efforts remains to be seen.

Asian Americans

The Asian population of the United States (Japanese, Chinese, Filipinos, Koreans, Laotians, and Vietnamese, among others) more than doubled between 1980 and the present but still constitutes only 4.3 percent of the total population. The Asian population can be broken roughly into three segments: descendants of nineteenth-century immigrants (Chinese and Japanese), post-World War II immigrants (Filipinos, Asian Indians, and Koreans), and recent refugees from Southeast Asia (Cambodians, Laotians, and Vietnamese).

A century ago, Asian immigrants were met with sharp and occasionally violent racism. Today, incidents of racial violence directed at Asians are rare but still occasionally make headlines. Despite these handicaps, Asian Americans have experienced high levels of social mobility. A higher percentage of Asian Americans than white Americans have college, doctoral, medical, and law degrees (Le 2006). Educational levels are especially high among Japanese and Chinese Americans, many of whom come from families that have lived in the United States for generations. Education levels are also high among Asian Indians and Filipinos, many of whom came to the United States to get a graduate education or with graduate degrees in hand. Current evidence suggests that the more recent streams of immigrants from Southeast Asia will follow the same path. For example, although many of the Southeast Asian refugees who came to the United States between 1975 and 1984 began their lives here on welfare, almost twice as many Vietnamese youths aged 20 to 24 are enrolled in school as are white youths of the same age.

Current Concerns

The high level of education earned by Asian Americans is a major step in opening doors to high-status occupations, and median income for Asian Americans is only slightly below that of whites (U.S. Bureau of the Census 2006). Yet, discrimination is not all in the past. Unofficial policies make it more difficult for Asian American applicants than for white Americans with the same credentials to gain admittance to elite colleges and universities. And highly educated Asians still earn significantly less than whites with the same professional credentials, primarily because they are less likely to move out of professional and technical positions into managerial and executive positions (Le 2006). Asians are often passed over for promotions because white employers assume Asians won't have the personality, social skills, or simply the "look" an executive is assumed to need. In addition, Asians less often even learn about available executive positions because they are less often accepted into the "old boy networks" in which most professional mentoring takes place, and in which individuals gain the contacts that can lead to higher-level jobs. Finally, Asian Americans— even those whose great-great-grandparents emigrated to this country—still are held back by others who assume they aren't "really" American. As one third-generation Japanese American said,

> I get real angry when people come up to me and tell me how good my English is. They say: "Oh, you have no accent. Where did you learn English?" Where did I learn English? Right here in America. I was born here like they were. [But] people see me now and they automatically treat me as an immigrant" (quoted in Zhou & Gatewood 2000, 18).

Native Americans

Native Americans (American Indians) are one of the smallest minority groups in the United States (about 1 percent of the entire population), and nearly half of their members live in just four states: Oklahoma, Arizona, California, and New Mexico. Native Americans are arguably our most disadvantaged minority group. Compared with African Americans, for example, Native Americans have lower median incomes and are more likely to live in houses without indoor plumbing, heating, or telephones (Table 8.2). Native Americans have the lowest rates of educational achievement and the highest rates of alcoholism and premature death of any U.S. racial or ethnic group (Kao & Thompson 2003). This situation exists despite hopeful new signs of economic vitality on some Indian reservations over the past 20 years, including the development of mineral reserves on the Navajo reservation and the advent of gambling casinos elsewhere.

Table 8.2 summarizes the situation on some of the larger Native American reservations. Keep in mind, though, that the status of Native Americans is highly diverse. Native Americans represent more than 200 tribal groupings, with different cultures and languages. Some have been successful: fish farmers in the Northwest, ranchers in Wyoming, and bridge builders in Maine. In urban areas, and east of the Mississippi where the impact of white society has been felt the longest, many Native Americans have blended into the majority culture and entered the economic mainstream. On isolated reservations with few economic resources (and little opportunity to draw crowds to casinos), on the other hand, socioeconomic conditions often are quite

TABLE 8.2
Social and Economic Characteristics of Selected Native American Reservations, 2000*

	Reservation					
	San Carlos Apache	Blackfeet	Wind River Cheyenne-Arapaho	Hopi	Navajo	Rosebud Sioux
Percent 65 years and over	5%	6%	12%	9%	7%	6%
Percent high school graduate (25 years or older)	58%	74%	83%	67%	56%	73%
Percent college graduate (25 years or older)	3%	13%	15%	10%	7%	11%
Percent families with female heads	17%	13%	8%	15%	14%	20%
Percent families below poverty level	48%	30%	16%	37%	40%	46%
Percent using wood to heat home	10%	14%	8%	37%	52%	8%
Percent lacking telephones	21%	11%	8%	32%	60%	25%
Percent with no vehicle	n/a	27%	12%	6%	24%	18%
Percent lacking indoor water or toilets	11%	2%	2%	27%	32%	3%
Median family income, 1999	$17,585	$26,832	$35,238	$22,989	$22,392	$18,673

SOURCE: U.S. Bureau of the Census 2000, American Factfinder: http://factfinder.census.gov/home/aian/aian_aff2000.html.
*Latest data available as of 2006.

poor. In addition, in white-dominated towns near large Native American reservations, prejudice and discrimination by whites remain major barriers.

Arab Americans

According to the 2000 U.S. census, Arab Americans comprise considerably less than 1 percent of the U.S. population. Because of recent world events, however, their status in this country is of particular political and sociological importance.

All Arab Americans are immigrants or children of immigrants from North Africa and the Middle East (including Morocco, Algeria, Saudi Arabia, and Iraq); Iran is not an Arabic country. The largest single group of Arab Americans is from Lebanon ("Arab American Demographics" 2006). Each of these countries has its own traditions, but they share common linguistic, cultural, and historical traditions. Some Arab Americans descend from families that emigrated to the United States in the late 1800s, some emigrated themselves only in the last few years. Two thirds of Arab Americans are Christian.

Arab Americans are a highly educated population. They are as likely as other white Americans to have graduated high school and are slightly more likely to have graduated college. As a result, the majority hold professional jobs, and their median incomes are somewhat above the U.S. average.

Current Concerns

Interestingly, the terrorist attacks of 9/11 have not harmed American opinion of Arabs. Many Americans who previously had no opinion of Arabs have since gotten a "crash course" in Arabic and Muslim history and culture, and have taken pains not to discriminate against all Arabs or Muslims because of the actions of a few. A Gallup Poll taken only a month after the 9/11 attacks found that attitudes had actually improved since

■ Arab Americans, like these Michigan schoolchildren, are an increasingly important minority group in the United States.

1993, the year of the first terrorist attack on the World Trade Center. In 2001, 54 percent reported very or mostly favorable opinions of Arabs, and 66 percent reported very or mostly favorable opinions of Muslims (Jones 2001). Similarly, a poll conducted in 2003 found that 51 percent reported favorable attitudes toward Muslim Americans—considerably lower than the approximately 70 percent who held favorable views of Protestants, Catholics, and Jews, but not much lower than the 58 percent who held favorable views of evangelical Christians (Pew Research Center 2003b).

Still, political changes since 9/11 raise questions about the current and future status of Arabs and Muslims in this country. One special concern is whether the sweeping federal USA Patriot Act, passed to help the government fight terrorism, has encouraged discrimination. During the first six months of 2003, the Justice Department received more than 1,000 reports from Arab immigrants alleging that Department employees had violated their civil rights or civil liberties, with violations ranging from verbal abuse to beatings. The Department also has received more than 500 reports of "hate crimes" against Arab and Muslim Americans (Shenon 2003). Negative attitudes toward Arabs (American or not) can be reinforced by newspaper and television shows that highlight Arab terrorists and Arab anti-Americanism, as well as by films that depict Arabs either as terrorists or not at all (Semmerling 2006).

The Future of Racial and Ethnic Inequality in the United States

The last few decades have witnessed considerable improvement in the social status of various minority groups. Yet inequality remains. What are some of the reasons for this, and what are some of the strategies that can help reduce inequality?

Global Perspective

focus on

Genocide in Darfur

Racial inequality is not solely an American problem. Discrimination and prejudice in other countries also deny minority groups their rights and opportunities. The genocide in Darfur offers a recent example. Although the political process that underlies Darfur's ethnic strife may be unique to that society, the economic processes appear to be typical of those accompanying ethnic conflict in societies throughout history: Racial and ethnic hostilities are most pronounced when economic resources are scarce and the majority group's economic advantage is threatened.

Darfur is a region in western Sudan, the largest country in Africa. Like most African countries, Sudan was cobbled together by a colonial power—in this case, Britain—during the nineteenth century. The country is overwhelmingly composed of Sunni Moslems who use Arabic as their *lingua franca*, but northern Sudan is primarily

Arab, while the rest of the country is now considered black African. Ironically, the physical differences between these two groups are slight enough that westerners typically cannot distinguish Sudanese Arabs from Sudanese Africans. Indeed, prior to this conflict, ethnic identity was fluid and relatively unimportant, intermarriage was common, and the distinction between "Arab" and "African" was rarely used. African farmers coexisted easily with Arab herders, since each benefited from trading with the other. Moreover, Arab herders sometimes became farmers, and African farmers sometimes became herders, depending on their shifting economic circumstances.

Since Sudan achieved independence in 1956, northern Arabs have thoroughly dominated the country's economy and government. Yet the north holds few of the country's natural resources, especially agricultural lands and oil deposits. Moreover, global warming, growing human and livestock populations, and damaging agricultural

practices are all contributing to the "desertification" of northern Sudan. To maintain their dominance over the country and its natural wealth, the Sudanese Arabs who run the country's government have used military repression, political repression, and economic strangulation against their perceived enemies. In response, since the 1980s armed resistance by Sudanese Africans, in both southern and western Sudan, has increased, as has repression by the central government.

Since early 2003, however, the Sudanese government has moved from repression to what most observers describe as genocide against the people of Darfur (and, increasingly, against Africans in neighboring Chad). To facilitate this policy, the government has supplemented military forces by forming and arming local Arab militias, known as "Janjaweed." Janjaweed members are recruited from nomadic and semi-nomadic Arab tribes who hope to gain not only war loot, but also access to increasingly scarce water sources, pasture for live-

The Persistence of Inequality

There is no question that traditional racism in the United States has declined over time. Few Americans now believe that African Americans earn less money than whites because of their innate inferiority, and few believe that people should be discriminated against because of their race and ethnicity. So why does racial and ethnic inequality continue? Answers include the effects of subtle racism and institutionalized racism.

Subtle racism reflects widespread belief in the ideology of the American Dream. Because of this ideology, many people continue to blame the poor for their poverty; the predominant white explanation for poverty stresses the lack of motivation among African Americans. As a result of such beliefs, whites are reluctant to support governmental policies designed to promote economic equality between minorities and whites (Bobo & Kluegel 1993; Quillian 1996).

Institutionalized racism is an even more important source of continued racial inequality, in the United States as elsewhere. Institutionalized racism refers to situations in which everyday practices and social arrangements are assumed to be fair, but in fact systematically reproduce racial or ethnic inequality. For example, almost all Gypsy children in Czechoslovakia are placed in special schools for the mentally handicapped, and almost all children in these schools are Gypsy (New York Times

Institutionalized racism occurs when the normal operation of apparently neutral processes systematically produces unequal results for majority and minority groups.

stock, and arable lands (Human Rights Watch 2006).

As of 2006, the Sudanese army and the Janjaweed have killed more than 100,000 civilians and produced almost 2 million refugees as individuals flee the warfare (U.S. State Department 2006). In addition, the Janjaweed have engaged in aerial carpet bombing, systematic torture, mass amputations with machetes, and mass rape—all aimed overwhelmingly at civilians rather than at resistance fighters. To justify these actions, the government and Janjaweed have encouraged racial stereotyping of African Sudanese as inferior. As a result, Sudanese civilians increasingly identify themselves as Arab or as African, rather than as Sudanese or as members of a specific tribe. Meanwhile, both intraethnic and interethnic violence is exploding.

Since 2003, and in a bid to control valuable lands and water supplies, the Sudanese government has encouraged racial stereotyping of African Sudanese as inferior and has promoted the slaughter of Sudanese Africans by Sudanese Arabs like these "janjaweed" militia members. Yet before the war ethnic identity in Sudan was fluid and relatively unimportant, intermarriage was common, and people rarely distinguished between "Arab" and "African" Sudanese.

2006). Czech school authorities argue that Gypsy children are placed in these schools based on standardized evaluations, but this policy effectively makes it impossible for Gypsy children to succeed in Czech society. Less extreme versions of school segregation and tracking reinforce racial inequality in the United States.

Combating Inequality: Race versus Class

In this chapter and Chapter 7, we have shown how both social class, on the one hand, and race or ethnicity, on the other hand, affect one's life chances. When a person has a lower status on both of these dimensions, we speak of **double jeopardy.** This means that disadvantages snowball. For example, poor African American, Hispanic, and Native American teenagers are more likely than poor white teenagers to be unemployed or to end up in prison.

Sociologists have hotly debated whether race or class is more important for understanding the structure of inequality in the United States today. The question most often asked is, "Is the status of lower-class African Americans due to the color-blind forces of class stratification, or is it due to class-blind racism?" A 1978 book titled *The Declining Significance of Race* set the tone for much of this debate. In it, African American sociologist W. J. Wilson argued that the status of African Americans had less to do

Double jeopardy means having low status on two different dimensions of stratification.

with racism than with the simple inheritance of poverty and the changing nature of the U.S. economy. As well-paying factory jobs disappeared and as other forms of employment shifted from the inner cities to the suburbs, the position of the poorest third of the African American population has disintegrated. Joblessness is up, the number of female-headed households is up, rates of drug use are up, and so on. For this reason, Wilson argued that the solutions to racial inequality are primarily economic. As a result, Wilson (1987) believes that African Americans can best be helped through strategies designed to create full employment and better jobs for *all* Americans, such as the movements for fair wages and for increasing educational opportunities described in Chapter 7.

Most sociologists disagree. They doubt that policies based on social class alone will be enough to resolve the problem of racial inequality in the United States. True, there are middle- and even upper-class minority group members, and it would be a serious mistake to assume that racism keeps all racial minorities poor and powerless. Nevertheless, race and ethnicity continue to be fundamental dividing lines in U.S. society. Membership in a minority group remains a handicap in social-class attainment and in social relationships. For example, the finding that middle-class African Americans are much more likely than are middle-class whites to live in poor neighborhoods suggests that the issue goes beyond class (Alba et al. 2000). Any successful strategy for combating inequality in the United States will have to address issues of race and ethnicity as well as social class.

The major strategies used in the United States to fight against racial and ethnic inequality are antidiscrimination and affirmative action laws. Since 1964, the United States has officially outlawed discrimination on the basis of race, color, religion, sex, and national origin. These laws have had considerable effect. States can no longer declare interracial marriage illegal or refuse to allow African Americans to vote or to attend state schools, and newspapers can no longer advertise that a job is open only to whites.

Whereas antidiscrimination laws make it illegal to discriminate, affirmative action rules require employers, schools, and others to actively work to increase the representation of groups that have historically experienced discrimination. This means, for example, that a college with very few minority faculty may be required to advertise new jobs through minority faculty organizations, as well as in regular employment bulletins. Affirmative action has proven much more contentious than antidiscrimination laws.

Where This Leaves Us

Racism and interethnic conflicts are problems worldwide, erupting in schoolyards, street corners, and courts of law. This does not mean, though, that these conflicts cannot be lessened or even eliminated. Irish people no longer are refused employment as was common in the nineteenth century, and Jews no longer are prohibited from living in certain neighborhoods or belonging to certain clubs as was common until the 1960s. Ideas about race and ethnicity are social constructions that change as societies change. To combat prejudice and discrimination, we will need to combat subtle and institutionalized racism, and we will need to address the social class inequalities that support racial and ethnic inequalities. Doing so will be both especially difficult and especially crucial if economic hard times continue in the United States.

Summary

1. A race is a category of people treated as distinct due to physical characteristics that have been given social importance. An ethnic group is a category whose members are thought to share a common origin and culture. Both race and ethnicity are socially constructed categories.

2. In the United States, the population is stratified by both race and class. These two factors work together to create greater advantages or disadvantages for different groups.

3. The concepts of majority and minority groups provide a general framework for examining structured inequalities based on ascribed statuses. Interaction between majority- and minority-group members may take the form of conflict, accommodation, acculturation, or social integration.

4. Prejudice and discrimination help create and maintain social distance. Segregation—caused both by economic differences across racial/ethnic groups and by prejudice and discrimination—also helps reinforce differences and inequalities.

5. In the United States, white ethnicity is now largely a symbolic characteristic. Its main consequence is that it has become the "standard" American ethnicity against which other groups are judged. White racial identity is typically invisible and carries considerable if unacknowledged privileges.

6. On many fronts, African Americans have improved their position in U.S. society. Nevertheless, African American families continue to have a median income that is far lower than that of white families. Major areas of continued concern are high rates of female-headed households, unemployment, and housing segregation.

7. Hispanics are the largest and fastest-growing minority group in the United States. Because many Hispanics are recent immigrants from less-developed countries, they generally have poor educations and low earnings. However, like their predecessors, current Hispanic immigrants likely will become socially integrated. Hispanic immigration helps the economy overall but may reduce income for the least-educated U.S. citizens.

8. Native Americans are the least prosperous minority group in the United States. Living conditions and economic prospects are the most difficult on geographically isolated reservations.

9. Asian Americans have used education as the road to social mobility. Even the newest immigrant groups outstrip white Americans in their pursuit of higher education. Despite some discrimination, Asian Americans have higher median family incomes than do white Americans and experience low levels of residential segregation.

10. Arab Americans are primarily middle class: well educated, with good jobs. Prejudice against Arab Americans is strong but did not increase substantially following the 9/11 terrorist attacks.

11. Prejudice and discrimination continue to be reinforced through both subtle and institutionalized racism. Effective efforts to counteract these forces will need to address not only racial and ethnic inequality but also broader economic inequality embedded in social class issues.

Thinking Critically

1. Within the next 50 years or so, non-Hispanic whites will be a *numerical* minority within the United States. In sociological terms, do you think they will be a minority group? What social, economic, or political changes do you expect as a result of changes in the relative size of the different U.S. racial and ethnic groups?

2. In thinking about the relationship between prejudice and discrimination, we generally assume that prejudice is the cause of discrimination. Can you think of a time or situation when the reverse might be true, that is, that prejudice would follow from discrimination?

3. Using the concepts in this chapter, and the information you see around you in the media, conversations, and so on, discuss whether Arab Americans are considered white. (Note: Do not discuss whether they *should* be considered white, just whether they are.)

4. List five things you typically do during the course of the week, such as going shopping or meeting with friends. How would that experience be different if you woke up tomorrow and found that your race had changed to African American or to white?

5. Some scholars contend that the major cause of racial/ethnic inequality in the United States today is institutionalized, not individual, racism. If this is so, what recommendations would you offer to policymakers who wanted to reduce racial or ethnic differences in quality of life?

6. When the Smithsonian Institution mounted an exhibition showing the horrors that resulted from the bombing of Hiroshima during World War II, some American veterans protested that the exhibit paid too much attention to Japanese suffering and portrayed the United States as an unfeeling aggressor. How should a multicultural society like the United States deal with such issues? What would you have done if you were in charge of the exhibit? Why?

7. What similarities and what differences do you see between the situation in Darfur and that of African Americans in the United States?

Companion Website for This Book

academic.cengage.com/sociology/brinkerhoff
Gain an even better grasp on this chapter by going to the Companion Website. This resource contains tutorial quizzes and flash cards to help you master key terms and concepts.

Suggested Readings

Correspondents of the *New York Times*. 2002. *How Race Is Lived in America: Pulling Together, Pulling Apart*. New York: Times Books. Based on an award-winning *New York Times* series, this book covers everything from white teenagers infatuation with hip-hop to the subtle and not-at-all subtle traumas of living with racism.

Lui, Meizhu, Robles, Barbara, and Leondar-Wright, Betsy. 2006. *The Color of Wealth: The Story Behind the U.S. Racial Wealth Divide*. Clearly explains the sources and consequences of the "wealth gap" between blacks and whites. New York: New Press.

Rusesabagina, Paul. 2006. *An Ordinary Man: An Autobiography*. Close to one million Rwandans were slaughtered in 1994 by their fellow citizens in genocidal attacks. Rusesabagina describes the sources of this violence, and how he came to take action against it (as depicted in the film *Hotel Rwanda*).

Morales, Ed. 2003. *Living in Spanglish: The Search for Latino Identity in America*. New York: St. Martin's. Explores how Latinos are creating a new culture, blending their Latino roots with Anglo-American culture.

Temple-Raston, Dina. 2003. *A Death in Texas: A Story of Race, Murder, and a Small Town's Struggle for Redemption*. New York: Henry Holt. On June 7, 1998, three white men from Jasper, Texas, chained an African American man to the back of a pickup truck and dragged him to his death. In this book Temple-Raston explores how the townspeople responded and uses this incident to illuminate America's racist legacy.

CHAPTER 9

Sex, Gender, and Sexuality

© Royalty-Free/Corbis

Outline

Sexual Differentiation

Men and women are different. Biology differentiates their physical structures, and cultural norms in every society differentiate their roles. In this chapter, we describe some of the major differences in men's and women's lives as they are socially structured in the United States. We will be particularly interested in the extent to which the ascribed characteristic of sex has been the basis for structured inequality.

Sex versus Gender

Sex is a biological characteristic, male or female.

Gender refers to the expected dispositions and behaviors that cultures assign to each sex.

Gender roles refer to the rights and obligations that are normative for men and women in a particular culture.

In understanding the social roles of men and women, it is helpful to make a distinction between sex and gender. **Sex** refers to the two biologically differentiated categories, male and female. It also refers to the sexual act that is closely related to this biological differentiation. **Gender,** on the other hand, refers to the normative dispositions, behaviors, and roles that cultures assign to each sex. (See the Concept Summary.)

Although biology provides two distinct and universal sexes, cultures provide almost infinitely varied **gender roles.** Each man is pretty much like every other man in terms of sex—whether he is upper class or lower class, African American or white, Chinese or Apache. Gender, however, is a different matter. The rights, obligations, dispositions, and activities of the male gender are very different for a Chinese man than for an Apache man. Even within a given culture, gender roles vary by class, race, and subculture. In addition, of course, individuals differ in the way they act out their

Concept Summary

Sex versus Gender

	Sex	Gender
Divides population into:	Male or female (or maybe intersex)	Masculine and feminine
Based on	Biological characteristics (chromosomes, sex hormones, penises or vaginas, etc.)	Cultural expectations regarding appropriate behaviors and attitudes for each sex
Consequences	On average, men have more upper body strength than women because of their hormones.	Men also have more upper body strength because women are warned that they will look "too masculine" if they lift weights too much.
	On average, men are taller than women because of their genes.	In poor countries, sex differences in height are amplified because boys receive more food than do girls.

expected roles: Some males model themselves after Brad Pitt and some after Johnny Depp or Will Smith.

Just how much of the difference between men and women in a particular culture is normative and how much is biological is a question of considerable interest to social and biological scientists. This question has led some biologists to investigate whether characteristics we typically think of as male and female also characterize nonhuman species. If they did, that would lend support to the idea that these male/female differences are biological. Results from these studies are decidedly mixed. Female goby fish sport bright colors to attract the opposite sex, but among birds it is the males who usually do so. Male baboons certainly dominate female baboons, but male marmosets (small monkeys) take care of the young, and male lions depend on the females to do all the hunting. Meanwhile, whales and elephants live in matriarchal families.

For the most part, social scientists are more interested in gender than in sex. They want to know about the variety of roles that have been assigned to women and men and, more particularly, about the causes and consequences of this variation. Under what circumstances does each gender have more or less power and prestige? How does having more or less power affect women's and men's everyday lives? And what accounts for the recent changes that have occurred in gender roles in our society?

Gender Roles across Cultures

A glance through *National Geographic* confirms that gender roles vary widely across cultures. The behaviors we normally associate with being female and male are by no means universal. Among the Wodaabe, a nomadic tribe of western Africa, boys carry mirrors with them from the time they can walk (Bovin 2001). Even when boys spend days alone in the bush herding cows, they begin each day by fixing their hair and putting on their jewelry, lipstick, mascara, and eye-liner. In contrast, because girls are primarily evaluated on their health and ability to work hard, they are expected to pay far less attention to their appearance than do boys. Wodaabe courtship mostly takes place during men's dance competitions, in which women judges select the winners based on the men's physical beauty and charm. Afterward, the women openly approach the men they find most attractive to be their romantic partners.

Despite cross-cultural variations such as these, and despite the fact that women do substantial amounts of work in all societies (often providing more than half of household food), in almost all societies women have less power and less value than men (Kimmel 2000). A simple piece of evidence is parents' almost universal preference for male children (Sohoni 1994), a preference which can be life threatening for girls. Currently, there are about 120 boys for every 100 girls in China and India—far higher than the natural ratio of about 105 to 100 (Hudson & DenBoer 2005). This difference is primarily due to the use of abortion to kill fetuses identified prenatally as female. Other girls are killed at birth or, more often, die because they receive less food and medical care than their brothers. The preference for boys is less strong in modern industrial nations, but parents in the United States nonetheless prefer their first child to be a boy by a two-to-one margin (Holloway 1994).

Another result of female power disadvantage is widespread violence toward women. According to the respected international organization Human Rights Watch, "Abuses against women are relentless, systematic, and widely tolerated, if not explicitly condoned. Violence and discrimination against women are global social epidemics" (Human Rights Watch 2004). For example:

- Each year, about 1.5 million American women are raped or physically assaulted by intimate partners (Tjaden & Thoennes 2000). Although men are also sometimes assaulted by their partners, they are more likely to be hit in self-defense and less likely to be seriously harmed or killed (Fox & Zawitz 2004). (Violence between intimates is further discussed in Chapter 11.)
- In Uganda, Darfur, Bosnia, and elsewhere, armies have used rape both as a systematic tool to subjugate the population and as a form of "sport" for soldiers.
- Between 100 and 140 million women, mostly in African countries but also in Asia, South America, and Europe, have undergone genital mutilation—removal of some or all of the clitoris and surrounding genitalia (World Health Organization 2000). Aimed at eliminating sexual desire in women, the practice is dangerous and even deadly.
- In India during 2004, more than 6,000 new brides were officially reported as murdered by their husbands or in-laws, with many more such murders unreported (U.S. Department of State 2005). These "dowry deaths," which are rarely punished, occur when a husband and his family consider a new wife's dowry inadequate.

At home and abroad, violence against women results from the lower status accorded to women. In growing numbers, women around the world are demanding equal rights. In some of the least-developed nations, this means changing cultural and legal values that treat women essentially as their husbands' or fathers' property. In the United States and the rest of the developed world, the problems are more subtle. Those problems lead sociologists to ask: How are gendered identities developed? And what are the institutional forces that maintain inequality, with or without overt violence and discrimination?

Theoretical Perspectives on Gender Inequality

Women rather than men bear children because of physical differences between the sexes. Most of the differences in men's and women's life chances, however, are socially structured. Different sociological theories offer different explanations for the persistence of this structured gender inequality.

Structural-Functional Theory: Division of Labor

The structural-functional explanation of gender inequality is based on the premise that a division of labor is often the most efficient way to get a job done. In the traditional sex-based division of labor, the man does the work outside the family and the woman does the work at home. According to this argument, a gendered division of labor is functional because specialization will (1) increase the expertise of each sex in its own tasks, (2) prevent competition between men and women that might damage the family, and (3) strengthen family bonds by forcing men and women to depend on each other.

Of course, as Marx and Engels noted, any division of labor has the potential for domination and control. In this case, the division of labor has a built-in disadvantage for women because by specializing in the family, women have fewer contacts, less information, and fewer independent resources. Because this division of labor contributes to family continuity, however, structural functionalists have seen it as necessary and desirable.

Conflict Theory: Sexism and Discrimination

According to conflict theorists, women's disadvantage is not an historical accident. Instead, it is designed to benefit men and to benefit the capitalist class.

Two major concepts employed by conflict theorists to explain how gender inequality benefits men and capitalists are sexism and discrimination. **Sexism** is the belief that women and men have biologically different capacities and that these differences form a legitimate basis for unequal treatment. Conflict theorists explain sexism as an ideology that is part of the general strategy of stratification. If others can be categorically excluded, the need to compete individually is reduced. Sexism, then, reduces women's access to scarce resources and allows men to keep those resources for themselves.

Discrimination is the natural result of sexism. If we believe that women are better suited to work with children and men are better suited for intellectual work, then we will be more likely to admit men to medical school than women, more likely to hire a man as a doctor than a woman, and more likely to hire a woman as a pediatrician than as a neurosurgeon.

> **Sexism** is a belief that men and women have biologically different capacities and that these form a legitimate basis for unequal treatment.

Symbolic Interactionism: Gender Inequality in Everyday Life

Symbolic interactionist theory is particularly useful for understanding the sources and consequences of sexism in everyday interactions. For example, sociologist Karin Martin (1998) was interested in understanding how boys and girls learn gender-normative ways of moving, using physical space, and comporting themselves. To do so, she studied 112 preschoolers in 5 different classrooms, at 2 different preschools, with 14 different teachers. She found that teachers routinely structure children's play and impose discipline in ways that reinforce gender differences. Little boys are actively discouraged from playing "dress-up" (even though many of them enjoy doing so), and little girls are discouraged from running, crawling, and lying on the floor. Whereas boys are allowed to have fun shouting, playing rough and tumble games, and moving about wildly, girls are disciplined to raise their hands, lower their voices, and restrict their movements. Likewise, boys' play was allowed to be more rough and tumble than that of girls. By the end of preschool, then, boys and girls are well on their way to learning the nonverbal behaviors and communication styles that are typical of, and seem so natural for, adult men and women. We will discuss these gendered differences in more detail later in this chapter.

A second study illustrating the symbolic interactionist perspective on gender inequality was based on observations conducted at a sleep-away camp during the course of one summer (McGuffey & Rich 1999). At this camp, high-ranking boys attained power and popularity primarily through athletic prowess. They bolstered their positions and won approval from other boys by acting aggressively toward lower-ranking boys and by sexually harassing girls. In addition, and most importantly, high-ranking boys led other boys in teasing, assaulting, or excluding any boys they deemed too "feminine" and any girls they deemed too "masculine." Interestingly, high-ranking boys were able to redefine "feminine" activities they enjoyed (such as hand-clapping games) into masculine activities. In these ways, high-ranking boys maintained their status and power over other boys, and almost all boys maintained greater status and power than girls.

Gender as Social Construction and Social Structure

To sociologists, gender is not simply something that individuals have—a biological given—but rather is something that is constantly re-created in individual socialization, in medical and cultural practices, and in social interaction. Similarly, sociologists describe gender as an attribute not only of individuals but also of social structures.

Developing Gendered Identities

From the time they are born, girls are treated in one way and boys in another—wrapped in blue blankets or pink ones, encouraged to take up sports or sewing, described as cute or as strong before they are old enough to truly exhibit individual personalities. In these ways, as symbolic interactionist studies illustrate, children learn their gender and gender role. By the age of 24 to 30 months, they can correctly identify themselves and others by sex, and they have some ideas about what this means for appropriate behavior (Cahill 1983).

Young children's ideas about gender tend to be quite rigid. They develop strong stereotypes for two reasons. One is that the world they see is highly divided by sex: In their experience, women usually don't build bridges and men usually don't crochet. The other important determinant of stereotyping is how they themselves are treated. Substantial research shows that parents treat boys and girls differently. They give their children "gender-appropriate" toys, they respond negatively when their children play with cross-gender toys, they allow boys to be active and aggressive, and they encourage their daughters to play quietly and visit with adults (Orenstein 1994). When parents do not exhibit gender-stereotypic behavior and do not punish their children for cross-gender behavior, their children are less rigid in their gender stereotypes (Berk 1989).

As a result of this learning process, boys and girls develop strong ideas about what is appropriate for girls and what is appropriate for boys. However, boys are punished more than girls for exhibiting cross-gender behavior. Thus, little boys are especially rigid in their ideas of what girls and boys ought to do. Girls are freer to engage in cross-gender behavior, and by the time they enter school, many girls are experimenting with boyish behaviors.

Reinforcing Biological Differences

Because of gender socialization, girls and boys and men and women understand quite well what a "proper" male or female should be like. These ideas can become self-fulfilling prophecies, as the *belief* that males and females are biologically different *keeps* males and females biologically different (Lorber 1994). To understand how this works, Shari Dworkin (2003) spent two

© Tony Freeman/PhotoEdit

■ Despite many changes in gender roles in the United States, boys and girls still tend to experience large doses of traditional gender socialization.

years doing participant observation at two gyms. She found that trainers at both gyms told women patrons that they could lift weights without fear because only men can "bulk up." Nonetheless, 25 percent of women didn't lift at all because they feared developing "masculine" muscles. Another 65 percent restricted their weight lifting to shorter periods or lighter weights after they *did* develop bigger muscles. By the end of two years training, these women remained relatively unmuscular. They lacked muscles not because they were inherently unable to develop them but because they chose not to do so, based on their beliefs about proper male/female differences.

Biological sex differences can also be reinforced by medical practices. Doctors sometimes prescribe hormones to keep girls from growing "too tall" and boys from being "too short" (Weitz 2007). Doctors also offer plastic surgery to women with small breasts and men with small pectoral muscles. In this way, the very bodies we see around us come to reinforce social ideas about male/female differences.

Our belief in the naturalness of biological differences is also reinforced when we are, in essence, kept from seeing how similar males and females can be. Television offers far more coverage of female cheerleaders and male football players than of male cheerleaders and female football players, reinforcing the idea that it is impossible for women to play strenuous sports and that no "real men" would be interested in cheerleading. Similarly, Olympic games that evaluate female figure skaters on their grace and male skaters on their speed and power force female and male skaters to develop different skills and leave audiences believing that female and male skaters naturally have quite different abilities. The same is true for athletic rules that limit the size of the basketball court on which girls can play or that forbid male and female athletes from competing together.

"Doing Gender"

Gender differences are also reinforced when we "do gender." Sociologists use the term "doing gender" to refer to everyday activities that individuals engage in to affirm their commitment to gender roles (West & Zimmerman 1987). Women who are professional body-lifters almost always wear long, blonde hair so that no one will question their femininity despite their muscles (Weitz 2004a), and male nurses sometimes talk about their athletic interests or heterosexual conquests to keep others from questioning their masculinity. Each of us does gender every day when we (whether male or female) choose to wear skirts or jeans, to speak softly or boldly, to get a butterfly tattoo or shark tattoo, and so on. In these ways we participate in the social construction of gender. Another way to think about this is that gender is not something that we innately have, but rather is something that we *do*.

Gender as Social Structure

Gender is also a social structure, a property of society (Risman 1998). Gender is built into social structure when workplaces don't provide day care; women don't receive equal pay; fathers don't receive paternity leave; basketballs, executive chairs, and power drills are sized to fit the average man; and husbands who share equally in the housework are subtly ridiculed by their friends. Importantly, this suggests that changing gender roles and attitudes will only produce social change if there are parallel changes in the social structure of gender. Equally important, when social structure changes, gender roles and attitudes change. For example, Barbara Risman found that fathers whose wives died or deserted them learned quickly how to be good "mothers," who could nurture their children as women would.

Connections

Personal Application

How are you doing gender right now? Are you sitting with your legs splayed apart or crossed at the ankles? Are you wearing makeup? What kind, and for what purposes? What color clothes are you wearing? (Probably not pink, if you are male.) If you are snacking on a muffin while reading this book, did you apologize beforehand or explain how you know you need to lose weight? These are all examples of doing gender.

Differences in Life Chances by Sex

In terms of race and social class, women and men start out equal. The nurseries of the rich as well as the poor contain about 50 percent girls. After birth, however, different expectations for females and males result in very different life chances. This section examines some of the structural social inequalities that exist between women and men.

Health

Women are at a substantial disadvantage in most areas of conventional achievement; in informal as well as formal interactions, they have less power than men. But men, too, face some disadvantages from their traditional gender roles.

Perhaps the most important difference in life chances involves life itself. Boys born in 2010 can expect to live 75.6 years, whereas girls can expect to live 81.4 years (U.S. Bureau of the Census 2006). On average, then, women live more than 6 years longer than men. Part of this difference is undoubtedly biological, with women's hormones offering them some protection. But men's gender roles also contribute to their lower life expectancies (Rieker & Bird 2000).

A major way male gender roles endanger men is by encouraging men to "prove" their masculinity through dangerous activities. As a result, compared with young women, young men are twice as likely to die in motor vehicle accidents and six times more likely to be killed by guns (Minino 2002). Similarly, men are far more likely than women to earn their living through dangerous jobs, such as fishing and lumbering.

But risk taking alone cannot explain all the difference between men's and women's life expectancies. For example, research suggests that men are at greater risk of dying from heart disease partly because the male gender role places little emphasis on nurturance and emotional relationships. Maintaining family and social relationships is usually viewed as women's work, and so men who stay single, get divorced, or are widowed often end up alone. Ultimately, this lack of social support leaves men especially vulnerable to stress-related diseases and may explain why their suicide rate is four times higher than women's (Minino 2002; Nardi 1992).

Education

Fifty years ago, few young women went to college. Those who did were encouraged to focus not on earning a B.A. but on earning an "MRS." (i.e., a marriage certificate). These days, women and men are about equally represented among high school graduates and among those receiving bachelor's and master's degrees. It is not until the level of the Ph.D. or advanced professional degrees (such as in architecture) that women are disadvantaged in quantity of education.

More important than the differences in level of education are the differences in *types* of education. From about the fifth grade on, sex differences emerge in academic aptitudes and interests: Boys take more science and math, whereas girls more often excel in verbal skills and focus their efforts on language and literature. In large part, these sex differences in aptitudes and interests are socially created (Sadker & Sadker 1994). In all subjects, but especially math and science, teachers typically assume that boys have a better chance of succeeding. One result is that teachers more often ask girls simple questions about facts and ask boys questions that require use of analytic skills. When boys have difficulty, teachers help them learn how to solve the problem, whereas when girls have difficulty, teachers often do the problem for them. By the

American Diversity

focus on

Gender Differences in Mathematics

Despite the many changes in gender roles over the last 50 years, boys still outperform girls on standardized math tests such as the SAT and ACT. By high school, boys substantially outnumber girls in advanced mathematics courses, and in adulthood women remain substantially underrepresented in occupations like engineering that depend heavily on mathematics. Are males actually better than females at math and, if so, is this difference based on nature or nurture?

Neuroscientists interested in this question have begun exploring the relationship between fetal exposure to sex hormones and characteristic differences in the brains of adult men and women. For instance, higher levels of fetal exposure to testosterone (a "male" hormone) are associated with right-brain dominance, while lower exposure levels are associated with left-brain dominance. This association may help explain why, compared with the opposite sex, men more often are left-handed with good visual-spatial skills (a "right-brain" trait) and women more often are right-handed with good verbal skills (a "left-brain" trait).

From findings such as these, some researchers reason that gender differences in mathematical performance are at least partially a result of hormonal differences. But just because hormonal differences are *associated* with mathematical performance does not mean that the hormonal differences *caused* the differences in performance. For one thing, gender differences in mathematical performance are considerably smaller in countries such as China that

less strongly consider mathematics a "male" field (Evans, Schweingruber, and Stevenson 2002.) At any rate, the gender differences in performance are small. Because the differences within each sex are so much larger than the differences between them, critics of the biological perspective argue that hormones can explain only a very small part of the overall variation in mathematical performance. This leaves a great deal of room for the influence of social factors. Evidence for this point of view comes from two lines of research.

Research suggests that the average test score for girls is lower than that for boys because girls 1) more often respond poorly to the stress of timed tests and 2) more often take the exam even if they are only average students in math (Lewin 2006). Girls' aver-

age score is lower than boys' because a broader pool of girls takes the test (Lewin 2006).

Second, research shows that the male advantage in mathematical performance is small, only emerges late in high school, and has declined steadily since the 1960s (Leahey and Guo 2001). One possible explanation for this pattern is that boys and girls are now being socialized more similarly, thereby reducing the traditional male advantage in math.

© Tom & Dee Ann McCarthy/CORBIS

■ Although boys substantially outnumber girls in advanced mathematics courses and in math-related careers, girls earn higher grades than do boys in most math courses and math tests. This suggests that socialization and social barriers, not inherent sex differences in mathematical ability, is the reason why so few girls become engineers or mathematicians.

time students arrive at college, girls often lack the necessary prerequisites and skills to major in physical sciences or engineering, even if they should develop an interest in them (Sadker & Sadker). As a result, women college graduates are overrepresented in education and the humanities, and men are overrepresented in engineering and the physical sciences—fields that pay considerably higher salaries.

Table 9.1 shows the proportion of bachelor's degrees earned by women in various fields of study in 1971 and in 2005. You can see from the table that there were changes over this period. Women comprised a far higher proportion of graduates in traditionally male fields in 2005 than in 1971. In fact, women now comprise about half of all graduates in business, pre-law, mathematics, and social sciences and history.

Still, striking differences between men and women remain. In 2006, only 19 percent of graduates in engineering and 25 percent of graduates in computer sciences were women. Meanwhile, 79 percent of graduates in education, 86 percent in health sciences (mostly nursing), and almost all graduates in home economics and library sciences were women (U.S. Department of Education 2006). Because engineers and computer scientists earn a great deal more than do home economics teachers, librarians, and nurses, these differences in college majors have implications for future economic well-being. This situation is an example of institutionalized sexism. (Recall that Chapter 8 discussed the parallel concept of institutionalized racism.)

Work and Income

Among Americans ages 25 to 44 in 2004, 92 percent of men compared with 75 percent of women were in the labor force (U.S. Bureau of the Census 2006). This gap is far smaller than it used to be and will likely continue to shrink (Figure 9.1). Although most young women nowadays still expect to be mothers, they also overwhelmingly expect to work full time after completing their education.

TABLE 9.1

Percentage of Bachelor's Degrees Earned by Women, by Field, 1971 and 2005

Between 1971 and 2005, the percentage of college degrees in traditionally male fields that were earned by women increased substantially. Nevertheless, engineering continues to be largely a male preserve, and education a female preserve. Because engineers earn roughly three times what teachers earn, this difference in majors is one reason why, on average, women earn less than men.

Field of Study	1971	2005
Business	9	50
Computer and information sciences	14	25
Education	75	79
Engineering	1	19
Health sciences	77	86
Home economics	97	97
Library and archival sciences	92	94
Pre-law	6	47
Mathematics	38	46
Social sciences and history	37	51

SOURCE: U.S. Bureau of the Census, 1996a; U.S. Department of Education 2006.

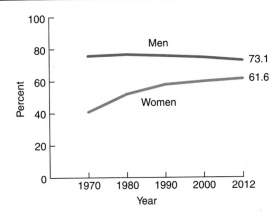

FIGURE 9.1
Labor-Force Participation Rates of Men and Women Aged 16 and Over, 1970-2012 (estimated)
In the 30 years between 1970 and 2000, women's participation in the labor force grew significantly, while men's barely changed. Sex differences in labor-force participation are expected to continue to decline slowly.
SOURCE: U.S. Bureau of the Census 2006.

Despite growing equality in labor-force involvement, major inequalities in the rewards of paid employment persist. In 2004, women who were full-time workers earned 76 percent as much as men (U.S. Bureau of Labor Statistics 2005). This percentage has not changed much since 1950.

Why do women earn less than men? The answers fall into two categories: differences in the types of occupations men and women have and differences in earnings of men and women in the same types of occupations.

Different Occupations, Different Earnings

A major source of women's lower earnings is that women are often employed in different occupations than are men, and women's occupations pay less than men's. The major sex difference as shown in Figure 9.2 is that women dominate sales, office, and service occupations, whereas men dominate blue-collar occupations. The proportion of men and women in professional and managerial occupations is equal. Generally, though, men professionals are doctors and women professionals are nurses; men manage steel plants and women manage dry cleaning outlets.

There are three major reasons why men and women have different occupations: gendered occupations, different qualifications, and discrimination.

1. *Gendered occupations.* Many occupations in today's segmented labor market are regarded as either "women's work" or "men's work." Construction is almost exclusively men's work; primary school teaching and day care are largely women's work. These occupations are so sex segregated that many men and women would feel uncomfortable working in a job where they were so clearly the "wrong" sex.

These stereotypes, combined with the low pay of traditionally female fields, keep most men out of these fields. However, growing numbers of women have moved into jobs that used to be reserved for men, such as insurance adjusting, police work, bus driving, and medicine. This does not, unfortunately, signal that women now have increased access to good jobs. Rather, women by and large are moving into jobs that men are abandoning because of deteriorating wages and working conditions (Reskin 1989).

Although sexism continues to have an impact, more and more women are finding employment in fields formerly open only to men.

FIGURE 9.2

Differences in Occupation by Sex, 2004

Men and women continue to be employed in different occupations in the United States. Women predominate in technical, sales, and clerical work, men in blue-collar jobs. Within each category, men typically hold higher-status positions than do women.
SOURCE: U.S. Bureau of the Census 2006.

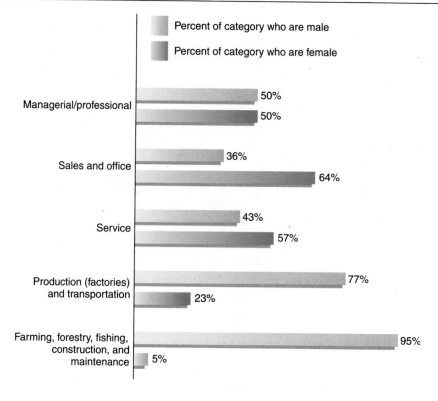

Source: U.S. Bureau of the Census 2006.

2. *Different qualifications.* Although the differences are smaller than they used to be, women continue to major in fields of study that prepare them to work in relatively low-paying fields, such as education, whereas men are more likely to choose more lucrative fields. More important than these differences in educational qualifications are disparities in experience and on-the-job training. Believing that women are likely to quit once they marry, have children, or lose interest, employers invest less in training and mentoring them (Tomaskovic-Devey & Skaggs 2002). As a result, women are less likely to be promoted to management positions—even if they have no intention of having children or marrying.

3. *Discrimination.* Although men and women have somewhat different occupational preparation, a large share of occupational differences is due to discrimination by employers (Hesse-Biber & Carter 2000). Employers reserve some jobs for men and some for women based on their own gender-role stereotypes. As a result, women remain nurses rather than nursing administrators, and salesclerks rather than store managers.

Same Occupation, Different Earnings

Not all occupations are highly sex segregated. Some, such as flight attendant, teacher, and research analyst, contain considerable proportions of both men and women. Within any given occupation, however, men typically earn substantially higher incomes (Table 9.2). There are two main explanations for this: different titles and discrimination.

TABLE 9.2

Sex Differences in Representation and Median Weekly Earnings, by Occupation*

Women are clustered in lower-paying occupations. But even when women have the same occupation as men, they tend to earn substantially less money. Women tend to be employed in lower-paying firms and subfields and to experience discrimination in hiring, raises, and promotion.

Occupation	Male Income	Female Income	% of workers who are women
Chief executives	$1875	$1310	24%
Lawyers	1,710	1255	33
Computer programmers	1151	1006	28
Elementary and middle school teachers	917	776	80
Retail salespersons	386	597	41

*Full time, year-round workers only.

SOURCE: U.S. Bureau of Labor Statistics 2005.

1. *Different titles.* Very often, men and women who do the same tasks are given different titles—women will be maids or executive assistants, and men doing the same work will be janitors or assistant executives. Simply because one job category is considered "male" and is occupied by males it is paid a higher wage.

2. *Discrimination.* Even when women and men have the same job titles, women tend to be paid less. One reason for this is that, within any given occupation, men tend to hold the more prestigious, better-paying positions (Hesse-Biber & Carter 2000; McBrier 2003). Male lawyers tend to be hired in large, high-paying firms to specialize in prestigious fields, whereas women tend to be hired in small, low-paying firms to specialize in prestigious fields, while women tend to work in small, low-paying firms, specializing in less prestigious fields. Male sales staff tend to be hired by stores and departments that offer better salaries or hefty commissions, female sales staff work in less remunerative areas. These differences reflect the segmented labor market (discussed in more detail in Chapter 13).

Even when women and men work in the same occupations and positions; work for the same employers; and have equal education, experience, and other qualifications, women earn less. The absence of any other explanations for this difference has led researchers to conclude that it must be caused by discrimination (Maume 2004).

This discrimination occurs in both female- and male-dominated fields. In female-dominated occupations, women's careers progress gradually. In contrast, men often encounter a "glass escalator" that invisibly helps them to move rapidly into administrative positions and prestigious specialties (Williams 1992; Hultin 2003). In male-dominated occupations, men's careers typically progress gradually, whereas women more often are pressured out of the occupation altogether (Maume 1999). This is often done through subtle discrimination such as exclusion from informal leadership and decision-making networks, sexual harassment, and other forms of hostility from male coworkers (Chetkovich 1998; Jacobs 1989b). This informal discrimination creates a "glass ceiling"—an invisible barrier to women's promotions (Freeman 1990).

Gender and Power

As Max Weber pointed out, differences in prestige and power are as important as differences in economic reward. When we turn to these rewards, we again find that women are systematically disadvantaged. In the family, business, the church, and elsewhere, women are less likely to be given positions of authority.

Unequal Power in Social Institutions

Women's subordinate position is built into most social institutions. In some churches, ministers quote the New Testament command, "Wives, submit yourselves unto your own husbands" (Ephesians 5:22). In colleges, women's basketball coaches are paid less than men's basketball coaches. In politics, prejudice against women leaders remains strong, and women still comprise only a minority of major elected officials in the United States and around the world. (See Map 9.1.)

Unequal Power in Interaction

As we noted in Chapter 4, even the informal exchanges of everyday life are governed by norms; that is, they are patterned regularities, occurring in similar ways again and again. Careful attention to the roles men and women play in these informal interactions shows clear differences—all of them associated with childhood socialization and with women's lower prestige and power.

Studies of informal conversations show that men regularly dominate women in verbal interaction (Tannen 1990). Men take up more of the speaking time, they interrupt women more often, and most important, they interrupt more successfully. Finally, women are more placating and less assertive in conversation than men, and women are more likely to state their opinions as questions ("Don't you think the red one is nicer than the blue one?"). This pattern also appears in committee and business meetings, which is one reason women employees are less likely than men to get credit for their ideas (Tannen 1994).

Laboratory and other studies show that this male/female conversational division of labor is largely a result of status differences (Kollock, Blumstein, & Schwartz 1985; Ridgeway & Smith-Lovin 1999; Tannen 1990). When women clearly have more status than men, such as when a female professor talks with a male student, women do not exhibit low-status interaction styles.

A Case Study: Sexual Harassment

Sexual harassment consists of unwelcome sexual advances, requests for sexual favors, or other verbal or physical conduct of a sexual nature.

The impact of women's relative lack of power becomes clear when we look at the topic of **sexual harassment**—unwelcome sexual advances, requests for sexual favors, and other unwanted verbal or physical conduct of a sexual nature. Although estimates vary widely depending on the definition and sample used, as many as half of all working women probably experience sexual harassment during their lifetime (Welsh 1998). Men also can be sexually harassed—by men as well as by women—but this occurs far less often.

There are two forms of sexual harassment (Shapiro 1994). By law, harassment exists when an employer, teacher, or other supervisor expects sexual favors (from inappropriate touching to sexual intercourse) in exchange for something else: keeping one's job, getting a good grade or letter of recommendation, and so on. Sexual harassment

MAP 9.1
Women in Political Office, 2006
In most countries, fewer than 10 percent of the legislators in the single or lower chamber of congress or parliament are female.
SOURCE: Inter-Parliamentary Union 2006.

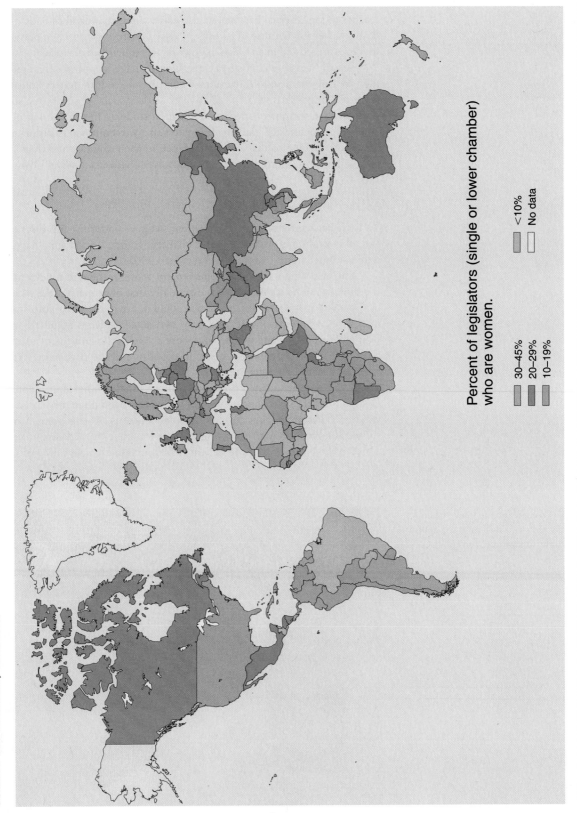

Percent of legislators (single or lower chamber) who are women.

30–45%
20–29%
10–19%
<10%
No data

ranges from subtle hints about the rewards for being more friendly with the boss or teacher to rape. Sexual harassment also exists when an individual finds it impossible to do his or her job because of a hostile sexual climate, such as when pornographic photographs are posted in an office or coworkers frequently make sexist or sexual jokes.

Sexual harassment exists because women have less power than men. (Similarly, men are only harassed in situations where they have little power.) But sexual harassment not only *reflects* women's relative powerless social position, it also helps to *keep* them in that position. For example, women students in engineering classes or firms who experience sexual harassment are less likely to continue to pursue a career in engineering. They also may lose confidence in their abilities and their judgment and may suffer long-lasting psychological troubles (Sadker & Sadker 1994).

Fighting Back: The Feminist Movement

To fight back against sexual harassment, woman battering, job discrimination, and the other problems discussed in this chapter, women—and men—have united in the feminist movement (Evans 2003; Freedman 2002).

The first American feminist movement arose in the mid-nineteenth century. At the time, women's legal status was essentially that of property. Like slaves (both male and female), women regardless of race could not own property, vote, make contracts, or testify in a court of law, and only two small colleges admitted women. Many women (both black and white) who were active in the movement to abolish slavery took from their experience a belief in equality and the organizing skills needed to start the feminist movement.

Because of feminist protest, by the end of the nineteenth century, the most egregious legal restrictions on women's lives had been lifted, and a growing (though still small) list of colleges accepted women students. At this point, feminist activity shifted almost entirely to obtaining the vote (suffrage) for women. In 1920, Congress adopted the Nineteenth Amendment, which granted female suffrage.

■ Women and men have taken to the streets—as well as to the courts—to fight to improve women's lives.

© Cleve Bryant/PhotoEdit

After the passage of the Nineteenth Amendment, feminist activity declined precipitously. In the 1960s, however, two groups of women began pressing for further change. The first group consisted of middle-aged, middle-class women who were brought together in mainstream political and professional organizations and who began organizing to end job discrimination and legal barriers that held women back. The second group consisted of young women activists brought together in the civil rights and anti-Vietnam War movements. These women quickly grew tired of making coffee and typing leaflets while male activists made all the decisions, and they quickly came to understand the parallels between racism and sexism.

At its core, the feminist movement holds that women and men are deserving of equal rights, and that women's lives, culture, and values are as important as are men's. Liberal feminists typically work to ensure that women have equal opportunity in all fields of life; because of feminist activism, academically select high schools and colleges no longer can refuse to admit female students. Radical feminists emphasize the need to change basic structures of society, not just ensure that women have equal access to existing structures. Radical feminists have, for example, led the battle against woman battering and for women's studies programs and research; 30 years ago, before the modern feminist movement took root, no sociology textbook would have included a chapter on sex and gender.

The Sociology of Sexuality

Like gender, sexuality is also a product of both biology and culture. Ideas about "proper" sexuality vary cross-culturally, and have varied historically. A hundred years ago, a woman who admitted to enjoying sexual pleasure could have been declared insane and locked in a mental hospital. Now, a woman who does *not* enjoy sexual pleasure may be labeled frigid and referred to a therapist. In ancient Greece, male youths were expected to engage in homosexual behavior with their adult male mentors; these days, adults (of either sex) who have sexual relations with minors can be imprisoned. In this section, we look at current sexual behavior in the United States.

Premarital Sexuality

In few areas of our lives are we free to improvise. Instead, we learn social scripts that direct us toward accepted behaviors and away from unaccepted ones. Sex is no exception.

One of the most important social scripts about sexuality has to do with when and under what circumstances it is okay to have sexual intercourse. Premarital intercourse has become increasingly accepted over the last few decades (Ku et al. 1998; Abma et al. 2004). Moreover, whereas in the 1950s couples typically only had sex if they intended to marry, now teens may "hook up" with no intention of even having a relationship.

Similarly, the proportion of never-married teenagers who say that they have had sexual intercourse increased from about 40 percent in the 1950s to about 50 percent for girls and 60 percent for boys by the late 1980s (Abma et al. 2004). Since then, however, rates of sexual intercourse among teens have declined slightly, to about 46 percent among both boys and girls (Abma et al. 2004). What explains this decline?

The answer is definitely *not* the abstinence-only sexual education programs that now dominate in the United States. Research consistently finds no credible evidence that such programs work except in the very short term (Dailard 2003).

More likely, the drop in teenage sexual activity reflects the growing awareness of the threats posed by AIDS and other sexually transmitted diseases. Not surprisingly, the percentage of teenagers who report using condoms the last time they had sexual intercourse has increased steadily since 1988. It is now common for young people to use condoms the first few times they have sexual relations with a new partner. After that, though, most conclude that they know and can trust their partners and so abandon condom use. Women are especially likely to believe that their partner loves them and wouldn't hurt them; men are especially likely to believe that they are invulnerable and don't need to worry. Unfortunately, it is usually impossible to know if someone has a sexually transmitted disease unless they admit it. But many individuals don't know they are infected, while others know but don't tell.

Marital Sexuality

In certain important ways, the sexual scripts followed by married couples have changed little over time. For example, frequency of sexual activity seems to have changed very little among married people over the years (Call, Sprecher, & Schwartz 1995; Laumann et al. 1994). And, now as in the past, most couples find that the frequency of intercourse declines steadily with the length of the marriage. The decline appears to be nearly universal and to occur regardless of the couple's age, education, or situation. After the first year, almost everything that happens—children, jobs, commuting, housework, finances—reduces the frequency of marital intercourse (Call et al. 1995). Nevertheless, satisfaction with both the quantity and the quality of one's sex life is essential to a good marriage (Blumstein & Schwartz 1983; Laumann et al. 1994).

Despite these historical continuities, the sexual scripts followed by married couples have undergone some important changes in recent decades. First, oral sex, a practice that was limited largely to unmarried sexual partners and the highly educated in earlier decades, is now more common. Second, women and men are now equally likely to have extramarital affairs. The double standard has disappeared in adultery, and recent studies suggest that as many as 50 percent of both men and women have had an extramarital sexual relationship (Laumann et al. 1994).

Sexual Minorities

Although the majority of the population is heterosexual—preferring sex and romance with the opposite sex—significant minorities diverge from this script. This section discusses homosexuals and transgendered persons.

Homosexuality in Society

The largest of the sexual minorities is homosexuals (also known as gays and lesbians). **Homosexuals** are people who prefer sexual and romantic relationships with members of their own sex. On well-regarded surveys, somewhere between 2 and 6 percent of Americans admit recent homosexual activity or describe themselves as homosexual, with rates about twice as high among men as among women (Binson et al. 1995; Lauman et al. 1994). Considerably more report ever engaging in homosexual activity, and it is likely that many more are unwilling to honestly report their behavior and preferences.

Homosexuals are people who prefer sexual and romantic relationships with members of their own sex.

Attitudes toward homosexuality have fluctuated greatly over time. During the last 50 years, however, American attitudes have become increasingly more positive. In a Gallup Poll conducted in 2006, 56 percent of surveyed Americans agreed that homosexual activity between consenting adults should be legal. Support for gay rights is highest among persons who are less religious, younger, urban dwellers, non-Southerners, more educated, and more liberal in general.

The Gay and Lesbian Rights Movement

Growing acceptance of homosexuality is a direct outgrowth of the gay and lesbian rights movement. The American gay and lesbian rights movement grew rapidly in the late 1960s and early 1970s, when gays and lesbians who had worked in the civil rights and feminist movements began questioning why they too should not have equal rights (Clendinen & Nagourney 2001; Marcus 2002).

The pivotal moment for the incipient gay rights movement was the Stonewall Riots, which began June 27, 1969. For many years before that date, the police had routinely raided gay bars in New York City. But something was different that night: This time the bar's patrons fought back. The police responded brutally, but the riot only grew, with about 2,000 people from the heavily gay and lesbian neighborhood joining in over the next few days. By the time the riots ended, the modern gay rights movement had come of age.

The AIDS epidemic also played an important role in the history of the movement. When AIDS was first identified in 1981, many erroneously labeled it a "gay plague," and both prejudice and discrimination increased. As gay men were forced by their illness to reveal their sexual identity or were identified as gay after they died of AIDS, heterosexuals came to realize how many of their friends, relatives, coworkers, neighbors, and favorite film stars (like Rock Hudson) were gay. As a result, stereotypes and prejudices often fell by the wayside.

The gay and lesbian rights movement has achieved some notable successes. The American Psychological Association no longer considers homosexuality *per se* an illness and, as of mid-2005, 30 states outlaw discrimination on the basis of sexual orientation at least in public employment (National Gay and Lesbian Task Force 2006); furthermore, acknowledged homosexuals have been elected to public office, including the U.S. senate. Most importantly, in 2003 the U.S. Supreme Court declared that states could no longer criminalize private, consensual, same-sex activities. Currently, the hottest battles are being fought over the right of gays to marry or enter into civil unions.

Transgender in Society

Transgendered persons are individuals whose sex or sexual identity is not definitively male or female. There are two main types of transgendered people: intersex persons and transsexuals.

Intersex persons are individuals who are born with ambiguous genitalia, such as a small penis as well as ovaries. Intersexuality is a naturally occurring, if rare, phenomenon. In the early stage of fetal development, all fetuses are sexually ambiguous. All fetuses (and adult humans) produce both male and female hormones (including estrogen and testosterone), and these hormones lead to sexual differentiation—the development of ovaries, penises, and so on—later in fetal development. Intersexuality occurs when that differentiation is incomplete. When such cases are identified, doctors typically use surgery or hormones to transform the individual's body into one that more closely matches our accepted ideas of what males or females should

Transgendered persons are individuals whose sex or sexual identity is not definitively male or female.

In February 2004, the city of San Francisco began issuing marriage licenses to gay and lesbian couples. This couple received one of the first licenses.

look like. As when plastic surgeons give women larger breasts, these medical interventions serve to reinforce social ideas about proper sexuality. In contrast, some other cultures recognize the existence of more than two sexes (Herdt 1994; Lorber 1994).

Unlike intersex persons, transsexuals' sex is not ambiguous: There are no observable biological differences between them and other heterosexual males or females. Instead, transsexuals are persons who psychologically feel that they are trapped in the body of the wrong sex. As with intersex persons, most doctors consider it appropriate to prescribe hormones or perform surgery (removing penises and constructing vaginas or vice versa) to give transsexuals the bodies they desire. Some observers, however, question the wisdom of these medical interventions (Meyerowitz 2002). They wonder whether, in a society that allowed both men and women more freedom, anyone would feel "trapped" in the wrong body, and they question whether there is really something so wrong with men who enjoy "chick flicks" and taking care of children, or with women who prefer wearing crew cuts and working on cars. To these observers, the medical treatment of transsexuality is another example of the social construction of both gender and sexuality.

Where This Leaves Us

Gender roles have changed dramatically over the last 30 years, in ways that have affected us deeply. As structural functionalists point out, traditional roles had their virtues. Everyone knew what was expected of them, and complementary male/female roles held families together by forcing each sex to depend on the other. In contrast, the decline in traditional gender roles has brought stress to many people—not only to men who lost rights and power and but also to women who found themselves caught between changing expectations.

But conflict theorists are also correct: Everyone did not benefit equally from traditional roles, and everyone paid some price for maintaining them. Women endured lower earnings, narrow educational and occupational opportunities, sexual harassment, sexist prejudice and discrimination, and, sometimes, physical violence. Men who held to traditional masculine gender roles experienced more stress, less nurturing relationships, and shorter lives.

Sex is a biological category, something we are born with. But sex, gender, and sexuality are also socially constructed. Doctors can change patients' physical bodies so that individuals' sex and gender better fit social expectations. Society, in general, continually evolves its ideas of what it means to be male and female, masculine and feminine, and all of us contribute to this process when we socialize our children, "do gender," and interact with each other. Creating a more just world will require that we change the social structure of gender and sexuality as well as its interpersonal aspects.

Summary

1. Although there is a universal biological basis for sex differentiation, a great deal of variability exists in the roles and personalities assigned to men and women across societies. In almost all cultures, however, women have less power than men.

2. Structural-functional theorists argue that a division of labor between the sexes builds a stronger family and reduces competition. Conflict theorists stress that men and capitalists benefit from sexism and a segmented labor market that relegates women to lower-status positions. Symbolic interactionism does not address why gender inequality arose but does help us understand how it is perpetuated in interaction.

3. Sex stratification is maintained through socialization. From earliest childhood, females and males learn ideas about sex-appropriate behavior and integrate them into their self-identities. Sex stratification is also maintained by medical and social practices that magnify biological differences between the sexes.

4. Gender is not simply an individual attribute. It is also a property built into social structures and built into our everyday actions. Sociologists use the term "doing gender" to refer to everyday activities that individuals engage in to affirm that they understand what is expected of them as male or female.

5. Men as well as women face disadvantages due to their gender roles. For men, these include higher mortality and fewer intimate relationships.

6. Women and men are growing more similar in their educational aspirations and attainments and in the percentage of their lives that they will spend in the work force.

7. Women who are full-time, full-year workers earn 76 percent as much as men. This is because they have different (poorer-paying) occupations and because they earn less when they hold the same occupations. Causes include different educational preparation and discrimination.

8. Women's subordinate position is built into all social institutions. Although some of this has changed, men disproportionately occupy leadership positions in social institutions. They also dominate women in conversation.

9. The feminist movement has fought for over 150 years to improve the position of American women and has had many notable successes.

10. Premarital sexuality is now widely accepted. However, it has declined in frequency since the late 1980s, primarily in response to the AIDS epidemic.

11. Homosexuality is growing more accepted in the United States. Some sociologists question whether the medical treatment of transgendered persons reflects and reinforces traditional ideas about gender roles.

Thinking Critically

1. Suppose you want your daughter to consider science as a future profession. How would you go about encouraging her to consider this career choice? As a member of the PTA at your daughter's school, what changes would you encourage her school to make in order to increase the chances of girls considering science as a profession?

2. Chapter 8 discussed institutionalized racism. Consider the parallels between racism and sexism. Can you think of some specific examples of how institutionalized sexism works against women in the workplace? Against men? How specifically might affirmative action pro-grams (also discussed in Chapter 8) help to alleviate this discrimination?

3. If men have more power, why do they die earlier and have higher rates of heart disease, suicide, and alcoholism? As women gain power, should we expect them to have similar health problems? Why or why not?

4. In TV commercials, males predominate about nine to one as the authority figure, even when the products are aimed at women. Using your sociological knowledge, how would you explain this?

Companion Website for This Book

academic.cengage.com/sociology/brinkerhoff
Gain an even better grasp on this chapter by going to the Companion Website. This resource contains tutorial quizzes and flash cards to help you master key terms and concepts.

Suggested Readings

Bingham, Clara and Leedy Gansler, Laura. 2003. *Class Action: The Story of Lois Jenson and the Landmark Case That Changed Sexual Harassment Law.* New York: Anchor books. Describes the brutal sexual harassment received by women miners and their 25-year legal battle for justice. This story was the inspiration for the film *North Country.*

Kimmel, Michael. 1995. *Manhood in America: A Cultural History.* New York: Free Press. An authoritative, entertaining, and wide-ranging history of men in the United States by a respected social scientist in the field of men, manhood, and masculinity.

Orenstein, Peggy. 1994. *School Girls: Young Women, Self-Esteem, and the Confidence Gap.* New York: Doubleday. A very good book, written in association with the American Association of University Women, on young women in today's middle schools. The book is based on participant observation in two California schools.

Weitz, Rose. 2004. *Rapunzel's Daughters: What Women's Hair Tells Us About Women's Lives.* New York: Farrar, Straus and Giroux. Weitz explores how young girls are socialized to see their appearance as central to their identities, how teenage girls use their appearance to manipulate their identities, and how women's appearances affect all aspects of their lives, from careers to romances.

Cross-Cutting Statuses:
Race, Sex, and Age

In the past two chapters on stratification, we have dealt with unequal life chance by race, ethnicity, and sex. For each of these characteristics, we have been able to demonstrate that there is a hierarchy of access to the good things in life and that some groups are substantially disadvantaged. In this section, we briefly discuss how disadvantage accumulates when we jointly consider race/ethnicity, sex, and age—a third cross-cutting status.

Race and Gender

Nonwhite women face a two-pronged dilemma. First, they have not benefited from the sheltered position of traditional white women's roles. Nonwhite women have always worked outside the home: For example, in 1900 married African American women were six times more likely to be employed than were married white women (Goldin 1992). Although they worked, they still had to face the economic and civic penalties of being women. Consequently, minority women traditionally have had less to lose and more to gain from abandoning conventional gender roles. On the other hand, nonwhite women face a potential conflict of interest: Is racism or sexism their chief oppressor? Should they work for an end to racism or an end to sexism? If they choose to work for women's rights, they may be seen as working against men of their own racial and ethnic group.

Nonwhite women face a potential conflict of interest: Is racism or sexism their chief oppressor? If they choose to work for women's rights, they may be seen as working against men of their own racial and ethnic group. If they don't, they face the economic and civic penalties of being women.

Current income figures indicate that sex is more important than race in determining women's earnings: The difference in earnings among Hispanic, African American, and European American women is relatively small compared with the difference between women and men. This suggests that fighting sex discrimination should be more important than fighting racial discrimination. But this conclusion overlooks the dependence of women and children on the earnings of their husbands and fathers. For example, because of the low earnings and limited employment opportunities of African American men, African American women and children are three times more likely than their white counterparts to live below the poverty level. As a result, nonwhite women have much to gain by fighting racism as well as sexism.

The dilemma remains. The women's rights movement is often seen as a middle-class white social movement; racial and ethnic movements have been seen as men's movements. Nevertheless, minority women have a long history of resistance to both racism and sexism.

Aging and Gender

Aging poses special problems for women. First is the problem of the double standard of aging: The signs of age—wrinkles, loose skin, and gray hair—are more stigmatizing on women than on men. In addition, men are assumed to gain maturity and authority with age,

■ Because more women than men survive to old age, older men need not live alone unless they want to.

whereas older women more often are assumed to have lost competence. Thus, age is associated with greater decreases in prestige and esteem for women than for men.

The life expectancy gap between men and women also makes the experience of old age very different for women than for men. On average, women live about 6 years longer than men. Taken together with the fact that women are usually 2 years or so younger than their husbands, this works out to an 8-year gap between when the average woman's husband dies and when she dies. As a result, above age 75, there are nearly twice as many women as men.

This mortality difference has enormous consequences for the quality of life. First, it means that most men will spend their old age married, with a spouse to care for them and to keep them from lingering in a nursing home before they die. The average woman, on the other hand, will spend the last years of her life unmarried and living alone, and will linger in a nursing home before dying (Federal Interagency Forum on Aging Related Statistics 2000). In addition, because women often rely on their husbands' retirement pay, often lose most of that pay when their husbands die, and often must use up their savings taking care of their husbands' medical needs, poverty is very common among older women.

Race and Age

Ethnic minority groups (other than Asians) earn substantially less during their peak working years than do non-Hispanic whites. This means that they are less likely to have accumulated assets such as home ownership to cushion income loss during retirement. In 1999, for instance, the median net worth of older African American households was $13,000 compared with a median net worth of $181,000 for older white households (Federal Interagency Forum on Aging Related Statistics 2000). But the most significant link between minority status and aging is that members of minority groups are less likely to live to experience old age.

Whereas 75 percent of white males can expect to survive until they are 65, only 58 percent of African American males will live that long. For white and African American women, the figures are 85 and 75 percent, respectively.

Another important issue in today's increasingly diverse society is how elderly immigrants experience U.S. culture. Many immigrants, such as those from Asian or Latino countries, have internalized their native culture's belief that age alone should command respect and prestige. However, U.S. culture places high value on activity, productivity, and individual achievement—values that are inconsistent with high prestige for old people. The assimilation of elderly immigrants and their children and grandchildren requires mutual adjustments in age roles and expectations (Lin & Liu 1993; Paz 1993).

CHAPTER 10

Health and Health Care

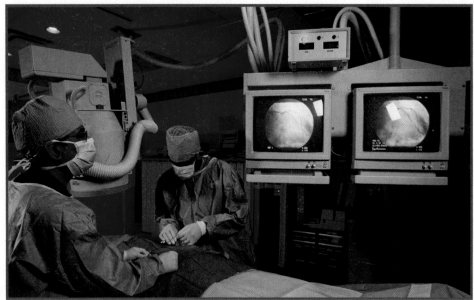

© Richard T. Nowitz/CORBIS

Outline

Health and Health Care as a Social Problem

At first glance, health seems a purely biological state, and health care a purely medical matter. Yet as this chapter will show, health, illness, and health care are deeply affected by social forces and social status.

Although it may seem that health and illness are not issues that need concern college-age students, this is far from true. Illness, disability, and traumatic injury can strike at any age. This is particularly important because the United States is alone among the industrialized nations in not providing access to health care to all citizens. As a result, close to 20 percent of Americans under age 65 lack health insurance, and health-related debt is a major cause of personal bankruptcy (Kaiser Commission 2005; Newman 1999b; Sullivan, Warren, & Westbrook 2000). Furthermore, health is the single most important factor that influences overall quality of life. Thus, we need to consider not only the social forces that affect health and illness but also why the U.S. health-care system has taken the particular form it has and the consequences of that system. We begin by looking at how sociologists think about illness itself.

Theoretical Perspectives on Illness

Because of their different approaches, each sociological theory of illness focuses on a different set of questions and offers a different set of answers. The classic structural-functionalist theory of illness looks at how (some) illness can help society run smoothly and how society limits illness that can interfere with that smooth flow. Conflict theory illustrates how competing interests lead to different definitions of illness, and symbolic interaction theory has been particularly useful for understanding the experience of illness.

Structural-Functionalist Theory: The Sick Role

The classic sociological theory of illness was first formulated by Talcott Parsons (1951). As a structural-functionalist, Parsons assumed that any smoothly functioning society would have ways to keep illness, like any other potential problem, from damaging it.

Parsons's most important contribution to sociology was the realization that illness is a form of deviance, in that it keep individuals from performing their normal social roles. The last time you were sick, for example, you might have taken the day off from work, asked your boyfriend or girlfriend to pick up groceries for you, or asked a professor to give you an extension on a paper. In fact, you might even have claimed to be sick just so you could get off from those obligations. To Parsons, therefore, illness (or claims of illness) is generally *dysfunctional* because it could threaten social stability.

Parsons also recognized, however, that allowing some illness was good for social stability. If no one could ever "call in sick" or take a "mental health day," no one

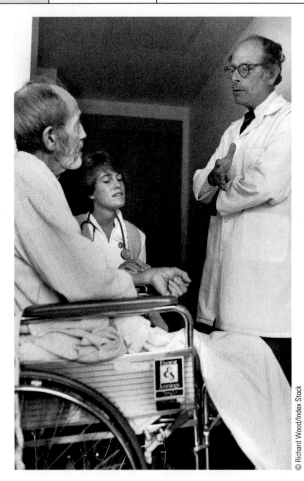

As the sick role describes, when we get sick we are expected to go to the doctor and do what the doctor tells us to get well.

The **sick role** consists of four social norms regarding sick people. They are assumed to have good reasons for not fulfilling their normal social roles and are not held responsible for their illnesses. They are also expected to consider sickness undesirable, to work to get well, and to follow doctor's orders.

would have the time needed to recuperate, and resentment would build among workers, students, and spouses who never got a break. In these ways, illness acts as a sort of "pressure valve" for society.

Defining the Sick Role

How does society control illness so that it increases rather than decreases social stability? The answer, according to Parsons, is the **sick role.** The sick role refers to four social norms regarding how sick people should behave and how society should view them. First, sick persons are assumed to have legitimate reasons for not fulfilling their normal social roles. This is why we give sick people time off from work rather than firing them for malingering. Second, cultural norms declare that individuals are not responsible for their illnesses. For this reason, we bring chicken soup to people who have colds rather than jailing them for stupidly exposing themselves to germs. Third, sick persons are expected to consider sickness undesirable and work to get well. This is why we sympathize with those who rest when they are ill and chastise those who don't. Finally, sick persons should seek and follow medical advice.

Critiquing the Sick Role

Parsons's concept of a sick role was a crucial step in beginning to think of illness sociologically. Subsequent research, however, has illuminated the limitations of the sick role model (Weitz 2007). This critique is highlighted in the Concept Summary.

First, in contrast to Parsons's analysis, ill persons sometimes *are* expected to fulfill their normal social roles. While no one expects persons dying of cancer to continue working, we often expect people with arthritis to do so, as well as those we suspect are malingerers or hypochondriacs because doctors have been unable to diagnose their condition. Similarly, regardless of illness, some professors expect students to turn papers in on time, some husbands expect their wives to cook dinner, and some employers expect their employees to come to work.

Second, sometimes people *are* held responsible for their illnesses. The last time you had a cold, did anyone chastise you for not taking care of yourself well enough? For not taking vitamin C, getting enough sleep, or eating healthy meals? Similarly, newspapers and television shows are full of stories that implicitly blame lung cancer on people who smoke, diabetes on people who eat too much, and so on.

Third, the sick role's assumption that sick individuals should work to get well simply doesn't fit those who have chronic illnesses that medicine can't cure. Similarly, the assumption that sick people should follow medical advice ignores those who can't afford or aren't helped by medical care.

Conflict Theory: Medicalization

Like other structural functionalists, Parsons assumed that social ideas about illness (in this case, the sick role) are designed to keep society running smoothly. In contrast, conflict theorists assert that, like other parts of social life, ideas about illness reflect competing interests among different social groups.

Strengths and Weaknesses of the Sick Role Model

Concept Summary

Elements of the Sick Role	Model fits well:	Model fits poorly:
Illness is considered a legitimate reason for not fulfilling obligations	Appendicitis, cancer	Undiagnosed chronic fatigue
Ill persons are not held responsible for illness	Measles, hemophilia	AIDS, lung cancer
Ill persons should strive to get well	Tuberculosis, broken leg	Diabetes, epilepsy
Ill persons should seek medical help	Strep throat, syphilis	Alzheimer's, colds

One of the major contributions of conflict theory to our understanding of illness is the concept of medicalization. As we saw in Chapter 6, **medicalization** refers to the process through which a condition or behavior becomes defined as a medical problem requiring a medical solution (Conrad 2005). One hundred years ago, masturbation, homosexuality, and, among young women, the desire to go to college were all considered symptoms of illness. These conditions are no longer considered illnesses not because their biology changed but because social ideas about them did. Similarly, one hundred years ago most women gave birth at home attended by midwives, few boys were circumcised, and plump people were considered attractive and lucky. Nowadays, pregnant women are expected to seek medical care, parents are expected to have their infant sons circumcised by doctors, and overweight people are considered to be at risk for illness or even to have the "illness" of obesity. These are all examples of medicalization.

For medicalization to occur, one or more organized social groups must have both a vested interest in it and sufficient power to convince others to accept their new definition of the situation. The strongest force currently driving medicalization is the pharmaceutical industry, which has a vested financial interest in enlarging the market for its products (Conrad 2005). For example, the pharmaceutical industry was the major force behind defining "male sexual dysfunction" as a disease—to be cured by Viagra (Loe 2004). Pressure for medicalization also can come from doctors who hope to enlarge their markets and from consumer groups who hope to stimulate research on or reduce the stigma of ambiguous conditions such as alcoholism or fibromyalgia (Barker 2005; Conrad 2005).

Conversely, doctors sometimes oppose medicalization because they don't want the responsibility for treating a condition (such as wife battering), and consumers sometimes oppose medicalization because they believe a condition is simply a natural part of life (such

Medicalization refers to the process through which a condition or behavior becomes defined as a medical problem requiring a medical solution.

■ Mass marketing of Viagra "sold" both the drug and the idea that impotence was a symptom of the disease "erectile dysfunction disorder."

as menopause). Insurers, too, may support or oppose medicalization, depending on their interests. For example, initially insurers rejected requests for expensive gastric bypass surgery for obese patients, arguing that obesity was not an illness. Now that most insurers have concluded that these surgeries reduce their long-term costs, they support diagnosing obesity as an illness and surgically treating it (Conrad 2005).

In each case, the battle over medicalization was won by the group that could bring the most money, influence, and other forms of power to bear.

Symbolic Interaction Theory: The Experience of Illness

The sick role model helps us understand cultural assumptions for how ill people should behave and how they should be treated by others, whereas conflict theory helps us understand how people come to be defined as ill in the first place. In contrast, symbolic interaction theory is particularly useful for understanding what it is like to live with illness on a day-to-day basis and, especially, what happens when doctors and patients have different definitions of the situation. This issue comes to the fore when doctors and patients disagree over treatment.

To doctors, any patient who does not follow their medical orders is engaging in *medical noncompliance*. Doctors typically assume that they know best how a disease should be treated, and therefore assume that any patient who does not follow their orders is either foolish or ignorant. Research by symbolic interactionists, however, suggests that the issue is far more complex. Some patients don't comply because health care workers offered only brief and confusing explanations of what to do and why. Other patients lack the money, time, or other resources needed to comply. Still others conclude that following medical advice is simply not in their best interests. They may decide, for example, against taking a drug that lowers blood pressure but leaves them unable to achieve erection, that reduces schizophrenic hallucinations but causes obesity, or that brings substantial side effects but seems to have no impact on their symptoms (Lawton 2003).

In sum, what doctors define as medical noncompliance, patients define as rational decision making. When doctors chastise patients for their noncompliance and fail to understand patients' perspectives, patients are likely to become even less willing to follow doctors' orders, creating a self-fulfilling prophesy.

The Social Causes of Health and Illness

In a widely cited article titled "A Case for Refocusing Upstream," sociologist John McKinlay (1994) offers the following oft-told tale as a metaphor for the modern doctor's dilemma:

> Sometimes it feels like this. There I am standing by the shore of a swiftly flowing river and I hear the cry of a drowning man. So I jump into the river, put my arms around him, pull him to shore and apply artificial respiration. Just when he begins to breathe, there is another cry for help. So I jump into the river, reach him, pull him to shore, apply artificial respiration, and then just as he begins to breathe, another cry for help. So back in the river again, reaching, pulling, applying, breathing, and then another yell. Again and again, without end, goes the sequence. You know, I am so busy jumping in, pulling them to shore, applying artificial respiration, that I have *no* time to see who the hell is upstream pushing them all in. (McKinlay 1994, 509–510)

Like the would-be rescuer in this story, doctors have few opportunities to focus upstream and ask why their patients get sick in the first place. Sociologists attempt to answer this question at two levels: the micro-level, in which individuals make choices about adopting behaviors that risk their health, and the macro-level, in which social structures limit the choices available to individuals.

But before we can ask why individuals' health is at risk, we need to know what those risks are. To do so, we need to look at the underlying causes of preventable death.

Underlying Causes of Preventable Death

In a highly-influential article published in the *Journal of the American Medical Association*, Mokdad and his colleagues (2004) reviewed all available medical literature to identify the underlying causes of preventable deaths (that is, deaths caused neither by old age nor by genetic disease). Nine factors—tobacco, poor diet and inadequate exercise, alcohol, bacteria and viruses, polluted workplaces and neighborhoods, motor vehicles, firearms, sexual behavior, and illegal drugs—emerged as underlying almost half of all preventable deaths in the United States (Table 10.1).

Of these nine factors, tobacco is clearly the most important—and is far more important than all illegal drugs combined. Whether smoked, chewed, or used as snuff, tobacco can cause an enormous range of disabling and fatal diseases, including heart disease, strokes, emphysema, and numerous cancers (World Health Organization 1998). About half of all smokers will die because of their tobacco use, with half of these dying in middle age and losing an average of 22 years from their normal life expectancy.

The second most common cause of premature deaths is a high-fat diet, sedentary lifestyle, and resulting obesity. Rates of obesity in the United States have skyrocketed since 1980 and show signs of continuing to increase (Centers for Disease Control and Prevention 2005). The combination of poor diet and insufficient exercise increases

TABLE 10.1
Underlying Causes of Preventable Death in the United States, 2000

Cause	Number	Percentage of All Deaths
Tobacco	435,000	18%
Poor diet and inadequate exercise[a]	100–400,000	5–17
Alcohol	85,000	4
Bacteria and viruses[b]	75,000	3
Polluted workplaces and neighborhoods	55,000	2
Motor vehicles[c]	43,000	2
Firearms	29,000	1
Sexual behavior	20,000	1
Illegal drugs	17,000	1

[a]Estimates vary.
[b]Not including deaths related to HIV, tobacco, alcohol, or illicit drugs.
[c]Includes motor vehicle accidents linked to drug use, but *not* to alcohol use.

SOURCE: Mokdad et al. 2004.

the risks of cardiovascular disease, strokes, certain cancers (of the colon, breast, and prostate), and diabetes, among other problems.

The remaining seven factors cause preventable deaths in a variety of ways. Alcohol and illegal drugs make unsafe sex more likely; alcohol, motor vehicles, firearms, and illegal drugs all contribute to deadly accidents; and alcohol, pollution, unprotected sex, and illegal drugs (when injected) can cause cancer, hepatitis, and other illnesses.

Micro-Level Answers: The Health Belief Model

Why do individuals engage in behaviors that endanger their health? Or, to ask the question more positively, why don't individuals adopt behaviors that will *protect* their health? Sociologists have identified four conditions—known collectively as the **health belief model**—that consistently predict whether individuals will do so (Becker 1974, 1993). These conditions are:

1. Individuals must believe they are at risk for a particular health problem.
2. They must believe the problem is serious.
3. They must believe that adopting preventive measures will reduce their risks significantly.
4. They must not perceive any significant financial, emotional, physical, or other barriers to adopting the preventive behaviors.

The experience of Pittsburgh Steelers quarterback Ben Roethlisberger illustrates this model. In June 2006, Roethlisberger suffered a concussion and numerous other injuries after crashing his motorcycle. He was not wearing a helmet at the time, even though helmets reduce the risk of dying in an accident by at least one third and reduce the rate of brain injury by two thirds (National Highway Traffic Safety Administration 2005).

Following his accident, Roethlisberger vowed never to ride a motorcycle without a helmet again. He now realized that the threat of a crash was real, and that the consequences of a crash could be serious or even fatal. Having crashed head-first into a car's windshield, it now made sense to him that wearing a helmet would signifi-

According to the **health belief model,** individuals will adopt healthy behaviors if they believe they face a serious health risk, believe that changing their behaviors would help, and face no significant barriers to doing so.

The Health Belief Model

Concept Summary

People Most Likely to Adopt Health-Protective Behaviors When They:	Example: Adopting Health Behaviors Likely	Example: Adopting Health Behaviors Unlikely
Believe they are susceptible	40-year-old smoker with chronic bronchitis who believes he is at risk for lung cancer.	16-year-old boy who believes he is too healthy and strong to contract a sexually transmitted disease.
Believe risk is serious	Believes lung cancer would be painful and fatal, and does not want to leave his young children fatherless.	Believes that sexually-transmitted diseases can all be easily treated.
Believe compliance will reduce risk	Believes he can reduce risk by stopping smoking.	Doesn't believe that condoms prevent sexual diseases.
Have no significant barriers to compliance	Friends and family urge him to quit smoking, and he can save money by so doing.	Enjoys sexual intercourse more without condoms.

cantly reduce his risk of death or brain injury. And when weighed against these potential benefits, the cost and discomfort of a helmet and the potential threat to his "tough guy" image if he wore one no longer seemed like important barriers.

Macro-Level Answers: The Manufacturers of Illness

At first glance, it's easy to conclude that poor individual choices explain most or even all preventable deaths. After all, like Ben Roethlisberger, other people also weigh their options and then choose to smoke tobacco, use firearms, engage in risky sex, and so on. But those choices are made in a broader social context. If we look more closely at that social context, we quickly come to what McKinlay (1994) describes as the **manufacturers of illness:** groups that promote deadly behaviors and social conditions. For example, cigarettes, beer, fast cars, good rifles, and sugary foods are inherently appealing to many people. But it is the manufacturers of these goods that largely determine how safe or dangerous their products will be, to whom and how they will be advertised, and where they will be sold. For example, car manufacturers have fought against bumpers that would make SUVs less dangerous to other cars, soda manufacturers have fought for the right to sell their high-calorie products in schools, and tobacco manufacturers have (implicitly) promoted smoking to teens and children through such tactics as the Joe Camel campaign and sponsoring youth-oriented concerts and music festivals.

Individual choice is even less a factor for the other underlying causes of death (Weitz 2007). People work with dangerous pesticides, inject illegal drugs that they don't know have been cut with dangerous chemicals, and live in apartments with lead in the water pipes because they lack alternatives. Manufacturers of illness in these circumstances include corporations that expose their workers to dangerous conditions, landlords who don't maintain their buildings, and politicians who oppose legalizing drugs so the drugs can be regulated. Finally, individuals are most likely to engage in unsafe behaviors—from eating doughnuts to shooting crack and having sex without condoms—if they feel they have nothing to look forward to anyway. These feelings are most common among those who are trapped at the bottom of the social class system.

As the health belief model suggests, these boys are unlikely to stop smoking because they are unlikely to believe—or even know—that smoking places them at risk for lung cancer and other serious diseases.

The **manufacturers of illness** are groups that promote and benefit from deadly behaviors and social conditions.

The Social Distribution of Health and Illness

Good health is not simply a matter of good habits and good genes. Although both elements play important parts, health is also strongly linked to social statuses such as gender, social class, and race or ethnicity. In this section, we provide an overview of how these statuses affect health in the United States and then briefly examine how changes in social structure have affected life expectancy in the former Soviet Union.

In the United States, the average newborn can look forward to 78 years of life (U.S. Bureau of the Census 2006). Although some will die young, the average person

TABLE 10.2
Life Expectancy by Sex, Race, and Family Income, United States, 2003
Generally, people of higher status live longer and have better health. White Americans report better health and live longer than do African Americans, and those with higher incomes report better health and live longer than those with lower incomes. The exception is sex: Despite their higher status, men have lower life expectancy than women. Men are as likely as women to report being in fair or poor health, but women report more disability and illness overall.

	Life Expectancy at Birth, 2003	Percentage Reporting Fair or Poor Health
Sex		
Male	75	9
Female	80	10
Race		
White	78	9
African Americans	73	14
Hispanic	NA	13
Asian	NA	7
Family income		
Poor	NA	20
Near poor	NA	15
Not poor	NA	6

SOURCE: U.S. Department of Health and Human Services 2005.

in the United States now lives to be a senior citizen. This is a remarkable achievement given that life expectancy was less than 50 years at the beginning of the twentieth century. Not everyone has benefited equally, however: Men, African Americans, and poorer people on average die younger than women, whites, and more affluent individuals (Table 10.2).

There is much more to health, of course, than just avoiding death. The distribution of illness and disability is at least as important as the distribution of mortality in evaluating a population's overall well-being. For every person who dies in a given year, many more experience serious illness or disability that affects the quality of their lives. In the following sections, we consider why and how gender, social class, and race/ethnicity are related to illness and mortality.

Gender

On average, U.S. women live about 5 years longer than U.S. men (see Table 10.2). Yet although women live longer, they also report significantly worse health than men at all ages: more arthritis, asthma, diabetes, cataracts, and so on (Lane & Cibula 2000; Rieker & Bird, 2000). These differences mean that women more often than men are plagued by disability and discomfort as they age. In part because of this combination of longer lives and more illnesses, women are considerably more likely than men to eventually enter a nursing home.

How can we explain why, as the saying goes, "women get sicker but men die quicker"? The answer lies in both biology and society. Probably because of their hormones, females are inherently stronger than males: As long as females receive sufficient food and caring, their chances of survival are greater than males at every stage of life from conception onward (Rieker & Bird 2000).

Social norms also protect women from fatal disease and injury (Rieker & Bird 2000). Odds are you know a lot more young men than young women who enjoy fast driving, dare-devil sports, slugging whiskey, or slugging others. These and other similar behaviors, all of which increase the chances of death, are socially encouraged for males but discouraged for females. Men are also more likely than women to use illegal drugs and to work at dangerous jobs. Finally, women are more likely than men to seek health care when they experience problems, although this has only a small impact on their overall health status.

It is less clear why, despite their lower chances of dying at any given age, women have higher rates of illness than do men (Barker 2005). Most likely, women's higher rates of illness stem from both their hormones (a biological effect) and the fact that, on average, they experience more stress than do men but have less control over the sources of that stress (a social effect). Because stress makes it more difficult for the body's immune system to function, it often leads to ill health.

Social Class

The higher one's income and social class, the longer one's life expectancy and the better one's health (see Table 10.2): Wealthy people live longer on average than do middle-class people, and middle-class people live longer than do poor people (Marmot 2004). This is true even in countries where everyone has access to health care, and even when we compare only people who have similar rates of smoking, obesity, and alcohol use (Banks et al. 2006). Moreover, these differences begin in infancy and childhood. For example, about 50 percent of children in New York City's homeless shelters have asthma, compared with 25 percent of children in the city's poorest neighborhoods and 6 percent of the city's children overall (Pérez-Peña 2004).

The reasons for the link between social class and illness are complex (Robert & House 2000; Marmot 2004; Wilkinson 2005). They are partially attributable to poorer people's inability to afford expensive medical care. However, environmental, economic, and psychosocial factors play even stronger roles in linking poverty with ill health. Lower-income people are more likely to live in unsafe and unhealthy conditions, near air-polluting factories, or in substandard housing. They are more likely to hold dangerous jobs and to lack sufficient, good-quality food. Low-income people also experience more stress than others but have less control over the causes of that stress. As a result, like women across income brackets, they are more likely to experience illness due to stress. In addition, whereas upper-income persons might cope with stress by taking a vacation or hiring a maid to help out at home, lower-income people have few such options. Instead, some will try to cope with stress through drinking, smoking, and other calming but health-risking behaviors.

Race and Ethnicity

Although income affects health more than does race and ethnicity (Weitz 2007), the latter nonetheless has a strong and independent effect. Asian Americans of Chinese, Japanese, Filipino, or Indian heritage typically are at least middle class and experience health at least as good as that of whites; the prognosis for recent, poorer immigrant groups from Southeast Asia remains unclear. In contrast, African Americans, Hispanic Americans, and Native Americans are on average poorer than non-Hispanic whites, and primarily as a result suffer disproportionately from the effects of low socioeconomic status on health. Because of lower incomes, these nonwhites are significantly more likely than whites or long-established Asian groups to lack health insurance (Kaiser Commission 2004). They are also more likely to experience stress and to live or work in areas

Hurricane Katrina vividly illustrated for all of us how race and social class combine to heighten the risk of illness, disability, and death for poor minority group members. This photo shows a man who needed medications waiting with other flood victims at the Convention Center in New Orleans.

contaminated by soot, carbon monoxide, ozone, sulfur, pesticides, and even radioactive wastes. For example, *60 percent* of all American children who have dangerously high levels of lead in their blood are African American, and only 17 percent are white non-Hispanic (Meyer et al. 2003).

In addition, regardless of income, the prejudice and discrimination experienced by minorities increases their rates of illness and death (Williams 1998; Williams & Jackson 2005). For example, because of racial segregation, even middle-class African Americans are more likely than whites to live in neighborhoods where violence and pollution threaten their health. Similarly, regardless of patients' symptoms or insurance coverage, doctors are more likely to offer white patients various life-preserving treatments (including angioplasty, bypass surgery, and the most effective drugs for HIV infection) and more likely to offer minorities various less desirable procedures, such as leg amputations for diabetes (Nelson, Smedley, & Stith 2002).

Taken together, these factors lead African Americans, Hispanics, and Native Americans to have significantly higher rates of illness and higher chances of dying at any given age than do whites.

Age

Not surprisingly, age is the single most important predictor of health, illness, and death. The two groups most at risk are the very young and the very old.

In poor countries, deaths are very common among infants and children younger than age 5. Some die because they are born prematurely, others because they do not get enough food, and still others because their immune systems are unable to fight disease, especially if they are malnourished.

Deaths of young children were also common in the Western world before the twentieth century. These days, such deaths are very rare in the United States, and young people are typically healthy. Compared with other developed nations, however, infant mortality remains shockingly high (Table 10.3). Infant mortality is especially high among African Americans, who (for all the reasons just discussed) are 2.5 times more likely to die in infancy than are whites (National Center for Health Statistics 2005).

Once past infancy, the chances of dying or developing a disabling illness only begin to rise gradually beginning at about age 40. By age 65, most people will have at least one long-lasting health problem, such as arthritis, hypertension, or hearing loss (Federal Interagency Forum on Aging Related Statistics 2000). Nevertheless, demographers estimate that three fourths of the years of life remaining after age 70 will be spent in good enough health to permit independent living in the community (Crimmins, Hayward, & Saito 1994). However, the odds of enjoying a healthy old age are significantly lower for racial and ethnic minorities: 74 percent of non-Hispanic whites aged 65 or older consider themselves healthy, compared with 65 percent of Hispanics and only 59 percent of African Americans (Federal Interagency Forum on Aging Related Statistics 2000).

A Case Study: Declining Life Expectancy in the Former Soviet Union

The single most important social factor affecting mortality is the standard of living—access to good nutrition, safe drinking water, and adequate housing free from environmental hazards. Differences in living standards help to explain why African American infants in the United States are more than twice as likely as white infants to die in their first year of life and why the average life expectancy of African American men is 7 years less than that of the average white non-Hispanic male. Differences in living standards also help to explain why, on average, Americans can expect to live 37 years longer than citizens of Sierra Leone (Population Reference Bureau 2006).

Throughout the world, improvements in living standards have been accompanied by increased life expectancy. Consequently, the precipitous decline in life expectancy in the former Soviet Union over the last 15 years is one of the most surprising current developments in world health; life expectancy for Russian men is now only 59 years—far lower than before the collapse of the Soviet Union and far lower than in other developed nations (Population Reference Bureau 2006).

What explains this shocking drop in life expectancy? First, during its decades as a dictatorship, the Soviet Union put industrial development above environmental protection. As a result, the countries of the former Soviet Union are now plagued by extensive environmental pollution. This has significantly raised rates of cancer and respiratory diseases, especially in the most industrialized regions (Cockerham 1997; Haub 1994). Second, as we've seen, stress is often an underlying cause of illness. After the collapse of the Soviet Union, incomes plummeted, social services ground to a halt, political uncertainty and corruption increased, and an entire way of life evaporated. The resulting rise in stress levels directly explains much of the increase in deaths. In addition, this stress also fostered sharp increases in smoking and drinking, with resulting deaths from disease, violence, and accidents. Finally, the former Soviet Union had never invested much in health care for the chronic illnesses, such as heart disease, that now cause most deaths. Since the collapse of the Soviet system, the quality of the health-care system has only worsened, further increasing death rates.

In sum, in the former Soviet Union as in the United States and elsewhere, social conditions are closely tied to health, illness, and death.

Mental Illness

So far we have talked about health and illness as if the only thing that matters is *physical* health. But mental health is also a crucial issue, affecting millions of people each year.

How Many Mentally Ill?

National random surveys of the U.S. population suggest that during the course of any given year, approximately 11 percent of working age adults experience a minor but still-diagnosable mental illness, and another 20 percent experience a moderate or severe illness (Kessler et al., 2005a). The most common illnesses are major depression and problems with alcohol use. These estimates, however, are probably a bit high, since they are based on reports of symptoms, not medical diagnoses of illnesses (Horwitz 2002). Survey researchers can't know, for example, if someone has lost

TABLE 10.3
Infant Mortality Rates per 1,000 Live Births

Singapore	1.9
Hong Kong	2.5
Japan	2.8
Sweden	3.1
Finland	3.1
Norway	3.2
Spain	3.6
Czech Republic	3.7
France	3.9
Netherlands	4.1
Portugal	4.1
Germany	4.3
Switzerland	4.3
Denmark	4.4
Belgium	4.4
Austria	4.5
Australia	4.5
Italy	4.8
Ireland	4.8
Israel	5.1
Greece	5.1
United Kingdom	5.2
Canada	5.4
New Zealand	5.6
U.S. white non-Hispanic	**5.8**
Cuba	5.8
U.S., all races	**6.6**
Hungary	6.6
Poland	6.8
Slovakia	7.8
Chile	7.8
Costa Rica	9.3
Kuwait	9.6
Puerto Rico	9.8
Azerbaijan	9.8
Bulgaria	11.6
Russia	12.4
U.S. blacks	**13.8**
Uruguay	15.0
Romania	16.7
Thailand	20.0
Mexico	24.9

SOURCE: Population Reference Bureau 2006.

American Diversity

focus on

Changing Populations, Changing Health Implications

With each passing year, fewer American babies are born and more Americans turn 65. At the same time, the white population is shrinking while the Hispanic and nonwhite populations are growing. What are the combined consequences of these two population trends for health and health care in America?

One obvious result of having more older Americans and more nonwhite Americans is that in the future there will be more Americans who are both older *and* nonwhite. This will have many important consequences, for the experience of old age is substantially different for nonwhite compared with white Americans (Takamura 2002). Most importantly, minority elderly are more likely than others to be poor: 26 percent of African American elderly live below the poverty line, compared with 21 percent of Latino elderly and only 8 percent of white non-Hispanic elderly. This has serious implications for health and health care, because poorer persons are both more likely to need services and less likely to have health insurance and less able to pay for them out of pocket.

But the problems extend beyond those who live in poverty. Even when incomes are equivalent, and after controlling for education, age, sex, marital status, and urban residence, minority elderly are still more likely than white elderly to lack health insurance. As a result, they find it more difficult to get the medical treatment they need, and their health problems are more likely to spiral out of control, making them more difficult (and expensive) to treat in the long run.

Finally, even if they are able to obtain health care, cultural barriers may make that health care less effective than it would otherwise be (Capitman 2002; Hayes-Bautista, Hsu, & Perez 2002; Takamura 2002). When health-care workers and health-care clients do not speak the same language, communication will necessarily be poor. Doctors and nurses may not understand what their patients need, and patients may not understand what their health-care providers want them to do. In these circumstances, patients easily become dissatisfied, do not follow instructions, and do not re-turn for follow-up care. In turn, care providers may come to regard patients as noncompliant and either unintelligent or unmotivated. This is obviously a problem in the current situation, in which most doctors are white non-Hispanics but increasing numbers of patients are nonwhite. Conversely, communication is also a problem for elderly white patients who now live in nursing homes. Although most doctors and nurses are white, day-to-day care in nursing homes is primarily left to poorly paid nurse's aides, virtually all of whom are nonwhite and most of whom are immigrants. For all these reasons, policymakers will need to pay close attention to both these population changes.

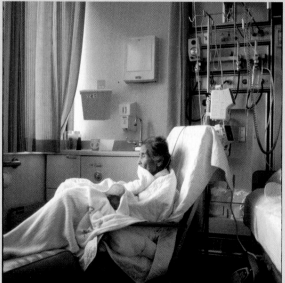
As the number of minority elderly increase, we are likely to see an increased number of elderly people who are poor, who lack health insurance, and who face cultural barriers in interacting with health care practitioners.

weight because of depression or because they are getting ready for a wrestling match.

Who Becomes Mentally Ill?

As with physical illness, social factors strongly predict mental illness. We focus here on two important factors: social class and gender.

Social Class Differences

Since the 1920s, when the first sociological studies of mental disorder were conducted, research has been clear: Poorer people experience more mental illness than wealthier people (Eaton & Muntaner 1999). Researchers are divided, however, about the reasons for this. Some argue that the social stress associated with lower-class life causes mental disorder. Others believe that the onset of mental illness causes people to lose their jobs and drift downward in social class.

Research clearly shows that the lower class does, in fact, experience more of the types of stress (such as job loss or chronic physical disabilities) that can cause mental disorders (Turner, Wheaton, & Lloyd 1995; Turner & Avison 2003; Ali & Avison 1997). The stresses of poverty and economic insecurity appear to be particularly important in understanding the causes of disorders such as major depression.

At the same time, research shows that the onset of disorders such as schizophrenia makes it difficult for people to keep a job. Not only may individuals lose their initial job, once potential employers discover that an individual has a history of mental disorder, they may be reluctant to hire him or her for anything other than a minimum wage job (Link et al. 1987, 1997). In these cases, a mental disorder clearly causes people to drift into a lower social class. Social drift, however, explains a lower proportion of mental illness than does the stress of lower social class life.

In addition to social class differences in *rates* of mental illness, there are also important differences in the *experience* of mental illness. Lower-class persons diagnosed with mental illness are hospitalized for longer periods of time and receive less effective types of treatment. In fact, most mental health treatment is given to middle-class persons experiencing short-term emotional problems, rather than to persons (of whatever social class) who are seriously mentally ill. Meanwhile, as funding for hospitals and health care has declined, lower-class mentally ill persons increasingly have been sent to jails or prisons rather than to clinics or hospitals when their behavior becomes socially unacceptable; experts estimate that between 10 and 20 percent of all U.S. prisoners are mentally ill (Human Rights Watch 2003).

Gender Differences

Depression is the most common form of mental illness, affecting about 17 percent of all adults living in the United States (Kessler et al. 2005). Because depression is so common and because it is much more commonly diagnosed in women, the overall rates of mental illness are higher for women than for men.

Why are women more likely than men to be diagnosed with depression? Most theorists hypothesize that women have higher rates of depression because they experience more stress *and* have less control over that stress (Horwitz 2002, 173–179). In fact, rates of depression are highest among those women with the least control over their lives: nonworking women and married mothers. A waitress with young children and a husband who expects a hot meal when he gets home, for example, has few means for controlling her life, schedule, or stress levels. By the same token, depression is especially common among men who have less power than their wives, have little control over their work, or lose their jobs.

In contrast, men are more likely than women to report substance abuse and "personality disorders" characterized by chronic maladaptive personality traits, such as compulsive gambling or violence (Kessler et al. 2005). Scholars theorize that because the traditional male role encourages men to respond to stress with aggression or substance abuse, those who experience stress and mental illness are more likely to develop these sorts of symptoms.

Working in Health Care

As in any other area of social life, health care has its own set of roles, statuses, and battles over power. In this section, we look at the two most important health care occupations, physicians and nurses, and discuss how each has attempted to obtain and maintain its position in the health-care hierarchy.

Physicians: Fighting to Maintain Professional Autonomy

Less than 5 percent of the medical workforce consists of physicians. Yet they are central to understanding the medical institution. Physicians are responsible both for defining ill health and for treating it. They set the standards for how patients should behave and play a crucial role in setting hospital standards and in directing the behavior of the nurses, technicians, and auxiliary personnel who provide direct care.

As will be described in Chapter 13, a profession is a special kind of occupation that demands specialized skills and permits creative freedom. No occupation better fits this definition than that of physician. Until about one hundred years ago, however, almost anyone could claim the title of physician; training and procedures were highly variable and mostly bad (Starr 1982). Some doctors were almost illiterate, many learned to doctor through apprenticeships, and most of the rest learned through brief courses where virtually anyone who could pay the fees could get certified. With the establishment of the American Medical Association in 1848, however, the process of professionalization began; the process was virtually complete by 1910, at which point strict medical training and licensing standards were adopted.

Understanding Physicians' Income and Prestige

The medical profession provides an example for stratification theories. As of 2004, family practitioners earned a median net salary of $156,000, and general surgeons earned an average of $283,000 (U.S. Bureau of Labor Statistics 2006a). Why are physicians among the highest-paid and highest-status professionals in the United States?

According to structural-functionalists, there is a short supply of persons who have the talent and ability to become physicians and an even shorter supply of those who can be surgeons. Moreover, physicians must undergo long and arduous periods of training. Consequently, high rewards must be offered to motivate the few who can do this work to devote themselves to it. The conflict perspective, on the other hand, argues that the high income and prestige accorded physicians have more to do with physicians' use of power to promote their self-interest than with what is best for society.

Central to the debate on whether physicians' privileges are deserved or are the result of calculated pursuit of self-interest is the role of the American Medical Association (AMA). The AMA sets the standards for admitting physicians to practice, punishes physicians who violate the standards, and lobbies to protect physicians' interests in policy decisions. Although less than half of all physicians belong to the AMA, it has enormous power. One of its major objectives is to ensure the continuance of the free market model of medical care, in which the physician remains an independent provider of medical care on a fee-for-service basis. In pursuit of this objective, the AMA has consistently opposed all legislation designed to create national health insurance, including Medicare and Medicaid. It has also tried to ban or control a variety of alternative medical practices such as midwifery, osteopathy, and acupuncture (Weitz 2007).

The Changing Status of Physicians

Although physicians have succeeded in maintaining high incomes, they have done less well in maintaining other professional privileges. Until the 1970s, most physicians worked as independent providers with substantial freedom to determine their conditions of work. They also benefited from high public regard; some patients considered them a nearly godlike source of knowledge and help. Much of this is changing. The many signs of changes include the following (Coburn & Willis 2000; Weitz 2007):

The popularity of doctor play sets and the large number of children who aspire to medical professions demonstrate the continuing prestige of doctors in contemporary society.

- A growing proportion of physicians work in group practices or for corporations, where fees, procedures, and working hours are determined by bureaucrats. As a result, physicians have lost a significant amount of their independence.
- The public has grown increasingly critical of physicians. Getting a second opinion is now general practice, and malpractice suits are about as common as unquestioning admiration. Patients are critical consumers of health care rather than passive recipients.
- Fees and treatments are increasingly regulated by government agencies and insurance companies concerned about reducing costs. The vast number of patients whose bills are paid by insurance companies or government agencies gives these groups some control over what treatments will be funded, for which patients, and at what fees.

Doctors have not accepted these changes lying down and have fought to maintain their professional autonomy in many ways. These include fighting for legal restrictions on who can bring malpractice lawsuits and how much money they can demand, fighting individually and collectively against insurance company regulations, and fighting in the court of public opinion to convince patients that physicians continue to have patients' best interests at heart.

Despite these problems, being a physician is still a very good job, offering high income and high prestige. But it is also part of an increasingly regulated industry that is receiving more critical scrutiny than ever before.

Nurses: Fighting for Professional Status

Of the nearly 10 million people employed in health care, the largest single category are the 1.8 million registered nurses. No hospital could run without nurses, and no doctor could function without them. Yet despite their great importance to the health-care system, their status remains far lower than we might expect. Why have nurses' attempts to improve their status achieved only modest success?

Nurses' Current Status

Nurses play a critical role in health care, but they have relatively little independence, either in their day-to-day work or in their training and certification. Physicians have a major voice in determining the training standards that nurses must meet and in enforcing these standards through licensing boards. On the job, even the most junior physician can give orders to experienced nurses. Reflecting this status difference, doctors' median net income is almost four times nurses' income (U.S. Bureau of Labor Statistics 2006a) (Table 10.4). Even when nurses have Ph.D.s or Masters degrees their salary remains a fraction of doctors' salaries. Finally, although the general public respects nurses for their dedication, it tends to discount nurses' specialized education. For all these reasons, nursing does not meet the sociological definition of a profession.

Why is nursing's status so low? The primary reason is its history as a traditionally female occupation. Before the twentieth century, most people believed that caring came naturally to women and, therefore, that mothers, daughters, cousins, and sisters should always be willing to help care for any sick family member (Reverby 1987). Nursing did not become a formal occupation until the mid-nineteenth century. Because of its historic roots, from the start it was considered a natural extension of women's character and duty rather than an occupation meriting either respect or rights (Reverby 1987). Nurses were encouraged to enter the field in a spirit of altruism and self-sacrifice and, as proper young women, to accept orders from doctors, hospital administrators, and their nursing superiors. This approach made it difficult for nursing as a field to fight for status, autonomy, or better working conditions. Moreover, the fact that the field was almost solely female in and of itself made it difficult for nursing to obtain the autonomy and public respect for its training and work that defines a profession.

Changing the Status of Nurses

To improve the status and position of nurses, nursing's leadership has worked for decades to raise educational levels (Weitz 2007). Until the 1960s, the standard nursing credential was an RN (registered nurse) diploma, obtained through a hospital-based training program. Now, almost all RNs hold 2- or 4-year nursing degrees from community colleges or universities. In addition, a small percentage of nurses obtain

TABLE 10.4
Physicians and Registered Nurses: Income, Sex, and Race
Nurses earn less than a quarter of what physicians earn. Critics wonder whether this reflects real differences in training and responsibility or whether it is another instance of traditional women's jobs being evaluated as less worthy than traditional men's jobs.

	Physicians	Registered Nurses
Median net income	$175,000*	$52,330*
Percentage female	31%**	93%**
Percentage African American	5%**	10%**

*2004 data.
**2005 data.

SOURCE: U.S. Bureau of Labor Statistics 2006a, 2006b.

graduate degrees and become nurse practitioners or nurse-midwives. These nurses enjoy considerably more autonomy, status, and financial rewards than do other nurses, including the right (in most states) to prescribe certain medications.

The drive to increase nurses' education and thus their status has succeeded only partially. Because many hospitals believe that associate-degree nurses receive the best practical training and make the best employees, associate-degree programs have remained more popular than higher-level training. Meanwhile, to control costs, hospitals have shifted many services to outpatient clinics where fewer RNs are needed, nurses' salaries are lower, and nursing jobs are less interesting and prestigious (Norrish & Rundall 2001). In addition, hospitals have reduced their nursing staffs and increased the work load of the remaining nurses (Gordon 2005). Finally, although more men now work as nurses, the field is still considered a "woman's profession," and for that reason, salaries and status remain relatively low.

The U.S. Health-Care System

Ensuring that people have access to health care is one of the most basic tasks of any society. The United States offers many ways through which people can get health care: private and publicly-funded insurance, private and public clinics and hospitals, or cash payments. Yet many Americans can obtain only low-quality care, many can obtain care only by making financial sacrifices, and many cannot afford care at all. How does health insurance in the United States work? Why do some people lack insurance, and what are the consequences of being uninsured? And why doesn't the United States have a national health care program?

Paying for Health Care

Medical care is the fastest-growing segment of the cost of living. In 1970, Americans spent an average of $340 per person (in 2003 dollars) on medical care. By 2003, they spent an average of $5,241, and by 2013 they will probably spend twice that (U.S. Bureau of the Census 2006).

Underlying many of the analyses of health care is one question: "Who pays?" There are three primary modes of financing health care in the United States: paying out of pocket, private insurance, and government programs. The cost of health care is so high that only the very rich can afford to pay out of pocket for anything beyond minor problems. As a result, most Americans must rely on private insurance or on government insurance programs. The remainder have no insurance and often are unable to pay for health care.

Private Insurance

Most insured Americans receive their health-care coverage from private insurance plans. Almost all of these obtain their insurance through their employer. Such insurance tends to be limited to employed adults (and their families) who work for large corporations or government agencies. Many jobs in small businesses and most minimum-wage jobs do not provide insurance benefits. In the past, the largest private insurers, like Blue Cross and Blue Shield, were nonprofit organizations that at least to some extent tried to keep costs down. These days, however, most private insurance providers are for-profit corporations. This is one reason why individual costs for health care have risen dramatically.

People who have no health insurance are particularly likely to receive their medical care in emergency-room settings where waits are long, care is costly, and follow-up treatment is rare.

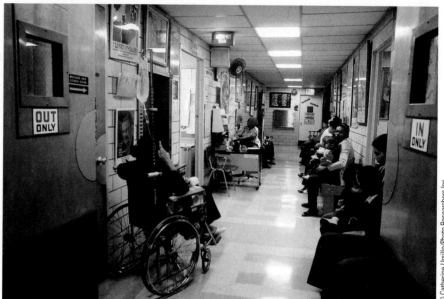

© Catherine Ursillo/Photo Researchers Inc.

Government Programs

The government has several programs that support medical care. The federal government provides some health care through its Veterans Administration hospitals, but its two largest programs are Medicaid and Medicare. In addition, local governments provide medical care through public health agencies and public hospitals.

Medicare is a government-sponsored health insurance program primarily for citizens older than age 65; premiums are deducted from Social Security checks. The enactment of this program in 1965 did a great deal to improve the quality of health care for the elderly. More than 95 percent of the elderly are now covered by health insurance. This is not a cheap program, however; in 2004, the government paid more than $301 billion in Medicare benefits (U.S. Bureau of the Census 2006).

Unlike Medicare, which is available to almost everyone older than age 65, *Medicaid* provides health insurance based on need. Funds come from both the federal government and state governments. Both eligibility and services are determined by states, some of which offer much more generous medical care than others. Generally speaking, though, you won't get Medicaid unless you are both very poor and either a child or pregnant.

The Uninsured

A significant portion of the U.S. population—about 18 percent of those younger than age 65—has no medical coverage (Figure 10.1) Thanks to Medicare, nearly 100 percent of the elderly are insured. Those who fall through the cracks are primarily young or middle-aged, unemployed, working poor, or employees of small businesses (Kaiser Commission 2005).

Every county in the United States does provide public hospital-based emergency care to those who cannot afford to pay for care. Once they seek care, however, uninsured individuals are often kept waiting for several hours by overworked

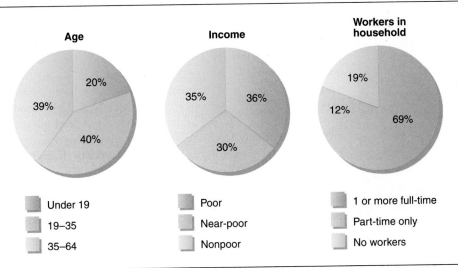

FIGURE 10.1

America's Uninsured Population, 2004

Approximately 18 percent of the U.S. population younger than age 65 has no health insurance. Most are working-age adults, poor or near-poor, and working full time.

SOURCE: Kaiser Commission on Medicaid and the Uninsured 2005.

and underpaid hospital staff before seeing a doctor. Once seen, they are more likely than others to receive substandard care. As a result of all these factors, uninsured persons are more likely than others with similar conditions to postpone needed medical care, to require hospitalization, and to die whether or not they are hospitalized (Kaiser Commission 2005; Weitz 2007).

Why Doesn't the United States Have National Health Insurance?

The United States is the only industrialized nation that does not guarantee health care to all of its citizens. Instead, health care is sold like any other commodity. Like dry cleaning, you get what you can afford, and if you can't afford it, you may have to go without. In contrast, in the rest of the industrialized world, medical care is like primary education—regarded as something that all citizens should receive regardless of ability to pay.

Why doesn't the United States have national health care? The answer, sociologists argue, lies in **stakeholder mobilization:** organized political opposition by groups with a vested interest in the outcome (Quadagno 2005).

Opposition to national health care has come from numerous sources (Quadagno 2005; Rothman 1997). In the past, labor unions opposed national health care because health insurance was one of the major benefits they could offer members. Opposition also came from the American Medical Association, which feared doctors might lose income or autonomy under a national health plan, and from middle- and upper-class Americans who had health insurance and were uninterested in paying taxes to support health care for others.

As the health-care crisis has worsened, affecting more and more middle-class Americans, support for national health care has grown among doctors, labor unions, the public, and even some major corporations who are tired of paying high prices for their employees' health insurance. The strongest opposition to national health care now comes from the pharmaceutical and health insurance industries. These industries poured millions into fighting former President Clinton's proposed health plan,

Connections

Example

How does lack of health insurance lead to illness or even death? Consider what happens to people with diabetes. Some diabetics can control their disease with exercise and diet, but the rest must rely on injections of insulin, which their body finds difficult to produce. Those who lack health insurance may have trouble affording insulin injections or affording the regular doctor visits needed to know how much insulin to take. As a result, their diabetes can spiral out of control, leading to strokes, gangrene, foot amputations, blindness, comas, kidney failure, and even death.

Stakeholder mobilization refers to organized political opposition by groups with vested interest in a particular political outcome.

China's government encourages the use of both Western and traditional medicine. This doctor works at an herbal pharmacy.

outspending those who favored it by a ratio of four to one (Quadagno 2005, 189). In addition, anti-tax sentiment and distrust of "big government" has became a powerful force in U.S. politics beginning in the 1980s, making it difficult to generate support for any governmental programs (Rothman 1997; Skocpol 1996). Nevertheless, polls consistently find that about two thirds of the U.S. public believe it is the federal government's responsibility to ensure health care for all Americans and are willing to pay more taxes to fund such services (Everybody In, Nobody Out 2005).

If Americans get national health insurance in the future, it will be because the middle class, labor unions, and corporations all find it increasingly difficult to pay their health-care bills and unite to fight against anti-tax lobbies, the health insurance industry, and the pharmaceutical industry.

An Alternative Model

Modern medical technology has enhanced our ability to extend and save lives. It is, however, extraordinarily expensive. In China, a largely rural society with a population in excess of 1 billion, Western-style medicine has taken a back seat to prevention. The focus has been on community organization for health-care delivery; improved sanitation, housing, and food; and traditional healing practices that Western physicians are only now coming to appreciate. China's strategies have effectively reduced the death rate at a relatively low cost: Life expectancy as of 2006 is only 6 years less than in the United States (Population Reference Bureau 2006).

Throughout the less-developed nations, in countries such as Costa Rica, Sri Lanka, Cuba, and Vietnam, the Chinese model of health care and illness prevention has proven more effective in reducing mortality than has building high-tech hospitals. To raise life expectancy in the United States to the level of other industrialized nations, we need to ensure that all Americans have access not only to good-quality health care, but also to nutrition and preventive health-care programs and to decent food, clothing, and shelter. Such strategies are likely to return more for each dollar invested than will money spent on expensive new biomedical technologies.

Where This Leaves Us

Sociological analysis suggests that health and illness are socially structured. To paraphrase C. Wright Mills once again, when one person dies too young from stress or bad habits or inadequate health care, that is a personal trouble, and for its remedy we properly look to the character of the individual. When whole classes, races, or sexes consistently suffer significant disadvantage in health and health care, then this is a social problem. The correct statement of the problem and the search for solutions require us to look beyond individuals to consider how social structures and institutions have fostered these patterns. The sociological imagination suggests that significant improvements in the nation's health will require changes in social institutions—increasing education, reducing poverty and discrimination, improving access to good-quality housing and food, and so on. Equalizing access to health care will also help but is considerably less important than making these social changes.

Summary

1. A major contribution of structural-functionalist theory to the study of health is the concept of a sick role. This concept explains how (some) illness can help society run smoothly and how society limits illness to keep it from interfering with that smooth flow.

2. From conflict theory we get the concept of manufacturers of illness: groups that benefit from promoting conditions that cause illness and disease. Conflict theory also helps us to understand how definitions of illness are developed in the process of medicalization and how competing interest groups battle over different potential definitions of illness.

3. Symbolic interaction theory has been particularly useful for understanding the experience of illness, including why patients sometimes do not follow doctors' orders.

4. Nine factors—tobacco, poor diet and inadequate exercise, alcohol, bacteria and viruses, polluted workplaces and neighborhoods, motor vehicles, firearms, sexual behavior, and illegal drugs—underlie almost half of all preventable deaths in the United States.

5. The sick role consists of four social norms regarding sick people. They are assumed to have good reasons for not fulfilling their normal social roles and are not held responsible for their illnesses. They are also expected to consider sickness undesirable, to work to get well, and to follow doctor's orders. The sick role model, however, fits some illnesses better than others.

6. The health belief model predicts that individuals will adopt behaviors that will protect their health if they believe they are at risk for a particular health problem, they believe the problem is serious, they believe that changing their behavior will reduce their risks, and no significant barriers keep them from changing their behavior.

7. Gender, social class, race/ethnicity, and age all help explain the patterns of health and illness in the United States. Men, racial and ethnic minorities, those with lower socioeconomic status, and the very young and old have higher mortality rates, largely due to social rather than biological forces.

8. The health disadvantage associated with lower socioeconomic status goes far beyond a simple inability to afford health care. Lower social class is associated with lower standards of living, more stress, lower education, and polluted environments, all of which increase the likelihood that individuals will have poor health.

9. Women and lower-class people have higher rates of mental illness than other groups. Although the reasons are complex, differences in exposure to stress appear to be the primary reason. Men have higher rates of substance abuse and personality disorders than women.

10. Physicians are professionals; they have a high degree of control not only over their own work but also over all others in the medical world. Structural functionalists argue that physicians earn so much because of scarce talents and abilities, whereas conflict theorists argue that high salaries are due to an effective union (the AMA). Physicians have less independence than they used to, due to increased corporate control, government oversight, and public criticism.

11. Nurses comprise the largest single occupational group in the health-care industry. Nurses earn much less than physicians, have less prestige, and take orders instead of giving them. The reasons for this primarily stem from its history as a "female" field.

12. Most insured Americans belong to private health insurance plans. Medicare is a government program that insures almost all senior citizens, and Medicaid is a government insurance program primarily for the very poor.

13. Eighteen percent of Americans are uninsured. Uninsured persons are more likely than others to postpone getting needed health care, to become ill, and to die if they become ill.

14. The United States is the only industrialized nation that does not make medical care available regardless of the patient's ability to pay. We do not have national health insurance because of stakeholder mobilization, which currently comes primarily from pharmaceutical and insurance corporations.

Thinking Critically

1. How have "manufacturers of illness" increased deaths caused by tobacco? By alcohol? By toxic agents? By diet?

2. How have social forces and political decisions increased deaths caused by sexual behavior? Caused by illicit drugs?

3. Think of a recent illness you or someone you know well experienced. Discuss each of the elements of the sick role, and whether or not it applied in this instance. Then think of someone you know who has a chronic illness, and do the same.

4. Who benefits when "male erectile dysfunction" is defined as an illness? How? Who loses? What do they lose?

5. Think of a friend of yours who smokes or engages in another unhealthy behavior. Use the health belief model to explain what else would have to change before your friend would be likely to change his or her behavior.

6. Why do so few men enter nursing? What could change this gender gap?

7. Who would gain if the United States adopted a national health care system? Who would lose, and what would they lose? Consider economic, social, political, and psychological costs.

Companion Website for This Book

academic.cengage.com/sociology/brinkerhoff

Gain an even better grasp on this chapter by going to the Companion Website. This resource contains tutorial quizzes and flash cards to help you master key terms and concepts.

Suggested Readings

Sered, Susan Starr and Fernandopulle, Rushika. 2005. *Uninsured in America: Life and Death in the Land of Opportunity.* Berkeley, CA: University of California Press. Explains in gripping detail who are the uninsured and what happens to individuals' health, income, and lives once they lose their health insurance.

Marmot, Michael G. 2004. *The Status Syndrome: How Your Social Standing Directly Affects Your Health and Life Expectancy.* London: Bloomsbury. Epidemiologist Michael Marmot, who received a knighthood for his research, explains why at each step up the social status ladder, persons live longer than those even one step below them.

Watts, David. 2005. *Bedside Manners: One Doctor's Reflections on the Oddly Intimate Encounters Between Patient and Healer.* New York: Three Rivers Press. In this book of short essays, poet, NPR-commentator, and doctor David Watts beautifully describes medicine, and the doctor/patient relationship, at its best.

Weitz, Rose. 2007. *The Sociology of Health, Illness, and Health Care: A Critical Approach.* 4th edition. Belmont, CA: Wadsworth. A sociological overview of how social factors affect who becomes ill or disabled, the experiences of illness and disability, and the health care world.

Family

© John & Yva Momatiuk/The Image Works

Marriage and Family: Basic Institutions of Society

Recent decades have seen many changes in American family life. Birth rates have declined sharply, divorce and single-parent families are now common, and the majority of women with small children work in the paid labor force. In addition to these statistical trends, major shifts in attitudes and values have occurred. Homosexuality, premarital sex, and extramarital sex have all become more acceptable. Related to many of these changes are the dramatic changes in the roles of women in our society.

These changes in family life have been felt, either directly or indirectly, by all of us. Is the family a dying institution, or is it simply a changing one? In this chapter, we examine the question from the perspective of sociology. We begin with a broad description of marriage and the family as basic social institutions.

To place the changes in the U.S. family into perspective, it is useful to look at the variety of family forms across the world. What is it that is really essential about the family?

Universal Aspects

In every culture, the family has been assigned major responsibilities, typically including the following (Seccombe & Warner 2004):

- Replacing the population through reproduction
- Regulating of sexual behavior
- Caring for dependents—children, the elderly, the ill, and the handicapped
- Socializing the young
- Providing intimacy, belongingness, and emotional support

Because these activities are important for individual development and the continuity of society, every society provides some institutionalized pattern for meeting them. No society leaves them to individual initiative. Although it is possible to imagine a society in which these responsibilities are handled by religious or educational institutions, most societies have found it convenient to assign them to the family.

Unlike most social structures, the family can be a biological as well as a social group. The **family** is a relatively permanent group of persons linked together in social roles by ties of blood, adoption, marriage, or quasi-marital commitments who live together and cooperate economically and in the rearing of children, if they have any. This definition is very broad; it would include a mother living alone with her child as well as a man living with several wives. The important criteria for families are that their members assume responsibility for each other and are bound together—if not by blood, then by some cultural markers such as marriage or adoption.

Marriage is an institutionalized social structure that is meant to provide an enduring framework for regulating sexual behavior and childbearing. Many cultures tolerate other kinds of sexual encounters—premarital, extramarital, or homosexual—but most cultures discourage childbearing outside marriage. In some cultures, the sanctions are severe, and almost all sexual relationships are confined to marriage; in others, nonmarital sexuality incurs little or no punishment.

The **family** is a relatively permanent group of persons linked together in social roles by ties of blood, adoption, marriage or quasi-marital commitment and who live together and cooperate economically and in the rearing of children.

Marriage is an institutionalized social structure that provides an enduring framework for regulating sexual behavior and childbearing.

Marriage is also a legal contract, specifying the obligations of each spouse. Until very recently, those obligations were sharply divided by sex: By law, husbands were obligated to financially support their wives, and wives were obligated to provide domestic services and sexual access to their husbands. These sex-specific obligations only started changing with the rise of the modern feminist movement in the 1970s.

Marriage is important for childbearing because it imposes socially sanctioned roles on parents and the kin group. When a child is born, parents, grandparents, and aunts and uncles are automatically assigned certain normative obligations to the child. This network represents a ready-made social structure designed to organize and stabilize the responsibility for children. Children born outside marriage, by contrast, are more vulnerable. The number of people normatively responsible for their care is smaller, and, even in the case of the mother, the norms are less well enforced. One consequence is higher infant mortality for children born outside of marriage in almost all societies, including our own.

Marriage and family are among the most basic and enduring patterns of social relationships. Although blood ties are important, the family is best understood as a social structure defined and enforced by cultural norms.

Cross-Cultural Variations

Families universally are expected to regulate sexual behavior, care for dependents, socialize the young, and offer emotional and financial security. The importance of these tasks, however, varies across societies. Offering economic security is more important in societies without government-provided social services; regulating sexual behavior is more important in cultures without contraception. In our own society, we have seen the priorities assigned to these family responsibilities change substantially over time. In colonial America, economic responsibility and replacement through reproduction were the family's primary functions; the provision of emotional support was a secondary consideration. More recently, however, some of the responsibility for socializing the young has been transferred to schools and day-care centers; financial responsibility for dependent elderly persons has been partially shifted to the government. At the same time, intimacy has taken on increased importance as a dimension of marital relationships.

Although all families share the same basic functions, hundreds of different family forms can satisfy these needs. This section reviews some of the most important ways cultures have fulfilled family functions.

Family Patterns

Throughout history and across cultures, people have typically lived with an assortment of kin: a husband and one or more wives; their children; and one or more grandparents, uncles, aunts, nephews, nieces, or cousins. This type of family is known as an **extended family.** Extended families have many benefits: There is always someone to hug or to talk with, finding a babysitter is easy, elderly and disabled relatives need not be left alone, and expenses can be shared.

In the United States, most Americans instead expect to live in a **nuclear family.** A nuclear family consists of a mother and father and their children. Nuclear families are valued by those who want their independence and are grateful not to have parents or in-laws looking over their shoulders.

In reality, only a small percentage of Americans live in nuclear families. Instead, most adults live either alone, with friends or lovers, or with their children only. Others live in extended families out of necessity. Still others (especially in immigrant families)

An **extended family** is a family in which a couple and their children live with other kin, such as the wife's or husband's parents or siblings.

A **nuclear family** is a family in which a couple and their children form an independent household living apart from other kin.

live in extended families because they enjoy doing so or because they consider it a moral obligation to take care of needy family members.

Marriage Patterns

Monogamy is a marriage in which there is only one wife and one husband.

Polygamy is any form of marriage in which a person may have more than one spouse at a time.

In the United States and much of the Western world, a marriage form called **monogamy** is practiced; each man may have only one wife (at a time), and each woman may have only one husband. Many cultures, however, practice some form of **polygamy**—marriage in which a person may have more than one spouse at a time. Most often, cultures allow men to have more than one wife, but a small percentage of cultures allow women to have more than one husband.

Even in cultures that allow—or even promote—polygamy, it has limits: Since there are nearly equal numbers of men and women in society, if some men (typically the wealthiest and most powerful) have more than one wife, other men have to do without. Consequently, even in societies where polygamy is accepted, most people actually practice monogamy, and young men may have to go elsewhere to find any wife at all. As this suggests, polygamy can only exist in societies where some men have considerably more power than do others and usually occurs only in societies where men in general have more power than do women.

The U.S. Family over the Life Course

Family relationships play an important role in every stage of our lives. As we consider our lives from birth to death, we tend to think of ourselves in family roles. Being a youngster usually means growing up in a family; being an adult usually means having a family; being elderly often means being a grandparent.

Because of the close tie between family roles and individual development, we have organized this description of the U.S. family into a life course perspective. This means that we will approach the family by looking at age-related transitions in family roles.

■ Some modern American families, like these fundamentalist Mormons, live a polygamous life despite legal and social opposition from most of their fellow citizens and from most other Mormons.

© AP/Wide World Photos

American Diversity:

Gay and Lesbian Families

What does it mean to be a family? As we have seen, a family is a relatively permanent group of persons linked together in social roles by ties of blood, adoption, marriage, or quasi-marital commitments who live together and cooperate economically and in the rearing of children, if they have any. By this definition, two men or two women who commit to each other, live together, and, if they have children, parent them together are a family.

Of course, gay or lesbian couples cannot biologically have children together through sexual intercourse. But the same is true of some heterosexual married couples, who also must rely on reproductive technologies or adoption if they want children. Similarly, many lesbians and gay men have children using artificial insemination or the like, and others have children from previous heterosexual relationships whom they raise together.

Issues related to gay families have become matters of fierce public debate in recent years. Should gays be allowed to adopt children? Is a gay man or lesbian inherently unfit for child custody or visitation rights? Should lesbians be allowed to use artificial insemination? Should gays and lesbians be allowed to marry so their partners can share their health insurance and Social Security benefits?

These are questions that go to the heart of the family. The traditional view is that homosexual unions are both unnatural and sinful. Others define the family by long-term commitment, and they are willing to tolerate or even encourage a variety of family forms—including gay and lesbian families—as long as they contribute to stable and nurturing environments for adults and children. In fact, research consistently finds that growing up with gay or lesbian parents has no measurable effect, other than perhaps increasing children's acceptance of nontraditional gender behavior (Freeman 2003).

There is no question that both homosexual activity and gay marriage are regarded more favorably now than in the past. A national survey conducted in 2006 found that 39 percent of Americans approve of gay marriages, up from 27 percent in 1996 (Pew Research Center 2006). Despite this growing support, however, few American lesbians and gays have the option of marrying their partners. Only a small number of U.S. jurisdictions allow same-sex couples to register their unions as "domestic partnerships," and even fewer permit same-sex marriages. Moreover, the federal Defense of Marriage Act prohibits any federal recognition of gay marriage. In contrast, a small number of other countries, including Canada, Spain, and South Africa, now recognize same-sex marriage. The question American society must now address is whether gay families should receive the same legal recognition and protection as other families in this country and as gay families in some other nations.

◼ Growing numbers of children in the United States are being raised and nurtured by gay or lesbian parents and their same-sex partners.

Childhood

U.S. norms specify that childhood should be a sheltered time. Children's only responsibilities are to accomplish developmental tasks such as learning independence and self-control and mastering the school curriculum. Norms also specify that children should be protected from labor, physical abuse, and the cruder, more unpleasant aspects of life.

Childhood, however, is seldom the oasis that our norms specify. A sizable number of children are physically or emotionally abused by their parents. For example, about 10 percent of girls experience rape or attempted rape during childhood (Tjaden & Thoennes 1998). In addition, nearly one fifth of all American children grow up in poverty. Although the proportion of children living in poverty in the United States has fallen slightly since 1993, the gap between rich and poor children appears to be growing and is now second highest among industrialized nations (Bennett & Lu 2000).

An important change in the social structure of the child's world is the sharp increase in the proportion of children who grow up in single-parent households: 35 percent of children are now born to single mothers (U.S. Bureau of the Census 2006). Many more experience the divorce of their parents and sometimes a second divorce between their parents and stepparents (Coleman, Ganong, & Fine 2000). Perhaps because single parents cannot provide as much money or time as married parents, studies show that, on average, children whose parents divorce have poorer self-esteem, academic performance, and social relationships than other children. These differences are slight, however, and stem primarily not from the divorce itself, but from the poverty and parental conflicts that precede or follow it (Coontz 1997; Demo & Cox 2000; Lamanna & Riedmann 2000). Consequently, some of these children would not have been any better off if their parents had remained married.

The increasing participation of women in the labor force has added another social structure to the experience of young children: day care. In 2004, about two

■ As increasing numbers of U.S. women, including those with infants younger than age 1, have entered the labor force, day-care centers have become much more important aspects of early childhood socialization.

© Gerd Ludwig/Woodfin Camp & Associates

thirds of mothers of children younger than age 6 were employed (U.S. Bureau of the Census 2006). About one third of preschool children with an employed mother attend a day-care center (Smolensky & Gootman 2003).

Research mostly supports the use of day-care centers. Some research suggests that day care can increase children's stress levels and behavioral problems, but this effect is not large and dissipates by third grade (NICHD Early Child Care Research Network 2005; Watamura et al. 2003). Other research suggests that day care increases children's math and reading skills and that any negative effects of child care are limited to certain types of children, families, or programs (Love et al. 2003). High-quality programs, which are most often attended by more affluent children, offer the most benefits overall. However, because lower-income children come from homes where they are less likely to get intellectual and social stimulation, they benefit considerably from day care, even in lower-quality programs. At any rate, many families cannot afford to have a parent stay home with the children. For these families, day care is far superior to leaving young children alone or in the care of older siblings.

Adolescence

Contemporary social structures make adolescence a difficult period. Because society has little need for the contributions of youth, it encourages young people to become preoccupied with trivial matters—such as eyebrow shaping or loading iPods. Yet, because adolescence is a temporary state, the adolescent is under constant pressure. Questions such as "What are you going to do when you finish school?", "What are you going to major in?", "What went wrong in Friday night's game?", and "How serious are you about that boy [girl]?" can create strain. That strain can be particularly high for gay and lesbian youth, who may find themselves interested in someone of the "wrong" sex, confused about their own feelings, and fearful over how their family might react.

Adolescents are supposed to become independent from their parents, acquiring adult skills and their own values. They are supposed to shift from the family to peer groups as a source of self-esteem. They must learn how to impress new people, and, last but not least, they are supposed to have fun (Gullotta, Adams, & Markstrom 2000). The average adolescent begins to date at about age 14, and an adolescent who is far behind may find that parents and friends are concerned. Thus, although society does not appear to expect much from them, adolescents experience a great deal of role strain. Many adults believe adolescence was the worst rather than the best time of their lives.

The Transition to Adulthood

Some societies have **rites of passage,** formal rituals that mark the end of one age status and the beginning of another. In our own society, there is no clear point at which we can say a person has become an adult. However, in the United States adulthood usually means that a person adopts at least some of the following roles: working, supporting oneself and one's dependents, voting, marrying, and having children. Some of these social roles are optional, and people may be considered adults who never vote, marry, or, in the case of women, hold a paid job. Nevertheless, the exit from adolescence always entails "escaping" from dependence on parents and family.

Making this escape, however, is now a harder and longer process than in the past (Settersten, Furstenberg, & Rumbaut 2006). In 1960, more than 80 percent of

Rites of passage are formal rituals that mark the end of one age status and the beginning of another.

30 year olds (male and female) had left home, finished school, and achieved financial independence. By 2000, the numbers who had done so had dropped by about 10 percent—not a huge drop, but still significant when compared with historical patterns (Furstenberg et al. 2004).

Why has the transition to adulthood slowed down? Changing attitudes have allowed women to extend their schooling and delay marriage and parenthood, and changing economic realities make it a wise choice for young men to do the same. Costs of living have risen rapidly, and paychecks have not kept pace, making it difficult for young people to strike out on their own. In addition, young people now graduate with more educational debts than was the case a generation ago. As a result, many young adults continue to live with their parents after leaving school, sometimes leaving home and returning several times before becoming independent. Some live with their parents to make ends meet, some to afford nice cars, cable television, fun vacations, and fast computers. Others can live on their own only because they receive substantial subsidies from their parents; about one third of young Americans between 18 and 34 receive such subsidies annually (Settersten, Furstenberg, & Rumbaut 2006).

Early Adulthood: Dating and Mate Selection

Nearly all Americans marry. In fact, the United States is the "marryingest" of industrialized nations. By the time they reach 30, a very high proportion of people in the United States have been married at least once. This strong cultural emphasis on marriage is one of the reasons that so many gays and lesbians also want to marry their life partners.

At first glance, it appears as if all persons are on their own in the search for a suitable spouse; few Americans (outside of certain religious and ethnic communities) rely on matchmakers or arranged marriages. On further reflection, however, it is clear that parents, schools, and churches are all engaged in the process of helping young people find suitable partners. Schools hold dances designed to encourage heterosexual relationships, churches have youth groups partly to encourage members to date and marry within their church, parents and friends introduce somebody "we'd like you to meet." Although dating may be fun, it is also a normative, almost obligatory form of social behavior.

The Changing Age of First Marriage

In the 1950s, teenagers dated in order to find a spouse. Many did so very quickly, and more than 50 percent of U.S. women were married before their twenty-first birthday. Times have changed. Teenagers no longer date with the expectation of settling down early. By their late twenties, forty percent of women and thirty percent of men still have never married. Some of them are not interested in marrying, but most are looking for at least a temporary partner. Courtship and dating are activities of individuals who are 35 and 55 as well as those who are 15 or 25.

Sorting through the Marriage Market

Over the course of one's single life, one probably meets thousands of potential marriage partners. How do we narrow down the marital field?

Obviously, you are unlikely to meet, much less marry, someone who lives in another community or another state. In the initial stage of attraction, **propinquity,** or spatial nearness, operates in this and a much more subtle fashion, by increasing the opportunity for continued interaction. It is no accident that so many people end up marrying coworkers or fellow students. The more you interact with others, the more

Propinquity is spatial nearness.

positive your attitudes toward them become—and positive attitudes may ripen into love (Homans 1950).

Spatial closeness is also often a sign of similarity. People with common interests and values tend to find themselves in similar places, and research indicates that we are drawn to others like ourselves. Of course, there are exceptions, but faced with a wide range of choices, most people choose a mate who is like them in many ways. This pattern is called **homogamy** (Kalmijn 1998). Most people marry within their social class (*social homogamy*), and most also marry within their racial, ethnic, or religious group, a type of homogamy known as **endogamy.** Intermarriages can only occur when individuals have contact with persons from other groups and accept those others as more or less equal. Although intermarriage remains rare, it has increased substantially in the last few decades, especially among Jews, Asian Americans, and more educated Americans (Kalmijn).

Physical attractiveness may not be as important as advertisers have made it out to be, but studies do show that appearance is important in gaining initial attention (Sullivan 2001). Its importance normally recedes after the first meeting.

Dating is likely to progress toward a serious consideration of marriage (or cohabitation) if the couple discover similar interests, aspirations, anxieties, and values (Kalmijn 1998; Seccombe & Warner 2004). When dating starts to get serious, couples begin checking to see if they share values such as the desire for children or commitment to an equal division of household labor. If he wants her to do all the housework and she thinks that idea went out with the hula hoop, they will probably back away from marriage.

Responding to Narrow Marriage Markets

Whether dating leads to marriage also depends on the local supply of "economically attractive" men. As early as 1987, William Julius Wilson noted that one of the reasons African American women were much less likely to marry than white women was the

Connections

Personal Application

Your college education is likely to affect whom you marry. Many people find a spouse in college classrooms or activities (based on propinquity). If you attend a college linked to your religion, race, or ethnic group, you are more likely to marry within your group (endogamy). If college throws you into contact with many others whose backgrounds are different from your own, you will be more likely to marry someone from a different background.

Homogamy is the tendency to choose a mate similar in status to oneself.

Endogamy is the practice of choosing a mate from within one's own racial, ethnic, or religious group.

Interracial marriage and dating have become far more common and socially accepted over the last few decades.

shrinking pool of African American men with good educations and jobs. Results from other researchers reinforce this conclusion: A shortage of males employed in good jobs with adequate earnings sharply reduces the likelihood that a woman will marry or cohabit (Lichter et al. 1992; Raley 1996; Teachman, Tedrow, & Crowder 2000). In fact, differences in the availability of marriageable men account for at least 40 percent of the racial difference in overall marriage rates. Similarly, local marriage markets affect the rates of intermarriage: Minority group members are significantly more likely to intermarry if they live in areas where members of their group are rare (Qian 1999).

In an interesting sidebar, researchers have found that "economically attractive" women are also more likely to marry. Their greater attractiveness to potential male partners apparently more than makes up for the fact that women with full-time employment and higher earnings tend to be choosier about the men they date and marry (Lichter et al. 1992).

Middle Age

The busiest part of most adult lives is the time between the ages of 20 and 45. There are often children in the home and marriages and careers to be established. This period of life is frequently marked by role overload simply because so much is going on at one time. Middle age, that period roughly between 45 and 65, is by contrast often a quieter time. Studies show that both men and women tend to greet the empty nest with relief rather than regret (Umberson et al. 2005).

Many middle-aged couples, however, look forward to the empty nest period only to find that they are sandwiched between the demands of their parents and those of their children. This generation squeeze has been fostered by two distinct trends. The first is rising life expectancy. Reflecting this trend, a 2003 national survey found that 17 percent of U.S. households include someone who is providing care for a relative older than age 65 (National Alliance for Caregiving 2003). At the same time, difficulties their children face in establishing themselves economically, rising age at marriage, and high rates of divorce have increased the likelihood that middle-aged adults still have children living at home. As a result of these pressures from both sides, many middle-aged people do not find the relief from family responsibility that they had hoped for. Not surprisingly, marital happiness is lower for parents whose adult children live with them (Umberson et al. 2005).

Age 65 and Beyond

One of the most important changes in the social structure of old age is that it is now a common stage in the life course—and often a long one. Almost all of us can count on living to age 65. Furthermore, if you live to age 65, you can expect to live an average of 18.4 more years (U.S. Department of Health and Human Services 2005). Most of these years will be healthy ones. Although most people age 70 and older experience some loss of stamina, 75 percent experience no major health limitations (Federal Interagency Forum on Aging Related Statistics 2000).

Family roles continue to be critical in old age. Having spouses, children, grandchildren, and brothers and sisters all contribute to well-being. Marriage is an especially important relationship, one that provides higher income, live-in help, and companionship. Because of men's shorter life expectancy and their tendency to marry younger women, however, marriage is not equally available: 79 percent of men aged 65 to 74 are still married compared with only 55 percent of women that age.

■ Both grandparents and grandchildren gain great personal satisfaction from their relationships.

Whether older people are married or not, relationships with children and grandchildren are typically an important factor in their lives. Most grandparents visit grandchildren every month and report very good relationships with them (American Association of Retired Persons 1999). Many children and families would have great difficulty without the help of grandparents, and many grandparents consider involvement with their grandchildren an important source of personal satisfaction (Allen, Blieszner, & Roberto 2000).

Grandparents are especially important when grandchildren are left parentless or effectively parentless, due to illness, disability, imprisonment, or substance abuse. More than a million children live with grandparents rather than with parents (Kreider & Fields 2005). These "skipped generation" households of grandparents and grandchildren are sharply at risk for poverty (Newman & Massengill 2006). Usually, though, the alternative is worse: sending the children to foster care, a system rife with problems.

The nature of intergenerational relationships depends substantially on the ages of the generations. When the older generation falls into the "young old" category, they are generally still providing more help to their children than their children are providing to them (Hogan, Eggebeen, & Clogg 1993). They are helping with down payments and grandchildren's college educations or providing temporary living space for adult children who have divorced or lost their jobs.

As the senior generation moves into the "old old" category, however, relationships must be renegotiated (Mutran & Reitzes 1984). Even in the "old old" category, most people continue to be largely self-sufficient, but they eventually will need help of some kind—for shopping, home repairs, and social support. Although these services are available from community agencies, most older people rely heavily on their families, especially their daughters (Kemper 1992; Lye 1996). Understandably, though, both older people and their adult children are happiest when these relationships are free of dependency. Elderly persons much prefer to live alone rather than with their children (Bayer & Harper 2000).

Roles and Relationships in Marriage

Marriage is one of the major role transitions to adulthood, and most people marry at least once. In fiction, the story ends with the wedding, and we are told that the couple lived happily ever after. In real life, though, the work has just begun. Marriage means the acquisition of a whole new set of duties and responsibilities, as well as a few rights. What are they and what is marriage like?

Gender Roles in Marriage

Marriage is a sharply gendered relationship. Both normatively and in actual practice, husbands and wives and mothers and fathers have different responsibilities. Although many things have changed, U.S. norms specify that the husband *ought* to work outside the home; it is still considered his responsibility to be the primary provider for his family—even though in about one quarter of dual-earner households, wives outearn their husbands (Winkler, McBride, & Andrews 2005). Similarly, although most Americans now believe that husbands and wives should share in household labor, most still expect that the wife will do the larger share. In fact, women currently perform two thirds to three quarters of household labor, and most women as well as men regard this as a fair arrangement (Coltrane 2000).

Although women who work outside the home typically do less housework than other women, this still leaves many working women (especially those with young children) subject to severe cases of role overload, or role strain. One adaptation women make to this overload is to lower their standards for cleanliness, meals, and other domestic services. They let their family eat at McDonald's and let the iron gather dust.

Another adaptation women make is to hire other women to perform domestic tasks. In this way, domestic labor remains a woman's job and the idea that women are responsible for this work is reinforced. In addition, since most employers of domestic help are white and middle class and most domestic workers are nonwhite and working class, paid domestic labor also reinforces class and race divisions within society (Hondagneu-Sotelo 2001; Parreñas 2000; Wrigley 1995).

The Parental Role: A Leap of Faith

The decision to become a parent is a momentous one. Children are extremely costly, both financially and in terms of emotional wear and tear—and the costs can continue for decades. It currently costs about $191,000 to raise a child to age 17, and another $42,000 by the time the child reaches age 34 (Settersten, Furstenberg, & Rumbaut 2006).

Parenthood is really the biggest risk most people will ever take. Few other undertakings require such a large commitment on so uncertain a return. The list of disadvantages is long and certain: It costs a lot of money, takes an enormous amount of time, disrupts usual activities, and causes at least occasional stress and worry. Also, once you've started, there is no backing out; it is a lifetime commitment. What are the returns? You hope for love and a sense of family, but you know all around you are parents whose children cause them heartaches and headaches. In fact, the presence of children in the home—especially infants and teenagers—seems particularly likely to reduce marital happiness, and happiness decreases with each additional child (Twenge, Campbell, & Foster 2003). Yet despite all this, most people want and have children.

Mothering versus Fathering

Despite some major changes, the parenting roles assumed by men and women still differ considerably (Cancian & Oliker 2000). Mothers are the ones most likely to drop out of the labor force to care for infants and young children; they are the ones most likely to care for sick children and to go to school conferences (Cancian & Oliker 2000). Fathers, on the other hand, are the ones likely to carry the major burden of providing for their families.

The overwhelming proportion of mothers who are employed—around 80 percent—has exerted pressure for fathers to increase their role in child care. Although research still finds that fathers "help" rather than "take responsibility" and that they are more likely to play with children than to change diapers, fathers have increased their role in child care. A growing proportion of fathers, however, do not live with their children. Among these fathers, contact tends to be low and child care virtually nonexistent.

Stepparenting

About 7 percent of American children currently live with a stepparent (Kreider & Fields 2005), and about one third will do so before they are 18—most often with a mother and a stepfather (Coleman, Ganong, & Fine 2000). If parenting is difficult, stepparenting is more so (Coleman, Ganong, & Fine 2000). Often stepparents are unsure what role they should take in their stepchildren's lives, and often their spouses and stepchildren are equally ambivalent. Older children, especially, are likely to reject stepparents and to discourage warm relationships, although many eventually develop close relationships with their stepparents. Stepmothers typically face more difficulties than stepfathers because stepmothers typically are more involved in their stepchildren's lives. In addition, stepmothers face more competition for the children's affections, since biological mothers are far more likely to remain involved in their children's lives than are biological fathers.

Although fathers now take more responsibility for child care and household tasks than they did in previous generations, mothers still bear far more of these burdens, leaving many feeling overworked and underappreciated.

Contemporary Family Choices

As discussed in Chapter 2, U.S. norms have changed over time to permit much wider variation in the way that people achieve core values. Although a happy family life remains at the center of things people in the United States value, the ways that individuals choose to meet this value have changed considerably. An increasing number of people will find themselves making three choices about marriage and the family: whether to marry or cohabit; whether to have children, either within or outside of marriage; and whether to give priority to work or to family.

Marriage or Cohabitation

During the last 30 years, the chances that an individual will *ever* cohabit has increased more than 400 percent for men and 1,200 percent for women. Cohabitation is also an increasingly common stage in the courtship process, with approximately half of all recently married couples cohabiting beforehand (Smock 2000).

Cohabitation is often but not always a prelude to marriage: Much of the decline in U.S. marriage rates is due to the increasing numbers of individuals who prefer cohabitation to marriage. The proportion of cohabiting couples that married within 3 years declined by half between the 1970s and the 1990s, and 40 percent of unmarried women who give birth these days are in cohabiting couples (Cherlin 2004). Indeed, Andrew Cherlin, a leading sociologist of the family, argues that we are now witnessing the **deinstitutionalization of marriage:** the gradual disintegration of the social norms that undergird the need for marriage, the meaning of marriage, and expectations regarding marital roles. This process has gained ground as cohabiting couples have won legal rights (such as the right to pass on property to each other or to sue for spousal maintenance if they split up). Conversely, the fight for (and against) gay marriage suggests that marriage still means a great deal to most Americans.

The **deinstitutionalization of marriage** is the gradual disintegration of the social norms that undergird the need for marriage, the meaning of marriage, and expectations regarding marital roles.

Having Children . . . or Not

Although most people in the United States plan to have children, increasingly they choose to do so outside of marriage. Others will choose to postpone parenthood, and increasing numbers will choose to remain childless. Still others will conclude that the best way to add children to their family is through adoption.

Nonmarital Births

About one third of all births in the United States are to unmarried women (Hamilton, Martin, & Sutton 2003). Most of these births (about three quarters) are to women 20 years of age and older. Now that most women participate in the labor force, many believe that they have the economic and psychological resources to tackle the tough job of parenting on their own. Some will decide against abortion if they become pregnant accidentally, and others will intentionally become pregnant or adopt even if they are not married (Hertz 2006). For the same reasons, births to unmarried women also have increased in Europe (Figure 11.1).

About one third of all births in the United States are to unmarried women. Girls who face bleak futures are especially likely to look to single motherhood as a way to find love and happiness.

© Mark Peterson/Corbis

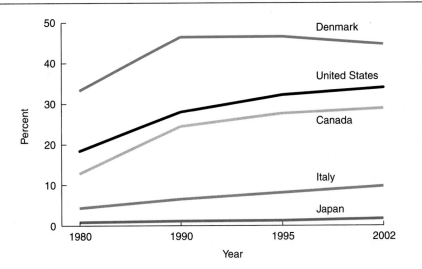

FIGURE 11.1
**Percentage of All Births to Un-
married Women, 1980-2002**
SOURCE: U.S. Bureau of the Census
2006.

Nonmarital childbearing among teenagers raises special concern, although the rate has declined steadily over the last decade (Hamilton, Martin, & Sutton 2003). Because teenage mothers are less likely to complete college or even high school, they are also likely to suffer economic hardship. In many instances, however, teenage pregnancy stems from poverty as well as causing it (Luker 1996; Newman & Massengill 2006; Edin & Kefalas 2006). Girls who face bleak futures sometimes conclude that single motherhood is a reasonable way to seek love and happiness. Other girls become pregnant because they fear that using contraceptives would suggest to a new boyfriend that they are "easy." Still others lack the power to insist that contraception be used.

Regardless of the mother's age, having a child outside marriage does not necessarily mean raising a child alone. About 40 percent of nonmarital childbirths are to women who live with the fathers of their babies (Smock 2000). Many of these women will eventually marry the father or another man; others will continue to share parenting outside of marriage.

Delayed Childbearing

Many married women are choosing to postpone childbearing until 5 or even 10 years after their first marriages. Today, 28 percent of U.S. women aged 30 to 34 are still childless as are 19 percent of those aged 40 to 44 (Dye 2005). Many of these intend to have children eventually, but have decided to wait until they are established in a career or in a stable marriage with someone who earns a good income.

Choosing Childlessness

While most women and men do eventually want children, increasing numbers have decided that they are uninterested in having children. Of course, this choice depends on access to effective contraception. But it also reflects social changes.

There have always been men who find sufficient satisfaction in their lives that they consider children both unnecessary for happiness and a hindrance to their work and other interests. As more women find satisfaction in their work and other aspects of their lives, they may come to adopt similar attitudes (Park 2005). These decisions are

Connections

Historical Note

One hundred years ago, unwed teen mothers were very rare. This does not mean, though, that teens were any less likely to get pregnant then than they are now. The difference is that now more pregnant teenagers keep their babies without getting married. Prior to the 1970s, this was an extremely rare choice, for the shame of raising a child alone was overwhelming. Girls who found themselves pregnant had three choices: getting an abortion (usually illegal and sometimes life-threatening), having a "shotgun" wedding, or giving up their babies. Those who chose the last option usually left their home towns and stayed in special institutions, where they could hide their pregnancies until their babies were born and given up for adoption.

© Big Cheese Photo LLC/Alamy

■ Now that birth control and abortion have significantly reduced the number of unwanted babies, and fewer single mothers give up their babies, it has become increasingly difficult to find babies to adopt. As a result, international adoption has become popular—at least among those who can afford it.

bolstered by the belief—backed by research—that having children *reduces* marital satisfaction and has little impact on happiness during middle age or later life (Umberson et al. 2005). Childlessness is also particularly common among women who were the eldest daughters in large families; these women often feel that they already raised several children and have no interest in doing so again.

Adoption

For those who want children but are single, lesbian, gay, or unable to bear or conceive children, adoption is often the best route to parenthood. In addition, about one quarter of those who adopt do so simply because they would like to give a needy child a home, while a small percentage adopt stepchildren or the children of relatives (Fisher 2003).

Finding a child to adopt is not easy, however. In 1963, when abortion was illegal and single motherhood was highly stigmatized among white Americans, about 40 percent of babies born to unwed white mothers were given up for adoption (Fisher 2003). These days, less than 1 percent are. (Single motherhood has consistently been more common and less stigmatized among African Americans.) As a result, increasing numbers of Americans now seek babies to adopt overseas, and obtaining a baby to adopt is difficult and expensive.

Adoption can be a wonderful way to create a family, and large, long-term studies find that the overwhelming majority of adoptions are highly successful for both parents and children. This is true even when the children are adopted after spending up to a few years in orphanages (Fisher 2003). But adoption also means the *disruption* of a family: One family lost a baby for another to get a baby. Even when mothers choose to give away a baby, they typically do so because they have no other viable choice: They cannot afford to feed a baby, they and their baby will suffer great stigma if they raise the child out of wedlock, or they lack the basic social support that anyone needs to raise a child. This is why about 50,000 children are adopted yearly from the U.S. child welfare system, whereas in Sweden, where mothers (whether married or not) receive extensive social services and support, fewer than a dozen children are put up for adoption each year (Rothman 2005).

Work versus Family

These days, whether or not a couple has children, it's likely that they are experiencing a time crunch at home. There are several reasons for this. First, 68 percent of married women aged 25 to 34 are now in the labor force (U.S. Bureau of the Census 2006). Second, for middle-class Americans, workdays and work weeks are growing longer. Individuals must work early, late, and on weekends and must take work home to demonstrate that they are serious players. Working-class Americans, on the other hand, increasingly can find only part-time employment. Those who have full-time jobs, meanwhile, often must work overtime to earn enough to make ends meet. Still others are pressured to work extra hours off the books and without pay in order to keep their jobs (Ehrenreich 2001). As a result, both working-class and middle-class parents can experience a time bind at home. Family meals are increasingly rare, and time at home becomes rigidly scheduled as parents try to get themselves to work, do the laundry, keep their home reasonably clean, and get their children to school or other activities on time.

This time bind is often explained as the inevitable result of decreasing real wages, global competitiveness in the workplace, and the growing taste for expensive consumer goods. In an influential study, however, sociologist Arlie Hochschild (1997) argues that many middle-class parents are choosing to spend more time at work because they find work more rewarding than being at home with their family. The more hectic it gets at home, the nicer the job looks. Bosses and coworkers hardly ever spill their juice, dirty their diapers, cry, or slam out of the house because they cannot use the car. Compared with home, the workplace tends to be relatively quiet and orderly and the work rewarding. For many, work rather than home is the place where you can put your feet up and drink a quiet cup of coffee, work is the place where you can get advice on your meddlesome mother-in-law or crumbling marriage, and work is the place where employers notice that you're under a lot of stress and provide free professional counseling. Plus, of course, at work there are paychecks, promotion opportunities, and recognition ceremonies.

More recent research, however, suggests that most work such long hours only because they have no choice (Jacobs & Gerson 2004). On a more positive note, recent research also suggests that, whatever the stresses of long work weeks, parents are finding ways to manage this time bind without cutting back on time spent with children (Bianchi, Robinson, & Milkie 2006). In fact, today's mothers spend as much time with their children as did mothers 40 years ago, and fathers spend considerably more time with children today. The difference is that today's mothers have cut back dramatically on the housework they do, whereas—today's fathers do a little more than they used to. In addition, parents preserve the time they can spend with their children by having fewer children and, if they can afford it, by hiring more outside help.

These solutions, however, are simply means to help parents work even longer hours. Real solutions would require a reduction in overtime work, a living wage that enabled individuals to work fewer hours, and a cultural shift that valued raising children as much as careers. For the time being, it seems likely that Americans will continue to be stressed by the competing demands of work and family.

Problems in the American Family

Some couples swear that they never have an argument and never disagree. These people are certainly in the minority, however, for most intimate relationships involve some stress and strain. We become concerned when these stresses and strains affect the mental and physical health of the individuals and when they affect the stability of society. In this section, we cover two problems in the U.S. family: violence and divorce.

Violence

Child abuse is nothing new, nor is wife battering. These forms of family violence, however, didn't receive much attention until recent years. In a celebrated court case in 1871, a social worker had to invoke laws against cruelty to animals in order to remove a child from a violent home. There were laws specifying how to treat your animals, but no restrictions on how wives and children were to be treated. In recent years, however, we have become both more aware and less tolerant of violence in the home.

The incidence of child abuse is particularly hard to measure, since it is difficult to obtain permission to interview children outside of their parents' presence. Surveys of child protective services professionals give us at least a starting point for estimating abuse. These surveys suggest that each year, 1.5 million children are known to be sexually, physically, or emotionally abused by their parents or caregivers, with about one third of these receiving serious physical injuries (Sedlak & Broadhurst 1996). This figure is obviously an underestimate, as it does not include those whose abuse remains hidden. For this reason, the best data currently available come from a national random survey of 16,000 Americans conducted for the National Institute of Justice (Tjaden & Thoennes 1998). Of women interviewed, 10 percent reported experiencing rape or attempted rape during their childhood, primarily at the hands of family members. These figures, too, are likely to be substantial underestimates, since they do not include those adults who refused to talk about their experiences. In addition, the survey did not include individuals who for whatever reason were in prison, a mental or general hospital, or some other institution at the time of the survey—all settings in which a disproportionate number of residents have experienced childhood abuse.

The same survey gives us our best measure of the extent of violence between adults in families (Tjaden & Thoennes 2000). The survey found that 22 percent of women and 7 percent of men have been physically assaulted by a spouse or cohabitant of the opposite sex (Table 11.1). Women were twice as likely as men to have required medical care after being assaulted. Violence was almost as common in male homosexual couples as in heterosexual couples, but was much rarer among lesbians.

Recently, concern also has been raised about violence directed at dependent, vulnerable, elderly parents by their adult children. Research has found, however, that most victims of elder abuse have been attacked by their spouses rather than by their children (Bergen 1998).

Family violence is not restricted to any class or race (Johnson & Ferraro 2000). It occurs in the homes of lawyers as well as the homes of welfare mothers. Violence is most likely to occur when individuals feel they are losing control, whether over their spouse or over other aspects of their lives. One reason men are more likely than women to beat their spouses is because they are more likely to believe that they *should* control their spouses (Johnson & Ferraro 2000).

Solutions to family violence are complex. The first step, however, is to make it clear that violence is inappropriate and illegal. New laws against spousal rape and

TABLE 11.1
Violence between Spouses and Cohabitants

	Percentage who have been assaulted
Men by female spouses or cohabitants	7%
Women by male spouses or cohabitants	22
Women by female cohabitants	11
Men by male cohabitants	23

SOURCE: Tjaden & Thoennes 2000.

other forms of family violence may clarify what used to be rather fuzzy norms about whether family violence was appropriate (Straus & Gelles 1986).

Divorce

In the United States, more than 2 million adults and approximately 1 million children are affected annually by divorce. The **divorce rate,** calculated as the number of divorces each year per 1,000 married women, rose steadily from the post-World War II period through the 1970s but has stayed at about 20 per 1,000 since then. That is, about 2 percent of all married women in the United States divorce annually.

A better way of looking at divorce is to calculate the **divorce probability:** the probability that a marriage will end in divorce within a given time period. Since most divorces occur within the first ten years of marriage, it is especially useful to look at the probability of divorce within those first 10 years. As Figure 11.2 indicates, the probability of divorce within the first 10 years increased steadily from the 1950s through the 1970s, but it has been dropping slowly since then (Kreider 2005). The same pattern holds true for the probability that a marriage will end in divorce by the time one partner dies. Of marriages begun in 1890 the proportion eventually ending in divorce was approximately 10 percent (Cherlin 1992). Today most researchers believe that between 40 and 50 percent of first marriages will eventually end in divorce.

What are the factors that make a marriage more likely to fail? Table 11.2 displays some of the predictors of divorce within the first 10 years of marriage. Research consistently finds six factors especially important (Bramlett & Mosher 2002; Teachman 2002):

- *Age at marriage.* Probably the best predictor of divorce is a youthful age at marriage. Marrying as a teenager or even in the early twenties doubles chances for divorce compared with those who marry later (see Table 11.2).
- *Parental divorce.* People whose parents divorced are themselves more likely to divorce.
- *Premarital childbearing.* Having a child before marriage reduces the stability of subsequent marriages. If an unwed woman marries before giving birth, however, that marriage is no more likely than others to end in divorce.
- *Education.* The higher one's education, the less likely one's marriage is to end in divorce. College graduates are only half as likely to divorce as are those without college degrees (Hurley 2005). Partly this is because people with higher educations are more likely to come from two-parent families, avoid premarital childbearing, and marry later. Independent of these other factors, however, higher education does reduce the chances of divorce.

divorce rate is calculated as the number of divorces each year per 1,000 married women

divorce probability is the estimated probability that a marriage will ever end in divorce within a given time period

FIGURE 11.2
Percentage of Women Divorced from Their First Husband within 10 Years, by Year of Marriage
The probability that a marriage will end in divorce within the first 10 years is much higher than it was 50 years ago, but has been declining since the late 1970s. Of course, even marriages that survive the first 10 years can end 15, 20, and even 30 years after marriage.

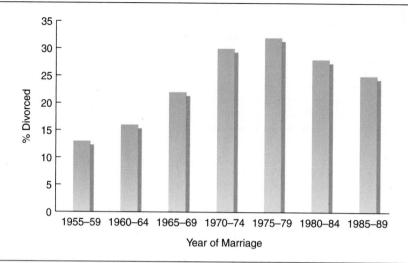

• *Race.* African Americans are substantially more likely than whites, Hispanics, or Asians to get divorced, although the difference has declined over time (Teachman 2002).
• *Religion.* Catholics are significantly less likely than others to get divorced, even after a variety of other demographic variables are taken into account.

Societal-Level Factors

Age at marriage, parental divorce, premarital childbearing, education, race, and religion affect whether a particular marriage succeeds or fails. These personal characteristics, however, cannot account for the fact that between 40 and 50 percent of couples now marrying are expected to eventually divorce (Kreider 2005). The shift from a lifetime divorce probability of 10 percent to one of 40 to 50 percent within the last century is a social problem, not a personal trouble, and to explain it we need to look at social structure.

The change in marital relationships is probably most clearly associated with changes in economic institutions. Rising divorce rates are not unique to the United States (Figure 11.3). Although divorce has always been more prevalent in the United States than elsewhere, most industrialized nations have experienced substantial increases in divorce. In France, Greece, and the United Kingdom, for instance, the divorce rate tripled between 1970 and 2000. These changes reflect changes in the economy. In agricultural societies, individuals' main assets were tools or land. Because divorce meant that one spouse would lose those assets, few could afford to consider it. In today's economy, middle-class individuals' main assets are their education and experience. Because people can walk away from a marriage and take these assets along, divorce no longer seems as risky. At the same time, changes in the economy have made it more difficult for lower-class men and women to support themselves or a family. The resulting economic hardships cause enormous stress within relationships, often resulting in divorce. Finally, now that women have greater opportunities to support themselves outside marriage, divorce can seem a more appealing option.

Because of these economic changes, women and men are now less impelled or less able to marry or to stay married. Because no new incentive for marriage has proven to be more effective than economic need, marriages have less institutional

TABLE 11.2

Factors Predicting Whether First Marriage Will Break Up within the First 10 Years

The probability that a first marriage will ever end in divorce is currently between 40 and 50 percent. Divorce is more likely for African Americans; children of divorced parents; and persons who marry young, have limited education, or have a child before marriage.

	% Ending in Divorce
Total	23%
Age at Marriage	
<18	48
18–19	40
20–24	29
25 or over	24
Education	
Less than 12 years	42
12 years	36
13 years or more	29
Children before marriage	
No	31
Yes	50
Race	
White	32
African American	47
Hispanic	34
Asian	20
Children of divorced parents	
Yes	43
No	29

SOURCE: Bramlett & Mosher 2002; U.S. Bureau of the Census 2006.

support than before. Nevertheless, it is important to note that most people whose marriages end in divorce eventually remarry, with remarriage especially common among young people, whites, and men (Coleman, Ganong, & Fine 2000).

How Serious Are These Problems?

In Chapter 4, we noted that some people view institutions as constraints that force people into uncomfortable and perhaps oppressive relationships; others see institutions as providing the stability and comfort frequently associated with old shoes. This difference of views is nowhere more present than in the case of the family.

Theoretical viewpoints sharply influence perceptions of the health of the modern family. If the family is an oppressive institution, then divorce is a form of liberation; if the family is the source of individual and community strength, then divorce undermines society. In this section, we briefly review some of the major criticisms of the contemporary family and conclude with a perspective on the future.

Loss of Commitment

Some critics argue that a major problem with the U.S. family—and, they would further argue, with U.S. culture—is an accent on individual growth at the expense of commitment to family and community (Bellah et al. 1985, 1991; Putnam 2000). This criticism is most likely to come from structural functionalists, who traditionally stress

FIGURE 11.3
Divorce Rates per 1,000 for
Selected Countries, 1980–2002
SOURCE: U.S. Bureau of the Census 2006.

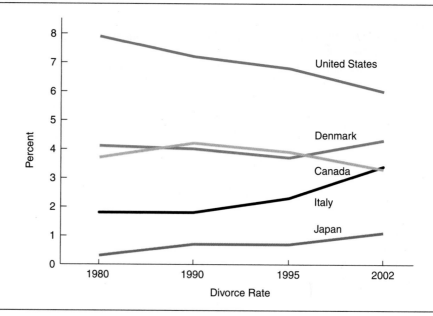

the subordination of individual to community needs, but it also comes from symbolic interactionists concerned with stable self-identity.

The most prominent symptom of the alleged emphasis on individual happiness and growth at the expense of long-term commitment is the rapid rise in the number of women who are raising children on their own. Because men don't want to be tied down by wives and children and because wives don't want to be tied down by husbands, fathers have walked away and mothers have let them, even encouraged them. Following divorce, many fathers all but sever ties to their children (Lye 1996; Seltzer 1994). The result is that the basic family unit, in the sense of long-term commitment to sharing and support, has become the mother-child pair. Husband-wife and father-child relationships are increasingly viewed as temporary and even optional. To many critics, this increasingly voluntary nature of family ties is dysfunctional, reducing the stability of the family and decreasing its ability to perform one of its major tasks: caring for children. Among the ill consequences they note is the increasing proportion of women and children living in poverty.

Inequality

Sociologists from a conflict perspective are more apt to criticize the family for its oppression of women and children. They point to the lower power of women and children relative to men and to the number of women and children who suffer abuse in the family. For example, Denzin (1984, 487–488) declares, "A patriarchal, capitalist society which promotes the ownership of firearms, women, and children; which makes homes men's castles; and which sanctions societal and interpersonal violence in the forms of wars, athletic contests, and mass media fiction (and news) should not be surprised to find violence in its homes. Violence directed toward children and women is a pervasive feature of sexual divisions of labor which place females and children in subordinate positions to adult males."

From this perspective, the family has not really changed for the worse: It has traditionally been an oppressive institution maintained by force and fraud. These critics do note the development of greater egalitarianism in the contemporary family but generally find that these changes are occurring too slowly.

■ Despite the very real problems many families face, family relationships continue to be a major source of satisfaction for most Americans throughout their lives.

© Richard Hutchings/Photo Researchers Inc.

One major problem from this perspective is that women's and children's independence from husbands and fathers is coming before their independence in the marketplace. The solution they recommend is equal opportunity and equal pay for women and state support for children in the form of family allowances. They note that in Scandinavian countries—where women have near equal opportunity in the job market and the state amply supports day care, education, and health care—single motherhood is not linked to poverty.

Another major problem noted by these critics is that even though women are increasingly encouraged to adopt what had formerly been viewed as "masculine" roles, the traditionally "female" role remains undervalued. As a result, few men are willing to fully share parenting or housework, and some will abandon their children if their marriages break up. Furthermore, neither women nor men who remain home with their children receive much respect or support for doing so. From this perspective, true solutions to problems within the family will require not only economic change but also change in the way we think about gender.

Where This Leaves Us

Despite all the changes and disruptions in families today, they remain the major source of economic support for children and of social support for people of all ages. Families also provide us with an important arena in which we can develop our self-concept, learn to interact with others, and internalize society's norms. Without the strong bonds of love and affection that characterize family ties, these developmental tasks are difficult if not impossible. Thus, the family is essential for the production of socialized members, people who can fit in and play a productive part in society.

The family is important not just in childhood but throughout the life course. Although we don't always get what we desire, our family members are still usually the ones we turn to when we need love, emotional support, financial assistance, and companionship. If you need an emergency loan to replace your car after an accident, or you need someone who will care for you for weeks on end while you recover from an illness, you are most likely to call on a close family member.

Given these benefits that the family gives to both the individual and society, it would seem to be a reasonable goal to support the family, while simultaneously reducing some of its more oppressive features. This goal is not impossible. Despite current rates of divorce, illegitimacy, childlessness, and domestic abuse, there are signs of health in the family: the durability of the mother–child bond, the frequency of remarriage, the number of stepfathers who willingly support other men's biological children, and the frequency with which elderly persons rely on and get help from their children.

There is no doubt that the family is changing. When you ask a young man what his father did when he was growing up, you are increasingly likely to hear, "What father?" or "Which father?" These recent changes must be viewed as at least potentially troublesome. At present we have no institutionalized mechanisms comparable to the family for giving individuals social support or for caring for children. The importance of these tasks suggests that the needs of families and especially children must be moved closer to the top of the national agenda.

Summary

1. Marriage and family are the most basic institutions found in society. In all societies, these institutions meet universal needs such as regulation of sexual behavior, replacement through reproduction, child care, and socialization.

2. Individual development over the life course is closely paralleled by changing family roles. The transition to adulthood is closely linked to adoption of marital and parental roles.

3. High rates of divorce and increases in the labor-force participation of women have led to major changes in the social structure of childhood. Nowadays, many U.S. children spend some time in a single-parent household before they are 18. About one quarter of preschoolers with employed mothers attend day-care centers. Although high-quality day-care programs can benefit children, low-quality programs are more common and may be detrimental.

4. Because of postponement of marriage and high rates of divorce, dating is no longer just a teen activity. Mate selection depends on love, but also on propinquity, homogamy, endogamy, and shared values.

5. Intergenerational bonds are important to individuals. Studies show that parent-child bonds remain strong over the life course and that sibling bonds remain important even in old age. The one exception is that at all ages, noncustodial fathers often have only weak ties to their children.

6. The increasing participation of wives in the breadwinning role is a major change in family roles. The change has been accompanied by a slight increase in husbands' housekeeping and child-care work and by a major decline in women's commitment to housework.

7. Cohabitation is now a common stage of courtship. Many cohabiting couples eventually marry, but a sizable minority are content to put off marriage indefinitely. The decline in marriage rates and increase in rates of cohabitation and divorce lead some to suggest that we are now experiencing the deinstitutionalization of marriage.

8. Growing numbers of women now choose to delay child-bearing or to forego having children altogether. One third of all births in the United States now occur outside of marriage. Teenage pregnancy stems from poverty, a lack of easily available contraception, and a lack of other ways to find meaning and personal satisfaction.

9. As the stigma against single motherhood has declined, it has become more difficult and expensive to find a child to adopt. Babies are most often available for adoption when single mothers are stigmatized and receive few social supports for raising a child on their own. The vast majority of adoptions are successful for both child and adoptive parents.

10. Violence against both children and intimate partners is relatively common in U.S. homes. In both homosexual and heterosexual relationships, men are more likely than women to batter their partners. Family violence is most likely when individuals feel they have lost control over their lives and their spouses and believe that they have a right to control their spouses.

11. It is estimated that 40 to 50 percent of first marriages will end in divorce. Factors associated with divorce include age at marriage, parental divorce, premarital childbearing, education, race, and religion. Reduced economic dependence on marriage underlies many of these trends.

12. Perceptions of the health of the family depend on the theoretical orientation of the viewer. Two problems are loss of commitment and inequality within the family.

Thinking Critically

1. What are the benefits of living in a nuclear family? What functions does it serve for individuals and society? What are its major dysfunctions?

2. Analyze the mate selection processes that you (or someone close to you) have undergone. Show how propinquity, homogamy, endogamy, and appearance were or were not involved. What role did parents play?

3. Do you know anyone who is taking care of an elderly parent or grandparent? Why do you think that person rather than some other family member has assumed that responsibility? What personal characteristics and what relational characteristics are involved?

4. How many children do you plan to have? What do you think the advantages and disadvantages will be? How and on what basis do you think you and your significant other should divide child-care responsibilities?

Companion Website for This Book

academic.cengage.com/sociology/brinkerhoff
Gain an even better grasp on this chapter by going to the Companion Website. This resource contains tutorial quizzes and flash cards to help you master key terms and concepts.

Suggested Readings

Coontz, Stephanie. 2006. *Marriage, a History: From Obedience to Intimacy, or How Love Conquered Marriage.* New York, NY: Penguin. A historian explores how marriage has evolved from an economic contract between families to a romantic quest between individuals, and the consequences of this change for all of us today.

Cosaro, William A. 1997. *The Sociology of Childhood.* Newbury Park, Calif.: Sage. A synthesis of theoretical and empirical work on children and how they shape and are shaped by their families and society. Also includes chapters on children as social problems and on the social problems of children.

Lipper, Joanna. 2003. *Growing Up Fast*. New York: Picador. A gripping portrayal of the lives and struggles of six teenage mothers in an economically depressed small northeastern city.

Savage, Dan. 2005. *The Commitment: Love, Sex, Marriage, and My Family*. New York: Dutton. Columnist Dan Savage writes of his 10-year relationship with another man, his mother's pressure for them to "get married," their 6-year-old son's conviction that men can't marry each other, and what the battles over gay marriage tell us about gays, straights, and American life.

CHAPTER 12

Education and Religion

© Will & Deni McIntyre/CORBIS

Educational and Religious Institutions

This chapter examines two institutions, education and religion. Both are central components of our cultural heritage and have profound effects on our society and on us as individuals. Most Americans are directly and personally affected by these institutions: almost all people in the United States have attended school, and a strong majority practice a religion. Even those who do not go to school or participate in a religion are affected by the omnipresence, norms, and values of these two institutions.

Theoretical Perspectives on Education

The **educational institution** is the social structure concerned with the formal transmission of knowledge.

The **educational institution** is the social structure concerned with the formal transmission of knowledge. It is one of our most enduring and familiar institutions. Nearly 3 of every 10 people in the United States are involved in education on a daily basis as students or staff. As former students, parents, or taxpayers, all of us are involved in education in one way or another.

What purposes are served by this institution? Who benefits? Structural-functional and conflict theories offer two different perspectives on these questions.

Structural-Functional Theory: Functions of Education

A structural-functional analysis of education is concerned with the consequences of educational institutions for the maintenance of society. Structural functionalists point out that the educational system has been designed to meet multiple needs. The major manifest (intended) functions of education are to provide training and knowledge, to socialize young people, to sort young people appropriately, and to facilitate positive and gradual change.

Training and Knowledge

The obvious purpose of schools is to transmit knowledge and skills. In schools, we learn how to read, write, and do arithmetic. We also learn the causes of the American War of Independence and the parts of a cell. In this way, schools ensure that each succeeding generation will have the skills needed to keep society running smoothly.

Socialization

In addition to teaching skills and facts, schools help society run more smoothly by socializing young people to conform. They emphasize discipline, obedience, cooperation, and punctuality. At the same time, schools teach students the ideas, customs, and standards of their culture. In American schools, we learn to read and write English, we learn the Pledge of Allegiance, and we learn the version of U.S. history that school boards believe we should learn. By exposing students from different ethnic and social class backgrounds across the country to more or less the same curriculum, schools help create and maintain a common cultural base.

Sorting

Schools are like gardeners; they sift, weed, sort, and cultivate their products, determining which students will be allowed to go on and which will not. Grades and test scores channel students into different programs—or out of school altogether—on the basis of their measured abilities. Ideally, the school system ensures the best use of each student's particular abilities.

Promoting Change

Schools also act as change agents. Although we do not stop learning after we leave school, new knowledge and technology are usually aimed at schoolchildren rather than at the adult population. In addition, schools can promote change by encouraging critical and analytic skills. Colleges and universities, are also expected to produce new knowledge.

Conflict Theory: Education and the Perpetuation of Inequality

Conflict theorists agree with structural functionalists that education reproduces culture, sorts students, and socializes young people, but they view these functions in a very different light. Conflict theorists emphasize how schools reinforce the status quo and perpetuate inequality.

Education as a Capitalist Tool

Some conflict theorists argue that one primary purpose of public schools is to benefit the ruling class. These theorists point to schools' **hidden curriculum,** the underlying cultural messages that schools teach. In public schools, this curriculum includes learning to wait your turn, follow the rules, be punctual, and show respect, as well as learning *not* to ask questions. All of these lessons prepare students for life in the working class (Gatto 2002). A different hidden curriculum in elite private schools trains young people to think creatively and critically and to assume that they are naturally superior and deserving of privilege. Conflict theorists note that both private and public schools teach young people to expect unequal rewards on the basis of differential achievement and so teach young people to accept inequality (Kozol 2005).

The **hidden curriculum** socializes young people into obedience and conformity.

Education as a Cultural Tool

Conflict theorists argue that, along with teaching skills such as reading and writing, children learn the cultural and historical perspective of the dominant culture (Spring 2004). For example, U.S. history texts describe the "Indian Wars" but rarely explain why Native American tribes resorted to warfare and give little or no coverage to the waves of anti-Chinese violence in the United States in the late nineteenth century or the removal of Japanese Americans to relocation camps during World War II. Art and music classes typically ignore the cultures of Latin America and Asia and gloss over the many contributions African Americans have made in the United States.

Education as a Status Marker

One supposed outcome of free public education is that merit will triumph over origins, that hard work and ability will be allowed to rise to the top. Conflict theorists, however, argue that basing decisions regarding who should get the best jobs and highest status on individuals' educational credentials does little to equalize economic

©Charles Gupton/Stock, Boston Inc.

■ In all societies, education is an important means of reproducing culture. In addition to skills such as reading and writing, children learn many of the dominant cultural values. In Japan, school uniforms emphasize group solidarity over individual achievement.

Credentialism is the assumption that some are better than others simply because they have a particular educational credential.

opportunity. Instead, a subtle shift has taken place. Instead of inquiring who your parents are, prospective employers ask what kind of education you have and where you got it. Because people from affluent families tend to end up with the best educational credentials—the median family income for Harvard students *who apply for financial aid* is about $150,000 (Leonhardt 2004)—the emphasis on credentials serves to keep "undesirables" out. Conflict theorists argue that educational credentials are mere window dressing; apparently based on merit and achievement, credentials are often a surrogate for race, gender, and social class (Brown 2001). In the same way that we use the term *racism* to refer to bias based on race, sociologists use the term **credentialism** to refer to bias based on credentials: *Credentialism* is the assumption that some are better than others simply because they have a particular educational credential.

Unequal Education and Inequality

The use of education as a status marker is reinforced by the very unequal opportunities for education available to different social groups and communities (Kozol 2005). In poor communities, students sit in overcrowded classrooms, where undertrained, substitute, or newly-graduated teachers are encouraged to focus on rote memorization rather than creative thinking skills. Students can choose to take auto mechanics or cosmetology, but their school is not likely to offer calculus, creative writing, or advanced placement (AP) classes. And regardless of which classes are offered, students find it difficult to learn when their classrooms lack proper heating or cooling and they must share outdated textbooks with other students. In contrast, in affluent communities, students sit in state-of-the-art classrooms and science laboratories and can choose from a variety of languages, challenging topics, and AP classes. A staff of advisors will help them gain admission to the most prestigious college that fits their needs and abilities; at the most selective U.S. colleges, 55 percent of freshmen come from families earning in the top 25 percent of income (Leonhardt 2004). Similarly, in mixed-income communities the

It is difficult for any children to learn in crowded classrooms that lack proper heating or cooling. It is even more difficult when students are taught by beginning or substitute teachers and must share out-dated textbooks with other students. Such conditions are considerably more common in poor and minority communities.

© Bob Daemmrich/The Image Works

wealthier students typically receive a far better education, with a very different range of classes, than do the poorer students (Bettie 2003).

Ethnic differences in access to educational opportunities mirror social class differences. Public school segregation was outlawed by the U.S. Supreme Court in 1954, and segregation did decline significantly over the next 30 years. Since the mid-1980s, however, judicial support for desegregation programs has declined, and school segregation has steadily increased for both Hispanic and African American students (Frankenberg & Lee 2002). Fewer than 15 percent of students are white in some public schools, from Boston to Birmingham. The higher the percentage of minority students at a school, the lower the chances that the school will offer students the opportunities they need to learn, to graduate high school, or to go on successfully to college. Within a given school as well, minority students are typically offered far fewer opportunities than are white students (Bettie 2003).

Symbolic Interactionism: The Self-Fulfilling Prophecy

In the modern world, the elite cannot directly ensure that their children remain members of the elite. To pass their status on to their children, they must provide their children with appropriate educational credentials. To an impressive extent, they are able to do so: Students' educational achievements are very closely related to their parents' social status.

How does this happen? Whereas conflict theorists emphasize how the *structure* of schools leads to these unequal results, symbolic interactionists focus on the processes that produce these results. Perhaps the most important such process is the self-fulfilling prophecy.

Self-Fulfilling Prophecy

One of the major processes that takes place in schools is, of course, that students learn. When they graduate from high school, many can type, write essays with three-part theses, and even do calculus. In addition to learning specific skills, they also undergo a process of cognitive development in which their mental skills grow and expand. In the ideal case, they learn to think critically, to weigh evidence, and to develop independent judgment.

An impressive set of studies demonstrates that cognitive development during the school years is greatest when teachers set high expectations for their students and, as a result, give their students complex and demanding work. Teachers are most likely to do this when students fit teachers' expectations for how "smart" students should look and behave. This is most likely when students are white and middle- or upper-class.

One explanation for this is that teachers share the racist and classist stereotypes common in our society. Another explanation is that white, well-off students are likely to have more cultural capital—attitudes and knowledge common in elite culture (Bourdieu 1984; Bettie 2003). They are more likely to have been introduced at home to the sort of art, music, and books that middle-class teachers value. They also are more likely to dress and behave in a way that teachers appreciate. This cultural capital helps them in their interactions with teachers and convinces teachers that they are worth investing time in (DiMaggio & Mohr 1985; Farkas et al. 1990; Kalmijn & Kraaykamp 1996; Teachman 1987).

In contrast, teachers (most of whom are white) are especially likely to assume that African American and Mexican American students are unintelligent and prone to trouble (Ferguson 2000; Bettie 2003). As a result, teachers often focus more on disciplining and controlling minority students than on educating them.

This process is a perfect example of a self-fulfilling prophecy. Those who are now teachers themselves grew up in a society still characterized by racist, sexist, and classist biases. When teachers biases' lead them to assume that certain students cannot succeed, the teachers give those students less opportunity to do so. So girls don't get taught calculus, boys (whether African American or white) don't learn how to cook, and working-class students (whether male or female, white or nonwhite) are encouraged to take cooking or auto mechanics rather than physics. This process helps to keep disadvantaged students from succeeding.

Current Controversies in American Education

In recent years, various proposals have emerged to improve the quality of education in the United States and to give young Americans the tools needed to be more competitive in an increasingly global job market. Three proposals that have been widely adopted are tracking, high-stakes testing, and school choice.

Tracking

Tracking occurs when evaluations made relatively early in a child's career determine the educational programs the child will be encouraged to follow.

Tracking is the use of early evaluations to determine the educational programs a child will be encouraged or allowed to follow. When students enter first grade, they are sorted into reading groups on the basis of ability. By the time they are out of elementary school, some students will be directed into college preparatory tracks, others into general education (sometimes called vocational education), and still others into remedial classes or "special education" programs. At all levels, and regardless of their actual abilities, minority and less affluent students are more likely to be put into lower tracks (Bettie 2003; Kao & Thompson 2003; Harry & Klingner 2005).

Because students who are assigned to lower-track classes receive few rewards for their academic efforts from either parents or teachers, many will quit school and seek their rewards from delinquent peers.

© AP/Wide World Photos

Ideally, tracking is supposed to benefit both gifted and slow learners. By gearing classes to their levels, both groups should learn faster and should benefit from increased teacher attention. In addition, classes should run more smoothly and effectively when students are at a similar level. In some ways, this is indeed true. Nevertheless, one of the most consistent findings from educational research is that students are helped modestly by assignment to high-ability groups but hurt significantly if put in low-ability groups (Kao & Thompson 2003).

An important reason students assigned to low-ability groups learn less is because they are taught less. They are exposed to less material, asked to do less homework, and, in general, are not given the same opportunities to learn. Because teachers expect low-track students to do poorly, the students find themselves in a situation where they cannot succeed—a self-fulfilling prophecy.

Less formal processes also operate. Students who are assigned to high-ability groups, for instance, receive strong affirmation of their academic identity; they find school rewarding, have better attendance records, cooperate more with teachers, and develop higher aspirations. The opposite occurs with students placed in low-ability tracks. They receive fewer rewards for their efforts, their parents and teachers have low expectations for them, and there is little incentive to work hard. Many will cut their losses and look for self-esteem through other avenues, such as athletics or delinquency (Bettie 2003). However, these negative effects of tracking diminish in schools where mobility between tracks is encouraged, teachers are optimistic about the potential for student improvement, and schools place academic demands on students who are not in college tracks (Gamoran 1992; Hallinan 1994).

High-Stakes Testing

Both federal and many local laws now require schools to measure student performance using standardized achievement tests. In many school districts, students must now pass these "high stakes" tests before they can move on to a higher grade. In addition, teachers and schools increasingly are evaluated, punished, or rewarded based on results from standardized examinations.

The emphasis on documenting school achievement through standardized test performance has pressed schools to pay more attention to the quality of the education their students receive and has encouraged them to make sure that all students receive good training in basic skills such as reading, writing, and arithmetic.

But high-stakes testing also has had unanticipated negative consequences (Berliner & Biddle 1995). Few schools have received additional resources to meet these new goals. As a result, schools have dropped classes in art, music, physical education, foreign languages, and even history and science so they can use these teachers for classes in reading, writing, and arithmetic—even when the teachers lack the training to teach these subjects (Berliner & Biddle 1995). Furthermore, teachers can

American Diversity:

focus on

What Do IQ Tests Measure?

- How many legs does a Kaffir have?
- Who wrote *Great Expectations*?
- Which word is out of place? sanctuary—nave—altar—attic—apse
- If you throw the dice and 7 is showing on top, what is facing down? 7—snake eyes—boxcars—little joe's—11

If you answered two, Dickens, attic, and 7, then you get the highest possible score on this test. What does that mean? Does it mean that you have genetically superior mental ability, that you read a lot, or that you shoot craps? What could you safely conclude about a person who got only two questions right?

The standardized test is one of the most familiar aspects of life in American schools. Whether it is the California or the Iowa Achievement Test, the SAT or the ACT, students are constantly being evaluated. Most of these tests are truly achievement tests; they measure what has been learned and make no pretense of measuring the capacity to learn. IQ tests, however, are supposed to measure the innate capacity to learn—mental ability. On these tests, African American, Hispanic, and Native

American students consistently score below white students, and working-class students score substantially below middle-class students. The obvious question is whether these tests are fair measures. Do African American, Hispanic, Native American, and working-class youths have lower mental ability than middle-class or white youths?

Before we can answer this question, we must first ask another: What is mental ability? It is an aspect of personality, "the capacity of the individual to act purposefully, to think rationally, and to deal effectively with his environment" (Wechsler 1958, 7).

Do questions such as those that opened this section measure any of these things? No, they do not. We can all imagine people who act purposefully, think rationally, and deal effectively with the environment but do not know who wrote *Great Expectations* and are ignorant about dice or church architecture. These people may be foreigners, they may have lacked the opportunity to go to school, or they may have come from a subculture where dice, churches, and nineteenth-century English literature are not important.

For this reason, good IQ tests try to measure the ability to think and reason independently of formal education. Do these tests achieve their intention? Do they measure the ability

to reason independently of years in school, subcultural background, or language difficulties? Again, the answer seems to be no.

There are two ways in which these tests are not culture-free. The first is that they reflect not only reasoning and knowledge but also competitiveness, familiarity with and acceptance of timed tests, rapport with the examiner, and achievement aspiration. Students who lack these characteristics may do poorly even though their ability to reason is well developed.

The more serious fault with such nonverbal tests is their underlying assumption. Reasoning ability is not independent of learning opportunities. How we reason, as well as what we know, depends on our prior experiences. The deprivation studies of monkeys and hospitalized orphans (see Chapter 3) demonstrate that mental and social retardation occur as a result of sensory deprivation. Just as the body does not develop fully without exercise, neither does the mind. Thus, reasoning capacity is not culture-free; it is determined by the opportunities to develop it. For this reason, there will probably never be an IQ test that measures test takers' true *abilities*, rather than measuring their previous learning opportunities.

afford to spend time only on teaching those aspects of the subjects that appear on the tests. In addition, teachers now must devote time simply to teaching test-taking skills. Meanwhile, the testing process itself costs school districts considerable time, energy, and money.

High-stakes testing also means that some students will be held back a grade and thereby stigmatized as failures. At the end of the 2002/2003 school year, for example, 23 percent of Florida third graders were held back because they failed to score high enough on the state reading test (Winerip 2003). Yet research suggests that holding students back can *reduce* their long-term academic performance and *increase* their chances of dropping out. Moreover, those who fail are disproportionately lower class and minority, for a variety of reasons. Similarly, when standardized achievement exams are used to determine who should graduate, be admitted to college, or receive financial aid, they typically increase inequality between races and social classes (McDill, Natriello, & Pallas 1986). Finally, there is some evidence that, to artificially improve their schools' rankings on high-stakes tests, schools are encouraging or even forcing low-performing students to leave school before taking the tests—turning potential dropouts into "push-outs" (Lewin & Medina 2003).

School Choice

Concern about the quality of American public education has led to a variety of proposals and programs for increasing school choice. **School choice** refers to a range of options (including tuition vouchers, tax credits, magnet schools, charter schools, and home schooling) that enable families to choose where their children go to school. Tuition vouchers and income tax credits are designed to help families pay for private (and, in some cases, religious) schools. Magnet schools are public schools that try to attract students through offering high-quality special programs or approaches; most commonly these schools emphasize either basic skills, language immersion, arts, or math and science. Charter schools are similar to magnet schools but are privately controlled. Charter schools receive some public funding and are subject to some public oversight, such as requirements that they offer certain courses and that their students meet specified measures of academic performance.

Proponents of school choice argue that when schools compete with each other for students, they provide better quality services, in the same way that Ford and Chevrolet compete to provide better cars (Chubb & Moe 1990; Schneider, Teske, & Marschall 2000). The school choice movement reflects the animosity toward "big government" that has been building in the United States for the last quarter century and is part of a broader movement toward **privatization:** the process of taking goods and services out of governmental control and instead treating them like any other marketable commodity. School choice has found supporters on the left as well as the right: black separatists, liberal believers in free-form "alternative schools," and Evangelical Christians all may prefer that their children attend schools where their own values are reinforced.

Although there is some merit to the arguments for school choice, it is difficult to scientifically document its benefits. The problem is that students who participate in school choice programs differ from other students from the beginning. Their parents are often more educated than other parents. More importantly, by definition their parents are committed to seeking out the best education for their children, knowledgeable about the options available, and willing to invest time and effort in obtaining the best options for their children. As a result, no matter what schools their children attend, they will likely do well.

School choice refers to a range of options (vouchers, tax credits, magnet and charter schools, home schooling) that enable families to choose where their children go to school.

Privatization is the process of taking goods and services out of governmental control and instead treating them like any other marketable commodity—something to be bought and sold in a competitive market.

Opponents of school choice identify several unintended negative consequences of these programs. First, these programs reinforce social inequality. Because tuition vouchers and tax credits do not cover the full cost of tuition and transportation, only middle- and upper-income children can afford to use them. Second, because white and affluent parents typically prefer not to send their children to schools with many poor or nonwhite students, school choice programs unintentionally increase segregation (Saporito 2003). Third, school choice programs reduce Americans' commitment to public education and to maintaining high-quality schools in all neighborhoods.

College and Society

Before World War II, college and even high school graduation was only common among the elite. Since then, however, there has been a tremendous growth in high school and college education, and today almost half of recent high school graduates ages 18 to 21 are enrolled in college. As Figure 12.1 shows, all segments of the population have been affected by this expansion in education, but significant differences still remain (Kao & Thompson 2003).

Who Goes?

Until recently, non-Hispanic white males were the group most likely to be enrolled in college, but this has changed (Figure 12.2). Because young men can earn good incomes right out of high school, many decide against going to college—even though in the long run they would earn far more money if they did so (Lewin 2006). Young women, on the other hand, have little chance of earning a good income unless they go to college. As a result, rates of college attendance for women in all ethnic groups have increased steadily, while rates among men have stayed stable. However, white men are still the most likely to receive professional and doctoral degrees and to graduate in the fields that promise the highest incomes.

Overall, though, sex differences in college attendance are fairly small compared to ethnic and social class differences (Lewin 2006; Mead 2006). Native Americans are the least likely to graduate high school. African Americans are still slightly less likely than whites or Asians to do so, and Hispanics are considerably less likely to do so, partly because many emigrated here as adults (U.S. Bureau of the Census 2006).

Why Go?

There is no question that a college education pays off economically. As Table 12.1 shows, college graduates are more likely to get satisfying professional jobs with good benefits and are less likely to be unemployed. They also earn nearly double the income of high school graduates.

A college education also offers many less tangible benefits. At its best, college teaches students not only specific skills in math, science, and other fields, but also how to think logically and critically about all aspects of the world. Research shows that students also emerge from college more knowledgeable about the world around them, more active in public and community affairs, and more open to new ideas than those who don't have a college degree (Funk & Willits 1987; Weil 1985); they also lead healthier lives and live longer (Ross & Mirowsky 1999).

College conveys psychological and social benefits as well (Kaufman & Feldman 2004). During college, students learn to talk and behave in ways that older adults

FIGURE 12.1

Educational Achievement of Persons 25 and Older by Race and Ethnicity, 1960–2004

Among whites, the proportion of adults graduating from high school has almost doubled since 1960; among African Americans, the proportion has quadrupled. Nevertheless, African Americans and Hispanics continue to have less education than do whites.

SOURCE: U.S. Bureau of the Census 2006.

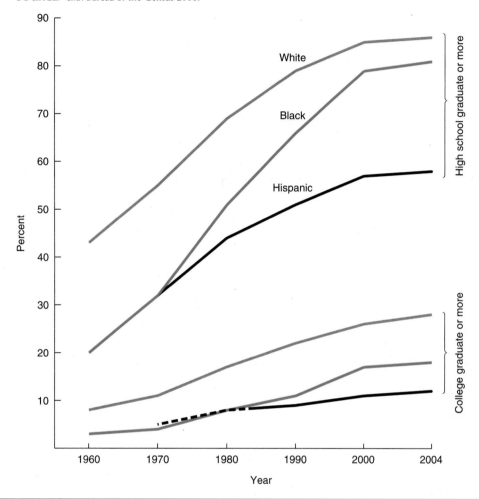

will interpret as smart and middle class (such as substituting "How are you?" for "Yo, whas up?"). College also teaches students to believe they are intelligent and are entitled to middle-class jobs. As a result, college graduates are more confident and more likely to apply for such jobs. At the same time, because American culture stresses that college graduates are more likely than others to have the skills needed for prestigious, high-paying jobs, college graduates are more likely to receive such jobs even if their actual skills are questionable (Brown 2001).

Understanding Religion

Unlike education, which we are forced by law to take part in, we have a choice about participating in religious organizations. Nevertheless, most people in the United States choose to participate, and religion is an important part of social life.

FIGURE 12.2

Percentage of Recent High School Graduates Enrolled in College, by Race, Ethnicity, and Sex, 1975 and 2003

Comparisons by sex, race, and ethnicity show increasing similarity in the likelihood that high school graduates from each category will attend college.

SOURCE: U.S. Bureau of the Census 2006.

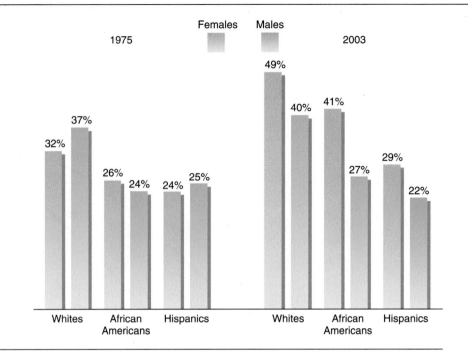

TABLE 12.1

Socioeconomic Consequences of Higher Education, 2003–2004

Going to college pays off—literally. Those who graduate college earn nearly twice as much as high school graduates, are more likely to be employed, and are more likely to have a professional job.

Education	Median Annual Income*	% with Managerial/Professional Job**	% Unemployed**
9–12 years, no degree	$18,990	6%	8.5%
High school graduate	28,763	16	5.0
Less than 4 years college	39,015	33	4.2
College graduate	55,751	72	2.7

*2003 data
**2004 data

SOURCE: U.S. Bureau of the Census 2006.

It is intertwined with politics and culture, and it is intimately concerned with integration and conflict.

What Is Religion?

How can we define *religion* so that our definition includes the contemplative meditation of the Buddhist monk, the speaking in tongues of a modern Pentecostal, the sacred use of peyote in the Native American Church, and the formal ceremonies of the Catholic Church? Sociologists define **religion** as a system of beliefs and practices related to sacred things that unites believers into a moral community (Durkheim [1915] 1961, 62). Religion includes belief systems (such as native African religions) that invoke supernatural forces as explanations of earthly struggles. It does *not* include belief systems such as Marxism and science that do not emphasize the sacred.

Religion is a system of beliefs and practices related to sacred things that unites believers into a moral community.

At its best, college encourages creative and critical thinking and broadens one's view of the world.

Sociologists who study religion treat it as a set of values. They do not, however, ask whether the values are true or false: whether God exists, whether salvation is really possible, or which is the true religion. Rather sociologists examine the ways in which culture, society, and other social forces affect religion and the ways in which religion affects individuals and social structure.

Why Religion?

Religion is a fundamental feature of all societies; Figure 12.3 shows the distribution of the world's population into the major religions as well as the distribution of religions

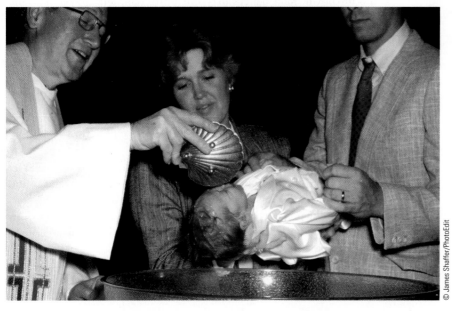

Ritual occasions such as baptisms help individuals and families to affirm their shared values and their membership in a community of believers.

FIGURE 12.3

Religious Affiliation in the United States and Worldwide

Seventy-six percent of Americans are Protestant or Catholic. Worldwide, Christianity remains the largest religion, but it is shrinking while Islam is growing.

SOURCE: Ontario Consultants on Religious Tolerance 2006; U.S. Bureau of the Census 2006.

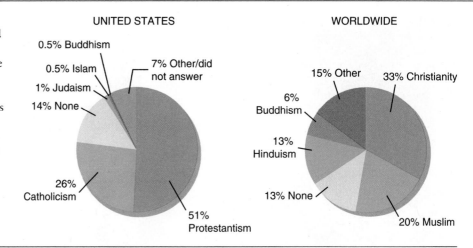

UNITED STATES

0.5% Buddhism
0.5% Islam
1% Judaism
14% None
7% Other/did not answer
26% Catholicism
51% Protestantism

WORLDWIDE

15% Other
6% Buddhism
13% Hinduism
13% None
33% Christianity
20% Muslim

Religious rituals help individuals cope with events that are beyond human understanding, such as death, illness, drought, and famine.

in the United States. Whether premodern or industrialized, every society has forms of religious activity and expressions of religious behavior.

Why is religion universal? One answer is that every individual and every society must struggle to find explanations for, and meaning in, events and experiences that go beyond personal experience. The poor man suffers in a land of plenty and wonders, "Why me?" The woman whose child dies wonders, "Why mine?" The community struck by flood or tornado wonders, "Why us?" Beyond these personal dilemmas, people may wonder why the sun comes up every morning, why there is a rainbow in the sky, and what happens after death.

Religion helps us interpret and cope with events that are beyond our control and understanding; tornadoes, droughts, and plagues become meaningful when they are attributed to the workings of some greater force. Beliefs and rituals develop as a way to control or appease this greater force, and eventually they become patterned responses to the unknown. Rain dances may not bring rain, and prayers may not lead to good harvests, but both provide a familiar and comforting context in which people can confront otherwise mysterious and inexplicable events. Regardless of whether they are right or wrong, religious beliefs and rituals help people cope with the extraordinary events they experience.

Why Religion Now? The Rise of Fundamentalism

Until the 1970s, many scholars implicitly assumed that religion would decline in importance as science and technology increased society's ability to explain and control previously mysterious events (Emerson & Hartman 2006). As a result, they assumed that **secularization**—the process of transferring things, ideas, or events from sacred authority (the clergy) to nonsacred, or secular, authority (the state, medicine, and so on)—would gradually increase.

Certainly science now explains many phenomena—illness, earthquakes, solar eclipses—that previously had been the territory only of religion. And compared with 40 years ago, many more Americans neither belong to religions, consider religion

Secularization is the process of transferring things, ideas, or events from the sacred realm to the nonsacred, or secular, realm.

TABLE 12.2
Changing Religious Commitment, 1962–2006

During the last 40 years, there has been a drop in the proportion of Americans who belong to a religion and who say religion is very important in their lives, and a *sharp* drop in the proportion who think that the Bible is the actual word of God.

	1962–65	2004–2006
Belong to a religion	98	90*
Religion is very important to their own lives	70	55*
Believe Bible is actual word of God, to be taken literally word for word	65	28**

*2004 data
**2006 data

SOURCE: Gallup Poll 2001; Polling Report 2006.

important in their lives, or even believe in God, as Table 12.2 shows. (In northern and western Europe, especially, the proportion of nonbelievers is exceedingly high.) But despite this evidence of secularization, commitment to fundamentalist religions has increased substantially over the last 30 years, in the United States and elsewhere (Sherkat & Ellison 1999; Stark & Finke 2000; Emerson & Hartman 2006).

Fundamentalism refers to religious movements that believe their most sacred book or books are the literal word of God, accept traditional interpretations of those books, and stress the importance of living in ways that mesh with those traditional interpretations. Fundamentalism exists around the world among Catholics, Protestants, Jews, Moslems, and others. Their beliefs are so strong that a small minority of fundamentalists are willing to engage in violence against nonbelievers who they feel are threatening their religion and way of life. Fundamentalist violence is most common in situations in which fundamentalists believe their religion is being suppressed by the government (Emerson & Hartman 2006). Unfortunately, whereas political terrorists aim primarily to get media attention with the goal of promoting social change, religious terrorists (like the Oklahoma City and World Trade Center bombers) are motivated by a sense of divine duty and often feel that the societies they attack are too morally corrupt to change. As a result, they are quite willing to kill for their cause (Hoffman 2006).

Rather than modernization reducing religious commitment, as earlier scholars hypothesized, it appears to have *increased* it: As individuals around the world find their basic values about life, the family, gender relations, and society challenged by modernization, they seek out conservative and fundamentalist religions to fight those changes (Emerson & Hartman 2006). Some researchers regard the adamant rejection of modern, Western beliefs about egalitarian gender relations, family structures, and social order to be so important to fundamentalism that they include this rejection in their definition of fundamentalism (e.g., Marsden 2006).

In addition, other theorists argue, commitment to religion remains a *rational* choice for individuals when the time and money costs of commitment are outweighed by its benefits. Those benefits include explanations for otherwise inexplicable events, the promise of supernatural rewards, integration into a community of like-minded individuals, and the lending of supernatural authority to traditional values and practices (Stark & Finke 2000).

Fundamentalism refers to religious movements that believe their most sacred book or books are the literal word of God, accept traditional interpretations of those books, and stress the importance of living in ways that mesh with those traditional interpretations.

Theoretical Perspectives on Religion

With this as background, we can now look at how sociological theorists approach the study of religion. In the next section, we explore three distinct theoretical perspectives. Structural functionalists, not surprisingly, focus on the functions that religion serves for both individuals and societies. Conflict theorists focus on how religion can foster or repress social conflict. The third perspective, associated with the work of Max Weber, combines elements from these first two perspectives.

Durkheim: Structural-Functional Theory of Religion

The structural-functional study of religion begins, most importantly, with the work of Emile Durkheim. Durkheim began his analysis of religion by identifying the three elements shared by all religions, which he called the elementary forms of religion ([1915] 1961).

Elementary Forms of Religion

The **profane** represents all that is routine and taken for granted in the everyday world, things that are known and familiar and that we can control, understand, and manipulate.

The **sacred** consists of events and things that we hold in awe and reverence—what we can neither understand nor control.

The first of the three elementary forms is that all religions divide human experience into the sacred and the profane. The **profane** represents all that is routine and taken for granted in the everyday world—things that are known and familiar, that we can control, understand, and manipulate. The **sacred,** by contrast, consists of the events and things that we hold in awe and reverence—what we can neither understand nor control.

Second, all religions hold beliefs about the supernatural that help people explain and cope with the uncertainties associated with birth, death, creation, success, failure, and crisis. These beliefs form the basis for official religious doctrines.

Third, all religions have rituals. In contemporary Christianity, rituals are used to mark such events as births, deaths, weddings, Jesus's birth, and the resurrection. In earlier eras, many Christian rituals were closely tied to planting and harvest; these are still important ritual occasions in many religions.

The Functions of Religion

Durkheim argued that religion would not be universal if it did not serve important functions for society. At the societal level, the major function of religion is that it gives tradition a moral imperative. Most of the central values and norms of any culture are reinforced through its religions. These values and norms cease to be merely the *usual* way of doing things and become perceived as the *only* moral way of doing them. They become sacred. When a tradition is sacred, it is continually affirmed through ritual and practice and is largely immune to change.

For individuals, Durkheim argued that the beliefs and rituals of religion offer support, consolation, and reconciliation in times of need. On ordinary occasions, many people find satisfaction and a feeling of belongingness in religious participation. This feeling of belongingness creates the moral community, or community of believers, that is part of our definition of religion.

Marx and Beyond: Conflict Theory and Religion

Like Durkheim, Marx saw religion as a supporter of tradition. This support ranges from injunctions that the poor and oppressed should endure rather than revolt (blessed be the poor, blessed be the meek, and so on) and that everyone should pay taxes (give unto Caesar) all the way to the endorsement of inequality implied by a belief in the divine right of kings.

Marx differed from Durkheim by interpreting the support for tradition in a negative light. Marx saw religion as the "opiate of the masses"—a way the elite kept the eyes of the downtrodden happily focused on the afterlife so that the poor would not notice their earthly oppression. This position is hardly value-free, and much more obviously than structural-functional theory, it makes a statement about the truth or falsity of religious doctrine.

Modern conflict theory goes beyond Marx's view. Its major contribution is in identifying the role that religions can play in fostering or repressing conflict between social groups. Religion has certainly *contributed* to conflict between Sunni and Shiite Moslems in Iraq and between Protestants and Catholics in Ireland, as well in many other countries. On the other hand, religion has *reduced* conflict when Muslim, Christian, Hindu, and other clergy have taught impoverished people to accept their fate as God's will or have preached that we are all God's children.

Whether it increases or reduces conflict, religion can and has served as a tool for groups to use in their struggles for power. Interestingly, although Marx believed that religion always helps to keep down the oppressed, we now know that oppressed groups can use religion to better their social position. One example of this is the powerful role the African American Church and leaders such as the Reverend Martin Luther King, Jr., played in fighting for civil rights in the United States.

Another contribution of conflict theory to the analysis of religion is the idea of the dialectic, that is, that contradictions build up between existing institutions and that these contradictions lead to change. Specifically, conflict theorists suggest that social change in the surrounding society can foster change in that society's religions. For example, changes in attitudes toward women have led Reform Jews, Methodists, and others to allow women to serve as ministers or rabbis. Conflict theorists also argue that changes in religion can lead to broader social change. For example, the rise of evangelical churches in (traditionally Catholic) U.S. Hispanic communities is playing a substantial role in organizing Hispanics into an effective political lobby. In March 2006, more than 500,000 people, most of them Hispanic and disproportionately evangelical Christians, marched in protest against proposed anti-immigration legislation. Many of these protesters had learned of the march through evangelical ministers.

Weber: Religion as an Independent Force

Max Weber's influential theory of religion combines elements of structural functionalism and conflict theory. Like Durkheim and other structural functionalists, Weber was interested in the forms of religion and their consequences for individuals and society. But Weber was also concerned with the processes through which religious answers are developed and how their content affects society. In his focus on the relationship between social and religious change, Weber shared interests similar to those of conflict theorists. Whereas conflict theorists typically focus on the ways that social conflict leads to religious change, however, Weber was most concerned with the way that changes in religious ideology could promote other types of social change.

For most people, religion is a matter of following tradition; people worship as their parents did before them. To Weber, however, the essence of religion is the search for knowledge about the unknown. In this sense, religion is similar to science: It is a way of coming to understand the world around us. And as with science, the answers religion provides may challenge the status quo as well as support it.

Charisma refers to extraordinary personal qualities that set an individual apart from ordinary mortals.

Where do people find the answers to questions of ultimate meaning? Often they turn to a charismatic religious leader. **Charisma** refers to extraordinary personal qualities that set the individual apart from ordinary mortals. Because these extraordinary characteristics are often thought to be supernatural in origin, charismatic leaders can become agents for dramatic social change. Charismatic leaders include Christ, Muhammad, and, more recently, Joseph Smith (Latter Day Saints), David Koresh (Branch Davidians), and the Ayatollah Khomeini (Iranian Islam). Such individuals give answers that often disagree with traditional answers. Thus, Weber sees religious inquiry as a potential source of instability and change in society.

In viewing religion as a process, Weber gave it a much more active role than did Durkheim. This is most apparent in Weber's analysis of the Protestant Reformation.

The Protestant Ethic and the Spirit of Capitalism

In his classic analysis of the influence of religious ideas on other social institutions, Weber ([1904–1905] 1958) argued that the Protestant Reformation paved the way for capitalism. The values of early Protestantism, which Weber called the *Protestant Ethic*, included the belief that work, rationalism, and plain living are moral virtues, whereas idleness and indulgence are sinful. What happens to a person who follows this ethic— who works hard, makes business decisions on rational rather than emotional criteria (for example, firing inefficient though needy employees), and is frugal rather than self-indulgent? Such a person is likely to grow wealthier. According to Weber, it was not long before wealth became an end in itself. At this point, the moral values underlying early Protestantism became the moral values underlying early capitalism.

In the century since Weber's analysis, other scholars have explored the same issues, and many have come to somewhat different conclusions. Nevertheless, this research has not changed Weber's major contribution to the sociology of religion: that religious ideas can be the source of tension and change in social institutions.

Tension between Religion and Society

Each religion is confronted with two contradictory yet complementary tendencies: the tendency to reject the world and the tendency to compromise with the world (Troeltsch 1931). If a religion denounces adultery, homosexuality, and fornication, does it have to categorically exclude adulterers, homosexuals, and fornicators, or can it adjust its expectations to take common human frailties into account? If "it is easier for a camel to go through the eye of a needle than for a rich man to enter the kingdom of God," must a church require that all its members forsake their worldly belongings?

How religions resolve these dilemmas is central to their eventual form and character. Scholars distinguish two general types of religious organizations: church and sect. The *church* represents the established religion, and the *sect* represents those who challenge that religion on moral grounds (Sherkat & Ellison 1999).

The categories of church and sect are what Weber referred to as *ideal* types. The distinguishing characteristics of each type are summarized in the Concept Summary. Although no church or sect may have all of these characteristics, the ideal types serve as useful benchmarks against which to examine actual religious organizations. Keep in mind, though, that churches and sects fall on a continuum: Some churches are more "churchlike" than others, some sects are more "sectlike" than others, and many religions fall somewhere in between.

Distinctions between Churches and Sects

Church and sect are ideal types against which we can assess actual religious organizations. Many religious organizations combine some characteristics of both. Nevertheless, Catholicism and Lutheranism are obviously churches, whereas the Nation of Islam has many of the characteristics of a sect.

	Churches	**Sects**
Degree of tension with society	Low	High
Attitude toward other institutions and religions	Tolerant	Intolerant rejecting
Type of authority	Traditional	Charismatic
Organization	Bureaucratic	Informal
Membership	Establishment	Alienated
Examples	Catholics	Jehovah's Witnesses
	Lutherans	Amish, Nation of Islam

Churches

In everyday language, we use the term "church" to refer to Christian religious organizations or places of worship. Sociologists, on the other hand, use the term **church** to refer to any religious organization that has become institutionalized. Churches have endured for generations, are supported by and support society's norms and values, and have become an active part of society. In some cases the religion is supported by the state, and in other cases church membership is even mandated.

Churches' embeddedness in their societies does not necessarily mean that they have compromised essential values. They still retain the ability to protest injustice and immorality. From the abolition movement of the 1850s to the Civil Rights struggle of the 1960s to the demonstrations against U.S. military intervention in Iraq, churchmen and women have been in the forefront of social protest. Nevertheless, churches are generally committed to working with society. They may wish to improve it, but they have no wish to abandon it.

Although all churches are institutionalized religious organizations, some are more institutionalized than others. In some societies, a state church automatically includes virtually every member of the society. The fate of the church and the fate of the nation are wrapped together, and the church is vitally involved in supporting the dominant institutions of society. Examples include the Roman Catholic church in Europe during the Middle Ages and Islam in contemporary Iran. In other societies, religious organizations accommodate both to society and to other religions. In the United States, Jewish, Catholic, Lutheran, Methodist, and Episcopalian clergy meet together in ecumenical councils, pray together at commencements, and generally adopt a live-and-let-live policy toward one another.

Structure and Function of Churches

Churchlike religions tend to be formal bureaucratic structures with hierarchical positions, specialization, and official creeds specifying their religious beliefs. Leadership is provided by a professional staff of ministers, rabbis, imams, or priests, who have received formal training at specialized schools. These leaders are usually arranged in a

Churches are religious organizations that have become institutionalized. They have endured for generations, are supported by and support society's norms and values, and have become an active part of society.

hierarchy from the local to the district to the state and even the international level. Religious services almost always prescribe formal and detailed rituals, repeated in much the same way from generation to generation. Congregations often function more as audiences than as active participants. They are expected to stand up, sit down, and sing on cue, but the service is guided by ceremony rather than by the emotional interaction of participants.

Generally, people are born into churchlike religions rather than converting to them. People who do change churches often do so for practical rather than emotional reasons: They marry somebody of another faith, another church is nearer, or their friends go to another church. Individuals also might change churches when their social status rises above that of most members of their church: Baptists become Methodists, Methodists become Episcopalians (Sherkat & Ellison 1999). Most individuals who change churches have relatively weak ties to their initial religion. Nevertheless, few make large changes: Orthodox Jews become Conservative Jews and members of one small Baptist church join a different small Baptist church (Stark & Finke 2000).

Churchlike religions tend to be large and to have well-established facilities, financial security, and a predominantly middle-class membership. As part of their accommodation to the larger society, churchlike religions usually allow scriptures to be interpreted in ways relevant to modern culture. Because of these characteristics, these religions are frequently referred to as *mainline churches*, a term denoting their centrality in society.

Sects

Sects are religious organizations that reject the social environment in which they exist.

Sects are religious organizations that arise in response to changes that they find repugnant either in churches or in the broader society (Sherkat & Ellison 1999). They often view themselves as restoring true faith, which they believe has been put aside by religious institutions too eager to compromise with society. They see themselves as preservers of religious tradition rather than innovators. Like the Reformation churches of Calvin and Luther, they believe they are cleansing the church of its secular associa-

This young American belongs to the Hindu cult commonly known as Hare Krishna.

© Vince Streano/CORBIS

tions. Most modern sects have emerged as protests against liberal developments in mainstream churches, such as the acceptance of homosexuals, divorce, or abortion.

Some sects' rejection of society's norms is so great that the relationship between the sect and the larger society becomes fraught with tension and even hostility. Egypt routinely incarcerates members of Moslem sects that it considers too extreme, and the United States in the past incarcerated Quakers (officially, members of the Society of Friends) who refused to serve in the military because the Bible says "Thou shalt not kill."

Those sects that experience the most tension with the broader society are known as cults. **Cults** are religions that are independent of the religious traditions of the society in which they function. Examples of cults in the United States are the Church of Scientology, the Unification Church (pejoratively known as "Moonies"), and the International Society for Krishna Consciousness (popularly known as "Hare Krishna"). Each of these religions stands outside of the Judeo-Christian tradition: They have a different God or Gods or no God at all and they don't use the Old Testament as a text.

Cults tend to arise in times of societal stress and change, when established religions do not seem adequate for explaining the upheavals that individuals experience. Because they are so alien to society's institutions, however, cults are of little assistance in helping people cope with everyday lives in mainstream society. Instead, they often urge their members to alter their lives radically and to withdraw from society altogether. Because of the radical changes they demand, cults generally remain small, and few survive more than a few years.

In contrast to cults, other sects partly reject the social world in which they live, but still reflect the religious heritage of the surrounding society. If a U.S. religious group uses the Bible as its primary source of inspiration and guidance, then it is probably a sect rather than a cult. The Amish are an excellent example: They base their lives on a strict reading of the Bible and remain aloof from the contemporary world.

The Amish church is exceptional in that it has managed to maintain its distance from the surrounding social world for generations. In contrast, most sects either dissolve or become increasingly churchlike over time. For example, the Church of Latter Day Saints (Mormons) has over time increased its accommodation to the larger society (Arrington & Bitton 1992). The church has officially abandoned polygamy, opened its priesthood to African American men, and left the seclusion of a virtual state church in Utah. The church and its members continue to hold religious and social views that differ from those of many other Americans, but they are now actively involved in the country's political, economic, and educational institutions.

Structure and Function of Sects

The hundreds of sects in the United States exhibit varying degrees of tension with society, but all are opposed to some basic societal institutions. Not surprisingly, these organizations tend to be particularly attractive to people who are left out of or estranged from society's basic institutions—the poor, the underprivileged, the handicapped, and the alienated. Not all of the people who convert to sectlike religions are poor or oppressed. Many are middle-class people who seek spiritual rather than material wealth, find established churches too bureaucratic, or seek a moral community that will offer them a feeling of belongingness and emotional commitment (Barker 1986). Others join sects such as Hasidic Judaism or Christian fundamentalist groups because they want to hold on to traditional norms and values that seem to have fallen from favor (Davidman 1991).

Sect membership is often the result of conversion or an emotional experience. Instead of merely following their parents into a sect, they actively choose to join.

A **cult** is a sectlike religious organization that is independent of the religious traditions of society.

Connections
Example

Unlike mainstream Mormonism, fundamentalist Mormonism remains far on the sect end of the continuum. It functions outside the law as well as outside mainstream religious traditions. Members of various breakaway Mormon groups believe that God desires them to engage in polygamy and that the mainstream Mormon church moved away from God's will when it abandoned polygamy. Thousands are believed to practice polygamy in Utah and nearby states; most live in isolated communities, under close control of a charismatic leader. In contrast, the mainstream Mormon church excommunicates identified polygamists. Law enforcement officers in Arizona and Utah have begun actively prosecuting polygamists, primarily because of cases in which middle-aged and elderly men married girls as young as 13.

■ Muslims throughout the world believe that, if possible, they should make at least one pilgrimage to the Sacred Mosque in Mecca. Aside from this commitment and a shared belief in the four other pillars of Islamic faith, however, there is considerable variation both within and between countries in the way that Islam is practiced and in the relationship between the church and the state.

Religious services are more informal than those of churches. Leadership remains largely unspecialized, and there is little, if any, professional training for the calling. The religious doctrines emphasize otherworldly rewards, and the scriptures are considered to be of divine origin and therefore subject to literal interpretation.

Sects share many of the characteristics of primary groups: small size, informality, and loyalty. They are closely knit groups that emphasize conformity and maintain significant control over their members.

A Case Study: Islam

Islam was founded in the seventh century A.D. by an Arab prophet named Muhammad, near Mecca, in what is now Saudi Arabia. It is the fastest-growing religion in the world, encompassing one fifth of the world's population ("Islam" 2001). (In contrast, Christianity is shrinking.) No matter where they are found, Muslims share a set of common beliefs. All Muslims believe in a single all-powerful God whose word is revealed to the faithful in the Koran, a book similar to the Christian Bible or the Jewish Torah. All Muslims must follow the Five Pillars of Islam. They must (1) profess faith in one almighty God and Muhammad his prophet, (2) pray five times daily, (3) make charitable donations to the Muslim community and the poor, (4) fast during daylight hours during the month of Ramadan, the time when the Koran was revealed to Muhammad, and (5) try to make at least one pilgrimage to Mecca. Prayer is usually in a mosque (an Islamic house of worship) and is led by an imam (a religious scholar). Because there is no formal central authority, there is considerable variation across countries in the relationship between Islamic clergy and the government and between Moslems and non-Moslems. In some nations Islam more closely resembles a church, and in others it more closely resembles a sect.

Islam as a Sectlike Religion: The Nation of Islam

Islam in the United States can be traced back to the importation of African Muslim slaves in the eighteenth and nineteenth centuries. In the late 1800s, new Muslim immigrants from the Middle East arrived and began to settle primarily in the Midwest. Today there are an estimated 6 million Muslims in the United States, about 40 percent of whom are African Americans (Smith 1999).

Of the many varieties of Islam practiced in the United States, perhaps the most widely recognized is that of the Nation of Islam, popularly known as "Black Muslims." Although the religious beliefs of this group differ markedly from those of traditional Islam, the Nation of Islam has become more churchlike and more like other Islamic groups in recent years (Smith 1999). Initially linked to a radical rejection of Christianity and of white American society, its members and clergy now participate actively in social and political affairs and have good relationships with Christian leaders. (However, anti-Semitism remains deeply entrenched.)

Membership in the Nation of Islam is growing most rapidly among poor and disenfranchised African American inner city residents. For these individuals, Islam can provide a sense of hope, community, identity, and freedom from the white-dominated world around them. In addition, the Muslim emphasis on community activism—antidrug campaigns and economic development—and on discipline and modest dress provide the sense of order and belonging commonly provided by all sects.

Islam as a Churchlike Religion: Egypt and Afghanistan

Until the collapse of the fundamentalist Islamic Taliban government following the U.S. invasion of Afghanistan in 2001, Afghanistan offered a good example of a modern-day Muslim state religion. Church and state were intertwined, with every citizen bound by religious law. For instance, all Afghan women, regardless of religion or nationality, could receive up to 80 lashes with a whip if they failed to cover their whole bodies and could be executed if they had sex outside of marriage. As long as Islamic clergy held all political power, there was little tension between religion and the larger society.

In contrast, in Egypt the government is more or less secular, even though 90 percent of the nation is Moslem. Tension between Islam and the state is palpable (Rubin 2002). Radical Islamic fundamentalists periodically incite violent anti-government attacks, and the government uses terror and repression to keep fundamentalists under control. More moderate Islamic mosques and imams, however, are allowed to function openly, and the government works with them to provide social services to the poor. As a result, Islam remains a highly organized, accepted part of Egypt's culture and society. Thus despite tension between Islam and the government, the religion remains churchlike rather than sectlike.

Islamic Fundamentalism

Recent years have seen a worldwide increase in Islamic fundamentalism. As is true of Christian fundamentalist churches, fundamentalist Muslim sects appeal especially to individuals who lack economic and political power. But Islamic fundamentalism also appeals to educated Muslims who, as mentioned in Chapter 7, despair of Western political and cultural domination and of the decline in traditional morality that they believe accompanied that domination (Amanat 2001; Barber 2001; Jacquard 2002).

All Islamic fundamentalists call for a rejection of the excesses and corruption of modern, secular culture and a return to "true" religious principles. Only the most radical Islamic fundamentalist sects, however, advocate violence to achieve these goals. Most Muslims, in fact, say the concept of *jihad*—holy war—primarily refers not to actual warfare but rather to the need to defend social justice, first through spiritual, economic, and political means and only if that fails through military means (Lawrence 1998).

Religion in the United States

When asked what religion they belong to, only 6 percent of Americans say they belong to none. Most people identify themselves not only as religious but also as members of some particular religion. The majority call themselves Protestants, but 27 percent are Catholics and 2 percent are Jews (Figure 12.3). Despite their differences, these three religions embrace a common Judeo-Christian heritage. They accept the Old Testament, and they worship the same God. They rely on a similar moral tradition (the Ten Commandments, for example), which reinforces common values. This common religious heritage supplies an overarching sense of unity and character to U.S. society—providing a framework for the expression of our most crucial values concerning family, politics, economics, and education.

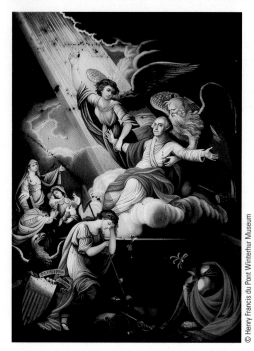

Within weeks of his appointment as commander of the army, Washington became an object of near worship (ART WORK: "Memorial to George Washington," painting on glass.)

U.S. Civil Religion

Americans also share what has been called a civil religion (Bellah 1974, 29; Bellah et al. 1985). **Civil religion** is a set of institutionalized rituals, beliefs, and symbols sacred to U.S. citizens. These include reciting the Pledge of Allegiance and singing the national anthem, as well as folding and displaying the flag in ways that protect it from desecration. In many U.S. homes, the flag or a picture of the president is displayed along with a crucifix or a picture of the Last Supper.

Civil religion has the same functions as religion in general: It is a source of unity and integration, providing a sacred context for understanding the nation's history and current responsibilities (Wald 1987). For example, shortly after the American colonies declared their independence from Britain, George Washington was declared commander of the U.S. army. With little military experience or charisma, Washington's major qualification for the job was that he didn't want it. Within weeks, he became an object of near worship. Why did this cult of Washington develop? It emerged, in part, because Washington symbolized the fledgling nation's unity and, in part, because his disdain for power made him a hero. In worshiping Washington, the colonists were worshiping their nation and the virtues they believed it embodied (Schwartz 1983).

Since then, we have made liberty, justice, and freedom sacred principles. We believe the American way is not merely the usual way of doing things but also the only moral way of doing them, a way of life blessed by God. The motto on our currency, our Pledge of Allegiance, and our national anthem all bear testimony to the belief that the United States operates "under God" with God's direct blessing.

Civil religion is the set of institutionalized rituals, beliefs, and symbols sacred to the U.S. nation.

Religiosity is an individual's level of commitment to religious beliefs and to acting on those beliefs.

Religious economy refers to the competition between religious organizations to provide better "consumer products," thereby creating greater "market demand" for their own religions.

Trends in Religiosity

Religiosity refers to an individual's level of commitment to religious beliefs and to acting on those beliefs. Membership in organized religions is considerably higher in the United States than in other developed nations, and reported rates of attendance at religious services have changed very little over the last several decades (although actual rates appear to have declined).

Why is religiosity so strong in the United States? According to sociologists Rodney Stark and Roger Finke (2000), the answer lies in our highly developed, competitive, and unregulated **religious economy.** They argue that because there are so many religious organizations in this country, each must compete with the others to provide better "consumer products," thereby generating greater "market demand" for them. An interesting example is the rise of emerging churches (described in the Focus on Media and Culture box), which now offer a new sort of religious perspective and experience to a growing number of American Christians.

But although most Americans believe in God, some are more involved in religion than others (Table 12.3). The most striking differences in religiosity are related to age and sex. Older people, women, Southerners, and African Americans are more likely than others to attend religious services regularly (Sherkat & Ellison 1999; General Social Survey 2004).

One interesting topic is the relationship between income, education, and religiosity. In the past, many scholars assumed that religion would appeal disproportionately to the poor, who were in greater need of hope, and to the uneducated, who

Media and Culture:

Emerging Churches

The newest movement within American Christianity is the rise of what have been called **emerging churches** (emergingchurch.info, Kimball 2003). As with other religious movements, these churches have emerged in response to growing discontent with the culture of traditional churches and culture of society at large. Specifically, emerging churches reflect dissatisfaction with the atomized, "inauthentic" life of modern Americans and the bureaucratization of religious belief in modern churches. Most who participate in emerging churches are young, white, and urban. Most also consider themselves evangelical Christians, but the appeal of these churches has spread beyond that core base.

Emerging churches promise an authentic religious experience closely shared with others in an informal space and relying on informal prac-

tices. Instead of sitting in pews, church members may sit on couches or bean bags. Instead of meeting in churches, they often meet in homes. Instead of reciting prayers or singing hymns out of hymnals, they talk about their feelings and beliefs, share their questions and tentative answers on matters of faith, and listen to music straight out of youth popular culture, often accompanied by video presentations. Emerging churches emphasize how individuals can live a life of mission, faith, and community—qualities many find sorely lacking in a broader culture that emphasizes working, consuming, and individual self-sufficiency. Finally, whereas traditional evangelical churches define themselves partly by their rejection of the moral values of contemporary American culture and hold that there is only one valid interpretation of scriptures, emerging churches have a more open perspective toward both scriptural interpretation and individual behavior. As a result,

they offer a better cultural fit for some young Americans.

Emerging churches would not have grown as large or as quickly if they had not taken advantage of new forms of media. Much of the early interest in this (highly amorphous) movement developed because a loose network of leaders shared their ideas with each other and with a growing audience on the Internet. The Internet has continued to play an important role in spreading the emerging church movement, especially through the use of blogs. Similarly, part of emerging churches' success stems from their willingness to adopt the ideas and forms of contemporary popular culture, from the emphasis on self-awareness to the use of rock videos. In sum, the rise of emerging churches illustrates how religious movements reflect the surrounding culture, respond to that culture, and can use available media and culture to increase their impact.

were more likely to lack "scientific" explanations for natural and human events. It is true that those with a college education are less likely, overall, to say that religion is important to them. However, as Table 12.3 indicates, college graduates are as likely to attend church as nongraduates. Moreover, among those who consider religion important in their lives, graduates and nongraduates are equally likely to hold conservative religious beliefs (Sherkat & Ellison 1999). In general, churchgoing appears to be more strongly associated with being conventional than with being disadvantaged. It is a characteristic of people who are involved in their communities, belong to other voluntary associations, and hold traditional values.

Emerging churches are linked by 1) the belief that American life and modern Christian churches are atomized, bureaucratic, and inauthentic and 2) an emphasis on informal rituals, a more open perspective toward scripture and behavior, and living a life of mission, faith, and community.

Consequences of Religiosity

Because religion teaches and reinforces values, it has consequences for attitudes and behaviors. People who are more religious tend to be friendlier, more cooperative, healthier, happier, and more satisfied with their lives and marriages (Ellison 1991, 1992; Sherkat & Ellison 1999; Thomas & Cornwall 1990). These benefits in large part stem from the social support and sense of belonging that individuals receive from their religious communities.

Persons who are more religious tend to have more conservative attitudes on sexuality and personal honesty; they also may have more conservative attitudes about family life, such as supporting the use of corporal punishment to discipline children

TABLE 12.3

Percentage of Americans Who Attend Religious Services Regularly, 2004

Southerners, older people, women, and African Americans are more likely than other Americans to attend religious services at least twice per month.

Total:	42%
Region	
East	37
Midwest	42
South	50
West	36
Age	
Below 29	33
30–49	43
50 and older	48
Sex	
Male	37
Female	48
Education:	
No college	41
Some college or college grad	43
Postgraduate education	48
Race	
White	39
African American	63
Other	44

SOURCE: General Social Survey 2004.

(Ellison, Bartkowski, & Segal 1996). Not surprisingly, some conservative religious groups have played significant roles in supporting conservative political movements, such as the antiabortion movement and certain right-wing hate groups.

Yet, we should not assume that church members necessarily adopt the attitudes of their churches. Primarily because of immigration, for example, the membership in the Roman Catholic church in the United States has increased by about 25 percent over the past 40 years (Monroe 1995, 383). However, these new American Catholics aren't as likely to obey church leaders or to attend Sunday Mass as Catholics used to be. Although the Pope sees artificial birth control as sinful, 82 percent of U.S. Catholics disagree; 64 percent disagree that abortion is always wrong. Just one third agree that only males should be priests. Nevertheless, 86 percent say they approve of the Pope (Sheler 1995), a sign, perhaps, that in the American Catholic church, members take what they like and leave the rest.

Moreover, even though religious training generally teaches and reinforces conventional behavior, religion and the church can be forces that promote social change. As noted earlier, African American churches and clergy played a significant role in the Civil Rights movement of the 1950s and 1960s and evangelical churches are playing a significant role in the current immigrant rights movement. In Latin America, liberation theology aims at the creation of democratic Christian socialism that eliminates poverty, inequality, and political oppression (Smith 1991). Conversely, church members don't always adopt their churches' liberal views: In recent years, some Baptists and Episcopalians, among others, have split from their central churches because they disapprove of growing church support for gay rights and other liberal agendas.

The Religious Right

One of the most striking changes in religion in the United States today is the vitality and growth of the fundamentalist churches compared with the relative decline of mainline churches (Woodberry & Smith 1998). One outgrowth of this is the development of the Religious Right.

The Religious Right is a loose coalition of Protestant, and on some issues Catholic, fundamentalists who believe that the U.S. government and social institutions should operate according to their interpretation of traditional Christian principles and practices (Woodberry & Smith 1998). They believe that it is a Christian obligation to be active politically in making the United States a Christian nation. They are united by opposition to abortion, liberal sexual mores, gay and lesbian rights, and the separation of church and state, among other things. The Religious Right is politically most powerful in the South and West.

The Religious Right is best understood as a political rather than a religious movement. It aims to influence public policy through normal political processes—lobbying, campaign contributions, and getting out the vote. In doing so, of course, it is building onto the existing foundations of civil religion.

Despite this link to established values and despite the growth in fundamentalist religions, the movement has had more publicity than power (Woodberry & Smith 1998). First, its power is limited because many fundamentalists believe that churches should save their energies and resources solely for religious endeavors. Second, fundamentalists do not necessarily agree with the Religious Rights' political agenda (Davis & Robinson 1996). In fact, with the exception of abortion, studies show that since 1980, the social attitudes of evangelical Protestants have become more like those of Roman Catholics, Jews, and mainstream Protestants (DiMaggio,

Evans, & Bryson 1996). Finally, the power of the Religious Right is limited because few of its leaders hold positions of power in the media (Woodberry & Smith 1998). The net result is that although the Religious Right exercises "veto power" over certain issues within the Republican Party (such as ensuring that no presidential candidate will actively support increased access to abortion), its overall power is relatively limited. For example, President George W. Bush's decision to propose, in 2006, a Constitutional amendment banning gay rights reflected the strength of the Religious Right, but the Senate's immediate vote against that proposal underscored its weakness.

Where This Leaves Us

Structural-functional theory and conflict theory are both right. On the one hand, schools and churches are preservers of tradition. Both institutions socialize young people to understand and accept traditional cultural values and to find their place in society. Occasionally schools and churches teach people to think for themselves, but more often both stress unquestioning acceptance of authority and of contemporary social arrangements, including social inequalities.

On the other hand, schools and churches are in the forefront of social change. Nowhere are the battles over oppression in the least-developed nations, abortion, or homosexuality fought more bitterly than in the councils of our major churches. Nowhere are the battles over race relations, sex and class equity, and clashing cultural values fought more bitterly than on school boards. Even if you are not religious and even after you finish your education, you cannot afford to ignore the vital roles education and religion play in creating or impeding social change.

Many observers, including Marx, have assumed that religion would appeal disproportionately to the poor and disadvantaged. Although the poor and disadvantaged are more likely than others to be fervent in their beliefs, studies show that churchgoing is correlated more strongly with being conventional than with being disadvantaged.

Summary

1. The structural-functional model of education suggests that education meets multiple social needs. It socializes young people to the broader culture, provides knowledge and skills, and can promote social change.

2. Conflict theory suggests that education helps to maintain and reproduce the stratification structure through four mechanisms: training a docile labor force that accepts inequality (the hidden curriculum), using credentialism to save the best jobs for the children of the elite, perpetuating the dominant culture, and ensuring that disadvantaged groups receive inferior educational opportunities.

3. Symbolic interactionists explore some of the processes through which education can reproduce inequality.

Key elements of this process are self-fulfilling prophesies and differences in children's cultural capital, both of which keep disadvantaged students from improving their lot.

4. Tracking generally helps students in high-ability groups but hurts those in low-ability groups. High-stakes testing encourages schools to pay more attention to the quality of the education they provide but has forced schools to cut programs and to focus on teaching students how to take tests. School choice gives parents and students options but can reinforce inequality and reduce support for public education.

5. About half of U.S. high school graduates between 16 and 24 are enrolled in college. Women are more likely

than men to attend and graduate college, but class and race differences are much greater than gender differences. Men from poor, minority families are the least likely to attend college.

6. Education pays off handsomely in terms of increased income, better jobs, and lower unemployment. It also offers nonmonetary benefits such as the likelihood of a longer life.

7. The sociological study of religion concerns itself with the consequences of religious affiliation for individuals and with the interrelationships of religion and other social institutions. It is not concerned with evaluating the truth of particular religious beliefs.

8. Despite earlier predictions, secularization has not increased significantly in the United States. Rather, mainstream religious organizations remain strong, and fundamentalist groups are growing in popularity. Religious membership and attendance remain at stable levels and are far higher than in Europe.

9. Durkheim argued that religion is functional because it provides support for the traditional practices of a society and is a force for continuity and stability. Weber argued that religion generates new ideas and thus can change social institutions. In contrast, Marx argued that religion serves as a conservative force to protect the status quo. More recent conflict theorists have explored the role that religion can play in either fostering or repressing social conflict.

10. All religions are confronted with a dilemma: the tendency to reject the secular world and the tendency to compromise with it. The way a religion resolves this question determines its form and character. Those that adapt to the broader world are called churches, whereas those that reject the world are called sects. In reality, however, churches and sects are not two discrete entities but rather form a continuum.

11. U.S. civil religion is an important source of unity for the U.S. people. It is composed of a set of beliefs (that God guides the country), symbols (the flag), and rituals (the Pledge of Allegiance) that many people of the United States of all faiths hold sacred.

12. A major development in the contemporary church is the growth of fundamentalism and its political arm, the Religious Right.

Thinking Critically

1. How have you been helped or harmed by tracking? If you have not experienced it, answer this question based on someone you know.

2. How would you reorganize elementary and secondary classrooms to best meet the needs of all students? What would be the manifest functions of your system? The latent functions? The potential dysfunctions?

3. Given what you now know about the process of secularization and the rise of fundamentalism, do you expect fundamentalism to grow or to recede in coming years? Why? Base your argument on your understanding of sociology, *not* on your religious beliefs.

4. If the Religious Right were to gain power, what changes would you expect to occur in U.S. government? U.S. society? Do you think they would be good for the United States? Why or why not?

Companion Website for This Book

academic.cengage.com/sociology/brinkerhoff
Gain an even better grasp on this chapter by going to the Companion Website. This resource contains tutorial quizzes and flash cards to help you master key terms and concepts.

Suggested Readings

Kozol, Jonathan. 2005. *The Shame of the Nation: The Restoration of Apartheid Schooling in America*. New York: Crown. Based on visits he made to 60 schools in 11 states, Kozol argues that minority children continue to receive shamefully unequal educations in the nation's most crowded and worst-equipped schools.

Nathan, Rebekah. 2005. *My Freshman Year: What a Professor Learned by Becoming a Student*. Ithaca, N.Y.: Cornell University Press. An anthropology professor spends a year undercover as an undergraduate and dorm resident and reports on student culture today.

Smith, Jane I. 1999. *Islam in America*. New York: Columbia University Press. An excellent overview of the history, religious beliefs, and social life of American Muslims.

Armstrong, Karen. 2001. *The Battle for God: A History of Fundamentalism*. New York: Ballantine Books. Fascinating account of the rise of fundamentalism among Christianity, Islam, and Judaism.

CHAPTER 13

Politics and the Economy

© Royalty-Free/Corbis

Political and Economic Institutions

Political and economic institutions are so closely related that we sometimes refer to them as a single institution—the political economy. Although we can treat them separately to some extent, we will see that many issues, such as the political power of organized labor, cut across both institutions. Throughout this chapter, a primary concern is to provide a sociological perspective that will help you to interpret your own political and economic experiences as well as the headlines in the national news.

Power and Political Institutions

Lisa wants to watch *American Idol* and John wants to watch football; fundamentalists want prayer in the schools and the American Civil Liberties Union wants it out; state employees want higher salaries and other citizens want lower taxes. Who decides?

Whether the decision maker is Mom or the Supreme Court, those who are able to make and enforce decisions have power. As we discussed in Chapter 7, power is the ability to direct others' behavior, even against their wishes. Here we will describe two kinds of power: coercion and authority. Although both mothers and courts have power, there are obvious differences in the basis of their power, the breadth of their jurisdiction, and the means they have to compel obedience. The social structure most centrally involved with the exercise of power is the state, and that is the focus of this section.

Coercion

The exercise of power through force or the threat of force is **coercion.** The threat may be physical, financial, or social injury. The key is that we do as we have been told only because we are afraid not to. We may be afraid that we will be hit, beaten, or killed, but we may also be afraid of a fine or of rejection.

Authority

Threats are sometimes effective means of making people follow your orders, but they tend to create conflict and animosity. It would be much easier if people would just agree that they were supposed to do whatever you told them to do. This is not as rare as you might suppose. This kind of power is called **authority;** it refers to power that is supported by norms and values that legitimate its use. When you have authority, your subordinates agree that, in this matter at least, you have the right to make decisions and they have a duty to obey. This does not mean that the decision will always be obeyed or even that each and every subordinate will agree that the distribution of power is legitimate. Rather, it means that society's norms and values legitimate the inequality in power. For example, if a dad tells his teenagers to be home by midnight, the kids may come in later. They may even argue that he has no right to run their lives. Nevertheless, most people, including children, would agree that a father does have this right.

Coercion is the exercise of power through force or the threat of force.

Authority is power supported by norms and values that legitimate its use.

Connections

Social Policy

Although states have a legitimate monopoly on coercion, not all coercion engaged in by states is legitimate. According to Amnesty International (amnestyusa.org), torture is still widely used in Mexico, Russia, China, and elsewhere. Accused prisoners and political activists may be beaten, raped, shocked with electricity, held under water until they almost drown, subjected to sensory deprivation, and more. The United States has used torture in Afghanistan, Iraq, and Guantánamo Bay and for decades has trained its allies in the use of torture (McCoy 2006). To prevent torture, the United Nations has established an International Criminal Court to prosecute those who authorize torture and a network of independent observers to monitor prison conditions around the world.

Traditional authority, like that enjoyed by King Mohamed VI of Morocco, exists when an individual's right to make decisions for others is widely accepted based on time-honored beliefs.

Traditional authority is the right to make decisions for others that is based on the sanctity of time-honored routines.

Charismatic authority is the right to make decisions that is based on perceived extraordinary personal characteristics.

Rational-legal authority is the right to make decisions that is based on rationally established rules.

Because authority is supported by shared norms and values, it can usually be exercised without conflict. Ultimately, however, authority rests on the ability to back up commands with coercion. Parents may back up their authority over teenagers with threats to ground them or take the car keys away. Employers can fire or demote workers. Thus, authority rests on a legitimization of coercion (Wrong 1979).

In a classic analysis of power, Weber distinguished three bases on which individuals or groups can be accepted as legitimate authorities: tradition, extraordinary personal qualities (charisma), and legal rules.

Traditional Authority

When the right to make decisions is based on the sanctity of time-honored routines, it is called **traditional authority** (Weber [1910] 1970c, 296). Monarchies and patriarchies are classic examples of this type of authority. For example, a half century ago, the majority of women and men in our society believed that husbands ought to make all the major decisions in the family; husbands had authority. Today, most of that authority has disappeared. Traditional authority, according to Weber, is not based on reason; it is based on a reverence for the past.

Charismatic Authority

An individual who is given the right to make decisions because of perceived extraordinary personal characteristics is exercising **charismatic authority** (Weber [1910] 1970c, 295). These characteristics (often an assumed direct link to God) put the bearer of charisma on a different level from subordinates. Gandhi's authority was of this form. He held neither political office nor hereditary position, yet he was able to mold national policy in India. More recently and less positively, Osama bin Laden also is believed by his followers to exercise charismatic authority.

Rational-Legal Authority

When decision-making rights are allocated on the basis of rationally established rules, we speak of **rational-legal authority.** This ranges from a decision to take turns to a decision to adopt a constitution. An essential element of rational-legal authority is that it is impersonal. You do not need to like or admire or even agree with the person in authority; you simply follow the rules.

Our government is based on rational-legal authority. When we want to know whether Congress has the right to make certain decisions, we simply check our rule book: the Constitution. As long as Congress follows the rules, most of us agree that it has the right to make decisions and we have a duty to obey.

Combining Bases of Authority

Analytically, we can make clear distinctions among these three types of authority. In practice, the successful exercise of authority usually combines two or more types. An elected official who adds charisma to the rational-legal authority stipulated by the law will have more power; the successful charismatic leader will soon establish a bureaucratic system of rational-legal authority to help manage and direct followers. All types of authority, however, rest on the agreement of subordinates that someone has the right to make a decision about them and that they have a duty to obey it.

Political Institutions

Power inequalities are built into almost all social institutions. In institutions as varied as the school and the family, roles associated with status pairs such as student–teacher and parent–child specify unequal power relationships as the normal and desirable standard.

In a very general sense, **political institutions** are all those institutions concerned with the social structure of power, including the family, the workplace, the school, and even the church or synagogue. The most prominent political institution, however, is the state.

Political institutions are institutions concerned with the social structure of power; the most prominent political institution is the state.

Power and the State

The **state** is the social structure that successfully claims a monopoly on the legitimate use of coercion and physical force within a territory. It is usually distinguished from other political institutions by two characteristics: (1) its jurisdiction for legitimate decision making is broader than that of other institutions, and (2) it controls the use of legalized coercion in society.

The **state** is the social structure that successfully claims a monopoly on the legitimate use of coercion and physical force within a territory.

Jurisdiction

Whereas the other political institutions of society have rather narrow jurisdictions (over church members or over family members, for example), the state exercises power over the society as a whole.

Generally, states are responsible for gathering resources (taxes, draftees, and so on) to meet collective goals, arbitrating relationships among the parts of society, and

Concept Summary

Power

Concept	Definition	Example From Family
Power	Ability to get others to act as one wishes despite their resistance; includes coercion and authority	"I know you don't want to mow the lawn, but you have to do it anyway."
Coercion	Exercise of power through force or threat of force	"Do it or else."
Authority	Power supported by norms and values	"It is your duty to mow the lawn."
Traditional authority	Authority based on sanctity of time-honored routines	"I'm your father, and I told you to mow the lawn."
Charismatic authority	Authority based on extraordinary personal characteristics of leader	"I know you've been wondering how you might serve me, . . ."
Rational-legal authority	Authority based on submission to a set of rationally established rules	"It is your turn to mow the lawn; I did it last week."

FIGURE 13.1
Americans' Perceptions of Government Responsibilities
SOURCE: Calculations from General Social Survey, sda.berkeley.edu.

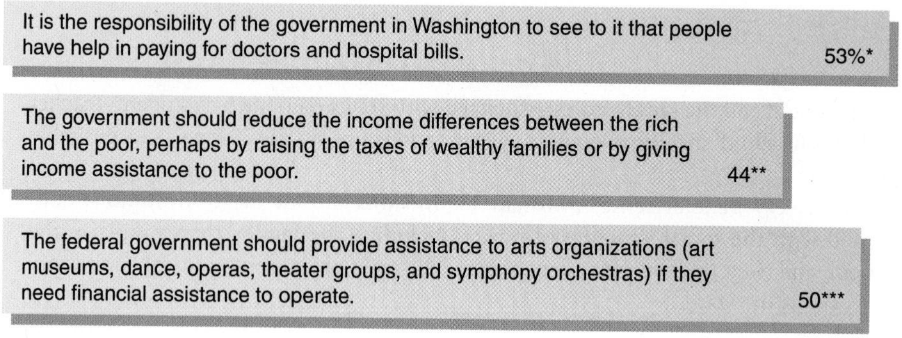

It is the responsibility of the government in Washington to see to it that people have help in paying for doctors and hospital bills.
53%*

The government should reduce the income differences between the rich and the poor, perhaps by raising the taxes of wealthy families or by giving income assistance to the poor.
44**

The federal government should provide assistance to arts organizations (art museums, dance, operas, theater groups, and symphony orchestras) if they need financial assistance to operate.
50***

*2004
**2002
***1998

maintaining relationships with other societies. As societies have become larger and more complex, the state's responsibilities have grown. Recent polls (Figure 13.1) indicate that about half of all Americans think the U.S. government is also responsible for supporting the arts, aiding the poor, and helping those who lack access to health care.

State Coercion

The state claims a monopoly on the legitimate use of coercion. To the extent that other institutions use coercion (for example, the family or the school), they do so with the approval of the state, and as the state gives, the state also takes away. Thus, the state has withdrawn approval from husbands who beat their wives, parents who beat their children, and teachers who hit their students.

The state uses three primary types of coercion. First, the state uses its police power to claim a monopoly on the legitimate use of physical force. It is empowered to imprison and even to kill people in certain circumstances. This claim to a monopoly on legitimate physical coercion has been strengthened in recent years by the declining legitimacy of coercion by other institutions, such as the home and the school. Second, the state uses taxation to legally take money from individuals. Finally, the state is the only unit in society that can legally maintain an armed force and that is empowered to deal with foreign powers.

A variety of social structures can be devised to fulfill these functions of the state. Most basically, states can be categorized into two basic political forms: authoritarian systems and democracies.

Authoritarian Systems

Authoritarian systems are political systems in which the leadership is not selected by the people and legally cannot be changed by them.

Most people in most times have lived under **authoritarian systems.** Authoritarian governments go by a lot of other names: dictatorships, military juntas, despotisms, monarchies, theocracies, and so on. What they have in common is that the leadership was not selected by the people and cannot be changed by them. In some of these countries, elections may be held, but the elections are rigged so that only certain individuals can win. Afghanistan under the Taliban was an authoritarian system, as is Cuba under Castro.

Authoritarian structures vary in the extent to which they attempt to control people's lives, the extent to which they use terror and coercion to maintain power, and the purposes for which they exercise control. Some authoritarian governments, such as monarchies and theocracies, govern through traditional authority; others have no legitimate authority and rest their power almost exclusively on coercion.

Democracies

There are several forms of **democracies,** many of them rather different from that of the United States. All democracies, however, share two characteristics: there are regular, constitutional procedures for changing leaders, and these leadership changes reflect the will of the majority.

In a democracy, two basic groups exist: the group in power and one or more legal opposition groups that are trying to get into power. The rules of the game call for sportsmanship on both sides. The losers have to accept their loss and wait until the next constitutional opportunity to try again, and the winners have to refrain from eliminating or punishing the losers. Finally, there has to be public participation in choosing among the competing groups.

Why are some societies governed by democracies and others by authoritarian systems? The answer appears to have less to do with virtue than with economics. Democracy is found primarily in the wealthier nations of the world. A large and relatively affluent middle class is especially important. Members of the middle class usually have sufficient social and economic resources to organize effectively; their economic power and organization enable them to hold the government accountable. The key factor, however, is not the overall wealth of the nation, or even the size of the middle class, but the way the wealth is distributed: Democracy is also found in poorer nations that do not have extremes of income inequality, such as Costa Rica and Sri Lanka. Still, democracy is possible even in the absence of these conditions. The largest democracy in the world, for instance, is India, which has a relatively small middle class and tremendous income inequality.

Democracy can also flourish in societies with many competing groups, each of which comprises less than a majority. In such a situation, no single group can win a majority of voters without negotiating with other groups; because each group is a minority, safeguarding minority political groups protects everybody.

Although democratic stability depends on competing interest groups, two additional conditions must be met. First, if minority political groups are so divided or

Democracies are political systems that provide regular, constitutional opportunities for a change in leadership according to the will of the majority.

Democracy is now taking root in South Africa, where the financial and political power of the white minority has been counterbalanced by the sheer numbers and political determination of the black majority.

© 1994 Tom Muscionico/Contact Press Images

ineffective that there is little chance they can win an election, the public may become disillusioned with the democratic process (Weil 1989). Second, if competing interest groups do not share the same basic values and interests, they are not likely to abide by the rules of the game. The repeated failure of peace talks and eruptions of violence between Israelis and Palestinians demonstrate how fundamental differences can make it difficult for democracy to flourish.

Globalization and State Power

As the Israeli government and the incipient Palestinian government have fought for land and autonomy, each has been both helped and hindered by organizations outside their borders. The United Nations and the European Union send diplomats and peacekeeping forces, the World Court judges whether either government is breaking international laws, the World Bank decides whether to extend low-interest loans to build the economy, and multinational oil companies pressure politicians in the United States, the Middle East, and elsewhere to safeguard their own interests. Each of these is an example of globalization—in this case, the globalization of the economy and law.

Some sociologists now argue that as a result of this globalization, the power of the nation-state has declined sharply (Sassen 2006; Appelbaum 2005). Multinational corporations, they argue, now have considerable ability to set the terms under which they operate and to limit the power of nation-states where they do business. For example, corporations have fought against minimum wage laws in the United States and against price controls designed to keep tortillas affordable in Mexico. Similarly, international regulatory organizations and associations such as the European Union and the International Monetary Fund also are challenging the power of the state. Others argue that globalization has been going on since the days of the great sailing ships, yet government regulation of the economy and citizenry is greater than ever. They therefore conclude that any threat to the nation-state has been greatly exaggerated (Wolf 2005).

Who Governs? Models of U.S. Democracy

Everyone agrees that the United States is a democracy. Political parties that have at least moderately different economic and social agendas vie for public support, and every 2, 4, or 6 years there are opportunities to "turn the rascals out" and replace the leadership. There is substantial debate, however, about whether the decisions made by our leaders really reflect the will of the majority. This section outlines the two major sociological models of how these decisions are made: the pluralist model and the power-elite model. Although there are some important differences between these two models (see the Concept Summary), it is notable that in both the key actors are organized entities—businesses, unions, political action committees, or government agencies—rather than individuals.

Structural-Functional Theory: The Pluralist Model

Like all structural-functionalist models, the pluralist model of political power begins from the assumption that when things are working properly, the different parts of our political process will run smoothly and harmoniously, for the good of all. The pluralist model focuses on the processes of checks and balances within the U.S. government

Two Models of American Political Power

	Pluralist	Power-Elite
Basic units of analysis	Interest groups	Power elites
Source of power	Situational; depends on issue	Inherited and positional; top positions in key economic and social institutions
Distribution of power	Dispersed among competing diverse groups	Concentrated in relatively homogeneous elite
Limits of power	Limited by shifting and cross-cutting loyalties	Potentially limited when other groups can unite in opposition
Role of the state	Arena where interest groups compete	One of several sources of power

and on coalition and competition among governmental and nongovernmental groups. This model argues that the system of checks and balances built into the U.S. Constitution makes it nearly impossible for either the judicial, legislative, or executive branch of government to force its will on the other branches. Similarly, the model argues that different groups with competing vested interests hold power in different sectors of American life. Some groups have economic power, some have political power, and some have cultural power. Because each group has some power, all are reasonably content and no extreme group can force its views on the others. (See Focus on Media and Culture: Blacklisting the Dixie Chicks for an interesting case study.)

A vital part of this model is the hypothesis of shifting allegiances. According to pluralist theorists, different coalitions of interest groups arise for each decision. For example, labor unions and automakers fight over wages, but collaborate to fight against Japanese auto imports. This pattern of shifting allegiances keeps any interest groups from consistently being on the winning side and keeps political alliances fluid and temporary rather than allowing them to harden into permanent and unified cliques (Dahl 1961, 1971). As a result of these processes, pluralists see the decision-making process as relatively inefficient but also relatively free of conflict, a process in which competition among interest groups (each of which has some sources of power) keeps any single group from gaining significant advantage over the others.

Research suggests the limits of the pluralist model. Typically, the power elite stick together, while other groups lack the resources to successfully challenge the elite (Burris & Salt 1990; Clawson & Su 1990; Korpi 1989). In the United States, programs designed to share wealth or access to opportunities more equitably—such as civil rights or Social Security—have had the best chance of success in two situations: 1) when a crisis caused the elite to favor at least some change and 2) when the elite disagreed among themselves (Jenkins & Brent 1989).

Conflict Theory: The Power-Elite Model

In contrast to the pluralist model, the power-elite model, which is based on conflict theory, contends that a relatively unified elite group makes all major decisions—in its own interests (Domhoff 1998). In his classic work, *The Power Elite*, C. Wright Mills

Media and Culture:

focus on

Blacklisting the Dixie Chicks

On March 10, 2003, while performing in London, England, lead singer Natalie Maines of the Texas-based Dixie Chicks told the audience, "Just so you know, we're ashamed the president of the United States is from Texas."

At the time, the Dixie Chicks had two hit recordings: "Landslide" was number one on adult contemporary radio stations and "Travelin' Soldier" was number one on country music stations. Within days, however, both songs virtually disappeared from the radio.

Why was the response to Maines's comments so swift and so complete? Was it because the mass media (in this case, radio stations) are tightly controlled by a handful of politically conservative corporations? Or was it because of widespread popular outrage?

To answer this question, sociologist Gabriel Rossman (2004) collected and analyzed data on how stations across the nation responded to the controversy. For this analysis, he divided radio stations first by music format (adult contemporary versus country) and then by ownership (large corporate chains versus smaller chains and independent). Since 1996, there has been tremendous consolidation of ownership of radio stations; Clear Channel alone now owns more than 1,000 stations across the country, and many observers have raised concern that this level of consolidation both forces out politically independent stations and pressures chain-owned stations to adopt political stances that reflect the conservative views of their corporate owners.

As it happened, corporate ownership did not increase the likelihood that a station would drop the Dixie Chicks's songs. Corporate chains make their decisions through a highly bureaucratic process, which was both too slow to respond to this brouhaha and too focused on maintaining a profit to care particularly about a political con-

troversy. In contrast, music format had a significant effect. Country music is known for its conservative values and is popular with conservative audiences. It is therefore not surprising that, regardless of ownership, country music stations were significantly more likely than adult contemporary stations to drop the Dixie Chicks. Finally, regardless of music format, stations located in politically conservative communities were more likely to drop the Dixie Chicks. In sum, and supporting the pluralist model, corporate interests proved generally unwilling to rock the boat and drop a popular music group *unless* they knew it was what their audience wanted them to do. Power to the people, indeed.

Four years later, the Dixie Chicks got their revenge: With popular sentiment swinging strongly against the war in Iraq, their new, openly rebellious CD, *Taking the Long Way*, became a smash hit, selling more than a million copies within the first month after its release and sweeping the Grammy Awards.

The **power elite** comprises the people who occupy the top positions in three bureaucracies—the military, industry, and the executive branch of government—and who are thought to act together to run the United States in their own interests.

(1956) defined the **power elite** as the people who occupy the top positions in three bureaucracies: the military, industry, and the executive branch of government. Through a complex set of overlapping cliques, these people share decisions having national and international consequences (Mills 1956, 18).

There is no question that the power elite has become more diverse since Mills's day. The independent power of the military has declined, whereas that of the cultural elite—which includes both movie stars and religious leaders—has grown. Increasing numbers of African Americans, Hispanics, and women hold high corporate positions and elected office, especially at local levels. On the other hand, women and minorities are still greatly outnumbered in positions of power by white males. Moreover, most "outsiders" who become part of the power elite come from at least middle-class homes, attend elite schools, and are willing and able to fit in: light-skinned minorities, Jews who marry Christians, and women who learn to play golf and even to smoke cigars (Zweigenhaft & Domhoff 1998).

The power-elite theory argues that individuals have power by virtue of the positions they hold in key institutions. If the interests of these individuals and institutions were in competition with one another, this model would not be significantly different from the pluralist model. The key factor in elite theory is the unity of purpose and outlook that top position holders have as a result of common membership in the upper class.

In sum, the power-elite model shares with the pluralist model a belief that different groups with different types of power can come together in shifting allegiances to create social change. The models differ in that the pluralist model assumes that power is generally distributed in a relatively equitable fashion and, as a result, that virtually all groups can change their situations if they so desire. The power elite model, on the other hand, argues that power is very unevenly distributed, and therefore creating meaningful social change is difficult, unless people organize together in unions, social movements, and the like.

Individual Participation in U.S. Government

Democracy is a political system that explicitly includes a large proportion of adults as political actors. Yet it is easy to overlook the role of individual citizens while concentrating on leaders and organized interests. This section describes the U.S. political structure and process from the viewpoint of the individual citizen.

Who Votes?

The average citizen is not politically oriented. About one third of the voting-age population does not even register to vote, and almost half (42 percent in 2004) does not vote even in presidential elections (U.S. Bureau of the Census 2006). An astonishing 75 to 80 percent do not vote in typical local elections.

This low level of political participation poses a crucial question about power in U.S. democracy. Who participates? If they are not a random sample of citizens, then some groups must have more influence than others. Studies show that voters differ from nonvoters by social class and age, but not by race.

Although all U.S. citizens over the age of 18 have the right to vote, middle-aged, better off, and better-educated citizens are most likely to do so.

Social Class

One of the firmest findings in social science is that political participation (indeed, participation of any sort) is strongly related to social class. Whether we define participation as voting or letter writing, people with more education, more income, and more prestigious jobs are more likely to be politically active. They know more about the issues, have stronger opinions, are much more likely to believe they can influence political decisions, and thus are more likely to try to do so. This conclusion is supported by data on voting patterns from the 2004 presidential election (Table 13.1). The higher the level of education, the greater the likelihood of registering to vote and of actually voting—those who have graduated from college are more than twice as likely to vote as those who have not completed high school.

Age

Another significant determinant of political participation is age. As Table 13.1 shows, older persons are considerably more likely than younger persons to register to vote and to vote (U.S. Bureau of the Census 2006). Even in the turbulent years of

TABLE 13.1

Participation in the 2004 Election, Among Voting Age Population

The likelihood of registering to vote and actually voting is greater among people who are older, better educated, employed, and non-Hispanic.

	Percentage Registered	Percentage Who Voted
Total	65.9	58.3
Education		
8 years or less	32.5	23.6
Some high school	45.8	34.6
High school graduate	61.5	52.4
Some college	73.7	66.1
College graduate or more	78.1	74.2
Race/ethnicity		
White	67.9	60.3
African American	64.4	56.3
Hispanic	34.3	28.0
Age		
18–20	50.7	41.0
21–24	52.1	42.5
25–34	55.6	46.9
35–44	64.2	56.9
45–64	72.7	66.6
65 and older	76.9	68.9
Employment status		
Employed	67.1	60.0
Unemployed	56.3	46.4

SOURCE: U.S. Bureau of the Census 2006.

the Vietnam War, when young antiwar demonstrators were so visible, young adults were significantly less likely to vote than were middle-aged individuals. In that period, many young adults engaged in other forms of political participation that did, in fact, influence political decisions. In most time periods, however, the low participation of younger people at the polls is a fair measure of their overall participation.

Race and Ethnicity

Racial differences in political participation have virtually disappeared. African Americans and white non-Hispanics are about equally likely to vote. Hispanics as a group are less likely to vote. Partly this is because of their lower average socioeconomic status but largely it is because many are not citizens and so are not eligible to vote.

Which Party?

Unlike the United States, most European nations have parliamentary governments. In these nations, seats in Parliament are assigned based on proportional representation: The proportion of votes received by each party more or less matches the proportion of seats it receives in Parliament. As a result, many different parties can have members in Parliament.

In contrast, seats in the U.S. Congress (and other U.S. political offices) are won through a "winner take all" process: In each election, whoever receives the most votes wins. As a result, although in theory the United States could have many political parties, in practice political power in this country is divided between two parties, the Democratic Party and the Republican Party. The odds of anyone from another party winning an election are so small that there is little point in supporting other parties.

Although both major political parties in the United States are basically centrist, there are philosophical distinctions between them. For the last century, the Democratic Party has been more associated with liberal morality; support for social services; and the interests of the poor, the working class, and minorities. The Republican Party has been more associated with conservative morality, tax cuts, and the interests of industry and the affluent. Because of these characteristics, the Democrats tend to attract more voters who are female, younger, minority, and less-educated than do the Republicans (Table 13.2).

A growing proportion of voters align themselves with neither party but declare an intention to vote on the basis of issues rather than party loyalty. When the 10 percent (or more) of voters who call themselves independent go to the polls, however, they usually have to choose between a Democratic and a Republican candidate.

Why So Few Voters?

The United States prides itself on its democratic traditions. Yet U.S. citizens are only half as likely to vote as are citizens of other Western nations. Moreover, studies consistently find that persons are most likely to vote if they are more affluent

TABLE 13.2
Percentage Voting Democrat or Republican in 2004 Presidential Election
Although American parties are not closely tied to social class, better-off individuals tend to vote Republican. Poor and nonwhite individuals tend to vote Democrat as, increasingly, do women.

	Democrat	Republican
Race/Ethnicity		
White	42	58
African American	90	10
Education		
Grade school	69	31
High School	45	54
College	50	50
Sex		
Male	46	54
Female	53	47
Age		
Under 30	66	34
30–45	45	55
46–61	44	56
62–77	51	49
78 and older	52	48
Total	50	50

SOURCE: U.S. Bureau of the Census 2006.

and educated, among other things. Yet rates of voting in the United States have declined steadily for the last century, even though both income and educational achievement have increased. This paradox has led some scholars to argue that political participation has declined because a growing number of Americans believe that the political process is corrupt, that the Democrats and Republicans are more similar than different, and that it makes little difference who gets elected (Southwell & Everest 1998).

Other scholars argue that voting rates are so low at least in part because politicians (sometimes intentionally, sometimes unintentionally) have made it difficult for people to vote (Piven & Cloward 1988, 2000). Until only a few years ago, both registering to vote and voting were more cumbersome in the United States than in any other Western democracy. In many states, individuals had to register annually, pass literacy tests, or pay special taxes. They also had to both register and vote in specific locations during specific limited hours, which was especially difficult for persons who held strictly scheduled, working-class jobs.

Voter registration has increased significantly since passage of the National Voter Registration Act of 1993. However, as was visibly evident in the 2000 presidential election—when, among other things, students at Ohio's Kenyon College had to wait for as much as 8 hours before they could vote—barriers to voting still remain. In the 2004 election, potential voters (especially in poor and minority districts) were still hampered by transportation difficulties, polling places that closed too early, and problems with processing voter registrations and absentee ballots (People for the American Way 2004).

But perhaps the major reason that voting turnout has been so low is that no major political party has sought to involve marginalized Americans nor to actively address their concerns; voting rates have been highest when social movements have energized these constituencies to believe that they can make a difference and when political parties have reached out to them (Winders 1999).

A Case Study: Ex-Felon Disenfranchisement

As we've seen, a surprising number of Americans choose not to participate in the democratic process. An even more surprising number of Americans are *kept* from participating. It is estimated that approximately 4.7 million individuals are barred from voting—disenfranchised—because they were once convicted of a felony (Ugger & Manza 2002; Hull 2005). In some states, only those still in prison are forbidden from voting; in other states, a felony conviction brings lifelong **ex-felon disenfranchisement.** Because the United States has both a high rate of felony convictions (primarily for drug-related crimes) and unusually restrictive laws on the voting rights of ex-felons, the United States has a higher rate of ex-felon disenfranchisement than almost any other country. In essence, the very possibility of rehabilitation is ignored: Someone convicted at age 20 of selling marijuana, for example, might be ineligible to vote for the rest of his or her life, even if he or she never again commits a crime and becomes a successful worker, parent, and community citizen.

Ex-felon disenfranchisement is the loss of voting privileges suffered by those who have been convicted of a felony. In some states, ex-felon disenfranchisement applies only to those in prison; in other states, it is lifelong.

Importantly, because poverty sometimes pushes individuals to commit crimes, and because the criminal justice system more often convicts poor criminals than equally guilty wealthy criminals, those subject to felon disenfranchisement overwhelmingly are poor. The number of disenfranchised poor people is high enough to significantly decrease the chances of electing politicians who favor helping the poor (Ugger & Manza 2002).

Modern Economic Systems

Our description of the U.S. political process has crossed over into discussions of economic processes again and again. From the role of the working class to the role of the power elite, we find that understanding government requires understanding the economic relationships that underlie it. At this point, we turn to an explicit assessment of economic relationships and how they affect the individual worker and the economy as a whole.

Economic institutions are social structures concerned with the production and distribution of goods and services. Issues such as scarcity or abundance, guns or butter, and craftwork or assembly lines are all part of the production side of economic institutions. Issues of distribution include what proportion goes to the worker versus the manager, who is responsible for supporting nonworkers, and how much of society's production is distributed on the basis of need rather than effort or ability. The distribution aspect of economic institutions intimately touches the family, stratification systems, education, and government.

In the modern world, there are basically two types of economic systems: capitalism and socialism. Because economic systems must adapt to different political and natural environments, however, we find few instances of pure capitalism or pure socialism. Most modern economic systems represent some variation on the two and often combine elements of both.

Economic institutions are social structures concerned with the production and distribution of goods and services.

Capitalism

Capitalism is the economic system in which most wealth (land, capital, and labor) is private property, to be used by its owners to maximize their own gain. Capitalism is based on competition. Each of us seeks to maximize our own profits by working harder or devising more efficient ways to produce goods. Such a system encourages hard work, technical innovation, and a sharp eye for trends in consumer demand. Because self-interest is a powerful spur, such economies can be very productive.

Even when it is very productive, though, a capitalist economy has drawbacks. These drawbacks all center around problems in the distribution of resources. First, the capitalist system at its most ideal represents a competitive bargain between labor (workers) and capital (owners of industries), both of whom control a necessary resource. But this is not a bargain between equals: almost always, capital has more power to obtain the bargain it wants than does labor. As a result, workers earn only a fraction of what capitalists earn. Second, those who have neither labor nor capital to bargain with (children, stay-at-home moms, the elderly, the disabled, and workers whose jobs have disappeared) always lose out, for with nothing to exchange, they are outside the market. They must rely on aid from others, which is not always forthcoming. Third, because public goods such as streets, watersheds, sewers, or defense offer no profit, pure capitalism has no interest in providing them. Yet society cannot function without these services. Thus capitalist systems must have some means of distribution other than the market.

Capitalism is the economic system, based on competition, in which most wealth (land, capital, and labor) is private property, to be used by its owners to maximize their own gain.

Socialism

If capitalism is an economic system that maximizes production at the expense of distribution, socialism is a system that stresses distribution at the expense of production. As an ideal, **socialism** is an economic structure in which productive tools are owned and managed by the workers and used for the collective good.

Socialism is an economic structure in which productive tools (land, labor, and capital) are owned and managed by the workers and used for the collective good.

In theory, socialism has several major advantages over capitalism. First, societal resources can be used for the benefit of society as a whole rather than for individuals. This advantage is most apparent in regard to common goods such as protecting the environment. A related advantage is that of central planning. Because resources are controlled by the group, they can be deployed to help reach group goals. This may mean diverting them from profitable industries such as television production to industries that are considered more likely to benefit society in the long run, such as education, agriculture, or steel. The major advantage claimed for socialism, however, is that it produces equitable (although not necessarily equal) distribution.

The creed of pure socialism is *from each according to his or her ability, to each according to his or her need.* Under socialism, everyone should receive what they need to survive, and everyone should work their best to achieve that common goal. Workers are expected to be motivated by loyalty to their community and their comrades. In reality, the hardworking woman with no children is not likely to work her hardest when the lazy worker next to her takes home a larger paycheck simply because she has several children and thus a greater need. Nor is the farmer as likely to make the extra effort to save the harvest from rain or drought if his rewards are unrelated to either effort or productivity. Because of this factor, production is usually lower in socialist economies than in capitalist economies.

Mixed Economies

Most Western societies today represent a mixture of both capitalist and socialist economic structures. In many nations, services such as the mail and the railroads and key industries such as steel and energy are socialized. This socialism rarely results from pure idealism. Rather, public ownership is often seen as the best way to ensure continuation of vital services that are not profitable enough to attract private enterprise. Other services—for example, health care and education—have been partially socialized because societies have judged it unethical to deny these services to the poor and too inefficient to provide them on the open market.

In the case of many socialized services, general availability and progressive tax rates have gone far toward meeting the maxim "from each according to his or her ability, to each according to his or her need." There are still inequalities in education and health care, but many fewer than there would be if these services were available on a strictly cash basis. The United States has done the least among major Western powers toward creating a mixed economy, and our future direction is unclear. By and large, the Republican Party has pushed to reduce government provision of social services and the Democratic Party has pushed to increase such services. The future mix of socialist and capitalist principles will reflect political rather than strictly economic conditions.

The Political Economy

Political economy refers to the interaction of political and economic forms within a nation. Both capitalism and socialism can coexist with either authoritarian or democratic political systems. Socialism and democracy are combined in many Western European nations, such as the United Kingdom and Sweden. Other nations, such as China and Cuba, combine socialism with an authoritarian political system. We often use (and misuse) the term "communist" to refer to societies in which a socialist economy is guided by a political elite and enforced by a military elite. The goals of socialism (equality and efficiency) are still there, but the political form is authoritarian rather than democratic.

Political economy refers to the interaction of political and economic forms within a nation.

Global Perspective:

focus on

Democratic Socialism in Sweden

What would it be like to live and work in Sweden? You would have guaranteed access to quality public transportation; guaranteed income if you were ill, disabled, or elderly; guaranteed access to comfortable housing; and free education all the way through college, graduate school, or professional training. After you or your partner gave birth to or adopted a baby, you would be entitled to a full year of paid parental leave. Once you went back to work, you could use a free, high-quality, state-funded day-care center. In exchange for these benefits, you would pay about 25 percent of your paycheck in federal income taxes and almost as much in local taxes.

Sweden is a democratic socialist society with an economy that mixes corporate capitalism with significant welfare benefits for workers and nonworkers alike. Because Sweden is a democracy, the majority of Swedes have voted to receive these benefits and to pay high taxes for them. But Sweden's economy wasn't always arranged this way.

Sweden owes its economic organization in part to the rise of a strong labor movement (Koblik 1975). As industrialization began in Sweden in the 1870s, labor union members worked to create the Social Democratic Party, a political party dedicated to equitable wages, job security, and welfare programs for the entire society. While Communists in Russia were fighting and winning the Russian Revolution in 1914–1917, members of Sweden's Social Democratic Party were politicking for seats in parliament. After holding power on and off during the 1920s, the Social Democratic Party won an important election in 1932 and then retained virtually uninterrupted political power until the present day. This has allowed the welfare state established by the party to develop deep roots.

The welfare state's emergence and success also reflects the deeply held Swedish belief in the responsibility of the community to look out for all its members. This attitude, in turn, has been fostered by the cultural homogeneity of the Swedish population. Until about 1980, the population of Sweden was overwhelmingly ethnically Swedish. Currently, however, foreign immigrants and individuals who have at least one immigrant parent comprise close to 20 percent of the Swedish population. As a result, many question whether support for the welfare state will decline if ethnic Swedes become unwilling to extend the benefits of their social system to immigrants, and if immigrants bring with them a philosophy of individual rather than social responsibility.

Not everyone in Sweden is a member of the Social Democratic Party, of course. Conservative groups favor a freer market economy. Furthermore, Social Democrats today are worried about whether Sweden's welfare society can survive the current international economic downturn (Olsen 1996). Controlling Sweden's transnational corporations so that they do not export jobs and continue to pay high taxes at home is proving to be more and more difficult. Some economists are beginning to point out that Sweden's market socialism is based on an inherent irony: Strong and profitable capitalist businesses are necessary so that workers can be employed and taxes for welfare benefits can be collected. But insisting on generous worker benefits and full employment eats into capitalist profits (Olsen 1996).

■ Swedish day-care center.

© Joe Rodriguez/Black Star Publishign/PictureQuest

Likewise, some capitalist nations are democratic and some are authoritarian. The United States and Japan have both capitalist economies and democratic political systems. Singapore and Saudi Arabia, on the other hand, have capitalist economies but autocratic political systems, in which elections are either nonexistent or virtually meaningless. These examples remind us that both capitalism and socialism can coexist with authoritarian regimes.

Privatization and the U.S. Political Economy

As we've seen, the United States is a democracy based in capitalism. This capitalistic basis of our system is reflected in the recent trend toward the privatization of government services. **Privatization** refers to the process through which government services are increasingly "farmed out" to private, capitalistic firms. The term is also used to refer to the practice of redesigning government services to operate more like corporate businesses.

> **Privatization** refers to two processes: 1) the increasing "farming out" of government services to private, capitalistic firms and 2) the increasing redesign of government services to operate more like corporate businesses.

Privatization has affected many types of government services (Jurik 2004). Some cities and states contract out water testing and delivery to private bidders. Others deliver public water very cheaply to private bottlers, who earn extraordinarily high profits by filtering it and selling it as a luxury good. Yet public water supplies are both more heavily regulated and safer than are Perrier, Calistoga Springs, or other private waters (Public Citizen 2006). Similarly, health care in U.S. prisons is now primarily offered by doctors who work on contracts with private firms. Some states have gone even farther and have contracted out the running of the prisons themselves (Hallett 2002), as well as welfare and other government services.

Major public universities illustrate the second form of privatization: redesigning public services to mimic corporate processes. These universities are still owned and run by state and city governments but, like corporations, are increasingly focused on the bottom line (Washburn 2006). Professors are hired and fired not based on the quality of their teaching, or even the quality of their research, but on whether their research will bring grant dollars or remunerative patents to their university.

Supporters of privatization argue that it brings greater efficiency to water supplies, prisons, universities, and other government services by motivating individuals to work hard and keep a sharp eye on cost–benefit ratios. Opponents argue that professors, scientists who test our water supply, guards who staff our prisons, and the like should make decisions based on what is best for our society, not what will generate the greatest profit.

The U.S. Economic System

Why are lawyers paid more than schoolteachers? Why are so many small grocery stores in New York and Los Angeles run by Korean immigrants? Why do so few farm kids end up staying on the farm? And how much can you expect to earn after you graduate? To answer these questions, we need to first understand the economic "big picture." To do so, we need to address four topics: the postindustrial economy, the dual economy, segmented labor markets, and the "Wal-Mart Effect."

The Postindustrial Economy

> **Primary production** is extracting raw materials from the environment.

In a preindustrial economy, the vast majority of the labor force is engaged in **primary production,** extracting raw materials from the environment. Prominent among primary production activities are farming, herding, fishing, forestry, hunting, and min-

Primary production involves direct contact with natural resources—fishing, hunting, farming, forestry, and mining.

ing. Preindustrial economic structures were characteristic of Europe until 500 years ago and are still typical of many societies.

The Industrial Revolution brought a shift from primary to **secondary production,** the processing of raw materials. For example, ore, cotton, and wood are processed by the steel, textile, and lumber industries, respectively; other secondary industries turn these materials into automobiles, clothing, and furniture. The shift from primary to secondary production is characterized by enormous increases in the standard of living.

Secondary production is the processing of raw materials.

Postindustrial development rests on a third stage of productivity, **tertiary production.** This stage is the production of services. The tertiary sector includes a wide variety of occupations: physicians, schoolteachers, hotel maids, short-order cooks, and police officers. It includes everyone who works for hospitals, governments, airlines, banks, hotels, schools, or grocery stores. None of these organizations produces tangible goods; they all provide service to others. They count their production not in barrels or tons but in numbers of customers.

Tertiary production is the production of services.

The tertiary sector has grown very rapidly in the last half-century and is projected to grow still more. As Figure 13.2 illustrates, only 19 percent of the labor force was involved in tertiary production in 1920; by 1956, the figure had grown to 49 percent, and by 2010 it is expected to include 80 percent of the labor force. Simultaneously, the portion of the labor force employed in primary production has almost disappeared, and the proportion employed in secondary production has halved. These shifts do not mean that primary and secondary production are no longer important, however. A large service sector depends on primary and secondary sectors that are so efficiently productive that large numbers of people are freed from the necessity of direct production.

As we saw in Chapter 7, a major change since 1980 is that America's postindustrial economy is increasingly interwoven with that of other nations. This globalization is likely to increase in the near future.

FIGURE 13.2
Changing Labor Force in the United States
Since 1820, the labor force in the United States has changed drastically. The proportion of workers engaged in primary production has declined sharply, while the proportion engaged in service work has expanded greatly. SOURCE: U.S. Bureau of Labor Statistics 2002.

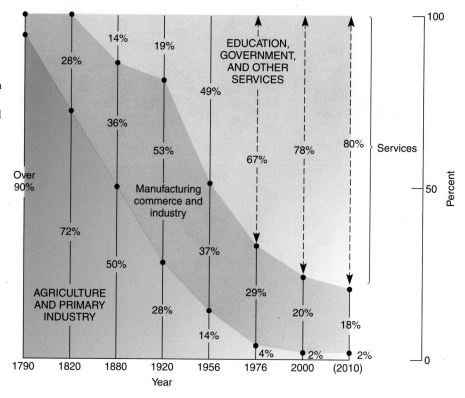

The Dual Economy

A **dual economy** consists of the complex giants of the industrial core and the small, competitive organizations that form the periphery.

The U.S. economic system can be viewed as a **dual economy.** Its two parts are the complex giants of the industrial core and the small, competitive organizations that form the periphery. They are distinguished from each other on two dimensions: the complexity of their organizational forms and the degree to which they dominate their economic environment (Baron & Bielby 1984).

The Industrial Core: Corporate Capitalism

Although there are more than 250,000 businesses in the United States, most of the nation's capital and labor are tied up in a few transnational giants that form the industrial core. The top 20 U.S. companies are large bureaucracies that control billions of dollars of assets and employ thousands of individuals. These giants loom large on both the national and international scene.

At the local level, we are familiar with the situation in which a region's one major employer holds city and county government hostage and bargains for tax advantages and favorable zoning regulations in exchange for increasing or retaining jobs. The growing size and interdependence of firms in the industrial core are now causing this scene to be reenacted at the federal level.

Wealthy capitalists are linked to each other by shared ownership of large firms; large firms are linked to one another by the members on their boards of directors, the businesses they invest in, and the businesses that invest in them. As a result of this interdependence, relations among large firms have become more cooperative than competitive. Although decreased competition reduces productivity and efficiency, it

Transnational corporations constantly seek new access to natural resources and new markets for their products.

increases joint political influence (Mizruchi 1989, 1990). For example, as the proportion of the nation's assets held by the top 100 firms increased, their political power increased and the taxes they were required to pay decreased—even while individual income tax rose (Jacobs 1988).

That power can extend to influencing U.S. foreign policy. A desire to protect the interest of transnational companies like Dole and United Fruit certainly played a role in the U.S. decision to lend support to dictatorships in Guatemala, Honduras, and other Latin American countries during the twentieth century. Similarly, to protect transnational oil companies, the United States covertly orchestrated the 1951 coup against democratically-elected Iranian Prime Minister Mohammad Mossadegh and subsequently propped up the authoritarian regime of the Shah of Iran (Kinzer 2003). (Popular resentment of the Shah's repressive regime eventually led to the Islamic Iranian revolution in 1979, which stimulated Islamic fundamentalism worldwide.) Some observers suspect that the U.S. conflicts with Iraq in 1991 and 2003 had more to do with protecting U.S. oil interests than with fighting terrorism.

The Periphery: Small Business

The periphery of the U.S. economy is made up primarily of small businesses that are owned and run by a family or small group of partners. They are usually characterized by few employees, economic uncertainty, and relatively little bureaucracy. The periphery can be divided into a formal and an informal economy.

THE FORMAL ECONOMY   The formal economy consists of small banks, farms, retail stores, restaurants, and repair services. Some small manufacturing companies are also part of this sector.

Although this segment of the periphery is smaller than it was 100 years ago, it continues to furnish jobs for a substantial portion of the population and to be an important part of both the industrial and service sectors. It has been an especially important avenue of opportunity for minorities in the United States. Koreans, Hispanics, and

■ The small businesses of the periphery provide important economic opportunities for minority Americans, such as these Korean American grocers.

African Americans who face discrimination in the corporate world can sometimes achieve moderate prosperity by operating neighborhood grocery stores, laundries, fast-food franchises, and the like.

The **informal economy** is the part of the economy that largely escapes state regulation; also known as the underground economy.

INFORMAL ECONOMY   An important sector of the periphery is the underground or **informal economy.** This is the part of the economy that largely escapes state regulation. It includes illegal activities such as prostitution, smuggling immigrants, and running numbers, as well as a large variety of legal but unofficial enterprises such as home repairs, house cleaning, and garment subcontracting. Often referred to disparagingly as "fly-by-night" businesses, enterprises in the informal sector are nevertheless an important source of employment. This is especially true for those segments of the population who would like to avoid federal record keeping: undocumented aliens, foreign students, senior citizens and welfare recipients who don't want their earnings to reduce their benefit levels, adolescents too young to meet work requirements, and many others.

The Segmented Labor Market

The **segmented labor market** parallels the dual economy. Hiring, advancement, and benefits vary systematically between the industry core and the periphery.

Parallel to the dual economy is a dual labor market, generally referred to as a **segmented labor market.** A segmented labor market is one in which hiring, advancement, and benefits vary systematically between the industrial core (the corporate sector) and the periphery (the competitive sector).

In the corporate sector, most hiring is done at the entry level, most upper-level positions are filled from below, and credentials are critical for hiring and promotion. Within core firms, there are predictable career paths for both blue- and white-collar workers. Although both job security and benefits have declined over the last 20 years (Newman 1999a), employment remains generally secure, and benefits still are relatively good. Wages and benefits are best in the very largest firms.

© Michael Newman/PhotoEdit

In the competitive sector, on the other hand, credentials are less important, career paths are short and unpredictable, security is minimal, and benefits are low or nonexistent. At the same time, there is less bureaucracy and red tape, and both workers and managers have more freedom in their work.

The competitive sector offers an employment haven for those who do not meet the demands of the corporate sector: those who do not have the required credentials, who have spotty work records, who want to work part time, or who have been "downsized" from the corporate world. As a result, a disproportionate number of youths, minorities, and women work in the competitive sector. Because keeping a job and getting a promotion are governed almost exclusively by personal factors rather than by seniority or even ability, however, this sector is less likely to promote minorities or women; there is no affirmative-action officer at Joe's Café. As a result, the gender gap in wages is significantly larger in the competitive sector than it is in core industries (Coverdill 1988).

The "Wal-Mart Effect"

So far we have talked about entire segments of the economy at a time—large corporations, informal businesses, and so on. The implicit message is that no one corporation or organization is that important on its own. This, however, is no longer the case. One corporation—Wal-Mart—is so large and so powerful that economists now identify a "Wal-Mart Effect" on the U.S. economy overall.

Until the 1980s, federal laws prohibited any corporation from becoming a monopoly. A monopoly is a corporation that holds so large a market share for a given good or service that it could drive any competitors out of business and then set any prices it wanted for its goods and services. These laws were substantially weakened by elected officials, beginning with Ronald Reagan, who were opposed to "big government" of all sorts. Wal-Mart is the result.

Wal-Mart earns its profit not by setting prices high, but by setting prices low and selling in vast quantities. There is no question that Wal-Mart benefits individual consumers via its low prices. But consumers and everyone else pay a high price in many other ways. Because Wal-Mart holds such a large share of any given market (for toys, for tires, for clothing), any manufacturer that doesn't sell its products

© AP/ Wide World Photos

■ Although individual consumers benefit from Wal-Mart's low prices, its low pay scale, ability to drive competitors out of business, and general unwillingness to pay benefits to its employers hurts communities in many ways. As a result, protests against Wal-Mart stores and policies have increased around the country.

through Wal-Mart risks being driven out of business. But manufacturers that work with Wal-Mart *also* can be driven out of business when Wal-Mart requires them to price their goods so low that the manufacturers no longer earn a profit. To avoid this fate, manufacturers have either cut wages to the bone or have moved jobs overseas where labor is cheaper.

Many stores that used to compete with Wal-Mart also have been driven out of business by Wal-Mart's predatory pricing; in towns across the United States, the arrival of Wal-Mart has quickly led entire downtowns to virtually shut down and led to significant drops in wages at stores that continue to compete with Wal-Mart. When small businesses go under, not only do individuals lose jobs, but towns lose a stable middle class with a vested interest in civic affairs. The resulting loss in jobs leaves many workers with no option other than to seek employment at Wal-Mart, where average salaries are below the poverty level (United Food and Commercial Workers 2006). In sum, in the new U.S. economy, Wal-Mart not only sets its own prices and employees' wages, but also effectively sets the prices for goods from its suppliers and for wages at both its suppliers and its competitors (Fichman 2006; Lynn 2006).

Work in the United States

From the individual's point of view, economic institutions mean jobs. For some, jobs are just jobs; for others, they are careers. But 40 years of involvement in the world of work is central to most people's lives.

Occupations

Aside from the simplest consequences of working (income and filling up much of your time), what you do at work is probably as important as whether you work. Here some of the important differences between the professions and white- and blue-collar work are described.

Professions

Professions are occupations that demand specialized skills and creative freedom.

The highest level of occupations are known as **professions.** Sociologists generally define an occupation as a profession when it meets three characteristics. First, it must have the autonomy to set its own educational and licensing standards and to police its members for incompetence or malfeasance. For example, doctors, rather than consumers, make up the licensing boards that judge doctors accused of incompetence. Second, a profession must have its own technical, specialized knowledge, learned through extended, systematic training. For example, both lawyers and car mechanics have specialized knowledge. But lawyers must study for years before entering the field, whereas mechanics need study only for months. As a result, sociologists consider lawyers to be professionals but do not consider car mechanics to be professionals. Third, a profession must be believed by the public to follow a code of ethics and to work more from a sense of service than from a desire for profit. So, for example, even though the public realizes that some individual ministers, doctors, and lawyers place personal profits above public service, it believes that most members of these professions do not.

The rewards that professionals achieve vary considerably. Physicians and lawyers can receive very high incomes, while schoolteachers typically earn more modest pay. The major reward that all professionals share, however, is substantial freedom from supervision. Because their work is nonroutine and requires personal judgments, professionals have been able to demand—and get—the right to do their work more or less their own way.

Freedom from supervision remains the most outstanding reward of professional work, but it is a reward that is being eroded. Increasingly, people in the professions work for others within bureaucratic structures that constrain many of the most characteristic aspects of professional life. Teachers must now spend considerable class time prepping students for standardized exams, and doctors must limit their prescriptions to drugs approved by managed care organizations.

What Color is Your Collar?

Fifty years ago, the color of your collar was a pretty good indication of the status of your job and your gender. Men who worked with their hands wore blue (or brown or flannel) collars; women who worked with their hands as beauticians, maids, or waitresses often wore pink. Managers and others who worked in clean offices wore white collars. Those days are past. The labor force is far more diversified, and some of the old guidelines no longer work. The bagger at Safeway wears a white shirt and tie; the librarian wears blue jeans and sandals. Yet the librarian is a white-collar worker and the bagger is not.

Traditional white-collar workers are managers, professionals, typists, and salespeople—those who work in clean offices and are expected to be able to think independently. Blue-collar workers are people in primary and secondary industry who work with their hands; they farm, assemble telephones, build houses, and weld joints. Although some blue-collar workers earn more than some white-collar workers, blue-collar jobs are characterized by lower incomes, lower status, lower security, closer supervision, and more routine.

■ Like most at the upper-end of the white-collar spectrum, this professional has a job that requires a college education, the ability to think independently, and excellent communication skills.

■ The fastest growing jobs in the United States today are minimum-wage service jobs that offer few benefits and fewer prospects for advancement.

FIGURE 13.3
Projected Fastest Growing Jobs, 2004–2014
The demand for labor is expected to grow between 2004 and 2014. But the ten jobs with the largest projected growth rates are all service jobs, and most are low-skill and low-wage jobs.
SOURCE: U.S. Bureau of Labor Statistics 2006a.

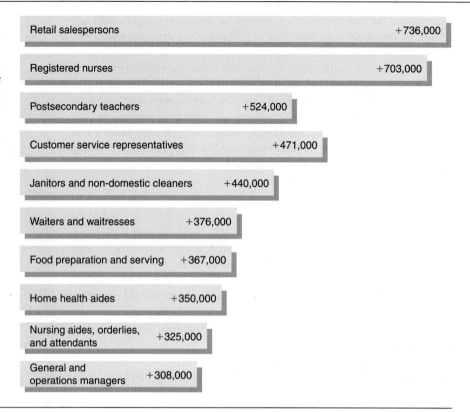

Retail salespersons +736,000
Registered nurses +703,000
Postsecondary teachers +524,000
Customer service representatives +471,000
Janitors and non-domestic cleaners +440,000
Waiters and waitresses +376,000
Food preparation and serving +367,000
Home health aides +350,000
Nursing aides, orderlies, and attendants +325,000
General and operations managers +308,000

Fifty years ago, this simple division of the labor force included most workers. These days it leaves out a growing category of low-skilled, low-status workers who often appear in company-supplied brown polyester suits or turquoise jackets and who fry hamburgers, stock K-Mart shelves, and collect money at the corner gas station. Characteristically, these workers hold jobs that have a short or nonexistent career ladder, and they earn the minimum wage or close to it.

Occupational Outlook

As the graph in Figure 13.2 indicated, the outlook for the future is for greater expansion of the tertiary sector. What will this mean for the kinds of jobs that are available in the future? As Figure 13.3 shows, all ten of the occupations with the greatest projected increase between now and 2014 are service jobs, and seven of the ten are low-wage and low-skill jobs.

In contrast, some traditional occupational categories are expected to suffer major declines. The occupations with the largest projected decreases include both blue- and white-collar jobs: Typists and word processors will have a harder time finding jobs, as will sewing-machine operators and farmers. The declining opportunities in these occupations reflect a variety of factors: changing age structure, loss of U.S. jobs due to migration of industry overseas, and new technology.

The most controversial issue is what kind of new jobs the economy will offer. Optimistic observers note that executive and professional jobs are growing faster than average and suggest that the high quality and good pay of these new jobs indicate what awaits today's college graduates. Others focus on the rapid increase in what one critic has called "McJobs" (Ritzer 1996). Although not all these jobs entail selling hamburgers, many are low-status jobs with low wages and no benefits: health aides, personal and home-care aides, and cashiers.

Both the optimists and the critics are correct in their expectations for the future. Although a four-year college degree will not guarantee a secure, well-paying job, good jobs for college graduates and those with technical training—computer engineers and scientists, registered nurses, and system analysts—nonetheless are growing rapidly. At the same time, however, bad jobs, often traditionally done by females and paying very poorly—home health aides, waiters and waitresses, and personal care assistants—are also growing rapidly (James, Grant, & Cranford 2000). Thus, the fastest-growing occupations require either years of advanced education or almost no skill at all, and the latter offer very little reward. The big losers in the transformation of the labor market are likely to be the traditional working class. Unlike their parents, who could find good, unionized jobs, young working-class people who do not obtain four-year college degrees will find a hard road ahead (Blau 2001; Perrucci & Wysong 2002).

The Meaning of Work

For most people, work is essential as the means to earn a livelihood. As noted in Chapter 7, one's work is often the most important determinant of one's position in the stratification structure and, consequently, of one's health, happiness, and lifestyle.

Work is more than this, however. It is also the way most of us structure our lives. It determines what time we get up, what we do all day, who we do it with, and how much time we have left for leisure. If we ourselves do not hold a job, our parents' or spouses' job may structure our lives: There's a big difference—one that goes beyond mere income—between being a preacher's kid or an army brat, a doctor's wife or a janitor's wife. Thus, the nature of our work and our attitude toward it can have a tremendous impact on whether we view our lives as fulfilling or painful. If we are good at it, if it gives us a chance to demonstrate competence, and if it is meaningful and socially valued, then it can be a major contributor to life satisfaction.

Work Satisfaction

U.S. surveys consistently find that the large majority (80 percent) of workers report satisfaction with their work. Although such a report may represent an acceptance of one's lot rather than real enthusiasm, it is remarkable that so few report dissatisfaction.

Studies of job satisfaction concentrate on two kinds of rewards that are available from work. **Intrinsic rewards** arise from the process of work; you experience them when you enjoy the people you work with and feel pride in your creativity and accomplishments. **Extrinsic rewards** are more tangible benefits, such as income and security; if you hate your job but love your paycheck, you are experiencing extrinsic rewards.

Generally, the most-satisfied workers are those in the learned professions, people such as lawyers, doctors, and professors. These people have considerable freedom to plan their own work, to express their talents and creativity, and to work with others; furthermore, their extrinsic rewards are substantial. The least-satisfied workers are those who work on factory production lines. Although their extrinsic rewards can be good, their work is almost completely without intrinsic reward; they have no control over the pace or content of the work and are generally unable to interact with coworkers. Between these extremes, professionals and skilled workers generally demonstrate the greatest satisfaction; semiskilled, unskilled, and clerical workers indicate lower levels of satisfaction. Nevertheless, even those who hold highly routine, physically demanding jobs such as cashiers and cooks at fast-food restaurants often

Intrinsic rewards are rewards that arise from the process of work; they include enjoying the people you work with and pride in your creativity and accomplishments.

Extrinsic rewards are tangible benefits such as income and security.

enjoy the satisfactions that come from doing a job well, earning a steady paycheck, and socializing with fellow workers (Newman 1999b).

Alienation

Alienation occurs when workers have no control over the work process or the product of their labor.

Another dimension of the quality of work life is alienation. **Alienation** occurs when workers have no control over their labor. Workers are alienated when they do work that they think is immoral (build bombs) or meaningless (push papers or brooms, or put together small pieces with no understanding of how those small pieces will become part of some larger whole). Work is also alienating when it takes physical and emotional energy without giving any intrinsic rewards in return. Alienated workers feel *used*.

The concept of alienation was first developed by Karl Marx to describe the factory system of the mid-nineteenth century. In 1863, a mother gave the following testimony to a committee investigating child labor:

> When he was seven years old I used to carry him [to work] on my back to and fro through the snow, and he used to work 16 hours a day. . . . I have often knelt down to feed him, as he stood by the machine, for he could not leave it or stop. (as quoted in Hochschild 1985, 3)

This child was truly an instrument of labor. He was being used, just as a hammer or a shovel is, to create a product that would belong to someone else.

Emotional labor refers to the work of smiling, appearing happy, or in other ways suggesting that one enjoys providing a service.

Although few Americans work on assembly lines any more, modern work can also be alienating. Service work, in fact, has its own forms of alienation, known as **emotional labor.** In occupations from nursing to teaching to working as flight attendants, not merely our bodies but also our emotions become instruments of labor. To turn out satisfied customers, we must smile and be cheerful in the face of ill humor, rudeness, or actual abuse. Studies of individuals in these occupations show that many have trouble with this emotional component of their work. After smiling for 8 hours a day for pay, they feel that their smiles have no meaning at home. They lose touch with their emotions and feel alienated from themselves (Hochschild 1985). This is especially true when workers feel that they have no control over their job conditions (Bulan, Erickson, & Wharton 1997).

Alienation is not the same as job dissatisfaction (Erikson 1986). Alienation occurs when workers lack control. It is perfectly possible that workers with no control, but with high wages and a pleasant work environment, will express high job satisfaction.

Technology and the Future of Work

The experience of working is tied directly to the technologies of the workplace. Many now argue that the shift toward more and more advanced technologies is inescapably anti-labor. These critics point out three negative effects of technology on labor: deskilling, displacing workers, and greater supervision.

1. *Deskilling.* In many occupational fields, the level of skill required has dropped to the point where it is difficult to take pride in craft or a job well done. Such *deskilling* can occur either when a job is cut into small pieces or mechanized (such as printing a newspaper) or when workers are limited to executing tasks which they did not participate in designing. Deskilling occurs at all levels of labor, not just on the assembly line (Burris 1998). For example, nurses and doctors these days often are required to follow set protocols for treating patients with little freedom to make independent decisions (Weitz 2007).

An important element of the deskilling process is that it reduces the scope for individual judgment. In hundreds of jobs across the occupational spectrum, computers make decisions for us. In the sawmill industry, for example, a computer now assesses the shape of a log and decides how it should be cut to get the most board feet of lumber from it. An important element of skill and judgment honed from years of experience is now made worthless (Braverman 1974).

2. *Displacement of the labor force.* One of the most critical complaints about automation is that it replaces people with machines. In the automobile industry, robots have replaced thousands of workers. In grocery stores, computerization has largely eliminated inventory clerks and pricing personnel and is increasingly replacing cashiers with "self-check-out" aisles. Meanwhile, in industry after industry, sophisticated technology has made sharp inroads into the number of hours of labor necessary to produce goods and services—and into the number of workers needed to get the work done. The resulting fear of job loss is one of the reasons employees seldom complain about deskilling. If they still have a job, they are happy (Vallas & Yarrow 1987).

3. *Greater supervision.* Computerization and automation give management more control over the production process. More aspects of the production process are determined by management through its computerized instructions, and fewer aspects are determined by the employees. Computers also keep more complex records on employees. For example, the scanner machines used in grocery stores do more than keep inventory records and add up your grocery bill. They also keep tabs on the checker by producing statistics such as number of corrections made per hour, number of items run through per hour, and average length of time per customer. It is not surprising, therefore, that studies show that computers have increased work alienation among the cashiers and typists who use them (Vallas 1987).

Whether new technologies are an enemy of labor may depend on which laborer we ask. Those persons, often women and less-skilled workers, whose jobs are being replaced by new technologies are unlikely to see anything good about them. On the other hand, most professionals in the knowledge industries (education and communications) regard these technologies as a boon. Computers have expanded their job opportunities and enhanced their lives. Even these workers, however, may occasionally wonder if they really benefit when technologies such as fax machines, cell phones, and e-mail allow—or even require—that they work at home, expanding work into a 24-hour-a-day job.

Technology by itself is a neutral force: It can aid management or it can aid the workers. Which technologies are implemented and the way they are implemented reflect a struggle between labor and management, and this struggle, not the technology itself, will determine the outcome.

Globalization and the Future of Work

The consequences of large-scale job loss are substantial. Not only individual workers but also entire communities are impoverished as new technologies facilitate corporate decisions to move factories to other countries where labor is cheaper. In fact, globalization is leading our national economy through a process of reverse development: Like a least-developed country, we export raw materials such as logs and wheat and import manufactured products such as VCRs and automobiles. People in Mexico, Japan, and Korea have jobs manufacturing products for the U.S. market while U.S. workers are making hamburgers.

Blue-collar workers are not the only ones affected by the loss of American jobs. Increasingly, white-collar jobs like computer programming and scientific research also are being moved overseas by corporations eager to find cheap workers whose scientific skills equal or surpass those of Americans. Other countries can now offer highly skilled workers, fluent in English, who are willing to work for far less than will U.S. workers. In one study conducted in 2006, 38 percent of surveyed multinational corporations reported plans to shift research centers from the United States and other developed nations to developing countries in the next three years, with China and India the most likely locations (Lohr 2006).

What can public policy do to protect jobs in the United States? There are three general policy options: the conservative approach, the liberal approach, and the social investment approach.

The Conservative Approach: Free Markets

Generally, business leaders and conservatives argue that the way to keep jobs in the United States is to reduce government oversight and leave wages and benefits up to market forces. In effect, this means reducing wages and benefits. If labor is cheap, conservatives argue, business will have less incentive to automate or to move assembly plants to Mexico or Indonesia.

By default, this policy has been implemented. In communities across the nation, management has used threats of plant closings to force wage concessions and reduce benefits. The power of labor unions is reduced to negotiating benefit protection in the face of wage reductions. Because so many workers have been afraid of losing their jobs, organized labor's power has been sharply reduced. Thus, one result of large-scale job loss is poorer working conditions for those who still have jobs.

The Liberal Approach: Government Policies

Liberals argue that private profit should not be the only goal of economic activity and that the state should see to it that economic decisions protect communities' and workers' interests. Among the specific policies recommended are (1) federal trade policies that make U.S.-made goods more competitive in international markets and that reduce the advantage that foreign-made products have in the United States, (2) vigorous state investment in industries that will provide the largest number of decent jobs, (3) government oversight of mergers and plant closings to make sure plants behave responsibly, and (4) state support for worker efforts to buy and manage their own industries.

In addition to policies such as these which are designed to keep people working, liberals also favor social welfare policies to protect those who are thrown out of work. Among the policies recommended are early notifications of proposed plant closings, retraining and relocation assistance for displaced workers, and more generous unemployment benefits. These are the sort of policies that some Western European nations have adopted. (See the Focus on a Global Perspective box in this chapter).

The Social Investment Approach

Finally, some observers (not easily categorized as either liberal or conservative) note that low-tech jobs move overseas solely to save money, but high-tech jobs move overseas both to save money *and* to seek intellectual talent (Lohr 2006). These observers argue that the best way for the United States to protect high-income jobs in information technology and scientific research is to make sure that American students receive the best possible education in reading, writing, science, and mathematics, from grade school through graduate school.

Where This Leaves Us

As you sit in the classroom to prepare yourself for a good position in the labor force, the economy itself is changing. Indeed, it is changing so fast that you may need to retool your skills several times before your work life is complete. Rapid developments in technology have dramatically changed the workplace from what it was 20 or even 10 years ago. The globalization and "Wal-Martization" of the economy has further changed the job situation for Americans. Although the specter of unemployment still haunts ethnic minorities and blue-collar workers the most, middle-class workers are also experiencing the pangs of job insecurity and alienation, as corporations strive to cut costs and jobs move overseas.

One political approach to the changing job situation is to adopt social welfare policies that would protect American workers both against overseas competition and against corporations focused on profitability. Yet in an era of political conservatism, when relatively few Americans vote—and voting is especially uncommon among those who need the government's help the most—it seems unlikely that the United States will expand its protections for average citizens. The growing concentration of money and power also works against any meaningful changes. In the meantime, American workers will increasingly need to learn new skills and be creative to find and keep jobs.

Summary

1. Power may be exercised through coercion or through authority. Authority may be traditional, charismatic, or rational-legal.

2. Any ongoing social structure with institutionalized power relationships can be referred to as a political institution. The most prominent political institution is the state. It is distinguished from other political institutions because it claims a monopoly on the legitimate use of coercion and it has power over a broader array of issues. Globalization, however, may be limiting this power.

3. Democracy is most likely to flourish in societies that have vibrant, competing interest groups, large middle classes, and relatively little income inequality.

4. Two major models are used to describe the U.S. political process: the pluralist model and the power-elite model. Although they disagree on whether power is centered in an elite or more broadly distributed, they agree that organized groups have far more power to influence events than do individuals.

5. Voting and other forms of political participation is especially low in the United States. Political participation is greater among those with high social status and among middle-aged and older people—establishment types who are more likely to support the status quo.

6. Although the Democratic Party tends to attract working-class and minority voters and the Republican Party tends to attract white and better-off voters, both U.S. political parties tend to have middle-of-the-road platforms with broad appeal.

7. Some sociologists believe that voting rates are lower in the United States than in other industrialized countries because potential voters are politically alienated. Others believe that rates are low because government policies make it difficult or impossible for Americans to vote and because no major political party has actively sought to involve marginalized groups.

8. Capitalism is an economic system that maximizes productivity but pays little attention to the equitable distribution of resources to the people; socialism emphasizes distribution of resources but neglects aspects of production. Most societies mix capitalist and socialist elements in their economies.

9. Each nation has its own political economy: the particular combination and interaction of political and economic forms within a nation. Both capitalist and socialist nations can be either democratic or dictatorial.

10. Changes from preindustrial to industrial to postindustrial economies have had profound effects on social organization. The tertiary sector of the economy occupies

about three quarters of the U.S. labor force; it includes doctors and lawyers as well as truck drivers and waitresses.

11. The United States has a dual economy containing two distinct parts: the industrial core and the periphery (or competitive sector). These are paralleled by a segmented labor market.

12. Currently, one corporation, Wal-Mart, has the power to affect organizations and individuals across the economic spectrum. Because of its great market share, it can affect prices for workers' labor, for its suppliers, and for its competitors.

13. For the near future, the largest number of new jobs will likely be low-status, low-wage service positions. The major losers will be those who have occupied traditional blue-collar jobs.

14. Although most U.S. workers report satisfaction with their work, many are nevertheless alienated because they are estranged from the products of their labor or from their emotions.

15. Critics argue that technology has had three ill effects on labor: deskilling jobs, reducing the number of jobs, and increasing control over workers.

Thinking Critically

1. The family and the classroom are more often authoritarian than democratic. Give examples of how this works, and explain the pros and cons of autocratic versus democratic approaches.

2. Keeping in mind what you just read about the factors associated with voting, what impact, if any, do you think the most recent presidential election will have on future voting rates? Why will it have this impact?

3. As an employee, what would you like about working in Sweden? What would you dislike? As an employer, what would you like and dislike about doing business in Sweden? How can Sweden's democratic socialist government continue to resolve these differences between the interests of workers and those of business?

4. How has technology affected your schoolwork in the last 10 years? How has it made your work easier? Harder? How has it made it easier or harder for teachers to monitor or control your behavior?

5. How do you think a postindustrial economy will affect your working and economic future? How will a globalized economy affect you?

6. Which of the three general policy options outlined in the text do you think the United States will follow and why? In your opinion, which would be the best one to follow and why?

Companion Website for This Book

academic.cengage.com/sociology/brinkerhoff
Gain an even better grasp on this chapter by going to the Companion Website. This resource contains tutorial quizzes and flash cards to help you master key terms and concepts.

Suggested Readings

Domhoff, G. William. 1998. *Who Rules America: Power and Politics in the Year 2000* (3rd ed.). Mountain View, Calif.: Mayfield. A fully updated edition of a classic text on the power elite.

Ehrenreich, Barbara. 2005. *Bait and Switch: The (Futile) Pursuit of the American Dream*. New York: Metropolitan Books. A 50-something journalist goes undercover as a job-seeker. Over the course of a year, she uncovers the barriers that keep middle-aged Americans who have been laid off or are trying to reenter the job market from finding decent employment.

Draut, Tamara. 2006. *Strapped: Why America's 20- and 30-Somethings Can't Get Ahead.* New York: Doubleday. Clearly explains the economic difficulties facing young Americans as, over the last 30 years, wages have decreased, income inequality has increased, and the cost of living—especially the cost of a college education and health care—has soared.

Newman, Katherine S. 1999. *No Shame in My Game: The Working Poor in the Inner City.* New York: Knopf. A finely detailed portrait of the lives of the urban working poor which documents both the difficulties and the rewards of low-wage work.

Piven, Frances Fox, and Cloward, Richard A. 2000. *Why Americans Still Don't Vote: And Why Politicians Want It That Way.* Boston: Beacon. Scholar-activists Piven and Cloward, who played key roles in passage of the "motor voter" National Voter Registration Act of 1993, argue that liberal politicians have done little to encourage Americans to vote while conservative politicians have worked tirelessly to ensure that the poor and minorities *don't* vote.

Politics, Religion, and the Culture War

Over the last decade, it has become commonplace for political commentators, theologians, scholars, and the popular press to lament the increasing polarization of American society. As examples of this polarization they cite:

- Heated debates among Episcopalians about the proper church stance on homosexual clergy
- Strongly divergent views toward abortion held by Right to Life and Freedom to Choose groups
- Conflict between supporters and opponents of capital punishment
- Debate over whether private and religious schools should be eligible for government-funded school vouchers.

The evidence seems to be mounting that Americans are no longer as tolerant of ideological differences as they once were. Some scholars have argued that the United States is increasingly split into two broad and antagonistic groups, each vying for control of American culture. Inflammatory titles like *Godless: The Church of Liberalism* (Coulter 2006) and *Lies and the Lying Liars Who Tell Them: A Fair and Balanced Look at the Right* (Franken 2003) scream out from book shelves, and talk show hosts left and right spew venom at each other. This divisiveness over important social values has appeared so extreme that it is frequently referred to as the "culture wars" (Hunter 1991; Wuthnow 1988). But what evidence is there that Americans have actually become much more deeply split in their attitudes?

The evidence seems to be mounting that Americans are no longer as tolerant of ideological differences as they once were. Inflammatory titles like *Godless: The Church of Liberalism* and *Lies and the Lying Liars Who Tell Them: A Fair and Balanced Look at the Right* scream out from book shelves, and talk show hosts left and right spew venom at each other.

Using data from the General Social Survey and the National Election Survey, researchers have found little evidence that attitudes have become more polarized since the 1970s, with the exception of attitudes toward abortion. Instead, Americans have become more unified in a number of respects—their support for racial integration and the rights of women to participate in public life, for example, as well as their tough stance on crime (Evans, Bryson, & DiMaggio 2001; Bolzendahl & Brooks 2005). Differences in attitudes between young and old; African Americans and whites; Southerners and Northerners; high school dropouts and college graduates; and Protestants, Catholic, and Jews persist, to be sure. But the polarization of attitudes has steadily declined among all of these groups, with the one exception being a widening gap between Republicans and Democrats.

If attitudes have not generally become more hostile and divisive, why do so many observers believe they have? According to Miller and Hoffmann (1999), the answer lies in a series of political events in the 1970s and 1980s that made the terms *liberal* and *conservative* much more value-laden.

The 1970s saw the emergence of many conservative Christian political groups. In 1974, Third Century Publishers began publishing a wide range of Christian literature aimed at political issues. In 1978, Pat Robertson founded the Christian Broadcasting Network, and in 1979 Jerry Falwell founded the Moral Majority. These groups have had only marginal electoral success. There is no doubt, however, that they have been effective in increasing the tendency to label certain positions on moral issues as either wholly "liberal" or "conservative." Also, Republicans since the 1980s have successfully used the term *liberal* to denigrate any groups who support higher taxes and more social services.

The rise of these conservative political groups was countered by the political activities of adults who had come of age in the protest movements of the late 1960s and early 1970s. As youths, these individuals had advocated the progressive social and political agenda that fostered the gay rights movement, political debates over the Equal Rights Amendment, and legal access to abortion. By the 1980s they were established adults, some of whom remained politically active. As a result, by this time not only were there more Christian political organizations with more visible and clearly defined conservative platforms, but competing organizations with alternative, liberal views also had a greater presence and greater visibility. The emergence of these groups and the political wrangling of their high-profile leaders no doubt fueled the increasingly negative view of liberals held by conservatives and vice versa (Miller & Hoffmann 1999; Wuthnow 1996).

Miller and Hoffmann (1999) argue that the end result of these historical, political events has been not only a polarization of the categories "liberal" and "conservative" but also an increased tendency for individuals to adopt whatever label is promoted by their religious denomination (or political party), no matter what their attitudes might be on any specific issue. If Miller and Hoffmann are right, the culture wars do not stem from fundamental differences in our attitudes but rather from a series of heated, high-profile, political campaigns.

CHAPTER 14

Population and Urban Life

© Alan Schein Photography/CORBIS

Populations, Large and Small

Birth and death—nothing in our lives quite matches the importance of these two events. Naturally, each of us is most intimately concerned with our own birth and death, but to an important extent, our lives are also influenced by the births and deaths of those around us. Do we live in large or small families, large or small communities? Is life predictably long or are families, relationships, and communities periodically and unpredictably shattered by death?

In this chapter we take a historical and cross-cultural perspective on the relationship between social structures and population. The study of population is known as **demography,** and those who study it are known as demographers. Demographers focus primarily on three processes: **fertility** (childbearing), **mortality** (deaths), and migration. Here we will look at these three demographic processes and also at the effect of population size on social relationships within communities. We are interested in questions such as how birth rates or community size affect social structures and, conversely, how changing social structures affect birth rates and community size.

Currently, the world population is 6.3 billion, give or take a couple hundred million. This is two and a half times as many people as lived in 1950. World population has grown primarily because fertility has declined only slowly while mortality among infants and very young children has declined precipitously. In part because of this population growth, millions are poor, underfed, and undereducated; pollution is widespread; and the planet's natural resources have been ransacked.

These problems are among the causes of migration, which has caused some villages, cities, and nations to grow in size while others shrink. Because some people may enter an area just as others leave it, demographers focus on **net migration:** the number of people who move into an area minus the number who move out. Some migrants, known as *immigrants*, move from one country to another, while others, known as *internal migrants*, move within a nation.

Migration, in turn, leads to another set of social concerns, as nations wrestle with how to respond to the newcomers in their midst. Some nations, like the United States, allow immigrants to eventually become citizens. Other nations refuse citizenship not only to almost all immigrants but also to their children and grandchildren. For example, Germany generally will not grant citizenship to the children of Turkish immigrants, even if these children are born, raised, and educated in Germany. Immigration has substantial consequences, then, not only for population growth and economic development, but also for issues such as the meaning of citizenship and nationality.

In sum, population size and change is vitally linked to many of our era's crises. The next section examines the process by which current world population was reached.

Demography is the study of population—its size, growth, and composition.

Fertility is the incidence of childbearing.

Mortality is the incidence of death.

Net migration is the number of people who move into an area minus the number who move out.

Understanding Population Growth

Although population is concerned with intimate human experiences such as birth and death, the big picture of population growth and change can be understood only if we use statistical summaries of human experience. Three measures are especially

The **crude birth rate** is the number of live births per 1,000 persons.

The **crude death rate** is the number of deaths per 1,000 persons.

The **rate of natural increase** is the crude birth rate minus the crude death rate, calculated as a percentage.

important: the crude birth rate, the crude death rate, and the rate of natural increase. The **crude birth rate** is the number of live births per 1,000 persons, and the **crude death rate** is the number of deaths per 1,000 persons. To find the **rate of natural increase,** we subtract the number of deaths from the number of births and then calculate the result as a percentage.

Table 14.1 shows these rates in 2006. Worldwide, the crude birth rate in 2006 was 21 births per 1,000 population; the crude death rate was a much lower 9 per 1,000. Because the number of births exceeded the number of deaths by 12 per 1,000, the rate of natural increase of the world's population was 1.2 per hundred, or 1.2 percent. If your savings were growing at the rate of 1.2 percent per year, you would undoubtedly think that the growth rate was very low. A growth rate of 1.2 percent in population, however, means that the planet will hold an extra 1.4 *billion* people by the year 2025.

The frightening prospect of welcoming another 1.4 billion people in less than 20 years is complicated by the fact that the growth is uneven. As Table 14.1 shows, growth rates are startlingly different across the areas of the world. Africa, for example, may double its population size in less than 30 years, whereas in Europe deaths actually exceed births and the population is shrinking.

These differentials in growth are of tremendous importance. Because most population growth is occurring in poor nations, the world will likely be poorer in 2025 than it is now. How did these different population patterns evolve?

Population in Former Times

For most of human history, fertility (childbearing) was barely able to keep up with mortality (death), and the population grew little or not at all. Historical demographers estimate that in the long period before population growth exploded, both the birth and death rates hovered around 40 to 50 per 1,000. (Birth rates are still almost 40 per 1,000 in Africa.) Translated into personal terms, this means that the average woman spent most of the years between the ages of 20 and 45 either pregnant or nursing. If both she and her husband survived until they were 45, she would produce an average of 6 to 10 children. The average life expectancy was perhaps 30 or 35 years. Such a low life expectancy was largely due to very high infant mortality: perhaps one quarter to one third of all babies died before they reached their first birthday. Both birth and death were frequent occurrences in most preindustrial households.

TABLE 14.1

The World Population Picture, 2006

In 2006, the world population was 6.6 billion and growing at a rate of 1.2 percent per year. Growth was uneven, however; the less developed areas of the world were growing much more rapidly than the more developed areas. As a result, most of the additions to the world's population were in poor nations.

Area	Crude Birth Rate (births/1,000 persons)	Crude Death Rate (deaths/1,000 persons)	Rate of Natural Population Increase (percent)	Projected Population Increase, 2006-2025
World	21	9	1.2%	1,385,000,000
More-developed nations	11	10	0.1	39,000,000
Less-developed nations	23	8	1.5	1,346,000,000

SOURCE: Population Reference Bureau 2006.

The Demographic Transition in the West

Beginning in the eighteenth century, a series of events occurred that revolutionized population in the West. First, mortality dropped; then, after a period of rapid population growth, fertility declined. Because studies of population are called demography, this process is called the **demographic transition.** Although this transition occurred at different times in different countries, the process was more or less similar across Europe and in the United States.

The Decline in Mortality

Prior to the demographic transition, widespread malnutrition was an important factor underlying high levels of mortality. Although few died of outright starvation, poor nutrition increased the susceptibility of the population to disease. Improvements in nutrition were the first major cause of the demographic transition's decline in mortality, beginning in the 1700s and continuing into the early twentieth century. New crop varieties from the Americas (corn and potatoes especially), new agricultural methods and equipment, and increased trade all helped improve nutrition in Europe and the United States. The second major cause of the decline in mortality was a general increase in the standard of living, as improved shelter and clothing left people healthier and better able to ward off disease. Changes in hygiene were vital in reducing communicable disease, especially those affecting young children, such as typhoid fever and diarrhea (Kiple 1993).

In the late nineteenth century, public-health engineering led to further reductions in communicable disease by providing clean drinking water and adequate treatment of sewage. For example, between 1900 and 1970, the life expectancy of white Americans increased from 47 to 72, and the life expectancy of African Americans increased from 33 to 64 (U.S. Bureau of the Census 1975, 2006). Thus, although life expectancy has been increasing gradually since about 1600, the fastest increases occurred in the first few decades of the twentieth century. Medical advances probably account for no more than one sixth of this overall rise in life expectancy (Bunker et al. 1994). Instead, public-health initiatives, better nutrition, and an increased standard of living are largely responsible for rising life expectancies (McKinlay & McKinlay 1977; Weitz 2007). Interestingly, once the standard of living in a nation reaches a certain point—approximately $6,400 per capita income—further increases in life expectancy depend less on increasing income than on reducing the income gap between rich and poor (Wilkinson 1996). This is one major reason why on average Cubans live almost as long as do Americans and Swedes live longer than Americans.

The Decline in Fertility

The Industrial Revolution also affected fertility, although less directly. Industrialization meant increasing urbanization, greater education, and the real possibility of getting ahead in an expanding economy. Pensions and other social benefits became more common with industrialization, so people no longer needed to have many children to care for them in their old age (Friedlander & Okun 1996). In addition, as mortality rates dropped, parents no longer needed to have eight children to count on two surviving. Perhaps even more important, industrialization created an awareness of the possibility of doing things differently from how they had been done by previous generations. As a result, the idea of controlling family size to satisfy individual goals spread even to areas that had not experienced industrialization, so that by the

---

The **demographic transition** is the process of moving from the traditional balance of high birth and death rates to a new balance of low birth and death rates.

Connections

Personal Application

Have you ever traveled to a less-developed country? If you did, the odds are that you got a nasty stomach virus for a day or two, but otherwise suffered no health problems. Yet malaria, cholera, dysentery, and the like kill millions in these countries each year. Why are American tourists virtually immune? Vaccinations, antibiotics, and access to soap and water help. But the most important reason is that, unlike many residents of less-developed countries, tourists start out healthy, well nourished, well sheltered, and well clothed. As a result, even if they come in contact with dangerous germs, their bodies most likely will be able to fight against infection.

end of the nineteenth century, the idea of family limitation had gained widespread popularity (van de Walle & Knodel 1980). Currently in Europe and North America, birth and death rates are about even, and there is little population growth.

The Demographic Transition in the Non-West

In the less-developed nations of the non-West, Africa especially, birth and death rates remained at roughly preindustrial levels until the first decades of the twentieth century. After that, in some areas such as Latin America, Taiwan, Singapore, and South Korea, economic development and improvements in the standard of living caused both death and birth rates to plummet, much as they had previously done in the West. It took a few more decades for death rates to begin falling in Africa, following improvements in public-health engineering and the introduction of very basic medical interventions (primarily childhood vaccinations and treatment for childhood diarrhea).

Unlike the mortality decline in the West and in the developed countries of Asia, in other countries decreasing death rates were not gradual. Nor were they caused by changes in social structure or standard of living. Instead, death rates dropped because of sudden change brought in from outside. Consequently, the changes in social structure and culture that are necessary to cause fertility decline are occurring more slowly, and populations continue to grow steeply. These patterns, however, are being affected dramatically by the AIDS epidemic. In African countries like Botswana, Swaziland, and Zimbabwe, where between 30 and 40 percent of the population is infected with HIV, death rates have soared, and population growth has slowed considerably (UNAIDS/WHO 2002).

Population and Social Structure: Two Examples

In this section we explore contemporary relationships between social structure and population in two societies: Ghana, where fertility is high, and Italy, where fertility is low.

Ghana: Is Fertility Too High?

Ghana is an example of a society in which traditional social structures encourage high fertility. It is also an example of a society in which high fertility may ensure continuing traditionalism—and poverty.

The Effects of Social Roles on Fertility

Fertility rates have declined in Ghana in recent years but remain high. Ghana still has a crude birth rate of 33 per 1,000 population. Mortality, however, is down to 10 per 1,000. This means that the rate of natural increase in Ghana is 2.3 percent per year (Population Reference Bureau 2006). If that rate continues, the population could double in less than 30 years. An aggressive family-planning program is unlikely to reduce this growth substantially because Ghanaian women, on average, still want 4.3 children while having 4.6 children (Ghana Statistical Service 1999).

One of the most important reasons for this high fertility is women's roles. In Ghana, children are an important—or even the only—source of esteem and power open to many women. Women who cannot bear children risk divorce or abandonment. This is especially true for the 23 percent of Ghanaian women who live in polygamous unions. The number of children a woman has—especially the number of sons—strongly affects her position relative to that of her co-wives. Moreover, because infant mortality remains relatively high, Ghanaian women believe they must have four or more children to ensure that two survive to adulthood.

Another important cause of high fertility is the need for economic security. Most Ghanaians work in subsistence agriculture. To survive, families need children as well as adults to work in the fields. In addition, when children grow up and marry, they can add to the family's economic and political security by creating political and social allegiances to other families. Finally, children are the only form of old-age insurance available to Ghanaians: Parents who grow old or ill must rely on their children to support them. Conversely, having children is relatively inexpensive: no expensive medical treatment for children is available, schooling is either inexpensive or unaffordable, and children don't expect to own designer jeans or $150 tennis shoes. With a cost/benefit ratio of this sort, it is not surprising that Ghanaians desire many children.

Many families in Africa, especially polygamous families, have numerous children, and overpopulation is a cause for some concern.

The Effects of High Fertility on Society

Although high fertility may appear to be in the best interests of individual women, its consequences for society are less beneficial. If Ghana's population continues to explode, the nation will have to dramatically increase its governmental expenditures just to maintain current levels of support for education, highways, agriculture, and the like. Thus, a decision that is rational on the individual level turns out to be less wise on the societal level.

This problem sometimes leads people in the West to ask: "Are they stupid? Can't they figure out they would be better off if they had fewer children?" Unfortunately for the argument, nations don't have children; people do. High fertility continues to be a rational choice for individual Ghanaians.

Policy Responses

To reduce its population growth, Ghana has established an excellent family-planning program that makes contraception available, convenient, and affordable to women who want it. When women want several children, however, access to contraception has limited impact. Currently, only 19 percent of all married women in Ghana use modern contraceptive methods (Population Reference Bureau 2006). Contraceptive use is considerably higher among younger, better educated, urban women. Study after study has found that the best way to reduce fertility is to combine access to contraception with educational and economic development and better status for women (Poston 2000).

Italy: Is Fertility Too Low?

In a world reeling from the impact of doubling populations in the less-developed world, it is ironic that many developed countries are worried that fertility is too low.

© Claudia Kunin/CORBIS

■ In Europe, many families have only one child, and underpopulation is increasingly a cause for concern.

Replacement level fertility requires that a woman bear an average of 2.1 children—one to replace herself, one to replace her partner, and a little extra to cover childhood mortality.

The Effects of Social Roles on Fertility

With modern levels of mortality, fertility must average 2.1 children per woman if the population is to replace itself: two children so that the woman and her partner are replaced and a little extra to cover unavoidable childhood mortality. This is called **replacement level fertility.** If fertility is less than this, the next generation will be smaller than the current one.

Currently in Italy, the average woman is having only 1.3 children (Table 14.2). This means that the next generation of Italians will be much smaller than previous ones, unless the country absorbs many new immigrants.

Why is fertility so low in Italy? In essence, the situation in Italy is the reverse of that in Ghana. Most women are educated and many hold paying jobs outside of the home. Women's social status is close to that of men, so women do not need to have children to have a purpose in life or to assure their social standing. Because few Italians work in agriculture, and all children are expected to be in school, having children doesn't add to a family's labor pool. Finally, the Italian government provides a good safety net in the form of disability insurance, health care, old-age pensions, and the like, which means that couples do not need to have children to take care of them in sickness or old age.

The Effects of Low Fertility on Society

Given the serious worldwide dilemmas posed by population growth and Italy's very high density, why should we consider low fertility a problem? There are two main concerns: the large numbers of old people compared with young people, and rising nationalistic fears resulting from the importing of immigrant labor.

Very low fertility creates an age structure in which the older generation is as large as or larger than the younger generation on whom it relies for support. As a result, it is increasingly difficult for Italy to fill all the occupations needed to keep the nation running, from taxi-drivers to doctors. At the same time, the cost of paying for old-age pensions and health care is growing rapidly. (The same is true of Social Security in the United States.) By 2030, the most-industrialized nations will need to spend an additional 9 to 16 percent annually of their national net incomes, just to continue current pension and health benefits for older persons (Peterson 1999).

To counteract this problem, Italy has imported workers from other countries, primarily neighboring Albania. This has led to nationalist fears of cultural dilution. A survey conducted in 2003 found that an astounding 80 percent of Italians believed that Albanian immigrants were bad for Italy (Pew Research Center 2003a). These feelings have provoked anti-immigrant violence in Italy and has led Italy to clamp down on immigration. In turn, the isolation and discrimination experienced by immigrants in Italy has also led to outbreaks of violence by immigrants themselves. Similar conditions elsewhere in western Europe have produced similar results, such as the riots that blazed across France's immigrant neighborhoods for three weeks in late 2005.

Policy Responses

In response to concerns like these, Italy and other European nations have established incentives to encourage fertility. Among them are paid, months-long maternity leave; cash bonuses and housing subsidies for having more children; and monthly

TABLE 14.2

Fertility and Population Growth in Europe, 2006

Overall deaths are now slightly exceeding births in Europe. Thus, some nations are already experiencing population decline. The last column in the table shows the combined impact of births, deaths, and migration into and out of a country.

Country	Crude Birth Rate (births/1,000 persons)	Crude Death Rate (deaths/1,000 persons)	Rate of Natural Population Increase (percent)	Average No. of Children per Woman	Projected Population Change, 2006–2050* (percent)
Denmark	12	10	0.2%	1.8	+1%
Germany	8	10	−0.2	1.3	−9
Hungary	10	13	−0.3	1.3	−11
Italy	10	10	0.0	1.3	−5
Romania	10	12	−0.2	1.3	−29
Spain	11	9	0.2	1.3	−4
United Kingdom	12	10	0.2	1.8	+14%

*Reflects the impact of immigration and out-migration as well as birth and death rates. (United Kingdom receives more immigrants and Romania loses more to immigration than do the other countries in this table.)

SOURCE: Population Reference Bureau 2006.

subsidies for children until age 3 (Oleksyn 2006). Nevertheless, the costs of raising children far outstrip these benefits. As a result, while these incentive plans have kept birth rates from falling drastically, they have not helped to raise birth rates in Italy or other countries where women have attractive alternatives outside the home (Gautier & Hatzius 1997).

Countries such as France are attempting to increase fertility above replacement levels through billboards such as this one, which declares "France Needs Children."

Population and Social Problems: Two Examples

Analysis of world population growth reveals a good news/bad news situation. The good news is that fertility is declining in every part of the world (Table 14.3). The bad news is that the population of the world will nonetheless increase dramatically over the next 50 years. The reason for this gloomy prediction lies in the age structure of the current population. The next generation of mothers is already born—and there are a lot of them. Thus, we must plan for a world that will soon hold 8 or 9 billion people.

Population pressures can contribute to numerous social problems. In this section, we address two of them—environmental devastation and poverty.

Environmental Devastation: A Population Problem?

■ Deforestation is devastating tropical rainforests in Brazil, the Philippines, and elsewhere.

All around the world, there are signs of enormous environmental destruction: In the developed world, we have acid rain and oil spills; in Africa, desert environments are spreading rapidly due to deforestation and overgrazing. Both of these pose serious threats to the environment, but only the latter is truly a population problem.

It is estimated that the United States, which contains only 5 percent of the world's population, consumes one quarter of the world's resources and produces nearly three quarters of the world's hazardous waste (Ashford 1995). Our affluent, throwaway lifestyle requires large amounts of petroleum and other natural resources. Obtaining these resources results in the destruction of wilderness, the loss of agricultural lands, and the pollution of oceans. Using these resources causes illness-inducing air pollution, acid rain, and smog that are killing our forests. Although these problems would be less severe if there were half as many of us (and hence half as many cars, factories, and Styrofoam cups), they are not really population problems. They stem from our way of life rather than our numbers.

TABLE 14.3
Fertility Decline in World Regions, 1950-2006
In the last half century, the average number of children per woman has declined worldwide.

Region	Average Number of Children per Woman	
	1950	2006
Africa	6.6	5.1
Asia	5.9	2.4
Europe	2.6	1.4
Latin America	5.9	2.5
North America	3.5	2.0
Oceania	3.8	2.1

SOURCE: Population Bulletin 1999; Population Reference Bureau 2006.

© Harold Castro/FPG

In sub-Saharan Africa, however, population pressure is a major culprit in environmental destruction. In rural areas, the typical scenario runs like this: Population pressure forces farmers to plow marginal land and to plant high-yielding crops in quick succession without soil-enhancing rotations or fallow periods. The marginal lands and the overworked soils produce less and less food, forcing farmers to push the land even harder. They cut down forests and windbreaks to free more land for production. Soon, water and wind erosion becomes so pervasive that the topsoil is borne off entirely, and the tillable land is replaced by desert or barren rock. This cycle of environmental destruction—which destroys forests, topsoil, and the plant and animal species that depend upon them—is characteristic of high population growth in combination with poverty. When one's children are starving, it is hard to make long-term decisions that will protect the environment for future generations.

In sum, reducing population growth would reduce future pressure on natural resources, but it would not solve the current problem. The solution rests in an international moral and financial commitment to reducing rural poverty, improving farming practices, reducing the foreign debt of the less- and least-developed nations, *and* curbing wasteful and destructive practices in the developed nations.

Poverty in the Least-Developed World

Perhaps 500 million people around the world are seriously undernourished, and each year outbreaks of famine and starvation occur in Africa and Asia. A billion more are poorly nourished, poorly educated, and poorly sheltered. These people live in the same nations that have high population growth.

Some observers blame poverty in the developing nations on high fertility. Yet high fertility is not the only or even the primary cause of this poverty. Poverty and malnutrition result primarily from war, corruption, and inequality in nondemocratic countries and from a world economic system that extracts raw goods and profits from poorer countries (Chase-Dunn 1989; Dreze & Sen 1989; Sen 1999). It is a terrible irony that most poor countries export more food than they import (Lappé, Collins, & Rosset 1998). Cuba, for example, became poorer in the 1990s not because of population growth but because its authoritarian government failed to develop a strong economy and instead relied heavily on subsidies from the now-defunct Soviet Union. People in the Democratic Republic of the Congo, meanwhile, are dying of starvation because of war rather than because of high fertility.

Policy Responses

Although many factors contribute to poverty, almost all world leaders agree that reducing fertility is an important step toward increasing the standard of living in the poorer nations of the world. The most successful programs to reduce fertility have combined an aggressive family-planning program, economic and educational development, and improvements in the status of women (Poston 2000):

1. *Family-planning programs.* These programs are designed to make modern contraceptives and sterilization available inexpensively and conveniently to individuals who desire to limit the number of their children. For example, between 1975 and 1991, an aggressive family-planning program increased contraceptive use in

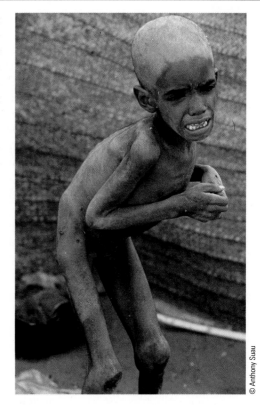

© Anthony Suau

■ Because of war, drought, corrupt governments, and income inequality within and between nations, about half a billion people in the world are seriously malnourished. Some, such as this child, are starving.

Bangladesh by 500 percent and decreased the average number of children per woman from 7 to 5 in just 16 years (Kalish 1994).

2. *Economic and educational development.* Experience all over the world shows that fertility declines as education increases and the country undergoes economic development. For example, South Korea's fertility has plummeted from 6.0 children per woman in 1960 to only 1.1 currently in the wake of its dramatic economic development (Population Reference Bureau 2006).

3. *Improving the status of women.* In countries where women have low status, having many children—especially sons—is the only way women can increase their social value and guarantee support in their old age. When women have greater education and can earn even a small income on their own, they gain greater power within the family. As a result, they typically marry later and have fewer children. In addition, they are better able to protect their daughters from being married off while still children. Consequently, the countries that have proven most successful in family planning and in economic growth are those, such as South Korea and Singapore, that have made particular efforts to increase education, economic options, and legal rights for women (United Nations Population Fund 2000).

Population in the United States

The U.S. population picture is similar to that in Italy, with low mortality and fertility rates, but there are also several differences. First, although fertility is at near replacement level, it has not dropped significantly below this level as has happened in Italy. Second, immigration continues to add substantially to the size of our population. Third, and partly because of this immigration, our population is younger than Italy's. In this section, we briefly describe fertility, mortality, and migration issues in the United States.

Fertility

For nearly 20 years, the number of children per woman in the United States has remained just around or just under 2.1—the level necessary to replace the population. This low fertility has been accompanied by sharp reductions in social-class, racial, and religious differences in fertility. Some women will have their children as teenagers and some when they are 40, but increasingly they will stop at two children.

Mortality

Death is almost a stranger to U.S. families. The average age at death is now in the late 70s, and many people who survive to age 65 live another 20 years. Parents can feel relatively secure that their infants will survive. If they don't divorce, young newlyweds can safely plan on a golden wedding anniversary.

Since 1970, we have added about 7 years to average life expectancy. These increases are due to better diagnosis and treatment of the degenerative diseases (such as heart disease and cancer) that strike elderly people. In addition, increases in life expectancy have been made possible by reducing (although not eliminating) racial and social-class differentials in mortality. In the early 1940s, African American women lived a full 12 years less than white women; today, the gap is down to less than 5 years (U.S. Bureau of the Census 2006).

On the other hand, the AIDS epidemic, first recognized in 1981, has given death a new face. Although death rates from AIDS have fallen in recent years, AIDS remains a leading cause of death for all persons ages 25 to 44, but especially for African Americans and Hispanics. Often spread through intravenous drug use (which has the most appeal for those who have the least to look forward to), AIDS is becoming a disease of the poor and disadvantaged.

Migration

Although it can safely be ignored as a factor in world population growth, migration often has dramatic effects on the growth of individual nations. The United States is one of the nations for which **immigration,** the permanent movement of people into another country, has had an important impact, particularly in Sunbelt states such as California, Arizona, and Florida.

Immigration is the permanent movement of people into another country.

Most U.S. citizens are descended from people who emigrated to the United States to improve their economic prospects, such as many recent migrants from Mexico. Other immigrants, such as those from Cambodia, Bosnia, and the Sudan, are primarily refugees driven from their homes by warfare or the economic destruction that often follows in its wake (see the Focus on a Global Perspective box in this chapter). Patterns of both internal migration and international immigration have created a unique set of problems in the United States and have dramatically changed our political landscape.

Immigration

The United States has always been a country of immigrants. Between 1870 and 1920, about 13 percent of the U.S. population was foreign born. That percentage dropped steadily until 1965 (when legal changes allowed more immigration) and has risen steadily ever since but still remains a little less than it was in 1920.

■ Hispanics make up a rising proportion of the U.S. population. This family is waiting to apply for legal residency in the United States.

© AP/Wide World Photos

Global Perspective:

focus on

International Migration

During 2005, approximately 9 million refugees fled their homes involuntarily, and another 3 million (who had fled earlier) remained outside their home countries as either stateless persons or asylum seekers (United Nations High Commissioner for Refugees 2006). Millions more chose voluntarily to seek new lives and new opportunities in other countries.

We often hear debate about immigrants and refugees in the United States, but what do we know about international migration? Map 14.1 shows recent migration patterns around the world. Most refugees flee from one developing nation to a neighboring developing nation, whereas many voluntary migrants move to industrialized nations in search of a better life.

Demographers believe that the economic and political turmoil of the last two decades, coupled with the opportunities presented by globalization, have substantially increased international migration. At least 160 million people lived outside their country of birth or citizenship in 2000, almost twice the number in 1990 (Martin & Widgren 2002). Although push factors such as war and famine account for much of this international migration, some migrants are also pulled by the economic growth and employment opportunities in newly industrializing nations, such as South Korea, Singapore, and Malaysia. Pull factors also account for much of the immigration from less developed to more developed countries. Strong European economies provide increasing numbers of jobs to a growing non-Western labor force. Migrants traditionally have been young men, but women and girls now make up 40 to 60 percent of the international migrant stream. Many of these are mothers who, in growing numbers, seek employment opportunities in more affluent neighboring countries in

MAP 14.1
Major Migration Patterns in the Early Twenty-first Century
SOURCE: Martin & Widgren 2002.

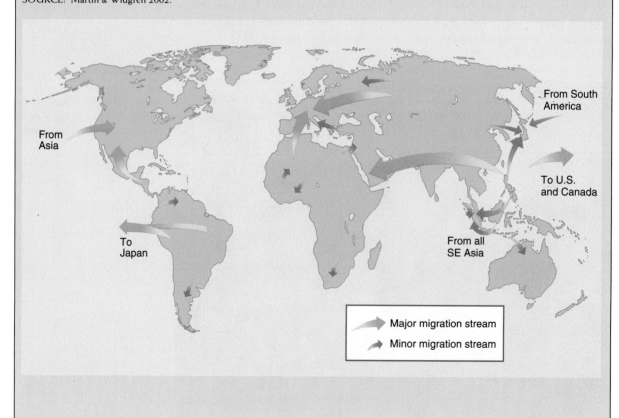

order to send money to the family, friends, or neighbors who are raising their children.

The money sent back home by migrants—both men and women—is a large and growing source of revenue for many nations. During 2005 alone, migrants sent $20 *billion* home to Mexico, three times the amount sent by migrants in 2000 (Hawley 2006). In rural areas across Latin America, money sent home by migrants far exceeds that received from the government or interna-

tional development agencies. Most of that money goes to help relatives, but increasingly migrants are pooling together their funds to build community projects or start industries in their home towns.

It is not yet clear, however, who profits most from the international migrant stream. Although countries such as Germany, France, and Italy face new challenges stemming from an ethnically diverse population, workers from developing nations help to sustain the

continued expansion of these nations' economies. Low birth rates have led to smaller labor forces and aging populations in Europe, Japan, and elsewhere. Thus, migrants from countries such as Turkey and Pakistan fill the demand for more workers, particularly at the low end of the labor hierarchy. Whether the money that migrants send home will significantly improve the quality of life in less developed nations remains an open question.

An estimated 1 million immigrants enter the United States each year. Almost all recent immigrants come from Latin America or Asia. Perhaps as many as half are illegal immigrants, most from Mexico or Central America.

Immigrants to the United States divide roughly into two very different groups. The first is skilled, well educated, able to speak English, and here legally, such as doctors and computer scientists from India. The second group is made up of low-skilled workers with little education or ability to speak English, many of whom are here illegally. Most of these workers come from Latin America. The experiences of these two groups, and their impact on the United States, differ markedly.

Because of immigration, the United States does not need to fear population decline. The racial and ethnic composition of the nation will change substantially, however. By 2050, it is estimated that the combination of Hispanic immigration and low fertility among whites will reduce the proportion of our population that is white non-Hispanic from 69 to 53 percent (Figure 14.1).

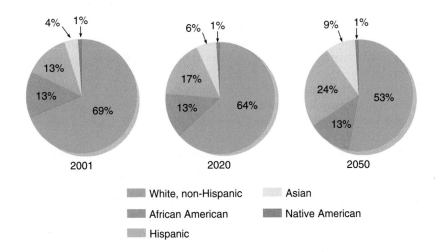

FIGURE 14.1
Changing Compositions of U.S. Population
If annual immigration remains at 1 million and if fertility remains low, the racial and ethnic composition of the U.S. population will change substantially in the decades ahead. The most noticeable effect will be a sharp rise in the proportion who are Hispanic and Asian and a corresponding decrease in the proportion who are non-Hispanic whites.
SOURCE: U.S. Bureau of the Census 2003b.

Most immigrants to the United States, both legal and illegal, are pushed from their native lands by poor local economies and are pulled by an unmet demand in the United States for low-skill, low-paid labor. In the past, many immigrants (especially from Mexico) would come to the United States to work briefly and then return to their home countries, a cycle they repeated whenever they needed to earn extra money. Ironically, recent efforts to clamp down on border crossings have made it too dangerous to cross the border repeatedly, and so many of these migrants instead have settled in the United States (Massey 2006).

The consequences of current immigration trends are likely to be both economic and cultural. From the standpoint of economics, research suggests that (1) immigrants are not taking jobs away from U.S. citizens but (2) the availability of low-wage illegal immigrants may depress wages for the least educated American citizens. Some economists believe immigrants have no effect on wages; others believe they reduce wages for high school dropouts by as much as 5 percent (Borjas & Katz 2006; Card 2005). From the standpoint of culture, it is likely that the United States will become a more pluralistic society in which salsa and soccer are as popular as hotdogs and baseball, but that the new immigrants will integrate into American society as did earlier waves of Hispanic and other immigrants (Alba & Nee 2003).

Internal Migration

The most striking and largest example of internal migration in U.S. history is the exodus of about one million people triggered by 2005's Hurricane Katrina. Most of New Orleans's population fled the hurricane and subsequent flooding. The long-term effect of this exodus will not be known for some time. On the negative side, New Orleans was already in economic and population decline before the hurricane, and this is likely to worsen, given the great damage suffered by its infrastructure and economy; as of 2006, its population is about 60 percent lower than before the hurricane. In addition, communities such as Houston may suffer because they now must pay to provide services to many poor former New Orleans residents. On the positive side, because many of those forced to leave New Orleans had been mired in poverty, some may in the long run benefit if they become integrated into more prosperous communities with better economic and educational opportunities.

Urbanization is the process of population concentration in metropolitan areas.

New Orleans, of course, is a unique case. More generally, the history of internal migration in the United States has been a story of **urbanization,** the process of population concentration in cities. For most of our history, urban areas grew faster than rural areas, with the largest urban areas growing the most. Since about 1970, however, this has all changed. Currently, the three major trends in internal migration are Sunbelt growth, migration from central cities to suburbs, and the resurgence of some nonurban areas.

Since 1970, there has been consistent movement of people from the Midwest and the northern states to the Sunbelt states of the Southeast and Southwest. Working people have followed jobs, and retirees have followed the sunshine. Most urban growth has occurred in these areas as well.

In the rest of the country, central cities have declined while their suburban rings have grown. Importantly, the middle class has disproportionately left the cities, so that increasingly cities are home to only the wealthy and the poor (Scott 2006). Urban poverty has sharply increased as jobs have moved to the suburbs, public transportation to the suburbs remains minimal, and the wealthy have driven up the cost of urban housing ("Out of Sight" 2000).

At the same time that central cities have been shrinking, nonurban areas have experienced some modest growth (Johnson 2003). Most of this growth, again, has occurred in Sunbelt states, especially in retirement destinations and in areas within a few hours' drive of a big city. But rising home prices, rising numbers of retirees, and rising numbers of workers who can live anywhere there is an Internet connection have led to a small but growing migration to more distant towns in more varied places, like northern Michigan and Archer County, Texas (Fessenden 2006).

Because suburban and nonurban life is so dependent on automobile transportation, migration to the suburbs and beyond has increased air pollution. In addition, much of the geographic relocation of the U.S. population since 1970 has been to those regions of the country that are least able to withstand the ecological impact of a large population. In many areas of Florida, California, and the Southwest, the demand for water already outstrips the supply. As states argue over water rights, political tensions are likely to increase; within states, competition for access to water may increase conflict between agricultural and urban interests.

Fertility, mortality, and migration patterns in the United States provide clear examples of the interrelationships between population and social institutions. Social class, women's roles, and race and ethnic relationships are all intimately connected to changes in population. One additional element of population that is especially important for social relationships is community size, an issue to which we now turn.

Urbanization

Most of our social institutions evolved in agrarian societies, where the vast bulk of the population lived and worked in the countryside. As late as 1850, only 2 percent of the world's population lived in cities of 100,000 or more (Davis 1973). Today, nearly a quarter of the world's population and more than two thirds of the U.S. population live in cities larger than 100,000. How did these cities develop and what are they like?

Theories of Urban Growth and Decline

Structural-functionalist theorists and conflict theorists hold very different views of the sources, nature, and consequences of urban life. Structural functionalists emphasize the benefits of urban growth and decline, while conflict theorists emphasize the political struggles that undergird these changes.

Structural-Functional Theory: Urban Ecology

Early structural-functional sociologists, many of whom lived in the booming Chicago of the 1920s and 1930s, assumed that cities grew in predictable ways. Some argued that (like Chicago), cities naturally grew outward in concentric circles from central business districts (Burgess 1925). Others believed that cities grew in wedge-shaped sectors, along transit routes, or in other patterns (Hoyt 1939). All structural functionalists, however, agreed that healthy and natural competition between economic rivals would lead cities to grow in whatever ways offered the most efficient means for producing and distributing goods and services. More recently, structural functionalists have assumed that urban decline and the growth of suburbs similarly reflect natural progress toward superior and more efficient ways of organizing economic and social life.

Conflict Perspectives: White Flight and Government Subsidies

In contrast, conflict theorists note that no patterns of urban growth have yet been discovered that hold across time and across different locations. Thus they conclude that there is nothing natural about urban growth or decline. Rather, they argue, each city grows or declines in its own unique way, depending on the relative power of competing economic and political forces (Feagin & Parker 1990).

These competing forces appear to have played an important role in drawing middle-class Americans from cities during the last half century. Western culture has long held an anti-urban bias, assuming that rural life is "purer" than city life. This view garnered strength during the early decades of the twentieth century, as first foreign immigrants and later African Americans moved in large numbers from the South to the cities of the Northeast and Midwest. These changes contributed greatly to white Americans' sense that the city was a dangerous place and encouraged middle- and upper-class Americans to flee the cities, a process known as "white flight." In contrast, throughout most of the world, the upper classes live in central cities and the poor are relegated to city outskirts and rural areas.

Suburbanization is the growth of suburbs.

The abandonment of American cities was greatly assisted by government subsidies for **suburbanization,** the growth of suburbs (Goddard 1994; Moe and Wilkie 1997). Since the 1930s, federal and local governments have responded to pressure from auto manufacturers and suburban developers by steadily reducing financial support for public transit while tremendously expanding subsidies for auto manufacturing, highways, road maintenance, and the like. As a result, people found it increasingly difficult to live, work, shop, or travel in dense cities with limited parking and decaying transit systems. In addition, since the 1950s the government has provided inexpensive home mortgages (along with tax breaks) to suburbanites while routinely denying mortgages to city dwellers. During the 1960s and 1970s, the government implemented a catastrophic "urban renewal" program that placed highways in the middle of stable, urban neighborhoods (most of which were minority and poor or working class) and moved dislocated residents to poorly constructed, public, high-rise housing. Finally, in the last two decades, local suburban governments have used tax subsidies to entice corporations to relocate to the suburbs.

All these changes pressured middle-class Americans to move to the suburbs, further contributing to the decay of our cities (Jackson 1985; Moe & Wilkie 1997). Of course, many people gratefully left their urban homes for suburbia and relished the freedom automobiles promised. But many others only reluctantly exchanged their close-knit urban neighborhoods, where they could read the newspaper while riding the bus to work, for sprawling suburbs where high walls separate neighbor from neighbor and long, nerve-wracking drives to work are the norm.

The Nature of Modern Cities

From the Industrial Revolution to the present, the modern city has grown in size and changed considerably in character. We look here at the development of industrial and postindustrial cities.

The Industrial City

With the advent of the Industrial Revolution, production moved from the countryside to the urban factory, and industrial cities, such as Boston, Detroit, and Pittsburgh, were born. These cities were mill towns, steel towns, shipbuilding towns, and, later, automobile-building towns; they were home to slaughterers, packagers, millers, processors, and fabricators. They were the product of new technologies, new forms of transportation, and vastly increased agricultural productivity that freed most workers from the land.

Fired by a tremendous growth in technology, the new industrial cities grew rapidly during the nineteenth century. In the United States, the urban population grew from 2 to 22 million in the half-century between 1840 and 1890. In 1860, New York was the first U.S. city to reach 1 million in population. The industrial base that provided the impetus for city growth also gave the industrial city its character: tremendous density and a central business district.

DENSITY   Until the middle of the twentieth century, most Americans walked to work—and everywhere else, for that matter. The result was dense crowding of working-class housing around manufacturing plants. Even in 1910, the average New Yorker commuted only two blocks to work. Entire families shared a single room, and in major cities such as New York and London, dozens of people crowded into a single cellar or attic. The crowded conditions, accompanied by a lack of sewage treatment and clean water, fostered tuberculosis, epidemic diseases, and generally high mortality.

CENTRAL BUSINESS DISTRICT   The lack of transportation and communication facilities also contributed to another characteristic of the industrial city, the central business district. The central business district is a dense concentration of retail trade, banking and finance, and government offices, all clustered close together so messengers could run between offices and businessmen could walk to meet one another. By 1880, most major cities had electric streetcars or railway systems to take traffic into and out of the city. Because most transit routes offered service only into and out of the central business district rather than providing cross-town routes, the earliest improvements over walking enhanced rather than decreased the importance of this district.

The Postindustrial City

The industrial city was a product of a manufacturing economy plus a relatively immobile labor force. Beginning about 1950, these conditions changed and a new type of city began to grow. Among the factors prominent in shaping the character of the postindustrial city are the change from secondary to tertiary production and greater ease of communication and transportation. These changes have led to the rise of urban sprawl and edge cities.

CHANGE FROM SECONDARY TO TERTIARY PRODUCTION   As we noted in Chapter 13, the last decades have seen a tremendous expansion of jobs in tertiary production and the subsequent decline of jobs in secondary production. The manufacturing plants that shaped the industrial city are disappearing. Many of those that remain have moved to the suburbs where land is cheaper and have taken working-class jobs, housing, and trade with them.

Instead of manufacturing, the contemporary central city is dominated by medical and educational complexes, information-processing industries, convention and entertainment centers, and administrative offices. These are the growth industries. They are also white-collar industries. These same industries, plus retail trade, also dominate the suburban economy.

© Peter Menzel

■ The suburbs are the fastest growing part of America, and the edge cities are the fastest growing cities.

EASIER COMMUNICATION AND TRANSPORTATION   Development of telecommunications and good highways has greatly reduced the importance of physical location. The central business district of the industrial city was held together by the need for physical proximity. Once this need was eliminated, high land values and commuting costs led more and more businesses to locate on the periphery, where land was cheaper and housing more desirable. Many corporate headquarters moved from New York or Chicago all the way to Arizona or Texas.

A key factor in increasing individual mobility was the automobile. Without the automobile, workers and businesses could not have moved to the city periphery, and space-gobbling single-family homes would not have been built. In this sense, the automobile and the automotive industry have been the chief architect of U.S. cities since 1950.

URBAN SPRAWL AND EDGE CITIES   These changes have led to the collapse of many central business districts. In their stead, urban sprawl and edge cities have emerged. Postindustrial cities, such as Atlanta, Las Vegas, and Miami, are much larger in geographical area than the industrial cities were. The average city in 1940 was probably less than 15 miles across; now many metropolitan areas are 50 to 75 miles across. No longer are the majority of people bound by subway and railway lines that only go back and forth to downtown. Retail trade is dominated by huge, climate-controlled, suburban malls. A great proportion of the retail and service labor force has also moved out to these suburban centers, and many of the people who live in the suburbs also work in them. Suburban centers that now have an existence largely separate from the cities that spawned them are known as **edge cities** (Garreau 1991). In 2006, the ten fastest growing cities were all edge cities.

Edge cities are suburban centers that now have an existence largely separate from the cities that spawned them.

Urbanization in the United States

What is considered urban in one century or nation is often rural in another. To impose some consistency in usage, the U.S. Bureau of the Census has replaced the common words *urban* and *rural* with two technical terms: *metropolitan* and *nonmetropolitan*.

A **metropolitan area** is a county that has a city of 50,000 or more in it *plus* any neighboring counties that are significantly linked, economically or socially, with the core county. Some metropolitan areas have only one county; others, such as New York, San Francisco, or Detroit, include half a dozen neighboring counties. In each case, the metropolitan area goes beyond the city limits and includes what is frequently referred to as, for example, the Greater New York area. A **nonmetropolitan area** is a county that has neither a major city in it nor close ties to such a city.

Currently, 78 percent of the U.S. population lives in metropolitan areas. This metropolitan population is divided between those who live in the central city (within the actual city limits) and those who live in the surrounding suburban ring. More than half of the metropolitan population live in the suburbs rather than in the central city itself. Although these people have access to a metropolitan way of life, they may live as far as 50 miles from the city center.

The nonmetropolitan population of the United States has shrunk to 22 percent of the U.S. population. Although there are nonmetropolitan counties in every state of the Union except New Jersey, the majority of the nonmetropolitan population lives in either the Midwest or the South. Only 5 percent are farmers, and many live in small towns rather than in purely rural areas.

A **metropolitan area** is a county that has a city of 50,000 or more in it plus any neighboring counties that are significantly linked, economically or socially, with the core city.

A **nonmetropolitan area** is a county that has no major city in it and is not closely tied to such a city.

Urbanization in the Less-Developed World

The growth of large cities and an urban way of life has occurred everywhere very recently; in the less- and least-developed nations, this growth is happening almost overnight (Figure 14.2). Mexico City, São Paulo, Bogotá, Seoul, Kinshasa, Karachi, Calcutta, and other cities in developing nations continue to grow rapidly. Their populations

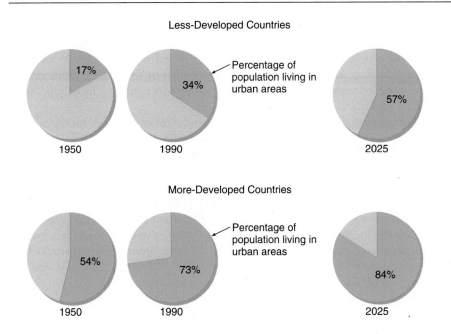

FIGURE 14.2
Urbanization Trends in the Developed and Less-Developed World, 1950-2025
Although the world is still more rural than urban, this is changing within our lifetime. Urbanization is growing particularly quickly in the less-developed world. In 1960, only 3 of the world's 10 largest cities were in developing countries: Shanghai, Buenos Aires, and Calcutta. In 1990, all except three (Tokyo, New York, and Los Angeles) were in the developing world. By the year 2025, three quarters of the world's urban population will live in less-developed countries. SOURCE: Haub 1993.

are likely to double in about a decade. The roads, the schools, and the sewers that used to be sufficient no longer are; neighborhoods triple their populations and change their character from year to year. These problems are similar to the problems that plagued Western societies at the onset of the industrial revolution, but on a much larger scale.

Urbanization in the less-developed world differs from that in the developed world, not only in pace, but also in causes. First, more than half of the growth in developing cities is due to a high excess of births over deaths, rather than to migration from the countryside. Second, many of the large and growing cities in the less-developed world have never been industrial cities. They are government, trade, and administrative centers. More than one third of the regular full-time jobs in Mexico City are government jobs. These cities offer few working-class jobs, and the growing populations of unskilled men and women become part of the informal economy—artisans, peddlers, bicycle renters, laundrywomen, and beggars.

Place of Residence and Social Relationships

Every year, new films and television shows depict the evils of city life, the boredom of suburbs, and the intolerance of small towns. How realistic are such images? This section explores the pleasures and perils of modern urban, suburban, and small-town life.

Urban Living

One of the primary questions raised by sociologists who study cities is the extent to which social relationships and the norms that govern them differ between rural and urban places. Here we look at sociological theories of urban life and research on the realities of urban living.

Theoretical Views

As we saw earlier, the Western world as a whole has an antiurban bias. Big cities are seen as haunts of iniquity and vice, corruptors of youth and health, and destroyers of family and community ties. City dwellers are characterized as sophisticated but artificial; rural people are characterized as unsophisticated, but warm and sincere. This general antiurban bias (which has been around at least since the time of ancient Rome), coupled with the very real problems of the industrial city, had considerable influence on early sociologists.

The classic statement of the negative consequences of urban life for the individual and for social order was made by Louis Wirth in 1938. Wirth argued that the greater size, heterogeneity, and density of urban living necessarily led to a breakdown of the normative and moral fabric of everyday life.

Gemeinschaft refers to society characterized by the personal and permanent ties associated with primary groups.

Gesellschaft refers to society characterized by the impersonal and instrumental ties associated with secondary groups.

Greater size means that many members of the community will be strangers to us. Rural society is characterized by what is called **gemeinschaft,** personal and permanent ties associated with primary groups, and urban society is characterized by **gesellschaft,** the impersonal and instrumental ties associated with secondary groups. Wirth postulated that urban-dwellers would still have primary ties but would keep their emotional distance from, for example, store clerks or strangers in a crowded elevator by developing a cool and calculating interpersonal style.

Wirth also believed that when faced with a welter of differing norms, the city dweller was apt to conclude that anything goes. Such an attitude, coupled with the lack of informal social control brought on by size, would lead to greater crime and deviance and a greater emphasis on formal controls.

Later theorists have had a more benign view of the city. Sociologists now suggest that individuals experience the city as a mosaic of small worlds that are manageable and knowable. Thus, the person who lives in New York City does not try to cope with 9 million people and 500 square miles of city; rather the individual's private world and primary ties are made up of family, a small neighborhood, and a small work group. In addition, sociologists point out, urban life provides the "critical mass" required for the development of tightknit subcultures, from gays to symphony orchestra aficionados to rugby fans. Wirth might interpret some of these subcultures as evidence of a lack of moral integration of the community, but they can also be seen as private worlds within which individuals find cohesion and primary group support.

Realities of Urban Living

Does urban living offer more disadvantages or advantages? This section reviews the evidence about the effects of urban living on social networks, neighborhood integration, and quality of life.

SOCIAL NETWORKS   The effects of urban living on personal integration are rather slight. Surveys asking about social networks show that urban people have as many intimate ties as rural people. There is a slight tendency for urban people to name fewer kin and more friends than rural people. The kin omitted from the urban lists are not parents, children, and siblings, however, but more distant relatives (Amato 1993). There is no evidence that urban people are disproportionately lonely, alienated, or estranged from family and friends.

NEIGHBORHOOD INTEGRATION   Empirical research generally reveals the neighborhood to be a very weak group. Most city dwellers, whether central city or suburban, find that city living has freed them from the necessity of liking the people they live next to and has given them the opportunity to select intimates on a basis other than physical proximity. This freedom is something that people in rural areas do not have. There is growing consensus among urban researchers that physical proximity is no longer a primary basis of intimacy (Flanagan 1993). Rather, people form intimate networks on the basis of kin, friendship, and work groups, and they keep in touch by telephone, e-mail, or instant messaging rather than by face-to-face communication. In short, urban people do have intimates, but they are unlikely to live near each other. When in trouble, they call on their good friends, parents, or adult children for help. In fact, one study of neighborhood interaction in Albany-Schenectady-Troy, New York, found that a substantial share—15 to 25 percent—of all interaction with neighbors was with *family* neighbors—parents or adult children who lived in the same neighborhood (Logan & Spitze 1994).

Neighbors are seldom strangers, however, and there are instances in which being nearby is more important than being emotionally close. When we are locked out of the house, need a teaspoon of vanilla, or want someone to accept a United Parcel Service

Many people enjoy the vibrant nature of life in cities, as seen in this Pittsburgh neighborhood.

package, we still rely on our neighbors (Wellman & Wortley 1990). Although we generally do not ask large favors of our neighbors and don't want them to rely heavily on us, most of us expect our neighbors to be good people who are willing to help in a pinch. This has much to do with the fact that neighborhoods are often segregated by social class and stage in the family life cycle. We trust our neighbors because they are people pretty much like us.

QUALITY OF LIFE   Big cities are exciting places to live. People can choose from a wide variety of activities, 24 hours a day, 7 days a week. The bigger the city, the more it offers in the way of entertainment, libraries, museums, zoos, parks, concerts, and galleries. The quality of medical services and police and fire protection also increases with city size. These advantages offer important incentives for big-city living.

On the other hand, there are also disadvantages: more noise, more crowds, more expensive housing, and more crime. The rate of both violent crimes and personal crimes are approximately 20 per 1,000 in suburban and rural areas but 30 per 1,000 in urban areas (Bureau of Justice Statistics 2006).

Because of these disadvantages, many people would rather live close to a big city than actually in it. For most Americans, the ideal is a large house on a spacious lot in the suburbs, but close enough to a big city that they can spend an evening or afternoon there. Some groups, however, prefer big-city living, in particular, childless people who work downtown. Many of these people are decidedly pro-urban and relish the entertainment and diversity that the city offers. Because of their affluence and childlessness, they can afford to ignore many of the disadvantages of city living.

Sociological attention has been captured by cities such as Manhattan and San Francisco with their bright lights, ethnic diversity, and crowding. Nevertheless, only one quarter of our population actually lives in these big-city centers. The rest live in suburbs and small towns. How does their experience differ?

Suburban Living

The classic picture of a suburb is a development of very similar single-family detached homes on individual lots. This low-density housing pattern is the lifestyle to which a majority of people in the United States aspire; it provides room for dogs, children, and barbecues. This is the classic picture of suburbia. How has it changed?

The Growth of the Suburbs

The suburbs are no longer bedroom communities that daily send all their adults elsewhere to work. They are increasingly major manufacturing and retail trade centers. Most people who live in the suburbs work in the suburbs. Thus, many close-in suburban areas have become densely populated and substantially interlaced with retail trade centers, highways, and manufacturing plants.

These changes have altered the character of the suburbs. Suburban lots have become smaller, and neighborhoods of townhouses, duplexes, and apartment buildings have begun to appear. Childless couples, single people, and retired couples are seen in greater numbers. Suburbia has become more crowded and less dominated by the minivan set.

With expansion, suburbia has become more diverse. Although each suburban neighborhood tends to have its own style, stemming in large part from each development including houses of similar size and price, there are a wide variety of styles.

In addition to classic suburban neighborhoods, there are now areas of spacious mini-estate suburbs where people ride horses and lawn mowers, as well as dense suburbs of duplexes, townhouses, and apartment buildings. Some of the first suburbs, which were built after World War II, are now more than 50 years old. Because people tend to age in place, these suburbs are often characterized by retirees living on declining incomes (Lambert & Santos 2006). Many houses are becoming rundown, and renting is becoming increasingly common.

Suburban Problems

Many of the people who moved to suburbia did so to escape urban problems: They were looking for lower crime rates, less traffic, less crowding, and lower tax rates. The growth of the suburbs, however, has brought its own problems (Langdon 1994). Three of the most important are weak governments, car dependence, and social isolation and alienation.

The county and municipal governments of suburban towns and cities are fragmented and relatively powerless. One result of this is the very haphazard suburban growth associated with weak and inadequate zoning authority. In addition, because there is rarely any governmental body that has the power to make decisions for a city and its suburbs as a whole, it is nearly impossible to coordinate decisions across a metropolitan region. This means, for example, that if one suburb or city decides to ban smoking in restaurants, business will simply move to the next suburb.

The lack of regional planning is particularly important when it comes to transportation. Without effective regional decision making, it is difficult to develop effective mass transit systems or even highways. This leaves suburban dwellers in the lurch, since most commute from suburb to suburb or suburb to city. It also makes suburban dwellers even more dependent than others on automobiles. People who don't have cars are basically excluded from the suburban lifestyle and from jobs in either suburb or city. If you can't afford a car or can't drive one due to disability, aging, or youth, your quality of life in suburbia plummets.

Long commutes leave individuals with little time to socialize with coworkers after work or with neighbors and family once they arrive home. In addition, suburban zoning laws that forbid businesses such as cafes, beauty parlors, and taverns in residential neighborhoods deprive people of the natural gathering places that foster social relationships and a sense of community. Similarly, suburban houses with high fences and no front porches make it nearly impossible for neighbors to meet informally (Oldenburg 1997). When people live in one community and work in another, they may end up feeling alienated from both.

Small-Town and Rural Living

Approximately 25 percent of the nation's population lives in rural areas or small towns (less than 2,500 people). Some live within the orbit of a major metropolitan area, but most live in nonmetropolitan areas, from Maine to Alabama, California to Florida. These areas vary greatly and include everything from millionaire second-home towns like Telluride, Colorado, to dying farm or mill towns in Kansas or Maine, flourishing Amish communities in Pennsylvania, and booming Nebraska poultry processing towns. Some rural areas are overwhelmingly white, some are overwhelmingly African American, and a growing number have substantial Hispanic populations.

Across the board, people find rural and small-town living attractive for a number of reasons (Brown & Swanson 2003). It offers lots of open space, low property taxes, and affordable housing (except in vacation areas). There's much less worry

Connections

Social Policy

Many urban planners and activists are working to redesign suburbs to reduce the social isolation and alienation that too often characterizes them. New suburbs are being designed, and old ones redesigned, with bike paths, neighborhood parks, and front porches. To stimulate a pedestrian atmosphere and encourage people to get out and interact, some towns have widened sidewalks and encouraged downtown restaurants to add street-side tables. Others have used financial incentives to encourage builders to place interesting stores on the first floor of new downtown buildings with apartments above. Through these actions, planners and activists hope to bring a sense of community to suburbs.

■ Suburbs are intensely car dependent. As suburbs have grown, so have traffic jams and long commutes.

© AP/ Wide World Photos

about crime and drugs, although alcohol and methamphetamine abuse are actually most common in rural areas. Many also appreciate the more conservative views on politics, premarital sex, religion, and the like that typify nonmetropolitan areas. In addition, community ties remain strong in the small minority of rural towns still characterized by deep family roots, family-run farms, civically engaged churches, and small rather than large manufacturing plants, and both children and adults benefit from the neighborliness and community sentiment. In a city, you might find a bar like Cheers "where everybody knows your name." In a rural area, practically everyone does.

Although young people who grow up in nonmetropolitan areas often must leave to get an education or a job, these areas continue to grow (Johnson 2003). Most of this growth, however, is in "recreational" areas that attract second-home owners and retirees, areas near large cities that attract long-distance commuters, and areas with large-scale food manufacturing plants (meat packing, canning, and so on).

The major problem with rural life is the dearth of jobs, especially good-paying jobs with benefits (Jensen, McLaughlin, & Slack 2003). Family farms have all but disappeared, driven out of business by global competition or bought out by huge agribusinesses (the only ones with the money and power to compete in this global market). Only 5 percent of nonmetropolitan dwellers still work in agriculture, while the majority now work in low-wage service jobs in prisons, casinos, fast-food restaurants, and the like (McGranahan 2003). Because of these problems, many nonmetropolitan dwellers must endure long commutes to jobs in distant metropolitan areas. Stress over low wages, underemployment, and unemployment coupled with the physical stresses of the available work, lack of social resources, and limited access to health care combine to leave nonmetropolitan residents, on average, in somewhat poorer physical and mental health than urban or suburban residents (Morton 2003).

In addition, nonmetropolitan areas that have experienced inflows of "city people" are experiencing new strains due to growing stratification: The economic and cultural differences between the upper and lower ends of the population are

In the most desirable rural areas, good-paying jobs are scarce and housing is expensive. As a result, many rural families must live in inexpensive manufactured homes.

far greater than in the past (Brown & Swanson 2003). Forty years ago, ski resort owners and ski resort workers all lived in Telluride, if in different conditions. Now, resort owners and their clients live in luxury homes in or near the center of town, while most workers can only afford to live far from town in "rural ghettoes" of mobile homes and concentrated poverty. This stratification is particularly hard on school children, who find themselves increasingly marginalized and stigmatized by teachers and wealthier children whose expectations for clothing, vacations, and academic preparation cannot be met by poorer children. In sum, although life in small towns and rural areas still brings benefits, it can bring high costs as well.

Where This Leaves Us

There's no question about it: Numbers matter. As the world's population grows—and, in places, shrinks—all of us are affected. Population growth in the United States has enormous consequences for the environment because of the huge amounts of natural resources Americans use. Population growth in the less-developed nations is especially important because it not only stems from poverty but also produces even more poverty. Meanwhile, population loss in Europe leaves nations grappling with problems brought on by having too few young people compared with the number of old people.

The problems of population growth are intimately connected to the problems of urbanization—and suburbanization. Cities emerged with the rise in industrialization, a process that is still continuing in the developing nations. In turn, problems with urban life, accentuated by various social policies, have stimulated the growth of suburbs and "edge cities." Each of these environments offers its own dangers and its own rewards.

Summary

1. For most of human history, fertility was about equal to mortality and the population grew slowly or not at all. Childbearing was a lifelong task for most women, and death was a frequent visitor to most households, claiming one quarter to one third of all infants in the first year of life.

2. The demographic transition—the decline in mortality and fertility—developed over a long period in the West. Mortality declined because of better nutrition, an improved standard of living, improved public sanitation, and to a much more limited extent, modern medicine. Somewhat later, changes in social structure associated with industrialization caused fertility to decline. In the developing nations, mortality has declined rapidly, and fertility levels are only slowly declining in response.

3. Social structure, fertility, and mortality are interdependent; changes in one affect the others. Among the most important causes and consequences of high fertility are the low status of women.

4. The level of fertility in a society is directly linked to the costs and rewards of childbearing. In traditional societies, such as Ghana, most social structures (the economy and women's roles, for example) support high fertility. In many modern societies, such as Italy and the United States, social structure imposes many costs on parents.

5. When a nation's fertility declines, the nation faces several problems. Among these are labor-force shortages, difficulties in funding health and pension benefits for a burgeoning number of older people, and nationalistic fears over growing numbers of foreign workers.

6. Population growth is an important cause of environmental devastation in the less-developed world, but not in the developed world (where most environmental resources are used). Although population growth does contribute to poverty in the less-developed world, other factors are much stronger causes of poverty.

7. In the United States, life expectancy is high and continues to increase. Childlessness is increasing and fertility is near replacement level. Because of high immigration rates, however, the U.S. population is unlikely to decline. Because many of the new Americans are Asian and Latino, the racial and ethnic composition of the U.S. population is likely to change substantially. Immigration has not taken jobs from U.S. citizens but may have reduced wages among the least-educated native-born Americans.

8. Since the 1970s, central cities in most of the nation have shrunk, and urban poverty has increased. Meanwhile, suburban towns and cities have grown significantly, and nonurban areas have experienced modest growth. This movement to suburbia and to nonurban locations raises serious environmental questions. Across categories—urban, suburban, and nonurban—the Sunbelt states have seen the most growth.

9. Structural functionalists argue that cities grow and decline in predictable and natural ways, reflecting the most efficient means for producing and distributing goods and services. Conflict theorists, on the other hand, argue that city growth and decline reflect the outcomes of economic and political struggles between competing groups. Government subsidies played a major role in the twentieth-century growth of suburbs and decline of central cities.

10. The industrial city has high density and a central-city business district. The postindustrial city reflects the shift to tertiary production and increased ease in communication and transportation and is characterized by lower density and urban sprawl.

11. Urbanization is continuing rapidly in the less-developed world; many of its large cities will double in size in a decade. This urban growth is less the result of industrialization than of high urban fertility.

12. There are competing theories about the consequences of urban living. Wirth's theory suggests that urban living will lead to nonconformity and indifference to others. Other theorists suggest that the size of the city is managed through small groups and allows for the development of unconventional subcultures.

13. Urban living is associated with less reliance on neighbors and kin and more reliance on friends, with greater risk of crime.

14. Suburban living has become more diverse. Retail trade and manufacturing have moved to the suburbs, and the suburbs are now more densely populated, more congested, and less dominated by the minivan set. Suburban living has its own problems, including weak governments, transportation problems, and social isolation and alienation.

15. Among the benefits of small-town and rural living are less crime, stronger community ties, more open spaces, and more affordable housing (except in vacation areas). The most serious problem is the dearth of good-paying jobs with benefits, which result in somewhat poorer physical and mental health.

Thinking Critically

1. If you are a typical college-age student, your generation is considerably smaller than your parents' generation. How will this affect you? Consider the impact on you now, as your parents and their generation retire, and as you approach retirement. Think about both personal finances and resources and government programs and spending.
2. How is dormitory life similar to urban living? Similar to small-town living?
3. Make a list of the environmental resources you use in a day. Consider "natural" products such as oranges, as well as manufactured products, such as computers. How would your list compare with that of someone in a developing nation?
4. What would the United States be like if all immigration ceased? What would be the benefits? The disadvantages?

Companion Website for This Book

academic.cengage.com/sociology/brinkerhoff
Gain an even better grasp on this chapter by going to the Companion Website. This resource contains tutorial quizzes and flash cards to help you master key terms and concepts.

Suggested Readings

Duany, Andres, Plater-Zyberk, Elizabeth, and Speck, Jeff. 2001. *Suburban Nation: The Rise of Sprawl and the Decline of the American Dream.* New York: North Point Press. Three renowned architects cogently argue that the typical American suburb reduces sense of community; isolates children, the elderly, and stay-at-home mothers; and contributes to environmental degradation. They make a strong case for new suburban developments that blend homes, schools, and workplaces.

Pipher, Mary. 2003. *The Middle of Everywhere: Helping Refugees Enter the American Community.* New York: Harvest Books. Family therapist and best-selling author Pipher vividly describes the heartbreaking experiences of refugees from around the world who have settled in Nebraska.

Ellingwood, Ken. 2005. *Hard Line: Life and Death on the US-Mexico Border.* New York: Vintage. Ellingwood, a journalist, captures the feelings and experiences of illegal immigrants, border agents, human rights workers, and ranchers living on, working at, and trying to cross the border.

Tobar, Hector. 2005. *Translation Nation.* New York: Riverhead. A fascinating portrayal of the diversity of Latino immigrants, how they are becoming part of the United States, and how both Latino and Anglo-American culture is changing in the process.

Fertility, Mortality, and Social Structure

The Effects of Social Structure on Fertility

In Ghana, the average woman has four or five children; in Italy, the average woman has only one or two. These differences are the product of values, roles, and statuses in very different societies. The average woman in Ghana wants four or five children, and the average woman in Italy wants only one or two.

The most important factors affecting mortality are public health measures and the standard of living: access to good nutrition, safe drinking water, and decent housing. These factors almost entirely explain why the average American will live 33 years longer than the average person living in Sierra Leone.

The level of fertility in a society is strongly related to the roles of women. Generally, fertility is higher in places where women marry at younger ages, where their value is measured by the number of children they bear, where they have less access to education and income, where their roles outside the household are limited, and where they are relatively powerless (United Nations Population Fund 2000). Fertility also reflects the development—or lack of development—of a society's institutions. When the family is the main source of security, income, social interaction, and even salvation, fertility is high.

The Effects of Social Structure on Mortality

As noted earlier, the most important factors affecting mortality are public health measures and the standard of living—access to good nutrition, safe drinking water, and protective housing. Differences in these factors almost entirely account for the expectation that the average American will be healthier and will live 33 years longer than the average person living in Sierra Leone; these differences also explain why in Britain, the United States, and elsewhere each social class has lower death rates and better health than the social class below it (Marmot, Kogevinas, & Elston 1987; Weitz 2007). As a result, by focusing on public-health measures and on equalizing income and education, countries such as China, Cuba, Costa Rica, and Sri Lanka have achieved life expectancies that are approaching those of the far-wealthier United States (Caldwell 1993).

More subtly, social structure affects mortality through structuring social roles and lifestyle (Weitz 2007). Race, socioeconomic status, and gender all affect exposure to unhealthy or dangerous lifestyles. People with less education, for example, get less physical exercise and are more likely to smoke than those with more education, and young men are more likely to die in automobile accidents than young women.

The Effects of Fertility on Social Structure

Fertility has powerful effects on the roles of women. The greater the number of children a woman has, the less likely she is to work, play, or study outside the home. When the average woman bears only two children, fertility places much less restriction on her social involvement.

In addition to affecting women's roles, fertility has a major impact on the age structure of the population: the higher the fertility, the younger the population. This is graphically

FIGURE 14.3
A Comparison of Age Structures in Low- and High-Fertility Societies, 2006

When fertility is high, the number of children tends to be much larger than the number of parents. When this pattern is repeated for generations, the result is a pyramidal age structure. When fertility is low, however, each generation has a similar size, and a boxier age structure results.

SOURCE: U.S. Census Bureau International Data Base 2006 (http://www.census.gov/ipc/www/idbpyr.html

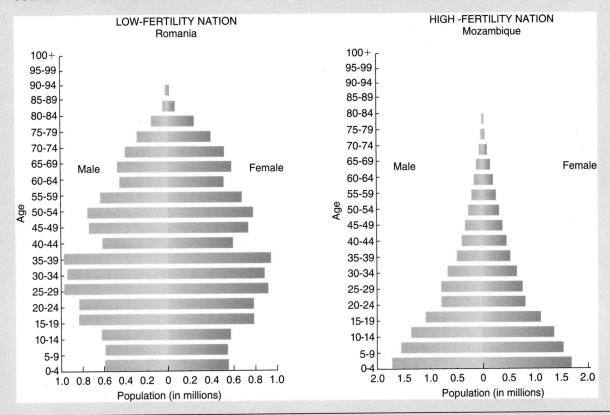

shown in the population pyramid in Figure 14.3. When fertility is low (as in Romania), the number of young people is about the same as the number of adults, and the population takes on a boxy shape when graphed; when fertility is high, as in Mozambique, there are many more children than adults, and the age structure takes on a pyramidal shape.

This pyramidal age structure has both short- and long-term consequences. In the short term, it means that a large proportion of the population is too young to work, and this reduces society's productivity. In the long run, it translates into explosive population growth. In those parts of sub-Saharan Africa that have not been hard hit by AIDS, the number of girls aged 0 to 4 exceeds the number of women aged 20 to 24 by a ratio of two to one. This means that in the next generation there will be twice as many mothers as there are today. In those parts of Africa that have been ravaged by AIDS, however, over the next decades the age structure will more closely resemble that of a low- than a high-fertility nation.

The Effects of Mortality on Social Structure

Like fertility, mortality has particularly strong effects on the family. A popular myth is that the preindustrial family was multiple-generational, what we call an extended family. A little reflection will demonstrate how unlikely it is that many children lived with their grandparents when life expectancy was only 30 to 35 years and when fertility was seven to eight children per woman. In fact, only a small percentage of families could have been three-generational. Quite often, if the children lived with their grandparents, it was because their parents were dead. The household of a high-mortality society was probably as fractured, as full of stepmothers, half-sisters, and stepbrothers, as is the current household of today's high-divorce society.

CHAPTER 15

Social Change

© Owen Franken/Stock, Boston Inc.

Outline

How Societies Change

Social institutions do not stand still. Often, things change without our knowing how or why. Immediately after the terrorist attacks of September 11, 2001, thousands of Europeans held candle-lit vigils to express their solidarity with the United States. One and a half years later, after U.S.-led forces invaded Iraq, even more Europeans demonstrated in protest against the United States. Meanwhile, Eastern Europe, South Africa, and Mexico lurch toward new economic and political forms, while the fortunes of all nations increasingly depend upon an international political and economic system.

What is going on? Many Americans shake their heads in confusion. The last decade has brought wonderful developments such as the fall of the Taliban, new drugs to treat AIDS, and wireless computer technologies. Balanced against these positive changes, however, are civil war and malnutrition in many developing nations, the destruction of the Amazon rainforest, and an epidemic of repetitive stress disorders linked to computer use.

All of these changes—both positive and negative—are referred to by sociologists as social change. **Social change** is defined as any significant modification or transformation of social structures or institutions over time. The rapid pace of social change and the complexity of twenty-first century problems lead many individuals to feel a sense of both urgency and helplessness. In this chapter, we describe three potential sources of social change: collective behavior, social movements, and technology.

> **Social change** is any significant modification or transformation of social structures and sociocultural processes over time.

Collective Behavior

- After the film Napoleon Dynamite appeared, "Vote for Pedro" T-shirts became suddenly popular around the country.
- In July 2006, during international soccer's World Cup, crowds stood up, cheered, and sang songs to spur on their national teams. Afterward, fans of the winning teams broke into spontaneous celebrations in the streets.
- In August 2005, after Hurricane Katrina devastated New Orleans, some informal groups of residents broke into stores to steal, while others banded together to rescue people from rooftops.

> **Collective behavior** is nonroutine action by an emotionally aroused gathering of people who face an ambiguous situation.

Despite the differences between these actions, all are examples of collective behavior. **Collective behavior** is spontaneous action by groups in situations where cultural rules for behavior are vague, inadequate, or debated (Marx & McAdam 1994). It includes such diverse actions as mob violence and spontaneous candle-lit vigils to protest mob violence, as well as the behavior of crowds surging into Wal-Mart for a sale or carousing in the streets during Mardi Gras. These are unplanned, more or less spur-of-the-moment actions, where individuals and groups improvise a joint response to an unusual or problematic situation. Collective behavior differs from social movements (discussed below) in that it is usually short lived, at least in part because participants lack a clearly defined social agenda and lack the resources and organization needed to affect public policy. (Some sociologists include social movements as part of collective behavior, but others, including this textbook's authors, prefer to keep the topics separate.)

Collective behavior, such as mosh pits and crowd-surfing at rock concerts, differs from social movement in being more spontaneous and relatively unplanned.

© Neal Preston/Corbis

As noted, collective behavior occurs when cultural rules are (1) vague, (2) inadequate, or (3) contested. Cultural rules are *vague* in many areas: Should a woman tattoo her whole arm? Should someone take a year off between high school and college? Cultural rules are often *inadequate* during crises or periods of rapid social change: Who should be rescued first during a disaster? What is appropriate—or safe—to post on a Facebook page? Cultural rules are *contested* when some social groups feel that the normal rules of the society work against them and decide to subvert or protest those rules.

Collective behavior can occur anywhere there is a group, from sidewalks, to prisons, to corporations (Marx & McAdam 1994). A rumor can lead illegal street vendors to quickly pack up their goods, and a prison may erupt in violence over squalid conditions. Within a corporation, a particular Windows desktop wallpaper may suddenly become popular on a floor, employees might help each other escape a disaster (as when the World Trade Center was attacked), or they might begin an informal work slowdown as a silent protest against low pay.

Even when collective behavior is not designed as protest, however, it can have the effect of challenging the status quo. For example, if enough college students post descriptions of drinking binges or wild sexual activity on Facebook or Myspace, then that behavior will likely come to seem more acceptable. The difference between collective behavior and social movements, however, is that social movements are organized, relatively broad based, long term, and intended to foster social change.

Social Movements

Social movements are individuals, groups, and organizations united by a common desire to change social institutions, attitudes, or ways of life (Tilly 2004). Examples include the immigrant rights and environmental movements, as well as the grassroots

A **social movement** is an ongoing, goal-directed effort to fundamentally challenge social institutions, attitudes, or ways of life.

Connections

Example

The word spread rapidly via text messaging and e-mail: Show up in New York City's Grand Central Station at 7 P.M., applaud wildly for 15 seconds, then disperse. Other "flash mobs" have gathered to duck-walk, spin in circles, or ask bookstore clerks for nonexistent titles. Although flash mobs may look like a social movement, they aren't yet. Participants are not motivated by anything other than the desire to have fun and have no program for social change. Flash mobs could, however, become a technique used by social movements or could grow into a movement if mobbers develop a more conscious critique of modern-day life.

struggle against drunk driving. A social movement is extraordinarily complex. It may include sit-ins, demonstrations, and even riots, but it also includes meetings, fundraisers, legislative lobbying, and letter-writing campaigns.

Both collective behavior and social movements challenge the status quo. As a result, they are related in at least two ways. First, social movements need and encourage some instances of collective behavior simply to keep issues in the public eye (Marx & McAdam 1994). There is nothing like a riot or police breaking up an illegal demonstration to get people's attention. Second, even though collective behavior is usually limited to a particular place and time, it can be part of a repeated mass response to problematic conditions. When this happens, collective behavior at a grassroots level may be a driving force in mobilizing social movements (Tilly 2004).

As we documented in Chapter 13, most people in the United States have relatively little interest in politics. Why, then, do some people shake off this lethargy and try to change the system? And under what circumstances do social movements succeed or fail?

Theoretical Perspectives on Social Movements

Three major theories explain the circumstances in which social movements arise: relative-deprivation theory, resource mobilization theory, and political process theory. All three theories suggest that social movements arise out of inequalities and cleavages in society, but they offer somewhat different assessments of the meaning, sources, and tactics of social movements.

Structural-Functional Theory: Relative Deprivation

Relative-deprivation theory argues that social movements arise when people experience an intolerable gap between their rewards and their expectations.

Poverty and injustice are universal phenomena. Why is it that they so seldom lead to social movements? According to **relative-deprivation theory,** social movements arise when we *believe* we should have more than we *actually* have—especially if we feel this deprivation is a result of unfair treatment (Walker & Smith 2002). Our expectations, in turn, are usually determined by comparing ourselves with others or with past situations. Because the theory refers to deprivation relative to other groups or times rather than to absolute deprivation, it is called *relative*-deprivation theory.

Figure 15.1 diagrams three conditions for which relative-deprivation theory would predict the development of a social movement. In Condition A, disaster or taxation suddenly reduces the absolute level of living. If there is no parallel drop in people's expectations, they will feel that their deprivation is illegitimate. In Condition B, both expectations and the real standard of living are improving, but expectations continue to rise

FIGURE 15.1
The Gap Between Expectations and Rewards
Relative-deprivation theory suggests that whenever there is a gap between expectations (E) and rewards (R), relative deprivation is created. It may occur when conditions are stable or improving as well as when the real standard of living is declining.

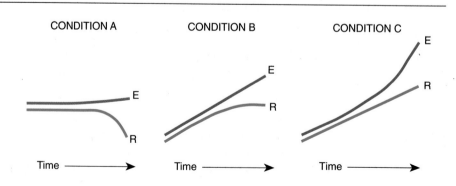

even after the standard of living has leveled off. Consequently, people feel deprived relative to what they had anticipated. Finally, in Condition C, expectations rise faster than the standard of living, again creating a gap between reality and expectations. Relative-deprivation theory has the merit of providing a plausible explanation for many social movements occurring in times when objective conditions are either improving (Condition C) or at least are better than in the past (Condition B).

Relative-deprivation theory is a structural-functional theory. Like other structural-functional theories, it assumes that in normal circumstances society functions smoothly. According to this theory, then, social movements arise only when social change occurs unevenly across social or cultural institutions or when the pace of change is simply too rapid.

There are two major criticisms of relative-deprivation theory. First, empirical evidence does not bear out the prediction that those who are most deprived, absolutely or relatively, will be the ones most likely to participate in social movement. Often, social movement participants are the best off in their groups rather than the worst off. For example, almost all of the 19 terrorists who destroyed the World Trade Center and attacked the Pentagon, as well as Osama Bin-Laden, were well educated and middle class or wealthy. In many other situations, individuals participate in and lead social movements on behalf of groups to which they do not belong, such as South African whites who fought against apartheid and people who fight for animal rights. Second, the theory fails to specify the conditions under which relative deprivation will lead to social movements. Why do some relatively deprived groups form social movements and others don't? Relative deprivation can play a role, but by itself it is not a good predictor of the development of social movements (Gurney & Tierney 1982).

Conflict Theory: Resource Mobilization

While structural functionalists assume that society generally works harmoniously, conflict theorists assume that conflict, competition, and, as a result, deprivation are common. If deprivation were all it took to spark a social movement, we would have active social movements all the time. Yet social movements only arise sporadically. Consequently, conflict theorists argue, relative deprivation by itself cannot explain why social movements emerge when they do. Rather, they argue, social movements emerge when individuals who experience deprivation can garner the resources they need to effectively mobilize for action. This theory is known as **resource mobilization theory** and is the most commonly used theory among American sociologists (McAdam & Snow 1997).

According to resource mobilization theory, then, the spark for turning deprivation into a movement is not anger and resentment but rather organization and re-

Resource mobilization theory suggests that social movements develop when individuals who experience deprivation can garner the resources they need to mobilize for action.

Theories of Social Movements

	Major Assumption	Causes of Social Movements
Structural-Functional Theory: Relative Deprivation	Social movements are an abnormal part of society	Social change produced disorganization and discontent
Conflict Theory: Resource Mobilization	Social movements are the normal outgrowth of competition between groups	Competition between organized groups
Symbolic Interaction Theory: Political Process	People join social movements because they have developed an "insurgent consciousness"	Political opportunities combined with an individual sense that change is needed and possible

sources. As a result, social movements will be more common in affluent societies than in poorer ones, since in affluent societies even the least well off may have access to the minimum resources needed for protest. Similarly, the building blocks of social movements are organized groups whose leaders are relatively well provided with resources, rather than discontented individuals from the economic bottom of a society. The final key is the ability of a movement to offer a frame for interpreting individual grievances and problems that will motivate individuals to participate.

Symbolic Interaction Theory: Political Process

Resource mobilization theory remains very important within sociology, but it has been criticized for two reasons. First, it downplays the importance of grievances and spontaneity as triggers for social movements (Klandermas 1984; Morris & Mueller 1992; Zygmunt 1986). Second, it overlooks the importance of the process through which vague individual grievances are transformed into new collective identities and organized political agendas (Jasper & Poulsen 1995; Williams 1995). **Political process theory** has arisen to fill this gap. According to political process theory, a social movement needs two things: political opportunities and an "insurgent consciousness." **Political opportunities** include preexisting organizations that can provide the new movement with leaders, members, phone lines, copying machines, and other resources. Whether or not political opportunities will exist depends on a number of factors, including the level of industrialization in a society, whether a war is going on, and whether other cultural changes are underway (Meyer 2004).

Insurgent consciousness is the individual sense that change is both needed and possible. In the same way that symbolic interactionism argues that individuals develop their identities and understanding of the social world through interactions with significant others, political process theory argues that individuals develop their sense of identity and of the possibility of change through interaction with others. For example, until the 1970s, newspapers regularly listed job ads in separate columns for men and for women, top universities refused to admit women as students, and some ministers told battered wives that they must have done something to cause their husbands to beat them. The growth of the women's movement depended upon convincing women that these were not merely personal problems but rather were problems they shared with other women *simply because they were women*. This point is neatly summed up in the feminist slogan "the personal is political."

Political process theory suggests that social movements develop when political opportunities are available and when individuals have developed a sense that change is both needed and possible.

Political opportunities are resources that allow a social movement to grow; they include preexisting organizations that can provide the new movement with leaders, members, phone lines, copying machines, and other resources.

Insurgent consciousness is the individual sense that change is both needed and possible.

Why Movements Succeed or Fail

Why do some movements succeed, while others disappear? Based on a historical review of 53 diverse social movement organizations (SMOs), sociologist William Gamson (1990) identified four possible outcomes of social movement activities. A fully successful SMO is one that wins both *new advantages* for those it aims to help and *acceptance* as a legitimate, reputable organization. Nelson Mandela's African National Congress, for example, now controls the government of the Republic of South Africa and has improved the situation of South Africa's black population enormously. Other SMOs, however, have not been as successful. Some SMOs are co-opted when their rhetoric and ideology gain nominal public approval, but the real social changes they had advocated have not occurred. Other SMOs are preempted when those in power adopt their goals and programs but continue to denigrate the organization and its ideology; many politicians, for example, now support the idea of equal pay for equal work but continue to belittle the feminists who brought the issue to public attention. Still others have little lasting effect on society. Table 15.1 outlines the four movement outcomes discussed by Gamson.

TABLE 15.1
Social Movement Outcomes
Based on his historical review of 53 social movement organizations, Gamson (1990) identified four possible movement outcomes. These four outcomes depend on a combination of two factors: whether or not the movement is able to win advantages for those it seeks to help and whether or not it is able to gain acceptance as a legitimate organization.

New Advantages	Acceptance	
	Full	None
Many	Full response	Preemption
None	Co-optation	Collapse

Empirical analysis of social movements in the United States and around the world suggests that a number of factors are important for movement success. Movements are most likely to succeed if they contain diverse organizations using diverse tactics, if they can garner sufficient resources, and if they can frame their goals and ideology in a way that attracts and keeps members.

Diverse Organizations and Tactics

A social movement is the product of the activities of dozens and even hundreds of groups and organizations, all pursuing, in their own way, the same general goals. For example, there are probably dozens of different SMOs within the environmental movement, ranging from the relatively conventional Audubon Society and Sierra Club to the radical Greenpeace organization and the ecoterrorists of the Earth Liberation Front (ELF). The organizations within a movement may be highly divergent and may compete with each other for participants and supporters. Because this assortment of organizations provides avenues of participation for people with a variety of goals and styles, however, the existence of diverse SMOs is functional for the social movement.

SMOs can be organized in one of two basic ways: as professional or as volunteer organizations. On the one hand, we have organizations such as the American Civil Liberties Union or the National Rifle Association, which have offices in Washington, D.C., and a relatively large paid staff, some of whom are professional fund-raisers or lobbyists who develop an interest in an issue only after being hired. At the other extreme is the SMO staffed on a volunteer basis by people who are personally involved—for example, neighbors who organize in the church basement to prevent a nuclear power plant from being built in their neighborhood. These two types of SMO are referred to, respectively, as the *professional* SMO and the *indigenous* SMO.

Evidence suggests that the existence of both types of organizations facilitates a social movement. The professional SMO is usually more effective at soliciting resources from foundations, corporations, and government agencies. It appeals to individuals who are ideologically or morally committed to the group's cause. On the other hand, because employees of professional SMOs are not themselves underprivileged and because they work daily with the establishment, professional SMOs sometimes lose the sense of grievance that is necessary to motivate continued, imaginative efforts for change. As a result, a social movement also requires sustained indigenous organizations (Jenkins & Eckert 1986). Indigenous organizations perform two vital functions. First, by keeping the aggrieved group actively supportive of the

■ During the spring of 2006, hundreds of thousands of demonstrators across the nation marched in support of immigrant rights. This was an indigenous social movement, mobilized through churches and by word of mouth rather than by professional social movement organizations.

© AP/ Wide World Photos

social movement, they help to maintain the sense of urgency necessary for sustained effort. Second, their anger and grievance propel them to more direct-action tactics (sit-ins, demonstrations, and the like) that publicize the cause and keep it on the national agenda.

The feminist movement is an excellent example of a social movement that combines both professional and indigenous SMOs. Informal networks continue to keep the discussion of equal rights and equal opportunities alive, even in periods when professional SMOs are nonexistent or marginalized. The most successful periods of feminist activism have been when professional SMOs, such as the National Organization for Women (NOW), worked in close cooperation with indigenous SMOs made up of informal networks and passionate individuals (Buechler 1993). In the absence of direct actions—candle-lit vigils for victims of wife abuse, boycotts of pornography stores, or equal rights rallies—pressure from both professional and indigenous SMOs can produce only modest results, at best.

When a movement is successful in reaching some of its major goals, it often becomes indistinguishable from any other humanitarian or political organization. For example, in the late nineteenth and early twentieth centuries, the League of Women Voters was considered a radical group because it fought for women's suffrage. Now it focuses mostly on getting out the vote and is considered middle-of-the-road. As new issues arise, however, a movement is likely to rely again on indigenous SMOs and on protests and demonstrations to gain new supporters or to reinvigorate existing rank-and-file members.

Mobilizing Resources

Mobilization is the process by which a unit gains significantly in the control of assets it did not previously control.

Mobilization is the process through which a social movement gains needed resources, of many types. These resources may be weapons, technologies, goods, money, or members. The resources available to a social movement depend on two factors: the amount of resources controlled by group members and the proportion of their resources that the members are willing to contribute to the movement. Thus, mobilization can proceed by increasing the size of the membership, increasing the proportion of assets that members are willing to give to the group, or recruiting richer members. Mobilization can also mean getting other organizations to work with a social movement. For example, the Civil Rights movement relied on aid from African American churches, and the antipornography movement has garnered support from both fundamentalist churches and feminist organizations.

Organizational factors also affect the odds that an SMO will succeed. Most importantly, SMOs must be able to mobilize sufficient resources to achieve their ends. Those resources can take many forms. During the spring of 2006, tens of thousands of high school students across the country walked out of their schools in protest against proposed anti-immigration legislation. These students were mobilized virtually overnight through text-messaging and cell phone calls—movement resources that Karl Marx never envisioned. In addition, SMOs are more likely to be successful when individuals must actively participate in the movement to derive any of the benefits from its victories. SMOs are also more likely to succeed if they have a centralized, bureaucratic structure; are able to avoid in-fighting; and cultivate alliances with other organizations (Gamson 1990; Marx 1971).

Frame Alignment

Political process theory has pointed to the importance of frame alignment for attracting and mobilizing new members. **Frame alignment** is the process that movements use to convince individuals that their interests, values, and beliefs are complementary to those of the SMO (Benford & Snow 2000; Snow et al. 1986). The Sierra Club, for example, might mail pamphlets to members of the Audubon Society in hopes of convincing them to join. It also might hold public meetings in a town plagued by pollution in hopes of convincing parents that their children's illnesses are caused by pollution, not by bad luck or bad genes. Other organizations, like cults and extremist groups, try to gain new members by convincing individuals that the way they have seen things is entirely wrong.

> **Frame alignment** is the process of convincing individuals that their interests, values, and beliefs are complementary to those of the social movement organization.

Who is most likely to be recruited through frame alignment? Studies of social movement activists show that, although ideology and grievances are important in bringing in new participants, the key factor is personal ties and networks. No matter how deeply committed an individual might be to a movement's ideology, they are not likely to become an active member unless they belong to a network of like-minded others. Conversely, they also are unlikely to become active if their friends, relatives, and acquaintances oppose the movement (McAdam 1986; McAdam & Paulsen 1993).

Effective Tactics

Finally, the particular tactics an SMO chooses can affect its chances of success. Researchers are divided on whether disruptive strategies such as strikes, boycotts, and violence can increase a movement's chances of success (Giugni 1998). However, innovative tactics are definitely useful for galvanizing public and media attention and sympathies (Marx 1971). For example, nonviolent sit-ins had a tremendous impact when first used during the Civil Rights movement. These days, however, they have little impact. Police now know how to deal with the tactic, and the media no longer find it newsworthy.

As this suggests, choice of tactics significantly affects media coverage, and media coverage can have a crucial impact on a movement's chances for success. The media can give or withhold publicity and can slant a story positively or negatively. Either

■ Although most of us view suicide bombings, like this attack in Jerusalem, as abhorrent, a small minority of social movements consider bombings a good way to gain attention and to spread their message.

© AP/ Wide World Photos

way, what the media choose to cover and how they cover it affect how the movement and its ideology are viewed by both participants and the public (e.g., Mulcahy 1995). As a result, SMOs work hard to get media coverage, through such tactics as staging dramatic events (such as hunger strikes, mass demonstrations, and suicide bombings) and timing those events to meet the deadline for the evening news on days when not much else is going on.

A particularly interesting example of this process occurred during China's Tiananmen Square uprising. Two major events in the spring of 1989 brought an unusually large contingent of international journalists to Beijing—the historic visit of Mikhail Gorbachev and the Asian Development Bank's first meeting in the People's Republic of China. Aware that the whole world would be watching and that state authorities would feel somewhat constrained from acting repressively, student activists used these journalists to create a global stage. Thus, foreign media played a crucial role in the movement, not only by reporting events to their audiences back home, but also by keeping Chinese citizens informed regarding movement developments (Zuo & Benford 1995). In both the West and East, then, media coverage appears to be a vital mechanism through which resource-poor organizations can generate public debate over their grievances.

Countermovements

A **countermovement** seeks to reverse or resist change advocated by an opposing social movement.

Countermovements are social movements that seek to reverse or resist changes advocated by an opposing movement (Lo 1982; Meyer & Staggenborg 1996). Countermovements can arise in response to any movement and can be either left-wing or right-wing.

Countermovements are most likely to develop if three conditions are met (Meyer & Staggenborg 1996). First, the original movement must have achieved moderate success. If the movement appears unsuccessful, then few will feel it worth their while to oppose it. Conversely, if the movement appears totally successful, then opposition will seem futile. Most tobacco smokers, for example, simply accepted new restrictions on smoking in the workplace rather than trying to resist them. On the other hand, when cities passed laws banning smoking in restaurants and bars, smokers realized that they had new allies: bar and restaurant owners who feared loss of customers. As a result, a countermovement has appeared to fight these laws.

Second, countermovements only arise when individuals feel that their status, power, or social values are threatened. This is most likely to happen if the original movement frames its goals broadly. The nineteenth-century temperance movement, which opposed all alcohol use, generated a strong countermovement. In contrast, the current movement against drunk driving, which identifies individual drunk drivers as the problem rather than alcohol consumption per se, has met almost no opposition.

Third, countermovements emerge when individuals who feel threatened by a new movement can find powerful allies. Those allies can come from within political parties, unions, churches, or any other important social group. Again, the alliance between smokers and bar owners is an example.

The conflict over abortion provides an excellent example of the interrelationship between movements and countermovements. The abortion rights movement of the 1960s was a quiet campaign, largely run by political elites—doctors, lawyers, and women active in mainstream political groups. For this reason, perhaps, it received little media coverage (Luker 1985). Its victory in the 1973 *Roe v. Wade* Supreme Court decision caught the country by surprise and galvanized the antiabortion movement (Meyer & Staggenborg 1996). That countermovement drew its supporters from women and men who believed that the legalization of abortion threatened religion, the stability of the family, and traditional ideas regarding women's nature and role. The antiabortion movement gained further support through highly visible, "newsworthy" actions that won media coverage for its views. In the years since *Roe v. Wade*, both the movement and the counter-movement have sought political allies—the pro-choice movement primarily within the Democratic Party and the antiabortion movement primarily within the Republican Party. Neither group, however, has yet achieved a decisive legal victory.

As these antiabortion and pro-choice protesters illustrate, whenever a social movement succeeds in creating social change, a countermovement is likely to develop.

Case Study: How the Environmental Movement Works

Being in favor of protecting the environment sounds like an innocuous position to take. After all, who is in favor of polluted air, dirty water, and disappearing species? Yet by default, nearly all of us are.

Ruining our environment is part of the status quo; it is part of our accepted way of life, of manufacturing and packaging merchandise, and of dealing with garbage. The average American produces 35 pounds of garbage each week, but recycles only a tiny fraction of this. Environmental protection, on the other hand, carries significant costs that few care to bear: higher-priced goods, more bother over recycling, more regulation, fewer consumer goods, and the loss of some jobs. Despite this apparent ill fit between environmentalism and modern life, the environmental movement continues to fight for its cause.

The Battle over Environmental Policy

This battle is being fought on many fronts—nuclear power, oil exploration in protected areas, hazardous wastes, forests, and suburban sprawl. Sometimes that battle takes extreme forms. "Mink liberators" in Utah have released animals from fur farms, bombed the fur breeder's cooperative that provides most of the food for the state's $20-million-a-year mink industry, and even set fire to a leather store. The Earth Liberation Front (ELF) announced that it fire-bombed and destroyed a $12-million mountaintop restaurant and ski-lift facility in 1998 to protect the last, best lynx habitat in Colorado (Glick 2001). Elsewhere, groups protesting suburban sprawl have set fire to sport utility vehicles and luxury home construction sites. Although many environmentalists disagree with this illegal sabotage, the spokesperson for one ELF cell says, "We know that the real 'ecoterrorists' are the white male industrial and corporate elite. They must be stopped" (Murr & Morganthau 2001).

Ecoterrorists who oppose suburban sprawl and the sale of gas-guzzling vehicles have taken actions such as spray-painting sport-utility vehicles and burning dealerships where SUVs are sold.

Although militants do much to publicize and galvanize the environmental movement, they cannot succeed by themselves. Arson, freeing animals, and bombing may buy time, but permanent victory in protecting forests, wildlife, and the rest of the environment involves court orders, legal battles, and other strategies. Thus, both professional and indigenous, conservative and radical SMOs help to push the movement forward.

The professional SMOs of the environmental movement—the Sierra Club, the Environmental Defense Fund, the National Audubon Society, and others—write letters to congressional representatives to urge support for clean-air laws or to lobby against dam projects or unrestrained suburban growth. They pay a battery of lawyers to get court injunctions when needed and to push for change in government policies. And, increasingly, they work with corporations to develop corporate policies that will protect the environment without hurting those corporations' bottom lines. For example, the Environmental Defense Fund prodded FedEx to use delivery trucks with hybrid fuel systems. This shift reduced air pollution, gasoline consumption, and FedEx's costs while burnishing the company's public image (Deutsch 2006; FedEx 2006).

The Environmental Movement Assessed

One reason corporations and federal agencies have adopted more environmentally friendly policies is that concern for the environment has increased markedly over the last two decades; most Americans now say they are willing to pay more taxes to clean up the environment.

Reflecting this growing public support, the environmental movement has had some notable successes. These include the rise in recycling, the establishment of new wilderness areas, and the passage of the Endangered Species Act. However, since the 1980s, increased antigovernment, antitax, and pro-business sentiment has dramatically limited economic and political support for environmental protection. Moreover, in a tight economic climate, Americans may be less willing to sacrifice economic growth for environmental benefits. If environmental protectionism starts to threaten the lifestyles and livelihoods of people other than mink farmers, resort owners, and luxury home builders, we will likely see more controversy rather than less (Cable & Cable 1995).

Technology

In social movements, individuals consciously aim to change their society. In other cases, people's intentions are more modest but may lead to great social change nonetheless. Such is the case with technology.

Technology is more pervasive than ever in our daily lives. Perhaps you woke up to an alarm this morning to find coffee already brewed in your preset electric coffeemaker, checked your cell phone for messages, and listened to MP3 files on your laptop, all before you made it to your first class. Technology is also more powerful and dangerous than ever before: The lethal power of a car or nuclear bomb is far greater than that of a horse-drawn cart or sword. It is vitally important, then, that we think about the social changes that technology can bring.

Technology is defined as the human application of knowledge to the making of tools and to the use of natural resources. It is important to note that the term *technology* refers not only to the tools themselves (material culture) but also to our beliefs, values, and attitudes toward them (nonmaterial culture). While we may be inclined

Technology involves the human application of knowledge to the making of tools and to the use of natural resources.

to think of technology in terms of today's "high-tech" advances, it also includes relatively simple tools such as pottery and woven baskets. Thus, technology has been a component of culture from the beginning of human society.

Because technology defines the limits of what a society can do, technological innovation is a major impetus to social change. As we saw in Chapter 4, technology helped to transform hunting, fishing, and gathering societies to horticultural, then agricultural, and then industrial societies. Currently, new technologies are developing to meet new needs created by a changing culture and society. The result is a never-ending cycle in which social change both causes and results from new technology. In this section, we briefly review two theories of technologically induced social change and present a case study of how information technology may change society. We then discuss the benefits and costs of two new technologies: information technology and reproductive technology.

Theoretical Perspectives on Technology and Social Change

Since the nineteenth century, sociologists have been interested in the link between technology and social change; as we saw in Chapter 1, many early scholars entered sociology because of their interest in the sources and consequences of the Industrial Revolution, an event that triggered dramatic social change. This section explores how structural functionalism and conflict theory explain the connections between technology and social change.

Structural-Functional Theory: Technology and Evolutionary Social Change

While structural-functional theory primarily asks how social organization is maintained in an orderly way, the theory does not ignore the fact that societies and cultures change. As pointed out in Chapter 1, according to the structural-functional perspective, change occurs through evolution: Social structures adapt to new needs and demands in an orderly way while outdated patterns, ideas, and values gradually disappear. Often, the new needs and demands that prompt this evolution are technological advances.

As noted in Chapter 2, sociologist William T. Ogburn (1922) added the concept of cultural lag to the idea of evolutionary change. Because the components of a society are interrelated, Ogburn reasoned, changes in one aspect of the culture invariably affect other aspects. The society will adapt, but only after some time has passed. As an illustration, Ogburn noted that by 1870, large numbers of U.S. industrial workers were being injured in factory accidents, but workers' compensation laws were not passed until the 1920s—a cultural lag of about 50 years. Ogburn pointed out that a society can hardly adapt to a new technology before it is introduced. Hence, cultural lag is a temporary period of maladjustment during which the social structure adapts to new technologies.

Conflict Theory: Technology, Power, and Social Change

While structural functionalism sees social change as orderly and generally consensual, conflict theorists contend that change—including the adoption of new technologies—results from conflict between competing interests. Furthermore, conflict theorists assert that those with greater power can direct technological and social change to their own advantage. In a process characterized by conflict and disruption, social structure changes (or does not change) as powerful groups act either to alter or to maintain the status quo.

According to Thorstein Veblen (1919), those for whom the status quo is profitable are said to have a vested interest in maintaining it. **Vested interests** represent stakes in either maintaining or transforming the status quo; people or groups who would suffer from social change have a vested interest in maintaining the status quo, while those who would profit from social change have a vested interest in transforming it. Electric companies have a vested interest in promoting electric cars; gas companies have a vested interest in impeding this. College students have a vested interest in downloading textbooks for free from the Internet; publishers have a vested interest in preventing this.

Just as the benefits of a particular technology are unevenly distributed, so also are the costs. Conflict theorists argue that costs tend to go to the less powerful. Pollution-producing factories, which can earn great profits for corporations, are typically located in poor neighborhoods and never located in places like Beverly Hills or Scarsdale.

Like evolutionary theories, the conflict perspective on social change makes intuitive sense to many, and there is empirical evidence to support it. A general assumption of the conflict perspective is that those with a disproportionate share of society's wealth, status, and power have a vested interest in preserving the status quo. In today's rapidly changing society, this may no longer be the case, as powerful factions may be just as likely to support as to oppose technological innovations. Microsoft, for example, is fully in favor of developing new technologies that it can profit from, like Windows XP, even while it works to impede innovations that others control, like Linux and Apple software and computers. Furthermore, some scholars have argued that technology is virtually "autonomous." That is, once the necessary supporting knowledge is developed, a particular invention—like the personal computer or the atomic bomb—will be created by someone. And once created, it will be used. In other words, technological changes may be put in motion by social forces beyond our effective control.

> **Vested interests** are stakes in either maintaining or transforming the status quo.

The Costs and Benefits of New Technologies

Almost all of us are glad that personal computers now exist: Their benefits are obvious, and the problems they create seem small by comparison. Far fewer of us are happy that the atomic bomb exists, although most Americans were happy that our government was able to use it during World War II.

As these examples suggest, new technologies always offer both benefits and costs, many of which are not immediately obvious. This section explores this idea through two examples of new technologies: new reproductive technologies and information technology. It then explores two general problems inherent in the expansion of technology: the technological imperative and "normal accidents."

New Reproductive Technologies

New reproductive technologies—some simple, some complex—have substantially expanded the options of women and men who want children genetically related to them. Men whose wives are infertile can have their sperm inseminated into another woman who agrees to serve as a "surrogate mother" (usually for a fee). Women whose husbands are infertile can be inseminated with another man's sperm. Women who cannot conceive can have their eggs surgically removed and mixed with sperm in a test tube. If those eggs become fertilized (a process known as in vitro fertilization), doctors can

Connections

Historical Note

In the same way that new reproductive technologies are changing society and social attitudes, older reproductive technologies deeply affected social life. Consider "the pill." Prior to the 1960s, having sex meant a good chance of having a baby. The development of birth control pills—inexpensive, reasonably safe, and very effective—meant that couples could now have sexual intercourse without fearing pregnancy. This brought both positive and negative changes: Individuals found it easier to relax and enjoy sex, and found it harder to justify saying no to sex.

surgically implant them in the woman's uterus in the hopes that one or more eggs will develop into a viable pregnancy. The same technology is used to enable women who lack viable eggs (including post-menopausal women) to bear children using another woman's eggs. These technologies are available not only to childless couples but also to single men and women and to gay and lesbian couples. In 2003, approximately 36,000 babies were born as a result of procedures in which both eggs and sperm were handled in the laboratory (Wright et al., 2006). An unknown additional number of babies were born when women were inseminated vaginally after taking prescription hormones or undergoing surgical procedures to restore their fertility

Although these reproductive technologies have increased childbearing options, some sociologists have raised concerns about their health, social, and ethical implications (Rothman 2000). The potential health problems are numerous. Women who take prescription hormones to increase their chances of conceiving risk breast cancer or ovarian cancer in the future. Other women face long, difficult, and potentially life-threatening pregnancies when these hormones leave them carrying twins, triplets . . . or even septuplets. The children they give birth to are disproportionately likely to be born prematurely, and as a result to have greater risks of a wide variety of lifelong cognitive and health problems. Finally, women who undergo surgical procedures for infertility face all the dangers inherent in any surgery.

The social and ethical problems implicit in new reproductive technologies are more subtle. Perhaps most important, some of these techniques have low success rates, especially with older women. Even those who eventually give birth typically have to endure several cycles of treatment costing around $10,000, with most of those cycles resulting in no pregnancies, miscarriages, or babies who die quickly. Yet, the constant development of new techniques makes it difficult for childless individuals and couples to decide to adopt or come to terms with childlessness. Finally, these technologies raise the question of whether we are turning children into commodities available to the highest bidder; they also may encourage a narrow definition of parenthood as having genetic ties to a child rather than a broader definition of parenthood as loving and raising a child.

Information Technology

Consider the college student in 1974 who is assigned the task of writing a term paper on the consequences of parental divorce. She goes to the library and walks through the periodicals section until she stumbles on the *Journal of Marriage and the Family*, in which she eventually finds five articles—the number her professor requires—on her topic. She takes notes on three-by-five-inch cards (there are no photocopying machines) and goes home to draft her paper on her new electric typewriter. She cuts and tapes together her draft copy, moving sections around until it looks good, checks words of dubious spelling in her dictionary, and then retypes a final copy. She uses carbon paper to make a copy for herself. (Ask your mom or dad to explain this to you.) When she makes a mistake, she erases it carefully and tries to type the correction in the original space.

Now consider the student in 2004. This student starts her paper by logging onto *Sociological Abstracts*, an online bibliography of more than 100,000 sociology articles. When she enters the keywords *divorce* and *parental*, the program prints out full citations and a summary for 41 articles. After identifying, downloading, and/or photocopying the 5 articles she wants, she drafts a report on her computer, edits it to her satisfaction, runs it through her spelling checker, and adjusts the vocabulary a bit by using the built-in thesaurus. She also runs the report through her grammar checker,

Because today's computers are better and cheaper than those of even 10 years ago, a very large portion of all college students bring their own to campus with them. In fact, some colleges now require students to have their own notebook computer.

which will catch errors in punctuation, capitalization, and so forth. Finally, she sends the whole thing to her mother (who lives 2,000 miles away) by e-mail and asks her to read it for logic and organization. She receives the edited version from her mother in an hour, prints two copies, and hands in the report. Or she may send the paper to her instructor via e-mail or fax.

Information technology—computers and telecommunication tools for storing, using, and sending information—has changed many aspects of our daily lives. Over the past few decades, the United States has become an "information society." More and more workers are employed in information acquisition, processing, and communication. Aside from enabling us to write term papers more easily, how will information technology change our lives? Will it reduce or increase social class inequality? Will it make life safer and better? Or will it make life more stressful and isolated?

The answer is likely to be some of each. As shown in Map 15.1, access to computers is spreading around the world—if unevenly. This means we can link via computer to distant family and friends, to doctors and medical information, to libraries and databanks, and to world events. During the war in Iraq, for example, U.S. soldiers stay in touch with their families via e-mail, and American citizens discuss events in Internet chat rooms and follow events on 24-hour satellite and cable news stations (some broadcast from Europe or the Arab world). Iraqi citizens, meanwhile, can use their computers to find out both where the most recent bombs exploded and who won the Academy Awards. Information technology also allows us to participate more fully in the political process by making it possible to communicate more effectively and directly with our elected officials. By linking us to distant work sites, computers and e-mail allow employees to work from their homes, reducing the time they spend commuting to work and increasing the time they can spend with their families.

On the downside, advances in information technology have introduced new forms of crime (hacking and electronic theft), new defense worries (breaches of defense data systems and faulty software programs that may inadvertently launch World War III), new health problems (eye strain and repetitive stress injuries), and new inefficiencies ("I'm sorry, the system is down"). They also have introduced new

Information technology comprises computers and telecommunication tools for storing, using, and sending information.

MAP 15.1
Personal Computers per 1,000 People
SOURCE: International Bank for Reconstruction and Development 2002.

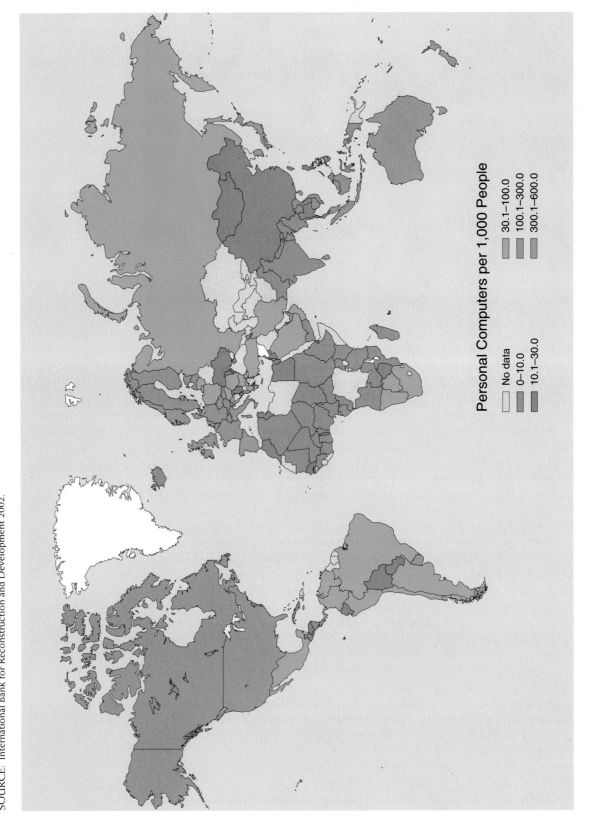

Personal Computers per 1,000 People

No data
0–10.0
10.1–30.0
30.1–100.0
100.1–300.0
300.1–600.0

Media and Culture:

focus on

Information Technology and Global Culture

No culture exists in isolation. Much of the content of U.S. culture, for instance, was acquired from successive waves of immigrants, each bringing their own language, values, and customs to American shores. In the early twenty-first century, one of the primary vehicles for cultural diffusion is the new information technology. The information revolution, which began in the United States, is rapidly spreading around the world. Although huge disparities persist, virtually every nation on every continent now has some access to the Internet and to e-mail technology. Between 1999 and 2002, the number of Internet users more than doubled, from 276 million to 605 million, and the number of people who had access to personal computers increased from 394 million to 550 million (Dutt, Lanvin, & Paua 2004).

Not only has the use of information technology spread throughout the world, but the fact that it has done so means that other aspects of cultural diffusion also take place much more rapidly. Ideas can sweep the world within days and be introduced into the most remote villages within weeks and months. A fervor for democracy, for example, swept the world in 1989. The year began with prodemocracy student protests in Tiananmen Square in Beijing, China, and ended with the fall of the Berlin Wall and the toppling of communist governments in Eastern Europe. Many of those seeking freedom and democracy relied on the ideals and symbols of the American Revolution; the Statue of Liberty lent symbolic support to the demonstration at Tiananmen Square. By the same token, Islamic fundamentalism is spreading globally in part through Internet chat rooms, Web sites, and the like.

The spread of information technology not only means more rapid change in all areas of life, it also means growing international similarity. In Moscow, Beijing, Nairobi, and Boston, business leaders are wearing the same kinds of suits. In the western United States and in Africa, plans to develop virtual universities are well underway. Throughout the world, young people are listening to the same kinds of music. White Americans now enjoy hip hop and salsa dancing, and Russian youth listen to *khard-roka* and *khevimetallu* (i.e., hard rock and heavy metal).

Information technology spreads not only information, but also ideas. Student protests in China's Tiananmen Square, for example, drew on ideals and symbols of American democracy such as the Statue of Liberty.

forms of social control. Information technology has given corporations, the police, lawyers, and government bureaucrats, among others, greater ability to build databases about you, combining information on the cars you buy, the Websites you visit, and the type of music you like with whether or not you have recently married, moved, had a child, or gotten a speeding ticket. Similarly, others now can obtain access to your computer files, deleted e-mail messages, and phone logs. One survey of 1,000 major corporations showed that almost two thirds engage in some form of "electronic surveillance" of their employees (Rosen 2000). Finally, new technologies have lengthened the number of working hours in a day, as notebook computers, e-mail, faxes, and cell phones increasingly invade our homes and even vacations.

The long-term effect of information technology on society will depend as much on social institutions as it does on the technological capacities of computers and telecommunications. Information technology offers us more freedom of residence and more input into local and federal legislative bodies, but we simultaneously lose some privacy and autonomy. Whether the blessings or costs will predominate will depend on how these technologies are implemented in schools, workplaces, and government bureaucracies. To the extent that they affect relationships among work, class, neighborhood, and family, the new technologies are of vital interest to those concerned with social institutions.

Making the best use of advancing technology and helping ensure that advances prompt desirable social changes requires social planning—the conscious and deliberate process of investigating, discussing, and coming to agreement about desirable actions based on common values.

The Technological Imperative

As we've already noted, once the knowledge needed to devise a certain technology is available, that new technology is likely to appear and to gain adherents. But we can make an even stronger statement: Once that technology is available, it becomes more and more difficult for anyone to decide against using it.

Whether working at home or relaxing on a park bench, technologies such as Palm Pilots keep us connected to our office and the world of work.

© Joel Gordon

Consider the automobile. In 1925, any city-dweller who had enough money could choose to commute to work by car. But if he chose not to do so, he could rely on a broad network of trolleys running on a frequent schedule to get him to his destination. He almost certainly lived fairly close to where he worked and could also choose to enjoy the walk instead. These days, the automobile has become completely enmeshed in our way of life. Billions of dollars in public subsidies pay for road building and parking lots and keep down the price of oil and gas for consumers. Meanwhile, public transportation has been cut to the bone. In many cities, walking or bicycling is dangerous or unpleasant because of high-speed traffic or freeways that divide neighborhoods.

This situation is an example of the **technological imperative:** the idea that once a technology is available, it becomes difficult to avoid using it. Think how annoyed people sometimes feel when their friends don't use cell phones, e-mail, or instant messaging and the pressures on holdouts to get these technologies.

Technological imperative refers to the idea that once a technology becomes available, it becomes difficult to avoid using it.

Normal Accidents

As our lives come increasingly to depend on highly complex and interconnected technologies, our vulnerability to technological problems increases exponentially. In the nineteenth century, most people got water from wells and used candles for lighting. If a well dried up or a house burned down, the disaster was limited to no more than a few households. Now we get our water from municipal water systems and our electricity from electric companies. When things go wrong, they go wrong big time.

The blackout of August 2003 provides a perfect example of this vulnerability. Electricity is provided to American households by a network of cooperating utility companies sharing a vast grid of electric cables. This grid is made possible by complex computerized technologies, designed to spread the demand over a broad region and reduce the chance of overloading the system in any one region. But because of its complexity and interconnectedness, a small problem can quickly mushroom to a huge problem for a huge area. This is what happened in 2003, when overloaded circuits in the Midwest resulted in 50 million people throughout the Midwest, Northeast, and even Canada losing electric power for up to several hours.

The blackout of August 2003, which shut down most train service in the northeast and left commuters reliant on busses, illustrated how dependent we have become on technology and how vulnerable we are when technology fails.

© Reuters/CORBIS

Normal accidents are accidents that can be expected to happen sooner or later, no matter how many safeguards are built into a system, simply because the system is so complex.

The blackout left people with a new awareness of how dependent we have become on technology, and how vulnerable we are when that technology fails. Because the system for distributing water to consumers runs on electricity, the blackout left thousands without water. Flashlight batteries ran down, leaving people with only candles for lighting. Many found themselves with no way of communicating with friends and relatives. Laptops and PDAs quickly ran out of power, while cell-phone networks either lost power or became overloaded, so people could neither phone nor e-mail. Even those who had working phones could not telephone others if they kept their phone directories on computers.

This process is an example of a "normal accident." **Normal accidents** are accidents that can be expected to happen sooner or later, no matter how many safeguards are built into a system, simply because the system is so complex (Perrow 1984). Normal accidents such as space shuttle crashes, accidental releases of radiation from nuclear power plants, and electric blackouts are the price we pay for modern technology.

Where This Leaves Us

Whether it originates in a social movement or in a new technology, any social change will have opponents. Every winner potentially produces a loser. This means that change creates a situation of competition and conflict.

In Chapter 1, we discussed the appropriate role of sociologists in studying social issues. Should they be value free, or should they take a stand? Issues of social change and conflict bring this question into sharp focus. Although most sociologists restrict their work to teaching and researching, a vocal minority argue that sociologists should take a more active role in monitoring and even creating social change. They believe that sociologists should be actively involved in helping individuals understand and resolve the conflicts that arise from competition, inequality, and social change.

What can sociologists contribute to ensure that social changes enhance social justice? A few of the areas that can be pursued include the following:

1. *The study of conflict resolution.* A growing number of universities have special courses or programs on conflict resolution. These courses are concerned with the development of techniques for handling disputes and negotiating peaceful settlements that can lead to positive social changes. Sociological research on topics such as small-group decision making and organizational culture are relevant here.

2. *Developing social justice perspectives.* At its core, sociology is concerned with the interaction of social groups and the role that power plays in those interactions. In their research and teaching, sociologists can explore how individuals, groups, and nations obtain and use power and how that power can be distributed and used more equitably.

2. *Modeling social change strategies.* Sociological research may lead to the development of more effective programs for improving the well-being of individuals and social groups, from Head Start programs to transnational investments.

The involvement of sociologists in issues of conflict resolution, social justice, and social change is not likely to be the crucial factor that creates a better world. We can be sure, however, that scholarly neglect of these issues is both shortsighted and immoral. To the extent that developing knowledge of the principles of human behavior will help us reduce social conflict, we have an obligation—as scholars, students, and citizens—to seek out knowledge and to apply it. Our future depends on this.

Summary

1. Collective behavior and social movements, although related, are distinct activities. Collective behavior is spontaneous and unplanned; a social movement is organized, goal-oriented, and long term.

2. According to relative-deprivation theory, social movements arise when individuals experience an unacceptable gap between what they have and what they expect to have. Expectations are derived from comparisons with other groups and other points in time.

3. Resource mobilization theory argues that social movements emerge when individuals are able to bring the necessary resources to effectively work together toward social change.

4. Political process theory builds on resource mobilization theory by recognizing that in addition to access to political opportunities and resources, successful movements must build a sense among participants that change is both needed and possible.

5. A successful movement needs a diverse range of organizations to accomplish different goals. It also must be able to mobilize needed resources of all sorts. To get new members, it must frame its ideology in ways that convince individuals that a problem is serious, that taking action on a problem is both proper and effective, and that individuals' interests, values, and beliefs mesh well with those of the movement. Regardless of ideology, however, individuals are most likely to be recruited when they have social ties to movement members and lack ties to movement opponents. Finally, successful movements need innovative tactics that will garner media attention.

6. Countermovements are social movements that seek to resist or reverse changes advocated by other social movements. A countermovement is most likely to develop if the original movement achieves modest success, if some individuals feel that their social position or values are threatened by changes achieved by the original movement, and if potential countermovement participants believe that they will have powerful allies.

7. In its effort to affect public policy, the environmental movement uses a variety of tactics, ranging from courtroom battles to sabotage. Among the reasons for the movement's growing successes are the wide variety of SMOs within the movement.

8. Technology is the human application of knowledge to the making of tools and hence to humans' use of natural resources. The term refers not only to the tools themselves (aspects of material culture) but also to people's beliefs, values, and attitudes regarding those tools (aspects of nonmaterial culture).

9. Social change is any significant modification or transformation of social structures or institutions over time. Technology is one important type and cause of social change.

10. Structural-functional theory primarily asks how technology contributes to orderly and positive social change. Cultural lag can be a serious problem when a technology enters a society too quickly for the culture to adapt to the changes it brings.

11. Conflict theorists contend that technological change results from and reflects conflict between competing interests. People or groups who would either suffer or profit from social change have vested interests—stakes in either maintaining or transforming the status quo.

12. Information technology has changed many aspects of our daily lives. It links us to people and information but has also created new defense worries, new inefficiencies, new forms of social control, and new illnesses and injuries. Similarly, new reproductive technologies have expanded the options of those who want children genetically related to them. At the same time, they have raised serious health, social, and ethical questions, such as whether we are turning children into commodities available to the highest bidder.

Thinking Critically

1. What social structural conditions in the larger society do you think helped spark the environmental movement? What countermovements do you know of that may impact the movement's success?
2. Suppose you were interested in mobilizing public opinion against the death penalty. What kind of activity or event would you try to use to get the media's attention?
3. How would you analyze the current debate over affirmative action policies and programs in terms of various groups' vested interests?
4. Europeans have opposed genetically modified plants and food much more vigorously than have Americans. How would you explain this difference based on your understanding of the factors that make societies more or less

likely to adopt new technologies and attitudes (see Chapter 2) and on your understanding of how social movements are able to successfully mobilize?

5. If you were to run for office, how would you use e-mail in your campaign? Which groups of your constituents would you be more likely to hear from via the Internet? How would you know whether they were actually U.S. citizens with the legal right to vote—or would it matter? How might you make sure that other voices, those without high-speed data ports and modems, were heard as well?

Companion Website for This Book

academic.cengage.com/sociology/brinkerhoff
Gain an even better grasp on this chapter by going to the Companion Website. This resource contains tutorial quizzes and flash cards to help you master key terms and concepts.

Suggested Readings

Cable, Sherry, and Cable, Charles. 1995. *Environmental Problems, Grassroots Solutions: The Politics of Grassroots Environmental Conflict.* New York: St. Martin's Press. A description of the environmental movement with details on how various regional, grassroots organizations address environmental issues and concerns.

Marx, Gary T., and McAdam, Douglas. 1994. *Collective Behavior and Social Movements: Structure and Process.* Englewood Cliffs, N.J.: Prentice-Hall. A concise, well-written review of the major ideas and issues in the study of collective behavior and social movements.

Morris, Aldon. 1984. *The Origins of the Civil Rights Movement.* New York: Free Press. A powerful account of the modern civil rights movement. Utilizing personal interviews and original documents, Morris draws particular attention to the importance of local African American community groups, in addition to the efforts of national leaders and organizations.

Rothman, Barbara Katz. 1998. *Genetic Maps and Human Imaginations: The Limits of Science in Understanding Who We Are.* New York: W. W. Norton. A thoughtful, provocative, and passionate collection of essays on the social implications of the Human Genome Project.

Stoll, Clifford. 1999. *High Tech Heretic.* New York: Doubleday. Soll, a legendary Internet pioneer and committed computer geek, challenges the idea that computers can replace good classroom teaching. Stoll also argues that the emphasis on computers has deprived teachers and students of support and funds for the things they really need.

Glossary

A

Accommodation occurs when two groups coexist as separate cultures in the same society.

Accounts are explanations of unexpected or untoward behavior. They are of two sorts: excuses and justifications.

Acculturation is the process of adopting some parts of the majority group culture while still retaining valued aspects of a minority culture.

Achieved status is a status that one works to obtain, such as being college-educated.

Alienation occurs when workers have no control over the work process or the product of their labor.

Anomie is a situation in which the norms of society are unclear or no longer applicable to current conditions.

Anticipatory socialization is role learning that prepares us for roles we are likely to assume in the future.

An **ascribed status** is fixed by birth and inheritance and is unalterable in a person's lifetime.

Assimilation is the process through which individuals learn and adopt the values and social practices of the dominant group within a society.

Authoritarian systems are political systems in which the leadership is not selected by the people and legally cannot be changed by them.

Authoritarianism is the tendency to be submissive to those in authority, coupled with an aggressive and negative attitude toward those lower in status.

Authority is power supported by norms and values that legitimate its use.

B

The **bourgeoisie** is the class that owns the tools and materials for their work—the means of production.

Bureaucracy is a special type of complex organization characterized by explicit rules and hierarchical authority structure, all designed to maximize efficiency.

C

Capitalism is the economic system, based on competition, in which most wealth (land, capital, and labor) is private property, to be used by its owners to maximize their own gain.

Caste systems rely largely on ascribed statuses as the basis for distributing scarce resources.

Charisma refers to extraordinary personal qualities that set an individual apart from ordinary mortals.

Charismatic authority is the right to make decisions based on perceived extraordinary personal characteristics.

Churches are religious organizations that have become institutionalized. They have endured for generations, are supported by and support society's norms and values, and have become an active part of society.

Civil religion is the set of institutionalized rituals, beliefs, and symbols sacred to the U.S. nation.

Class, in Marxist theory, refers to a person's relationship to the means of production.

Class consciousness occurs when people understand their relationship to the means of production and recognize their true class identity.

Class systems rely largely on achieved statuses as the basis for distributing scarce resources.

Coercion is the exercise of power through force or the threat of force.

Cohesion in a group is characterized by high levels of interaction and by strong feelings of attachment and dependency.

Collective behavior is nonroutine action by an emotionally aroused gathering of people who face an ambiguous situation.

Collective efficacy refers to the extent to which individuals in a neighborhood share the expectation that neighbors will intervene and work together to maintain social order.

A **community** is a collection of individuals characterized by dense, cross-cutting social networks.

Competition is a struggle over scarce resources that is regulated by shared rules.

Complex organizations are large formal organizations with complex status networks.

Conflict is a struggle over scarce resources that is not regulated by shared rules; it may include attempts to destroy, injure, or neutralize one's rivals.

Conflict theory addresses the points of stress and conflict in society and the ways in which they contribute to social change.

Consumerism is the philosophy that says "buying is good" because "we are what we buy."

A **control group** is the group in an experiment that does not receive the independent variable.

Cooperation is interaction that occurs when people work together to achieve shared goals.

Core societies are rich, powerful nations that are economically diversified and relatively free of outside control.

Correlation refers to a statistical relationship between two variables (for example, income increases when education increases).

Countercultures are groups that have values, interests, beliefs, and lifestyles that are opposed to those of the larger culture.

Countermovements are movements that seek to reverse or resist change advocated by another social movement.

Credentialism is the assumption that some are better than others simply because they have a particular educational credential.

Crime is behavior that is subject to legal or civil penalties.

Cross-sectional design uses a sample (or cross section) of the population at a single point in time.

Crude birth rate refers to the number of live births per 1,000 persons.

Crude death rate refers to the number of deaths per 1,000 persons.

A **cult** is a sectlike religious organization that is independent of the religious traditions of its society.

Cultural capital refers to attitudes and knowledge that characterize elite culture.

Cultural diffusion is the process by which aspects of one culture or subculture enter and are incorporated into another.

Cultural lag occurs when one part of culture changes more rapidly than another.

Cultural relativity requires that each cultural trait be evaluated in the context of its own culture.

Culture is the total way of life shared by members of a community. It includes not only language, values, and symbolic meanings but also technology and material objects.

The **culture of poverty** is a set of values that emphasizes living for the moment rather than thrift, investment in the future, or hard work.

Culture shock refers to the discomfort that arises from exposure to a different culture.

D

Deduction is the process of moving from theory to data by testing hypotheses drawn from theory.

The **deinstitutionalization of marriage** refers to the gradual erosion of social norms that stress the need for marriage and dictate how spouses should behave.

Democracies are political systems that provide regular, constitutional opportunities for a change in leadership according to the will of the majority.

Demographic transition refers to the shift from a society characterized by high birth rates and low life expectancies to one characterized by low birth rates and high life expectancies.

Demography is the study of population—its size, growth, and composition.

The **dependent variable** is the effect in cause-and-effect relationships. It is dependent on the actions of the independent variable.

Deterrence theories suggest that deviance results when social sanctions, formal and informal, provide insufficient rewards for conformity.

Development refers to the process of increasing the productivity and standard of living of a society—longer life expectancies, more adequate diets, better education, better housing, and more consumergoods.

Deviance refers to norm violations that exceed the tolerance level of the community and result in negative sanctions.

Dialectic philosophy views change as a product of contradictions and conflict between the parts of society.

A **differential** is a difference in the incidence of a phenomenon across social groups.

Differential association theory argues that people learn to be deviant when more of their associates favor deviance than favor conformity.

Disclaimers are verbal devices used to ward off doubts and negative reactions that might result from one's conduct.

Discrimination is the unequal treatment of individuals on the basis of their membership in categories.

Divorce probability is the estimated probability that a marriage will ever end in divorce within a given time period.

The **divorce rate** is calculated as the number of divorces each year per 1,000 married women.

Double jeopardy means having low status on two different dimensions of stratification.

Dramaturgy is a version of symbolic interaction that views social situations as scenes manipulated by the actors to convey the desired impression to the audience.

A **dual economy** consists of the complex giants of the industrial core and the small, competitive organizations that form the periphery.

Dysfunctions are consequences of social structures that have negative effects on the stability of society.

E

Economic determinism means that economic relationships provide the foundation on which all other social and political arrangements are built.

Economic institutions are social structures concerned with the production and distribution of goods and services.

Edge cities are suburban cities that now have an existence largely separate from the cities that spawned them.

The **educational institution** is the social structure concerned with the formal transmission of knowledge.

Emerging churches are characterized by 1) the belief that American life and modern Christian churches are atomized, bureaucratic, and inauthentic and 2) an emphasis on informal rituals, a more open perspective toward scripture and behavior, and living a life of mission, faith, and community.

Emotional labor refers to the work of smiling, appearing happy, or in other ways suggesting that one enjoys providing a service.

Empirical research is research based on critical and systematic examination of the evidence.

Endogamy is the practice of choosing a mate from within one's own racial, ethnic, or religious group.

Environmental racism refers to the disproportionately large number of health and environmental risks faced by minorities.

An **ethnic group** is a category whose members are thought to share a common origin and to share important elements of a common culture.

Ethnocentrism is the tendency to view the norms and values of our own culture as standards against which to judge the practices of other cultures.

Ex-felon disenfranchisement is the loss of voting privileges suffered by those who have been convicted of a felony. In some states, felon disenfranchisement applies only to those in prison; in other states, it is lifelong.

Exchange is voluntary interaction from which all parties expect some reward.

Excuses are accounts in which one admits that the act in question is wrong or inappropriate but claims one couldn't help it.

The **experiment** is a method in which the researcher manipulates independent variables to test theories of cause and effect.

An **experimental group** is the group in an experiment that experiences the independent variable. Results for this group are compared with those for the control group.

An **extended family** is a family in which a couple and their children live with other kin, such as the wife's or husband's parents or siblings.

Extrinsic rewards are tangible benefits such as income and security.

F

False consciousness is a lack of awareness of one's real position in the class structure.

The **family** is a relatively permanent group of persons linked together in social roles by ties of blood, adoption, marriage or quasi-marital commitment and who live together and cooperate economically and in the rearing of children.

Fertility is the incidence of childbearing.

Folkways are norms that are the customary, normal, habitual ways a group does things.

Formal social controls are administrative sanctions such as fines, expulsion, or imprisonment.

A **frame** is an answer to the question, What is going on here? It is roughly identical to a definition of the situation.

Frame alignment is the process of convincing individuals that their interests, values, and beliefs are complementary to those of a social movement.

Functions are consequences of social structures that have positive effects on the stability of society.

Fundamentalism refers to religious movements that stress traditional interpretations of religion and the importance of living in ways that mesh with those traditional interpretations.

G

Gemeinschaft refers to society characterized by the personal and permanent ties associated with primary groups.

Gender refers to the expected dispositions and behaviors that cultures assign to each sex.

Gender roles refer to the rights and obligations that are normative for men and women in a particular culture.

The **generalized other** is the composite expectations of all the other role players with whom we interact; it is Mead's term for our awareness of social norms.

Gesellschaft refers to society characterized by the impersonal and instrumental ties associated with secondary groups.

Globalization refers to the process through which ideas, resources, practices, and people increasingly operate in a worldwide rather than local framework.

Globalization of culture is the process through which cultural elements (including musical styles, fashion trends, and cultural values) spread around the globe.

A **group** is two or more people who interact on the basis of shared social structure and recognize mutual dependency.

The **guinea-pig effect** occurs when subjects' knowledge that they are participating in an experiment affects their response to the independent variable.

H

The **health belief model** proposes that individuals will be most likely to adopt healthy behaviors if (1) they believe their health is at risk, (2) they believe the risk is a serious one, (3) they believe that changing their behaviors would significantly reduce those risks, and (4) they face no significant barriers that would make changing their behaviors difficult.

The **hidden curriculum** consists of the underlying cultural messages taught by schools. Both public and private schools teach young people to accept inequality.

High culture refers to the cultural preferences associated with the upper class.

Homogamy is the tendency to choose a mate similar in various ways to oneself. Marrying someone of your social class is known as social homogamy; marrying within your ethnic or religious group is a type of homogamy known as endogamy.

Homosexuals (also known as gays and lesbians) are people who prefer sexual and romantic relationships with members of their own sex.

A **hypothesis** is a statement about relationships that we expect to find if our theory is correct.

I

The **I** is the spontaneous, creative part of the self.

The **id** is the natural, unsocialized, biological portion of self, including hunger and sexual urges.

Identity work is the process of managing identities to support and sustain our self-esteem.

An **ideology** is a set of beliefs that strengthen or support a social, political, economic, or cultural system.

Immigration is the permanent movement of people into another country.

Incidence is the frequency with which an attitude or behavior occurs.

The **independent variable** is the cause in cause-and-effect relationships.

The **indirect inheritance model** argues that children have occupations of a status similar to that of their parents because the family's status and income determine children's aspirations and opportunities.

Induction is the process of moving from data to theory by devising theories that account for empirically observed patterns.

The **informal economy** is the part of the economy that largely escapes state regulation; also known as the underground economy.

Informal social control is self-restraint exercised because of fear of what others will think.

Information technology comprises computers and telecommunication tools for storing, using, and sending information.

Institutionalized racism occurs when the normal operation of apparently neutral processes systematically produces unequal results for majority and minority groups.

Institutions are enduring social structures that meet basic human needs.

Insurgent consciousness is the individual sense that change is both needed and possible.

Internalization occurs when individuals accept the norms and values of their group and make conformity to these norms part of their self-concept.

Intrinsic rewards are rewards that arise from the process of work; they include enjoying the people you work with and pride in your creativity and accomplishments.

J

Justifications are accounts that explain the good reasons the violator had for choosing to break the rule; often they are appeals to some alternate rule.

L

Labeling theory is concerned with the processes by which labels such as *deviant* come to be attached to specific people and specific behaviors.

Latent functions or dysfunctions are consequences of social structures that are neither intended nor recognized.

Laws are rules that are enforced and sanctioned by the authority of government. They may or may not be norms.

Least-developed countries are those nations that are characterized by poverty and political weakness and that are considerably behind on every measure of development.

Less-developed countries are those nations that have lower living standards than the most-developed countries but are substantially better off than the least-developed nations.

The **linguistic relativity hypothesis** argues that the grammar, structure, and categories embodied in each language affect how its speakers see reality.

Longitudinal research is any research in which data are collected over a long period of time.

The **looking-glass self** is the process of learning to view ourselves as we think others view us.

M

Macrosociology focuses on social structures and organizations and the relationships between them.

A **majority group** is a group that is culturally, economically, and politically dominant.

Manifest functions or dysfunctions are consequences of social structures that are intended or recognized.

The **manufacturers of illness** are groups that promote and benefit from deadly behaviors and social conditions.

Marriage is an institutionalized social structure that provides an enduring framework for regulating sexual behavior and childbearing.

McDonaldization is the process by which the principles of the fast-food restaurant—efficiency, calculability, predictability, and control—are coming to dominate more sectors of American society.

The **me** represents the self as social object.

The **medical model of mental illness** holds that mental illness can be objectively defined, has identifiable causes, and if left untreated will become worse.

Medicalization refers to the process through which a condition or behavior becomes defined as a medical problem requiring a medical solution.

The **medicalization of deviance** refers to the process of redefining "badness" or oddness into illness.

A **metropolitan area** is a county that has a city of 50,000 or more in it plus any neighboring counties that are significantly linked, economically or socially, with the core county.

Microsociology focuses on interactions among individuals.

A **minority group** is a group that is culturally, economically, and politically subordinate.

Mobilization is the process by which a unit gains significantly in the control of assets it did not previously control.

Modernization theory sees development as the natural unfolding of an evolutionary process in which societies go from simple to complex economies and institutional structures.

Monogamy is the term for marriages in which there is only one wife and one husband.

Moral entrepreneurs are people who attempt to create and enforce new definitions of morality.

Mores are norms associated with fairly strong ideas of right or wrong; they carry a moral connotation.

Mortality is the incidence of death.

Most-developed countries are those rich nations that have relatively high degrees of economic and political autonomy.

N

Net migration is the number of people who move into an area minus the number who move out.

A **nonmetropolitan area** is a county that has no major city in it and is not closely tied to such a city.

The **norm of reciprocity** is the expectation that people will return favors and strive to maintain a balance of obligation in social relationships.

Normal accidents are accidents that can be expected to happen sooner or later, no matter how many safeguards are built into a system, simply because the system is so complex.

Norms are shared rules of conduct that specify how people ought to think and act.

A **nuclear family** is a family in which a couple and their children form an independent household living apart from other kin.

O

An **operational definition** describes the exact procedure by which a variable is measured.

Organizational culture refers to the pattern of norms and values that structures how business is actually carried out in an organization.

P

Participant observation includes a variety of research strategies—participating, interviewing, observing—that examine the context and meanings of human behavior.

Peripheral societies are poor and weak, with highly specialized economies over which they have relatively little control.

Political economy refers to the interaction of political and economic forms within a nation.

Political institutions are concerned with the social structure of power; the most prominent political institution is the state.

Political opportunities are resources that allow a social movement to grow; they include preexisting organizations that can provide the new movement with leaders, members, phone lines, copying machines, and other resources.

Political process theory suggests that social movements develop when political opportunities are available and when individuals have developed a sense that change is both needed and possible.

Polygamy is any form of marriage in which a person may have more than one spouse at a time.

Popular culture refers to aspects of culture that are widely accessible and commonly shared by most members of a society, especially those in the middle, working, and lower classes.

Power is the ability to direct others' behavior even against their wishes.

The **power elite** comprises the people who occupy the top positions in three bureaucracies—the military, industry, and the executive branch of government—and who are thought to act together to run the United States in their own interests.

Prejudice is an irrational, negative attitude toward a category of people.

Primary groups are groups characterized by intimate, face-to-face interaction.

Primary production is extracting raw materials from the environment.

Primary socialization is personality development and role learning that occurs during early childhood.

Privatization refers to two processes: 1) the increasing "farming out" of government services to private, capitalistic firms and 2) the increasing redesign of government services to operate more like corporate businesses.

The **profane** represents all that is routine and taken for granted in the everyday world, things that are known and familiar and that we can control, understand, and manipulate.

Professional socialization is role learning that provides individuals with both the knowledge and a cultural understanding of their profession.

Professions are occupations that demand specialized skills and creative freedom.

The **proletariat** is the class that does not own the means of production. They must support themselves by selling their labor to those who own the means of production.

Propinquity is spatial nearness.

R

A **race** is a category of people treated as distinct because of *physical* characteristics to which *social* importance has been assigned.

Racism is the belief that inherited physical characteristics associated with racial groups determine individuals' abilities and are a legitimate basis for unequal treatment.

Random samples are samples chosen through a random procedure, such as tossing a coin, that ensures that every individual within a given population has an equal chance of being selected for the sample.

The **rate of natural increase** is the crude birth rate minus the crude death rate, calculated as a percentage.

Rational-legal authority is the right to make decisions based on rationally established rules.

Relative-deprivation theory argues that social movements arise when people experience an intolerable gap between their rewards and what they believe they have a right to expect; it is also known as breakdown theory.

Religion is a system of beliefs and practices related to sacred things that unites believers into a moral community.

Religiosity is an individual's level of commitment to religious beliefs and to acting on those beliefs.

Religious economy refers to the competition between religious organizations to provide better "consumer products," thereby creating greater "market demand" for their own religions.

Replacement level fertility—about 2.1 births per woman—is the rate of fertility needed to maintain a population at any given size.

Replication is the repetition of empirical studies with another investigator or a different sample to see if the same results occur.

Reproductive labor refers to traditionally female tasks such as cooking, cleaning, and nurturing that make it possible for a society to continue and for others to work and play.

Resocialization is the process of learning to replace one's self-concept and way of life with radically different ones.

Resource mobilization theory suggests that social movements develop when individuals who experience deprivation pull together the resources they need to mobilize for action.

Rites of passage are formal rituals that mark the end of one age status and the beginning of another.

A **role** is a set of norms specifying the rights and obligations associated with a status.

Role conflict is when incompatible role demands develop because of multiple statuses.

Role exit is the process by which individuals leave important social roles.

Role identity is the image we have of ourselves in a specific social role.

Role strain is when incompatible role demands develop within a single status.

Role taking involves imagining ourselves in the role of others in order to determine the criteria others will use to judge our behavior.

S

The **sacred** consists of events and things that we hold in awe and reverence—what we can neither understand nor control.

Sampling is the process of systematically selecting representative cases from the larger population.

Sanctions are rewards for conformity and punishments for nonconformity.

Scapegoating occurs when people or groups blame others for their failures.

School choice refers to a range of options (vouchers, tax credits, magnet and charter schools, home schooling) that enable families to choose where their children go to school.

Secondary groups are groups that are formal, large, and impersonal.

Secondary production is the processing of raw materials.

Sects are religious organizations that reject the social environment in which they exist.

Secularization is the process of transferring things, ideas, or events from the sacred realm to the nonsacred, or secular, realm.

A **segmented labor market** is an economy in which hiring, advancement, and benefits vary systematically between the industrial core and the periphery.

Segregation refers to the physical separation of minority- and majority-group members.

The **self** is a complex whole that includes unique attributes and normative responses. In sociology, these two parts are called the *I* and the *me*.

The **self-concept** is the self we are aware of. It is our thoughts about our personality and social roles.

Self-esteem is the evaluative component of the self-concept; it is our judgment about our worth compared with others' worth.

Self-fulfilling prophecies occur when something is *defined* as real and therefore *becomes* real in its consequences.

Sex is a biological characteristic, male or female.

Sexism is a belief that men and women have biologically different capacities and that these form a legitimate basis for unequal treatment.

Sexual harassment consists of unwelcome sexual advances, requests for sexual favors, or other verbal or physical conduct of a sexual nature.

The **sick role** consists of four social norms regarding sick people. They are assumed to have good reasons for not fulfilling their normal social roles and are not held responsible for their illnesses. They are also expected to consider sickness undesirable, to work to get well, and to follow doctor's orders.

Significant others are the role players with whom we have close personal relationships.

Social change is any significant modification or transformation of social structures and sociocultural processes over time.

Social class is a category of people who share roughly the same class, status, and power and who have a sense of identification with each other.

The **social construction of race and ethnicity** is the process through which a culture (based more on social ideas than on biological facts) defines what constitutes a race or an ethnic group.

Social control consists of the forces and processes that encourage conformity, including self-control, informal control, and formal control.

Social-desirability bias is the tendency of people to color the truth so that they sound nicer, richer, and more desirable than they really are.

Social distance is the degree of intimacy and equality in relationships between two groups.

Social integration means that belonging to a particular race, ethnic group, or religion does not affect individuals social position, social relationships, or how they are viewed by members of the dominant group.

Social mobility is the process of changing one's social class.

A **social movement** is an ongoing, goal-directed effort to fundamentally challenge social institutions, attitudes, or ways of life.

A **social network** is an individual's total set of relationships.

Social processes are the forms of interaction through which people relate to one another; they are the dynamic aspects of society.

A **social structure** is a recurrent pattern of relationships.

Socialism is an economic structure in which productive tools (land, labor, and capital) are owned and managed by the workers and used for the collective good.

Socialization is the process of learning the roles, statuses, and values necessary for participation in social institutions.

A **society** is the population that shares the same territory and is bound together by economic and political ties.

Sociobiology is the study of the biological basis of all forms of human (and nonhuman) behavior.

Socioeconomic status (SES) is a measure of social class that ranks individuals on income, education, occupation, or some combination of these.

The **sociological imagination** is the ability to see the intimate realities of our own lives in the context of common social structures; it is the ability to see personal troubles as public issues.

Sociology is the systematic study of human society, social groups, and social interactions.

The **sociology of everyday life** focuses on the social processes that structure our experience in ordinary face-to-face situations.

Stakeholder mobilization refers to organized political opposition by groups with vested interest in a particular political outcome.

The **state** is the social structure that successfully claims a monopoly on the legitimate use of coercion and physical force within a territory.

Status is an individual's position within a group relative to other group members; also social honor, expressed in lifestyle.

Strain theory suggests that deviance occurs when culturally approved goals cannot be reached by culturally approved means.

Stratification is an institutionalized pattern of inequality in which social statuses are ranked on the basis of their access to scarce resources.

Strong ties are relationships characterized by intimacy, emotional intensity, and sharing.

Structural-functional theory addresses the question of social organization (structure) and how it is maintained (function).

Subcultures are groups that share in the overall culture of society but also maintain a distinctive set of values, norms, and lifestyles and even a distinctive language.

Suburbanization is the growth of suburbs.

The **superego** is composed of internalized social ideas about right and wrong.

Survey research is a method that involves asking a relatively large number of people the same set of standardized questions.

Symbolic interaction theory addresses the subjective meanings of human acts and the processes through which people come to develop and communicate shared meanings.

T

Technological imperative refers to the idea that once a technology is available, it becomes difficult to avoid using it.

Technology involves the human application of knowledge to the making of tools and to the use of natural resources.

Tertiary production is the production of services.

A **theory** is an interrelated set of assumptions that explains observed patterns.

Total institutions are facilities in which all aspects of life are strictly controlled for the purpose of radical resocialization.

Tracking occurs when evaluations made relatively early in a child's career determine the educational programs the child will be encouraged to follow.

Traditional authority is the right to make decisions for others that is based on the sanctity of time-honored routines.

Transgendered persons are individuals whose sex or sexual identity is not definitively male or female. Some are hermaphrodites, some are transsexuals.

Transnational corporations are large corporations that produce and distribute goods internationally.

A **trend** is a change in a variable over time.

U

Urbanization is the process of population concentration in metropolitan areas.

V

Value-free sociology concerns itself with establishing what is, not what ought to be.

Values are shared ideas about desirable goals.

Variables are measured characteristics that vary from one individual or group to the next.

Vested interests are stakes in either maintaining or transforming the status quo.

Victimless crimes such as drug use, prostitution, gambling, and pornography are voluntary exchanges between persons who desire illegal goods or services from each other.

Voluntary associations are nonprofit organizations designed to allow individuals an opportunity to pursue their shared interests collectively.

W

Weak ties are relationships characterized by low intensity and intimacy.

White-collar crime refers to crimes committed by respectable people of high status in the course of their occupation.

World-systems theory is a conflict perspective of the economic relationships between developed and developing countries, the core and peripheral societies.

References

Abma J.C., Martinez, G.M., Mosher, W.D., Dawson, B.S. 2004. "Teenagers in the United States: Sexual activity, Contraceptive Use, and Childbearing, 2002." National Center for Health Statistics. Vital and Health Statistics 23(24):1–48.

Alba, Richard D., Logan, John R., and Stults, Brian J. 2000. "How Segregated Are Middle-Class African Americans?" Social Problems 47:543–558.

Alba, Richard, and Nee, Victor. 2003. Remaking the American Mainstream: Assimilation and Contemporary Immigration. Cambridge, MA: Harvard University Press.

Albas, Daniel, and Albas, Cheryl. 1988. "Aces and Bombers: The Post-Exam Impression Management Strategies of Students." Symbolic Interaction 11:289–302.

Alcock, John. 2001. The Triumph of Sociobiology. New York: Oxford University Press.

Ali, Jennifer, and Avison, William R. 1997. "Employment Transitions and Psychological Distress: The Contrasting Experiences of Single and Married Mothers." Journal of Health and Social Behavior 38:345–362.

Allan, Emilie, and Steffensmeier, Darrell. 1989. "Youth Unemployment and Property Crime." American Sociological Review 54:107–123.

Allen, Katherine R., Blieszner, Rosemary, and Roberto, Karen A. 2000. "Families in the Middle and Later Years: A Review and Critique of Research in the 1990s." Journal of Marriage and the Family 62:911–926.

Altheide, David. 2002. Creating Fear: News and the Construction of Crisis. Piscataway, NJ: Aldine Transaction.

Amanat, Abbas. 2001. "Empowered Through Violence: The Reinventing of Islamic Extremism." In Strobe Talbott and Nayan Chanda (eds.), The Age of Terror. New York: Basic Books.

Amato, Paul R. 1993. "Urban-Rural Differences in Helping Friends and Family Members." Social Psychology Quarterly 56:249–262.

American Association of Retired Persons. 1999. The AARP Grandparenting Survey: Sharing and Caring between Mature Grandparents and Their Grandchildren. Washington, D.C.: American Association of Retired Persons.

Anderson, David C. 1998. Sensible Justice: Alternatives to Prison. New York: Norton.

Appelbaum, Richard. 2005. Critical Globalization Studies. New York: Routledge.

"Arab American Demographics." 2006. Arab American Institute. www.aaiusa.org/arab-americans/22/demographics. Accessed May 2006.

Arana-Ward, Marie. 1997. "As Technology Advances, a Bitter Debate Divides the Deaf." The Washington Post. May 11, p. A11.

Arrigo, Bruce A. (ed.). 1998. Social Justice/Criminal Justice. Belmont, CA: Wadsworth.

Arrington, Leonard J., and Bitton, Davis. 1992. The Mormon Experience: A History of the Latter-Day Saints (2nd ed.). Urbana, IL: University of Illinois Press.

Asch, Solomon. 1955. "Opinions and Social Pressure." Scientific American 193:31–35.

Ashford, Lori S. 1995. "New Perspectives on Population: Lessons from Cairo." Population Bulletin 50 (1, March). Washington D.C.: Population Reference Bureau.

Atchley, Robert C. 1982. "Retirement as a Social Institution." American Review of Sociology 8:263–287.

Austin, Roy, and Allen, Mark. 2000. "Racial Disparity in Arrest Rates as an Explanation of Racial Disparity in Commitment to Pennsylvania's Prisons." Journal of Research in Crime and Delinquency 37:200–220.

Babbie, Earl. 2004. The Practice of Social Research (10th ed.). Belmont, CA: Wadsworth.

Bai, Ruoyun. 2003. "Chicken Wings." Studies in Symbolic Interaction 26:263-265.

Banks, James, Marmot, Michael, Oldfield, Zoe, and Smith, James P. 2006. "Disease and Disadvantage in the

United States and in England." Journal of the American Medical Association 295:2037–2045.

Barber, Benjamin R. 2001. Jihad vs. McWorld: How Globalism and Tribalism are Reshaping the World (rev. ed.). New York: Ballantine.

Barker, Eileen. 1986. "Religious Movements: Cult and Anticult Since Jonestown." Annual Review of Sociology 12:329–346.

Barker, Kristin. 2005. The Fibromyalgia Story: Biomedical Authority and Women's Worlds of Pain. Philadelphia, PA: Temple University Press.

Baron, James N., and Bielby, William T. 1984. "The Organization of a Segmented Economy." American Sociological Review 49:454–473.

Bayer, Ada-Helen, and Harper, Leon. 2000. Fixing to Stay: A National Study on Housing and Home Modification Issues. Washington, D.C.: American Association of Retired Persons.

Bearman, Peter S. and Moody, James. 2004. "Adolescent Suicidality." American Journal of Public Health 4: 89–95.

Beauvais, Fred. 1998. "American Indians and Alcohol." Alcohol Health and Research World 22:253–259.

Becker, Howard S. 1963. Outsiders: Studies in the Sociology of Deviance. New York: Free Press.

Becker, Marshall H. (ed.). 1974. The Health Belief Model and Personal Health Behavior. San Francisco: Society for Public Health Education.

———. 1993. "A Medical Sociologist Looks at Health Promotion." Journal of Health and Social Behavior 34:1–6.

Beckwith, Carol. 1983. "Niger's Wodaabe: 'People of the Taboo.'" National Geographic 164 (October):482–509.

Bell, Derrick. 1992. Race, Racism, and American Law. Boston: Little, Brown.

Bellah, Robert N. 1974. "Civil Religion in America." In Russell B. Richey and Donald G. Jones (eds.), American Civil Religion. New York: Harper & Row.

Bellah, Robert N., Madsen, Richard, Sullivan, William M., Swidler, Ann, and Tipton, Steven M. 1985. Habits of the Heart: Individualism and Commitment in American Life. Berkeley: University of California Press.

———. 1991. The Good Society. New York: Knopf.

Benford, Robert D. 1993. "'You Could Be the Hundredth Monkey': Collective Action Frames and Vocabularies of Motive within the Nuclear Disarmament Movement." Sociological Quarterly 34:195–216.

Benford, Robert D., and Snow, David A. 2000. "Framing Processes and Social Movements: An Overview and Assessment." Annual Review of Sociology 26:611–639.

Bennett, Neil G., and Lu, Hsien-hen. 2000. "Child Poverty in the States: Levels and Trends from 1979 to 1998." Child Poverty Research Brief 2. New York: National Center for Children in Poverty.

Berezin, Mabel. 1997. "Politics and Culture: A Less Fissured Terrain." Annual Review of Sociology 23:361–383.

Bergen, Raquel Kennedy. 1998. Issues in Intimate Violence. Thousand Oaks, CA: Sage.

Berger, Peter L., and Luckmann, Thomas. 1966. The Social Construction of Reality: A Treatise in the Sociology of Knowledge. Garden City, NY: Doubleday.

Berliner, David C., and Biddle, Bruce J. 1995. The Manufactured Crisis: Myths, Fraud, and the Attack on America's Public Schools. New York: Longman.

Bettie, Julie. 2003. Women Without Class: Girls, Race, and Identity. Berkeley: University of California Press.

Bian, Yanjie. 1997. "Bringing Strong Ties Back In: Indirect Ties, Network Bridges, and Job Searches in China." American Sociological Review 62:366–385.

Bian, Yanjie, and Soon, Ang. 1997. "Guanxi Networks and Job Mobility in China and Singapore." Social Forces 75:981–1005.

Bianchi, Suzanne M., Robinson, John P. and Milkie, Melissa A. 2006. Changing Rhythms of American Family Life. ASA Rose Monograph Series. New York: Russell Sage.

Binson, D., Michaels, S., Stall, R., Coates, T.J., Gagnon, J.H., and Catania, J.A. 1995. "Prevalence and Social Distribution of Men Who Have Sex with Men: United States and its Urban Centers." Journal of Sex Research 32:245–254.

Blau, Joel. 2001. Illusions of Prosperity: America's Working Families in an Age of Economic Insecurity. New York: Oxford University Press.

Blau, Peter M. 1987. "Contrasting Theoretical Perspectives." In J. Alexander, B. Giesen, R. Munch, and N. Smelser (eds.), The Micro-Macro Link. Berkeley: University of California Press.

Blendon, Robert, and Young, John T. 1998. "The Public and the War on Illicit Drugs." Journal of the American Medical Association 279:827–832.

Blumer, H. 1969. Symbolic Interactionism: Perspective and Method. Englewood Cliffs, NJ: Prentice-Hall.

Blumstein, Phillip, and Schwartz, Pepper. 1983. American Couples. New York: William Morrow.

Bobo, Lawrence, and Hutchings, Vincent. 1996. "Perceptions of Racial Group Competition: Extending Blumer's Theory of Group Position to a Multiracial Social Context." American Sociological Review 61:951–972.

Bobo, Lawrence, and Kluegel, James R. 1993. "Opposition to Race Targeting: Self-Interest, Stratification Ideology, or Racial Attitudes?" American Sociological Review 58:443–464.

Bolzendahl, Catherine, and Brooks, Clem. 2005. "Polarization, Secularization, or Differences as Usual? The Denominational Cleavage in U.S. Social Attitudes Since the 1970s." Sociological Quarterly 46:47–78.

Borjas, George J. and Katz, Lawrence F. 2007 (forthcoming). "The Evolution of the Mexican-Born Workforce in the United States." In George J. Borjas (ed.), Mexican Immigration. Chicago: University of Chicago Press.

Bonacich, Edna. 1972. "A Theory of Ethnic Antagonism: The Split Labor Market." American Sociological Review 37:547–559.

Booth, Alan, and Osgood, D. Wayne. 1993. "The Influence of Testosterone on Deviance in Adulthood: Assessing and Explaining the Relationship." Criminology 31: 93–117.

Bose, Christine E., and Rossi, Peter H. 1983. "Gender and Jobs: Prestige Standings of Occupations as Affected by Gender." American Sociological Review 48:316–330.

Bourdieu, Pierre. 1984. Distinction: A Social Critique of the Stratification of Taste. Cambridge, MA.: Harvard University Press.

Bourgois, Philippe. 1995. In Search of Respect: Selling Crack in El Barrio. New York: Cambridge University Press.

Bovin, Mette. 2001. Nomads Who Cultivate Beauty: Wodaabe Dances and Visual Arts in Niger. Uppsala, Sweden: Nordiska Afrikainstitutet.

Boyle, John P., and O'Connor, Liz Clapp. 1996. "Leveraging Technology and Partnerships in Criminal Investigations." Police Chief 63:19–24.

Braithwaite, John. 1981. "The Myth of Social Class and Criminality, Reconsidered." American Sociological Review 46:36–58.

———. 1985. "White Collar Crime." Annual Review of Sociology 11:1–25.

Bramlett, M. D., and Mosher, W. D. 2002. "Cohabitation, Marriage, Divorce, and Remarriage in the United States." Vital Health Statistics 23 (22):1–87.

Brannon, Robert L. 1996. "Restructuring Hospital Nursing: Reversing the Trend Toward a Professional Work Force." International Journal of Health Services 26:643–654.

Braverman, Harry. 1974. Labor and Monopoly Capital. New York: Monthly Review Press.

Breuninger, Paul. 1995. "Crime Scene Reconstruction Using 3D Computer-Aided Drafting." Police Chief 62:61–62.

Brinkman, Richard L. and Brinkman, June E. 1997. "Cultural Lag: Conception and Theory." International Journal of Social Economics 24:609–627.

Brooks, David. 2000. Bobos in Paradise: The New Upper Class and How They Got There. New York: Simon & Schuster.

Brown, David K. 2001. "The Social Sources of Educational Credentialism: Status Cultures, Labor Markets, and Organizations," Sociology of Education Extra Issue: 19–34.

Brown, David L. and Swanson, Louis E. (eds.). 2003. Challenges for Rural America in the Twenty-First Century. University Park, PA: Pennsylvania State University Press.

"Brundtland Report." 2000. Encyclopedia of the Atmospheric Environment. http://www. doc.mmu.ac.uk/aric/eae/Sustainability/Older/Brundtland_Report.html. Accessed May 2006.

Bryson, Bethany. 1996. "'Anything but Heavy Metal': Symbolic Exclusion and Musical Dislikes." American Sociological Review 61:884–899.

Buechler, Steven M. 1993. "Beyond Resource Mobilization? Emerging Trends in Social Movement Theory." Sociological Quarterly 34:217–235.

Bulan, Heather Ferguson, Erickson, Rebecca J., and Wharton, Amy S. 1997. "Doing for Others on the Job: The Affective Requirements of Service Work, Gender, and Emotional Well-Being." Social Problems 44:235–255.

Bullard, Robert D. 1993. Confronting Environmental Racism: Voices from the Grassroots. Boston: South End Press.

Bullard, Robert D., Warren, Rueben C., and Johnson, Glenn S. 2001. "The Quest for Environmental Justice." In Ronald L. Braithwaite and Sandra E. Taylor (eds.), Health Issues in the Black Community (2nd ed.). San Francisco: Jossey-Bass, pp. 471–488.

Bumpass, Larry L., Raley, R. Karen, and Sweet, James. 1995. "The Changing Character of Stepfamilies: Implications of Cohabitation and Nonmarital Childbearing." Demography 32:425–436.

Bunker, John P., Frazier, Howard S., and Mosteller, Frederick. 1994. "Improving Health: Measuring Effects of Medical Care." Milbank Quarterly 72:225–258.

Bureau of Justice Statistics. 2005. Prisoners in 2004. Washington D.C.: U.S. Government Printing Office.

———. 2006. Criminal Victimization in the United States, 2004. http://www.ojp.usdoj.gov/bjs/pub/pdf/cvus0403.pdf. Accessed June 2006.

Burgess, Ernest W. 1925. "The Growth of the City: An Introduction to a Research Project." In Robert E. Park, Ernest W. Burgess, and Roderick D. McKenzie (eds.), The City. Chicago: University of Chicago Press, pp. 47–62.

Burke, Peter J. 1980. "The Self: Measurement Requirements from the Interactionist Perspective." Social Psychological Quarterly 43(1):18–29.

Burman, Patrick. 1988. Killing Time, Losing Ground: Experiences of Unemployment. Toronto: Thompson Educational Publishing.

Burris, Beverly H. 1998. "Computerization of the Workplace." Annual Review of Sociology 24:141–157.

Burris, Val, and Salt, James. 1990. "The Politics of Capitalist Class Segments: A Test of Corporate Liberalism Theory." Social Problems 37:341–359.

Cable, Sherry, and Cable, Charles. 1995. Environmental Problems, Grassroots Solutions: The Politics of Grassroots Environmental Conflict. New York: St. Martin's.

Cahill, Spencer E. 1983. "Reexamining the Acquisition of Sex Roles: A Social Interactionist Perspective." Sex Roles 9:1–15.

Caldwell, John C. 1993. "Health Transition: The Cultural, Social, and Behavioral Determinants of Health in the Third World." Social Science and Medicine 36:125–135.

Call, Vaughn, Sprecher, Susan, and Schwartz, Pepper. 1995. "The Incidence and Frequency of Marital Sex in a National Sample." Journal of Marriage and the Family 57:639–652.

Camacho, David E. (ed.). 1998. Environmental Injustices, Political Struggles: Race, Class, and the Environment. Durham, NC: Duke University Press.

Cancian, Francesca M. and Oliker, Stacey J. 2000. Caring and Gender. Thousand Oaks, CA: Pine Forge Press.

Cancio, A. Silvia, Evans, T. Davic, and Maume, David J., Jr. 1996. "Reconsidering the Declining Significance of Race: Racial Differences in Early Career Wages." American Sociological Review 61:541–556.

Capitman, John. 2002. Defining Diversity: A Primer and a Review. Generations 26(3):8–14.

Card, David. 2005. "Is the New Immigration Really So Bad?" NBER Working Paper No. 11547. http://www.phil.frb.org/econ/conf/ immigration/card.pdf. Accessed June 2006.

Center for American Women and Politics. 2006. Women in Elective Office 2006. Fact Sheet. New Brunswick, NJ: Rutgers University.

Centers for Disease Control and Prevention. 2005. "Overweight and Obesity: Home." http://www.cdc.gov/nccdphp/dnpa/obesity. Accessed June 2006.

Central Intelligence Agency. 2002. World Factbook 2002. Washington, D.C.: Central Intelligence Agency.

Chapkis, Wendy. 1997. Live Sex Acts: Women Performing Erotic Labor. New York: Routledge.

Charon, Joel M. 2006. Symbolic Interactionism: An Introduction, An Interpretation (9th ed.). Englewood Cliffs, NJ: Prentice-Hall.

Chase-Dunn, Christopher. 1989. Global Formation: Structure of the World Economy. London: Basil Blackwell.

Cherlin, Andrew. 1992. Marriage, Divorce, and Remarriage (rev. ed.). Cambridge, MA: Harvard University Press.

Cherlin, Andrew J. 2004. "The Deinstitutionalization of American Marriage." Journal of Marriage and Family 66:848–861.

Chesney-Lind, Meda, and Shelden, Randall G. 2004. Girls, Delinquency, and Juvenile Justice (3rd ed.). Belmont, CA: Wadsworth.

Chetkovich, Carol. 1998. Real Heat. New Brunswick, NJ: Rutgers University Press.

Chirot, Daniel. 1977. Social Change in the Twentieth Century. San Francisco, CA: Harcourt Brace Jovanovich.

———. 1986. Social Change in the Modern Era. San Diego, CA: Harcourt Brace Jovanovich.

Chodak, Symon. 1973. Societal Development: Five Approaches with Conclusions from Comparative Analysis. New York: Oxford University Press.

Chodorow, Nancy. 1999. The Power of Feelings: Personal Meaning in Psychoanalysis, Gender, and Culture. New Haven, CT: Yale University Press.

Christenson, James A. 1984. "Gemeinschaft and Gesellschaft: Testing the Spatial and Communal Hypothesis." Social Forces 63:160–168.

Chubb, John E., and Moe, Terry M. 1990. Politics, Markets, and America's Schools. Washington, D.C.: Brookings Institute.

Clawson, Dan, and Su, Tie-ting. 1990. "Was 1980 Special? A Comparison of 1980 and 1986 Corporate PAC Contributions." Sociological Quarterly 31:371–387.

Clendinen, Dudley, and Nagourney, Adam. 2001. Out for Good: The Struggle to Build a Gay Rights Movement in America. New York: Simon and Schuster.

Coburn, David, and Willis, Evan. 2000. "The Medical Profession: Knowledge, Power, and Autonomy." In Gary L. Albrecht, Ray Fitzpatrick, and Susan C. Scrimshaw (eds.), Handbook of Social Studies in Health and Medicine. Thousand Oaks, CA: Sage, pp. 377–393.

Cockerham, William C. 1997. "The Social Determinants of the Decline of Life Expectancy in Russian and Eastern Europe: A Lifestyle Explanation." Journal of Health and Social Behavior 38:117–130.

Cohen, Philip N., and Huffman, Matt L. 2003a. "Individuals, Jobs, and Labor Markets." American Sociological Review 68:443–463.

———. 2003b. "Occupational Segregation and the Devaluation of Women's Work Across U.S. Labor Markets." Social Forces 81:881–907.

Coleman, Marilyn, Ganong, Lawrence, and Fine, Mark. 2000. "Reinvestigating Remarriage: Another Decade of Progress." Journal of Marriage and the Family 62:1288–1307.

Collins, Patricia Hill. 1991. Black Feminist Thought: Knowledge, Consciousness, and the Politics of Empowerment. New York: Routledge.

Collins, Sharon M. 1993. "Blacks on the Bubble: The Vulnerability of Black Executives in White Corporations." Sociological Quarterly 34:429–447.

———. 1997. "Black Mobility in White Corporations: Up the Corporate Ladder but Out on a Limb." Social Problems 44:55–67.

Coltrane, Scott. 2000. "Research on Household Labor: Modeling and Measuring the Social Embeddedness of Routine Family Work." Journal of Marriage and the Family 62:1208–1233.

Conley, Dalton. 2004. The Pecking Order: Which Siblings Succeed and Why. New York: Pantheon.

Conrad, John P. 1983. "Deterrence, the Death Penalty, and the Data." In Ernest van den Haag and John P. Conrad (eds.), The Death Penalty: A Debate. New York: Plenum.

Conrad, Peter. 2005. "The Shifting Engines of Medicalization." Journal of Health and Social Behavior 46:3–14.

Conrad, Peter, and Schneider, Joseph W. 1992. Deviance and Medicalization: From Badness to Sickness. Philadelphia: Temple University Press

Cook, Philip J., and Laub, John H. 1998. "The Unprecedented Epidemic in Youth Violence." In Michael Tonry and Mark H. Moore (eds.), Youth Violence. Chicago: University of Chicago Press.

Cooley, Charles Horton. 1902. Human Nature and the Social Order. New York: Scribner's.

———. 1967. "Primary Groups." In A. Paul Hare, Edgar F. Borgotta, and Robert F. Bales (eds.), Small Groups: Studies in Social Interaction (rev. ed.). New York: Knopf. (Originally published 1909.)

Cooney, Mark. 1997. "The Decline of Elite Homicide." Criminology 35:381–407.

Coontz, Stephanie. 1997. The Way We Really Are: Coming to Terms with America's Changing Families. New York: Basic Books.

Corcoran, M. 1995. "Rags to Rags: Poverty and Mobility in the United States." Annual Review of Sociology 21:237–267.

Corsaro, William. 2003. "We're Friends, Right?": Inside Kids' Culture. Washington, DC: National Academies Press.

———. 2004. Sociology of Childhood (2nd ed.). Thousand Oaks, CA: Sage Press.

Coser, Lewis A. 1956. The Functions of Conflict. New York: Free Press.

Coulter, Ann. 2006. Godless: The Church of Liberalism. New York: Crown Forum.

Coverdill, James E. 1988. "The Dual Economy and Sex Differences in Earnings." Social Forces 66:970–993.

Cowan, Ruth Schwartz. 1993. "Less Work for Mother?" In Albert H. Teich (ed.), Technology and the Future (6th ed.). New York: St. Martin's, pp. 329–339.

Crary, David. 2006. "Number of Women Behind Bars Surges." Arizona Republic. May 21:A5.

Crimmins, Ellen M., Hayward, Mark D., and Saito, Yashuhiko. 1994. "Changing Mortality and Morbidity Rates and the Health Status and Life Expectancy of the Older Population." Demography 31:168–169.

Cronon, William. 2003. Changes in the Land: Indians, Colonists, and the Ecology of New England (20th Anniv. ed.). New York: Hill and Wang.

Crowder, Kyle, South, Scott J., Chavez, Erick. 2006. "Wealth, Race, and Inter-Neighborhood Migration." American Sociological Review 71:72–94.

Crozier, Michael, and Friedberg, Erhard. 1980. Actors and Systems: The Politics of Collective Action. Chicago: University of Chicago Press.

Crutchfield, Robert D. 1989. "Labor Stratification and Violent Crime." Social Forces 68:489–512.

Culver, John H. 1992. "Capital Punishment, 1977–1990: Characteristics of the 143 Executed." Sociology and Social Research 76:59–61.

Cureton, Steven. 2000. "Justifiable Arrests or Discretionary Justice: Predictors of Racial Arrest Differentials." Journal of Black Studies 30:703–719.

Currie, Elliott. 1998. Crime and Punishment in America. New York: Henry Holt.

Curry, Theodore R. 1996. "Conservative Protestantism and the Perceived Wrongfulness of Crimes: A Research Note." Criminology 34:453–464.

Curtis, James E., Baer, Douglas E., and Grabb, Edward G. 2001. "Nations of Joiners: Explaining Voluntary Association Membership in Democratic Societies." American Sociological Review 66:783–805.

Dahl, Robert. 1961. Who Governs? New Haven, CT: Yale University Press.

———. 1971. Polarchy. New Haven, CT: Yale University Press.

Dailard, Cynthia. 2003. "Understanding 'Abstinence': Implications for Individuals, Programs and Policies." Guttmacher Report 6(5):4–6.

Dalby, Andrew. 2003. Language in Danger: The Loss of Linguistic Diversity and the Threat to Our Future. New York: Columbia University Press.

Daly, Martin, and Wilson, Margo. 1983. Sex, Evolution, and Behavior (2nd ed.). Boston: Willard Grant.

Davidman, Lynn. 1991. Tradition in a Rootless World: Women Turn to Orthodox Judaism. Berkeley: University of California Press.

Davis, James, and Stasson, Mark. 1988. "Small-Group Performance: Past and Future Research Trends." Advances in Group Processes 5:245–277.

Davis, Kingsley. 1961. "Prostitution." In Robert K. Merton and Robert A. Nisbet (eds.), Contemporary Social Problems. San Francisco: Harcourt Brace Jovanovich.

———. 1973. "Introduction." In Kingsley Davis (ed.), Cities. New York: W. H. Freeman.

Davis, Kingsley, and Moore, Wilbert E. 1945. "Some Principles of Stratification." American Sociological Review 10:242–249.

Davis, Nancy J., and Robinson, Robert V. 1996. "Are Rumors of War Exaggerated? Religious Orthodoxy and Moral Progressivism in America." American Journal of Sociology 102:756–787.

Death Penalty Information Center. Race and the Death Penalty. http://www.deathpenaltyinfo.org/article.php?did=105&scid=5. Accessed May 2006.

DeFrances, Carol J., and Smith, Steven K. 1998. Perceptions of Neighborhood Crime, 1995. Bureau of Justice Statistics Special Report, NCJ-165811. Washington, D.C.: U.S. Department of Justice.

Demo, David H., and Cox, Martha J. 2000. "Families with Young Children: A Review of Research in the 1990s." Journal of Marriage and the Family 62:876–895.

DeNavas-Walt, Carmen, and Cleveland, Robert. 2002. "Money Income in the United States: 2001." Current Population Reports P60–280. Washington, D.C.: U.S. Census Bureau.

Denzin, Norman K. 1984. "Toward a Phenomenology of Domestic, Family Violence." American Journal of Sociology 90:483–513.

Deutsch, Claudia H. 2006. "Companies and Critics Try Collaboration." New York Times. May 17:D1+.

Devine, Joel, Sheley, Joseph, and Smith, M. Dwayne. 1988. "Macroeconomic and Social Control Policy Influences in Crime Rate Changes, 1948–85." American Sociological Review 53:407–420.

DeWitt, J. L. 1943. Japanese in the United States. Final Report: Japanese Evacuation from the West Coast. Cited in Paul E. Horton and Gerald R. Leslie. 1955. Social Problems. East Norwalk, CT: Appleton-Century-Crofts.

Diamond, Jared M. 1997. Guns, Germs, and Steel: The Fates of Human Societies. New York: W.W. Norton.

———. 2005. Collapse: How Societies Choose to Fail or Succeed. New York: Viking.

DiMaggio, Paul, Evans, John, and Bryson, Bethany. 1996. "Have Americans' Social Attitudes Become More Polarized?" American Journal of Sociology 102:690–755.

DiMaggio, Paul, Hargittai, Eszter, Neuman, W. Russell, and Robinson, John P. 2001. "Social Implications of the Internet." Annual Review of Sociology 27:307–336.

DiMaggio, Paul, and Mohr, John. 1985. "Cultural Capital, Educational Attainment, and Marital Selection." American Journal of Sociology 90:1231–1261.

Dixon, William J., and Boswell, Terry. 1996. "Dependency, Disarticulation, and Denominator Effects: Another Look at Foreign Capital Penetration." American Journal of Sociology 102:543–562.

Dodds, Peter Sheridan, Muhamad, Roby, and Duncan, J. Watts. 2003. "An Experimental Study of Search in Global Social Networks." Science 301:827-829.

Dolnick, Edward. 1993. "Deafness as Culture." The Atlantic Monthly, September:37–53.

Domhoff, G. William. 1998. Who Rules America: Power and Politics in the Year 2000 (3rd ed.). Mountain View, CA: Mayfield.

Donnelly, Strachan, McCarthy, Charles R., and Singleton, Rivers, Jr. 1994. "The Brave New World of Animal Biotechnology." Special Supplement. Hastings Center Report 24 (January/February).

Douglas, Tom. 1983. Groups: Understanding People Gathered Together. London: Tavistock.

Drew, Rob. 2005. "'Once More, With Irony': Karaoke and Social Class." Leisure Studies 24:371–383.

Dreze, Jean, and Sen, Amartya. 1989. Hunger and Public Action. Oxford, England: Clarendon Press.

Durkheim, Emile. 1938. The Rules of Sociological Method. New York: Free Press. (Originally published 1895.)

———. 1951. Suicide: A Study in Sociology. New York: Free Press. (Originally published 1897.)

———. 1961. The Elementary Forms of the Religious Life. London: Allen & Unwin. (Originally published 1915.)

Dutta, Soumitra, Lanvin, Bruno, and Paua, Fiona. 2004. The Global Information Technology Report 2003–2004. New York: Oxford University Press.

Dworkin, Andrea. 1981. Pornography: Men Possessing Women. New York: Putnam.

Dworkin, Shari. 2003. "Holding Back: Negotiating a Glass Ceiling on Women's Muscular Strength." In Rose Weitz (ed.), Politics of Women's Bodies: Sexuality, Appearance, and Behavior. New York: Oxford University Press.

Dye, Jane Lawler. 2005. "Fertility of American Women: June 2004." Current Population Reports P20–555. U.S. Census Bureau, Washington, DC, 2005.

Eaton, William W., and Muntaner, Carles. 1999. "Social stratification and mental disorder." In Allan V. Horwitz and Teresa L. Scheid (eds.). A Handbook for the Study of Mental Health: Social Contexts, Theories, and Systems. Cambridge: Cambridge University Press.

Ebaugh, Helen Rose Fuchs. 1988. Becoming an Ex: The Process of Role Exit. Chicago: University of Chicago Press.

Eberhardt, Jennifer L., Davies, Paul G., Purdie-Vaughns, Valerie J., and Johnson, Sheri Lynn. 2006. "Looking Deathworthy: Perceived Stereotypicality of Black Defendants Predicts Capital-Sentencing Outcomes." Psychological Science 17:383–386.

Eckholm, Erik. 2006. "America's 'Near Poor' are Increasingly at Economic Risk, Experts Say." New York Times. May 8:A14.

Economic Policy Institute. 2006. Minimum Wage Policy Guide. Washington, DC: Economic Policy Institute.

Edin, Kathryn, and Kefalas, Maria. 2006. Promises I Can Keep: Why Poor Women Put Motherhood Before Marriage. Berkeley, CA: University of California Press.

Ehrenreich, Barbara. 2001. Nickel and Dimed: On (Not) Getting By in America. New York: Metropolitan.

Eichenwald, Kurt. 2005. Conspiracy of Fools: A True Story. New York: Broadway Books.

Ellison, Christopher. 1991. "Religious Involvement and Subjective Well-Being." Journal of Health and Social Behavior 32:80–99.

———. 1992. "Are Religious People Nice People? Evidence from the National Survey of Black Americans." Social Forces 71:411–430.

Ellison, Christopher G., Bartkowski, John P., and Segal, Michelle L. 1996. "Conservative Protestantism and the Parental Use of Corporal Punishment." Social Forces 74:1003–1028.

Ember, Lois. 1994. "Minorities Still More Likely to Live Near Toxic Sites." Chemical and Engineering News 72 (36, September 5):19.

Emerson, Michael O., and Hartman, David. 2006. "The Rise of Religious Fundamentalism." Annual Review of Sociology 32:127–144.

Entman, Robert M., and Rojecki, Andrew. 2000. The Black Image in the White Mind: Media and Race in America. Chicago: University of Chicago Press.

Entwisle, Doris E., and Alexander, Karl L. 1992. "Summer Setback: Race, Poverty, School Composition and Mathematics Achievement in the First Two Years of School." American Sociological Review 57:72–84.

Envirowatch. 2006. Environmental Contamination at Grassy Narrows. http://www .envirowatch.org/gnfnindex.htm. Accessed May 2006.

Erikson, Kai. 1986. "On Work and Alienation." American Sociological Review 51:1–8.

Evans, E. Margaret, Schweingruber, Heidi, and Stevenson, Harold W. 2002. "Gender Differences in Interest and Knowledge Acquisition: The United States, Taiwan, and Japan." Sex Roles 47:153–167.

Evans, Ellis D., Rutberg, Judith, Sather, Carmela, and Turner, Chari. 1991. "Content Analysis of Contemporary Teen Magazines for Adolescent Females." Youth and Society 23:99–120.

Evans, John H., Bryson, Bethany, and DiMaggio, Paul. 2001. "Opinion Polarization: Important Contributions, Necessary Limitations." American Journal of Sociology 106:944–959.

Evans, Sara. 2003. Tidal Wave: How Women Changed America at Century's End. New York: Free Press.

Everybody In, Nobody Out. 2005. http:// www.everybodyinnobodyout.org/DOCS/ Polls.htm#GovHI. Accessed May 2006.

Farkas, George, Grobe, Robert P., Sheehan, Daniel, and Shuan, Yuan. 1990. "Cultural Resources and School Success." American Sociological Review 55:127–142.

Farmer, Paul. 1999. Infections and Inequalities: The Modern Plagues. Berkeley: University of California Press.

———. 2003. The Uses of Haiti. Monroe, ME: Common Courage Press.

Feagin, Joe R., and Parker, Robert. 1990. Building American Cities: The Urban Real Estate Game. Englewood Cliffs, NJ: Prentice-Hall.

Feagin, Joe R., and Sikes, Melvin P. 1994. Living with Racism: The Black Middle-Class Experience. Boston: Beacon.

Federal Bureau of Investigation. 2006. Crime in the United States: 2004. /www.fbi.gov/ucr/ cius_04. Accessed June 2006.

Federal Interagency Forum on Aging Related Statistics. 2000. Older Americans 2000: Key Indicators of Well-Being. Washington, D.C.

FedEx. 2006. About FedEx: Hybrid Electric Vehicles. http://www.fedex.com/us/about/responsibility/environment/ hybridelectricvehicle.html. Accessed May 2006.

Feeley, Malcolm M., and Simon, Jonathan. 1992. "The New Penology: Notes on the Emerging Strategy of Corrections and Its Implications." Criminology 30:449–474.

Felson, Richard B. 1996. "Mass Media Effects on Violent Behavior." Annual Review of Sociology 22:103–128.

Felson, Richard B., and Trudeau, Lisa. 1991. "Gender Differences on Mathematics Performance." Social Psychology Quarterly 54:113–126.

Ferguson, Ann. 2000. Bad Boys: Public Schools in the Making of Black Masculinities. Ann Arbor, MI: University of Michigan Press.

Feshbach, Morris. 1999. "Dead Souls." Atlantic Monthly 283 (1):26–27.

Fessenden, Ford. 2006. "Farther Afield: Americans Head Out Beyond the Exurbs." New York Times. May 7:WK14.

Fine, Gary Alan. 1996. Kitchens: The Culture of Restaurant Work. Berkeley: University of California Press.

Firebaugh, Glenn. 1996. "Does Foreign Capital Harm Poor Nations? New Estimates Based on Dixon and Boswell's Measures of Capital Penetration." American Journal of Sociology 102:563–575.

Fisher, Allen P. 2003. Still "Not Quite as Good as Having Your Own"? Toward a Sociology of Adoption. Annual Review of Sociology 29:335–361.

Fishman, Charles. 2006. The Wal-Mart Effect: How the World's Most Powerful Company Really Works—and How It's Transforming the American Economy. New York: Penguin.

Flanagan, William G. 1993. Contemporary Urban Sociology. New York: Cambridge University Press.

"The Forbes 400." 2005. Forbes.com.http://www .forbes.com/lists/ 2005/54/Rank_1.html. Accessed June 2006.

Foster, John Bellamy. 2006. "The Household Debt Bubble." Monthly Review 58(1):1–11.

Foster, John L. 1990. Bureaucratic Rigidity Revisited. Social Science Quarterly 71:223–238.

Fox, James Alan, and Sawpits, Marianne W. 2004. Homicide trends in the United States. Washington, DC: Department of Justice.

Franken, Al. 2003. Lies and the Lying Liars Who Tell Them: A Fair and Balanced Look at the Right. New York: Dutton.

Frankenberg, Erika, and Lee, Chungme. 2002. Race in American Public Schools: Rapidly Resegregating School Districts. Cambridge, MA: Harvard University Civil Rights Project.

Freedman, Estelle B. 2002. No Turning Back: The History of Feminism and the Future of Women. New York: Ballantine.

Freeman, Sue J. M. 1990. Managing Lives: Corporate Women and Social Change. Amherst: University of Massachusetts Press.

Freud, Sigmund. 1925. The Standard Edition of the Complete Psychological Works of Sigmund Freud. Volume 19. London: Hogarth Press. (Republished in 1971.)

Friedlander, D., and Okun, B. S. 1996. "Fertility Transition in England and Wales: Continuity and Change." Health Transition Review Supplement 1–18.

Funk, Richard, and Willits, Fern. 1987. "College Attendance and Attitudinal Change: A Panel Study, 1970–81." Sociology of Education 60:224–231.

Furstenberg, Frank F. Jr., Kennedy, Sheela, McLoyd, Vonnie C., Rumbaut, Ruben G., and Settersten, Richard A., Jr. 2004. "Growing Up is Harder to Do." Contexts 3:33-41.

Gallup Poll. 2001. Religion after 9/11. Dec. 21. Princeton, NJ: Gallup Organization.

Gamoran, Adam. 1992. "The Variable Effects of High School Tracking." American Sociological Review 57:812–828.

Gamson, William. 1990. The Strategy of Social Protest (2nd ed.). Belmont, CA: Wadsworth.

Garfinkel, H. 1967. Studies in Ethnomethodology. Englewood Cliffs, NJ: Prentice-Hall.

Garreau, Joel. 1991. Edge City: Life on the New Frontier. New York: Doubleday.

Gatto, John Taylor. 2002. Dumbing Us Down: The Hidden Curriculum of Compulsory Schooling (10th ed.). New York: New Society.

Gautier, Ann H., and Hatzius, Jan. 1997. "Family Benefits and Fertility: An Econometric Analysis." Population Studies 51:295–306.

General Social Survey. 2004. University of California–Berkeley. http://sda.berkeley.edu.

Ghana Statistical Service. 1999. Ghana Demographic and Health Survey 1998. Accra, Ghana: Ghana Statistical Service.

Giddens, Anthony. 1984. The Constitution of Society. Cambridge, England: Polity Press.

Giugni, Marco G. 1998a. "Frontiers in Social Movement Theory." Sociological Forum 13:365–375.

———. 1998b. "Was It Worth the Effort? The Outcomes and Consequences of Social Movements." Annual Review of Sociology 24:371–393.

Glassner, Barry. 2000. The Culture of Fear: Why Americans Are Afraid of the Wrong Things. New York: Basic Books.

———. 2004. "Narrative Techniques of Fear Mongering." Social Research 71:819–827.

Glick, Daniel. 2001. "Web-Exclusive Excerpt: 'Powder Burn.'" http://www.msnbc.com/news/512636.asp?cp1=1. Accessed June 2003.

Goddard, Stephen B. 1994. Getting There: The Epic Struggle between Road and Rail in the American Century. Chicago: University of Chicago Press.

Goffman, Erving. 1959. The Presentation of Self in Everyday Life. New York: Doubleday.

————. 1961a. Asylums: Essays on the Social Situation of Mental Patients and Other Inmates. New York: Doubleday.

————. 1961b. Encounters: Two Studies in the Sociology of Interaction. Indianapolis, IN: Bobbs-Merrill.

————. 1963. Behavior in Public Places: Notes on the Social Organization of Gatherings. New York: Free Press.

————. 1974. Gender Advertisements. New York: Harper & Row.

Goldin, Claudia. 1992. Understanding the Gender Gap: An Economic History of American Women. New York: Oxford University Press.

Gordon, Suzanne. 2005. Nursing Against the Odds: How Health Care Cost Cutting, Media Stereotypes, and Medical Hubris Undermine Nurses and Patient Care. Ithaca, NY: Cornell University Press.

Gould, Roger V. 1991. "Multiple Networks and Mobilization in the Paris Commune, 1871." American Sociological Review 56:716–729.

Gouldner, Alvin. 1960. "The Norm of Reciprocity." American Sociological Review 25:161–178.

Granovetter, Mark. 1973. "The Strength of Weak Ties." American Journal of Sociology 78:1360–1380.

————. 1974. Getting a Job: A Study of Contacts and Careers. Cambridge, MA: Harvard University Press.

Grant, Don S., and Martinez-Ramiro, J. R. 1997. "Crime and restructuring of the U.S. economy: A Reconsideration of the Class Linkages." Social Forces 75:769–798.

Greenhouse, Steven. 2003. "I.B.M. Explores Shift of Some Jobs Overseas." New York Times. July 22:C11.

Grimes, Michael D. 1989. "Class and Attitudes Toward Structural Inequalities: An Empirical Comparison of Key Variables in Neo and Post-Marxist Scholarship." Sociological Quarterly 30:441–463.

Gullotta, Thomas P., Adams, Gerald R., and Markstrom, Carol A. 2000. The Adolescent Experience (4th ed.). San Diego, CA: Academic.

Gurney, Joan N., and Tierney, Kathleen, J. 1982. "Relative Deprivation and Social Movements: A Critical Look at Twenty Years of Theory and Research." Sociological Quarterly 23:33–47.

Hafferty, Frederic W. 1991. Into the Valley: Death and the Socialization of Medical Students. New Haven, CT: Yale University Press.

Hagan, Frank. 2002. Introduction to Criminology. Belmont, CA: Wadsworth.

Hagan, John, Gillis, A. R., and Simpson, John. 1985. "The Class Structure of Gender and Delinquency: Toward a Power/Control Theory of Common Delinquent Behavior." American Journal of Sociology 90:1151–1178.

Halle, David. 1993. Inside Culture: Art and Class in the American Home. Chicago: University of Chicago Press.

Hallett, M.A. 2002. "Race, Crime and For-Profit Imprisonment: Social Disorganization as Market Opportunity." Punishment & Society 4:369–393.

Hallinan, Maureen T. 1994. "School Differences in Tracking Effects on Achievement." Social Forces 72:799–820.

Hamilton, Brady E., Martin, Joyce A., and Sutton, Paul D. 2003. "Births: Preliminary Data for 2002." National Vital Statistics Reports 51(11):1–20.

Handel, Michael J. 2002. Sociology of Organizations: Classic, Contemporary and Critical Readings. Thousand Oaks, CA: Sage.

Harlow, H. F., and Harlow, M. K. 1966. "Learning to Live." Scientific American 1:244–272.

Harris, Irving B. 1996. Children in Jeopardy. New Haven, CT: Yale University Press.

Harris, Judith Rich. 1998. The Nurture Assumption: Why Children Turn Out the Way They Do. New York: Free Press.

Harris Poll. 2000. "The Public Tends to Blame the Poor, the Unemployed, and Those on Welfare for Their Problems." 24 (May 3).

Harry, Beth, and Klingner, Janette. 2005. Why are So Many Minority Students in Special Education? Understanding Race and Disability in Schools. New York: Teachers College Press.

Haub, Carl. 1994. "Population Change in the Former Soviet Republics." Population Bulletin, Vol. 49. Washington, D.C.: Population Reference Bureau.

Hawley, Chris. 2006. "Western Union Speaking Out on Immigration." Arizona Republic. March 19:A1+.

Hayes-Bautista, David, Hsu, Paul, and Perez, Aide. 2002. "The Browning of the Graying in America: Diversity in the Elderly Population and Policy Implications." Generations 26(3):15–24.

Haynie, Dana L., and Osgood, D. Wayne. 2005. "Reconsidering Peers and Delinquency: How Do Peers Matter?" Social Forces 84:1109–1130.

Hechter, Michael. 1987. Principles of Group Solidarity. Berkeley: University of California Press.

Heidensohn, Frances. 1985. Women and Crime: The Life of the Female Offender. New York: New York University Press.

Henley, Nancy M. 1985. "Psychology and Gender." Signs 11:101–119.

Herdt, Gilbert (ed.). 1994. Third Sex, Third Gender: Beyond Sexual Dimorphism in Culture and History. New York: Zone Books.

Hertz, Rosanna. 2006. Single by Chance, Mothers by Choice: How Women are Choosing Parenthood Without

Marriage and Creating the New American Family. New York: Oxford University Press.

Hertzman, Clyde, Frank, J., and Evans, Robert G. 1994. "Heterogeneities in Health Status and the Determinants of Population Health." In R. Evans, M. Barer, and T. Marmor (eds.), Why Are Some People Healthy and Others Not? The Determinants of Health of Populations. New York: Aldine de Gruyter, 67–92.

Hesse-Biber, Sharlene, and Carter, Gregg Lee. 2000. Working Women in America: Split Dreams. New York: Oxford University Press.

Hewitt, John, and Stokes, Randall. 1975. "Disclaimers." American Sociological Review 40:1–11.

Higley, Stephen R. 1995. Privilege, Power, and Place: The Geography of the American Upper Class. Lanham, MD: Rowman & Littlefield.

Hirschi, Travis, and Gottfredson, Michael. 1983. "Age and the Explanation of Crime." American Journal of Sociology 89:552–584.

Hitlin, Steven, and Piliavin, Jane Allyn. 2004. "Values: Reviving a Dormant Concept." Annual Review of Sociology 30:359–393.

Hochschild, Arlie R. 1985. The Managed Heart: The Commercialization of Human Feeling. Berkeley: University of California Press.

———. 1997. The Time Bind: When Work Becomes Home and Home Becomes Work. New York: Holt.

Hoffman, Bruce. 2006. Inside Terrorism. New York: Columbia University Press.

Hogan, Dennis P., and Astone, Nan Marie. 1986. "The Transition to Adulthood." Annual Review of Sociology 12:109–130.

Hogan, Dennis P., Eggebeen, David J., and Clogg, Clifford C. 1993. "The Structure of Intergenerational Exchanges in American Families." American Journal of Sociology 98:1428–1458.

Holloway, Marguerite. 1994. "Trends in Women's Health: A Global View." Scientific American. August:76–83.

Hondagneu-Sotelo, Pierrette. 2001. Domestica: Immigrant Workers Cleaning and Caring in the Shadows of Affluence. Berkeley: University of California Press.

Horwitz, Allan V. 2002. Creating Mental Illness. Chicago: University of Chicago Press.

Hout, Michael. 1986. "Opportunity and the Minority Middle Class: A Comparison of Blacks in the United States and Catholics in Northern Ireland." American Sociological Review 51:214–223.

Hoyt, Homer. 1939. The Structure and Growth of Residential Neighborhoods in American Cities. Washington, D.C.: Federal Housing Administration.

Hudson, Valerie M., and DenBoer, Andrea M. 2005. Bare Branches: The Security Implications of Asia's Surplus Male Population. Cambridge, MA: MIT Press.

Hull, Elizabeth A. 2005. The Disenfranchisement of Ex-Felons. Philadelphia, PA: Temple University Press.

Hultin, Mia. 2003. "Some Take the Glass Escalator, Some Hit the Glass Ceiling? Career Consequences of Occupational Sex Segregation." Work and Occupations 30:30–61.

Human Rights Watch. 2003. Ill-Equipped: U.S. Prisons and Offenders with Mental Illness. New York: Human Rights Watch.

———. 2004. Women's Rights Division. http://www.hrw.org/women/. Accessed June 2004.

———. 2006. Q & A: Crisis in Darfur. http://hrw.org/english/docs/2004/05/05/ darfur8536_txt.htm

Hunt, Darnell M. 1997. Screening the Los Angeles "Riots": Race, Seeing, and Resistance. New York: Cambridge University Press.

Hunter, James Davison. 1991. Culture Wars: The Struggle to Redefine America. New York: Basic Books.

Huntford, Roald. 2000. Scott and Amundsen. London, UK: Abacus History.

Hurley, Dan. 2005. "Divorce Rate: It's Not as High as You Think." New York Times. April 19:D7.

Iceland, John, and Wilkes, Rima. 2006. "Does Socioeconomic Status Matter? Race, Class, and Residential Segregation." Social Problems 53:248–273.

International Bank for Reconstruction and Development. 2002. World Development Indicators 2002. Washington, D.C.: World Bank.

International Centre for Prison Studies. 2002. "World Prison Brief." http://www.prisonstudies.org/. Accessed June 2003.

Inter-Parliamentary Union. 2006. Women in National Politics, 2006. http://www.ipu.org/wmn-e/ classif.htm. Accessed June 2006.

"Islam." 2001. Ontario Consultants on Religious Tolerance. http://www.religioustolerance .org/islam.htm. Accessed June 2003.

Jacobs, David. 1988. "Corporate Economic Power and the State: A Longitudinal Assessment of Two Explanations." American Journal of Sociology 93:852–881.

Jacobs, David, and Helms, Ronald E. 1997. "Testing Coercive Explanations for Order: The Determinants of Law Enforcement Strength over Time." Social Forces 75:1361–1392.

Jacobs, Jerry. 1989. Revolving Doors: Sex Segregation and Women's Careers. Stanford, CA: Stanford University Press.

Jacobs, Jerry A., and Gerson, Kathleen. 2004. The Time Divide: Work, Family, and Gender Inequality. Cambridge, MA: Harvard University Press.

Jacobson, Matthew Frye. 1998. Whiteness of a Different Color: European Immigrants and the Alchemy of Race. Cambridge, MA: Harvard University Press.

Jacquard, Roland. 2002. In the Name of Osama Bin Laden. Durham, NC: Duke University Press.

James, Angela D., Grant, David M., and Cranford, Cynthia. 2000. "Moving Up, but How Far? African American Women and Economic Restructuring in Los Angeles, 1970–1990." Sociological Perspective 43:399–420.

Janis, Irving. 1982. Groupthink: Psychological Studies of Policy Decisions and Fiascoes. Boston: Houghton Mifflin.

Jasper, James M., and Poulsen, Jane D. 1995. "Recruiting Strangers and Friends: Moral Shocks and Social Networks in Animal Rights and Anti-Nuclear Protests." Social Problems 42:493–512.

Jenkins, J. Craig, and Brent, Barbara. 1989. "Social Protest, Hegemonic Competition, and Social Reform." American Sociological Review 54:891–909.

Jenkins, J. Craig, and Eckert, Craig M. 1986. "Channeling Black Insurgency: Elite Patronage and Professional Social Movement Organizations in the Development of the Black Movement." American Sociological Review 51:812–829.

Jenness, Valerie. 1990. "From Sex as Sin to Sex as Work: COYOTE and the Reorganization of Prostitution as a Social Problem." Social Problems 37:403–420.

Jensen, Holger. 1990. "The Cost of Neglect." Maclean's. May 7:54–55.

Jensen, Leif, McLaughlin, Diane K., and Slack, Tim. 2003. "Rural Poverty: The Persistent Challenge." In David L. Brown and Louis E. Swanson (eds.), Challenges for Rural American in the Twenty-First Century. University Park, PA: Pennsylvania State University Press, pp. 118–131.

Johnson, Kenneth M. 2003. "Unpredictable Directions of Rural Population Growth and Migration." In David L. Brown and Louis E. Swanson (eds.), Challenges for Rural American in the Twenty-First Century. University Park, PA: Pennsylvania State University Press, pp. 19–32.

Johnson, Michael P., and Ferraro, Kathleen J. 2000. "Research on Domestic Violence in the 1990s: Making Distinctions." Journal of Marriage and the Family 62:948–963.

Johnson, Robert C. 1993. "Science, Technology, and Black Community Development." In Albert H. Teich (ed.), Technology and the Future (6th ed.). New York: St. Martin's, pp. 265–282.

Jones, Jeffrey M. 2001. "Americans Felt Uneasy Toward Arabs Even Before September 11." Gallup Poll Monthly, September:52–53.

Jones, Steve. 2003. Let the Games Begin: Gaming Technology and Entertainment among College Students. Washington, D.C.: Pew Internet and American Life Project.

Joseph, Brian D., DeStephano, Johanna, Jacobs, Neil G., and Lehiste, Ilse. 2003. When Languages Collide: Perspectives on Language Conflict, Language Competition, and Language Coexistence. Columbus, OH: Ohio State University Press.

Joynt, Jen, and Poe, Marshall. 2003. "Waterworld." Atlantic Monthly 292(1):42–43.

Jurik, Nancy. 2004. "Imagining Justice: Challenging the Privatization of Public Life." Social Problems 51:1–15.

Kaiser Commission on Medicaid and the Uninsured. 2005. The Uninsured and Their Access to Health Care. Washington, DC: Kaiser Family Foundation.

Kalab, Kathleen. 1987. "Student Vocabularies of Motive Accounts for Absence." Symbolic Interaction 10:71–83.

Kalish, Susan. 1994. "Culturally Sensitive Family Planning: Bangladesh Story Suggests It Can Reduce Family Size." Population Today 22 (February):5.

———. 1995. "Multiracial Births Increase as U.S. Ponders Racial Definition." Population Today 23(4):1–2.

Kalmijn, Matthijs. 1998. "Intermarriage and Homogamy: Causes, Patterns, and Trends." Annual Review of Sociology 24:395–421.

Kalmijn, Matthijs, and Kraaykamp, Gerbert. 1996. "Race, Cultural Capital, and Schooling: An Analysis of Trends in the United States." Sociology of Education 69:22–34.

Kao, Grace, and Thompson, Jennifer S. 2003. "Racial and Ethnic Stratification in Educational Achievement and Attainment." Annual Review of Sociology 29:417–441.

Katz, Jack. 1988. The Seductions of Crime: Moral and Sensual Attractions of Doing Evil. New York: Basic Books.

Kaufman, Peter, and Feldman, Kenneth A. 2004. "Forming Identities in College: A Sociological Approach." Research in Higher Education 45:463–496.

Keister, Lisa A., and Moller, Stephanie. 2000. "Wealth Inequality in the United States." Annual Review of Sociology 26:63–81.

Kemper, Peter. 1992. "Use of Formal and Informal Home Care by the Disabled Elderly." Health Services Research 27:421–451.

Kerbo, Harold R. 1991. Social Stratification and Inequality: Class Conflict in Historical and Comparative Perspective (2nd ed.). New York: McGraw-Hill.

Kessler, Ronald C., Berglund, Patricia, Demler, Olga, Jim, Robert, and Walters, Ellen E. 2005. "Lifetime prevalence and age-of-onset distributions of DSM IV disorders in the National Comorbidity Survey Replication." Archives of General Psychiatry 62:593–602.

Kimball, Dan. 2003. The Emerging Church. Grand Rapids, MI: Zondervan

Kimmel, Michael S. 2000. The Gendered Society. New York: Oxford University Press.

Kinzer, Stephen. 2003. All the Shah's Men: An American Coup and the Roots of Middle East Terror. New York: Wiley.

Kiple, Kenneth F. 1993. Cambridge World History of Human Disease. New York: Cambridge University Press.

Klandermas, Bert. 1984. "Mobilization and Participation: Social Psychological Expansion of Resource Mobilization Theory." American Sociological Review 49:583–600.

Kluegel, James R., and Smith, Eliot R. 1983. "Affirmative Action Attitudes: Effects of Self-Interest, Racial Affect, and Stratification Beliefs on Whites' Views." Social Forces 61:170–181.

Koblik, Steven. 1975. Sweden's Development from Poverty to Affluence 1750–1970. Minneapolis: University of Minnesota Press.

Kohn, Melvin, Schooler, Carmi, and Associates. 1983. Work and Personality: An Inquiry into the Impact of Social Stratification. Norwood, NJ: Ablex.

Kohut, Andrew, and Stokes, Bruce. 2006. America Against the World. New York: Henry Holt.

Kolko, Gabriel. 1999. "Ravaging the Poor: The International Monetary Fund Indicted by Its Own Data." International Journal of Health Services 29:51–57.

Kollock, Peter, Blumstein, Phillip, and Schwartz, Pepper. 1985. "Sex and Power in Conversation: Conversational Privileges and Duties." American Sociological Review 50:34–46.

Konig, René. 1968. "Auguste Comte." In David J. Sills (ed.), International Encyclopedia of the Social Sciences. Vol. 3. New York: Macmillan and Free Press.

Korpi, Walter. 1989. "Power, Politics, and State Autonomy in the Development of Social Citizenship." American Sociological Review 54:309–328.

Kozol, Jonathan. 2005. The Shame of the Nation: The Restoration of Apartheid Schooling in America. New York: Crown.

Kramer, Lisa A., and Lambert, Stephen. 2003. "Sex-Linked Bias in Chances of Being Promoted to Supervisor." Sociological Perspectives 44:111–127.

Kraska, Peter B., and Kappeler, Victor E. 1997. "Militarizing American Police: The Rise and Normalization of Paramilitary Units." Social Problems 44:1–18.

Krech, Shepard I. 1999. The Ecological Indians: Myth and History. W.W. Norton, New York.

Kreider, Rose M. 2005. "Number, Timing, and Duration of Marriages and Divorces: 2001." Current Population Reports, P70–97. U.S. Census Bureau, Washington, DC.

Kreider, Rose M., and Fields, Jason. 2005. "Living Arrangements of Children: 2001." Current Population Reports, P70–104. U.S. Census Bureau, Washington, DC.

Krysan, Maria. 2000. "Prejudice, Politics, and Public Opinion: Understanding the Sources of Racial Policy Attitudes." Annual Review of Sociology 26:135–168.

Ku, Leighton, Sonenstein, Freya, Lindberg, Laura, Bradner, Carolyn H., Boggess, Scott, and Pleck, Joseph. 1998. "Understanding Changes in Sexual Activity among Young Metropolitan Men: 1979–1995." Family Planning Perspectives 30(6):256–262.

Kunda, Gideon. 1993. Engineering Culture: Control and Commitment in a High Tech Corporation. Philadelphia: Temple University Press.

Kunstler, James Howard. 1994. The Geography of Nowhere: The Rise and Decline of America's Man-Made Landscape. New York: Simon & Schuster.

Lamanna, Mary Ann, and Riedmann, Agnes. 2000. Marriages and Families: Making Choices in a Diverse Society (7th ed.). Belmont, CA: Wadsworth.

Lambert, Bruce, and Fernanda Santos. 2006. "'First' Suburbs Growing Older and Poorer, Report Warns." New York Times. February 16:B1.

Lamont, Michelle, and Fournier, Marcel (eds.). 1992. Cultivating Differences: Symbolic Boundaries and the Making of Inequality. Chicago: University of Chicago Press.

Lane, Sandra D., and Cibula, Donald A. 2000. "Gender and Health." In Gary L. Albrecht, Ray Fitzpatrick, and Susan C. Scrimshaw (eds.), Handbook of Social Studies in Health and Medicine. Thousand Oaks, CA: Sage Publications, pp. 136–153.

Langdon, Philip. 1994. A Better Place to Live: Reshaping the American Suburb. New York: Harper.

Lappé, Frances Moore, Collins, Joseph, and Rosset, Peter. 1998. World Hunger: Twelve Myths. New York: Grove.

Lareau, Annette. 2003. Unequal Childhoods: Class, Race, and Family Life. Berkeley, CA: University of California Press.

Laumann, Edward O., Gagnon, John H., Michael, Robert T., and Michael, Stuart. 1994. The Social Organization of Sexuality: Sexual Practices in the United States. Chicago: University of Chicago Press.

Lavee, Yoar, McCubbin, Hamilton I., and Patterson, Joan M. 1985. "The Double ABCX Model of Family Stress and Adaptation: An Empirical Test by Analysis of

Structural Equations with Latent Variables." Journal of Marriage and the Family 47:811–825.

Lawrence, Bruce B. 1998. Shattering the Myth: Islam Beyond Violence. Princeton, NJ: Princeton University Press.

Lawton, Julia. 2003. "Lay Experiences of Health and Illness: Past Research and Future Agendas." Sociology of Health and Illness 25:23–40.

Le, C.N. 2006. "Employment & Occupational Patterns" Asian-Nation: The Landscape of Asian America. http://www.asian-nation.org/ index.shtml. Accessed June 2006.

Leahey, Erin, and Guo, Guang. 2001. "Gender Differences In Mathematical Trajectories." Social Forces 80:713-732.

Leblanc, Lauraine. 1999. Pretty in Punk: Girls' Gender Resistance in a Boys' Subculture. New Brunswick, NJ: Rutgers University Press.

Lee, Yun-Suk, Schneider, Barbara, and Waite, Linda J. 2003. "Children and Housework: Some Unanswered Questions." Sociological Studies of Children and Youth 9:105–125.

Lefkowitz, Bernard. 1997. Our Guys: The Glen Ridge Rape and the Secret Life of the Perfect Suburb. Berkeley: University of California Press.

Lemert, Edwin. 1981. "Issues in the Study of Deviance." Sociological Quarterly 22:285–305.

Lenski, Gerhard. 1966. Power and Privilege: A Theory of Social Stratification. New York: McGraw-Hill.

Leonhardt, David. 2004. "As Wealthy Fill Top Colleges, Concerns Grow Over Fairness." New York Times. April 22:A1+.

Levy, Emanuel. 1990. "Stage, Sex and Suffering: Images of Women in American Films." Empirical Studies of the Arts 8(1):53–76.

Lewin, Tamar. 2000. "Boom in Gene Testing Raises Questions on Sharing Results." New York Times. July 21:11.

———. 2006. "At Colleges, Women Are Leaving Men in the Dust." New York Times. July 9:A11.

Lewin, Tamar, and Medina, Jennifer. 2003. "To Cut Failure Rate, Schools Shed Students." New York Times. July 31:11.

Lewis, Oscar. 1969. "The Culture of Poverty." In Daniel P. Moynihan (ed.), On Understanding Poverty. New York: Basic Books.

Lichtenstein, Nelson. 2003. State of the Union: A Century of American Labor. Princeton, NJ: Princeton University Press.

Lichter, Daniel T., LeClere, Felicia B., and McLaughlin, Diane K. 1991. "Local Marriage Markets and the Marital Behavior of Black and White Women." American Journal of Sociology 96:843–867.

Lichter, Daniel T., McLaughlin, Diane K., Kephart, George, and Landry, David J. 1992. "Race and the Retreat from Marriage: A Shortage of Marriageable Men?" American Sociological Review 57:781–799.

Lin, Chien, and Liu, William T. 1993. "Intergenerational Relationships Among Chinese Immigrant Families from Taiwan." In Harriett Pipes McAdoo (ed.), Family Ethnicity: Strength in Diversity. Newbury Park, CA: Sage, pp. 271–286.

Lin, Nan. 1990. "Social Resources and Social Mobility: A Structured Theory of Status Attainment." In R. Breiger (ed.), Social Mobility and Social Structure. New York: Cambridge University Press, pp. 247–271.

Lincoln, James R., and McBride, Kerry. 1987. "Japanese Industrial Organization in Comparative Perspective." Annual Review of Sociology 13:289–312.

Link, Bruce G., Cullen, Francis T., Frank, James, and Wozniak, John F. 1987. "The Social Rejection of Former Mental Patients: Understanding Why Labels Matter." American Journal of Sociology 92:1461–1500.

Link, Bruce G., Struening, Elmer L., Rahav, Michael, Phelan, Jo C., and Nuttbrock, Larry. 1997. "On Stigma and Its Consequences: Evidence from a Longitudinal Study of Men with Dual Diagnoses of Mental Illness and Substance Abuse." Journal of Health and Social Behavior 38:177–190.

Linn, James Weber, and Scott, Anne Firor. 2000. Jane Addams: A Biography. Champlain, IL: University of Illinois Press.

Liska, Allen E., Chamlin, Mitchell B., and Reed, Mark. 1985. "Testing the Economic Production and Conflict Models of Crime Control." Social Forces 64:119–138.

Little, Ruth E. 1998. "Public Health in Central and Eastern Europe and the Role of Environmental Pollution." Annual Review of Public Health 19:153–172.

Lo, Clarence Y. H. 1982. "Countermovements and Conservative Movements in the Contemporary U.S." Annual Review of Sociology 8:10–34.

Loe, Meika. 2004. The Rise of Viagra: How the Little Blue Pill Changed Sex in America. New York: New York University Press.

Lofland, John. 1985. Protest: Studies of Collective Behavior and Social Movements. New Brunswick, NJ: Transaction.

Logan, John R., and Spitze, Glenna D. 1994. "Family Neighbors." American Journal of Sociology 100: 453–476.

Lohr, Steve. 2006. "Outsourcing Is Climbing Skills Ladder." New York Times. February 16:C1+.

London, Bruce, and Robinson, Thomas. 1989. "The Effect of International Dependence on Income Inequality and Political Violence." American Sociological Review 54:305–308.

Lorber, Judith. 1994. Paradoxes of Gender. New Haven, CT: Yale University Press.

Love, John M., Harrison, Linda, Sagi-Schwartz, Abraham, van Ijzendoorn, Marinus, Ross, Christine, Ungerer, Judy A., Raikes, Helen, Brady-Smith, Christy, Boller, Kimberly, Brooks-Gunn, Jeanne, Constantine, Jill, Eliason Kisker, Ellen, Paulsell, Diane, and Chazan-Cohen, Rachel. 2003. "Child Care Quality Matters: How Conclusions May Vary With Context." Child Development 74:1021–1033.

Lowney, Kathleen S. 2003. "Wrestling with Criticism: The World Wrestling Federation's Ironic Campaign against the Parents Television Council." Symbolic Interaction 26:427–446.

Luker, Kristen. 1985. Abortion and the Politics of Motherhood. Berkeley: University of California Press.

———. 1996. Dubious Conceptions: The Politics of Teenage Pregnancy. Cambridge, MA: Harvard University Press.

Lurigio, Arthur. 1990. "Introduction." Crime and Delinquency 36:3–5.

Luxembourg Income Study. 2002. http://www .lisproject .org/. Accessed June 2003.

Lye, Diane N. 1996. "Adult Child-Parent Relationships." Annual Review of Sociology 22:79–102.

Lynn, Barry C. "Breaking the Chain: The Antitrust Case Against Wal-Mart." Harpers 313(1874):29–36.

MacKenzie, Doris Layton, Wilson, David B., and Kider, Suzanne B. 2001. "Effects of Correctional Boot Camps on Offending." Annals of the American Academy of Political and Social Science 578:126–143.

Maher, Lisa, and Daly, Kathleen. 1996. "Women in the Street-Level Drug Economy: Continuity or Change?" Criminology 34:465–491.

Marcus, Eric. 2002. Making Gay History: The Half Century Fight for Lesbian and Gay Equal Rights. New York: Harper.

Marger, Martin. 2003. Race and Ethnic Relations (6th ed.). Belmont, CA: Wadsworth.

Marmot, M. G., Kogevinas, M., and Elston, M. 1987. "Social/Economic Status and Disease." Annual Review of Public Health 8:111–135.

Marmot, Michael G. 2004. The Status Syndrome: How Your Social Standing Directly Affects Your Health and Life Expectancy. London: Bloomsbury.

Marsden, George M. 2006. Fundamentalism and American Culture. New York: Oxford University Press.

Martin, Karin A. 1998. "Becoming a Gendered Body: Practices of Preschools." American Sociological Review 63:494–511.

Martin, Philip, and Widgren, Jonas. 2002. "International Migration: Facing the Challenge." Population Bulletin 57(1):1–40.

Martin, Teresa, and Bumpass, Larry. 1989. "Recent Trends in Marital Disruption." Demography 26:37–51.

Marx, Gary T. (ed.). 1971. Racial Conflict. Boston: Little, Brown.

Marx, Gary T., and McAdam, Douglas. 1994. Collective Behavior and Social Movements: Process and Structure. Englewood Cliffs, NJ: Prentice Hall.

Marx, Karl, and Engels, Friedrich. 1967. The Communist Manifesto. London: Penguin.

Massey, Douglas S. 1990. "American Apartheid: Segregation and the Making of the Underclass." American Journal of Sociology 96:329–357.

———. 2006. "The Wall that Keeps Illegal Workers In." New York Times. April 4:A23.

Matsueda, Ross L., and Kreager, Derek A., and Huizinga, David. 2006. "Deterring Delinquents: A Rational Choice Model of Theft and Violence." American Sociological Review 71:95–122.

Maume, David. 1999. "Occupational Segregation and the Career Mobility of White Men and Women." Social Forces 77:1433–1459.

———. 2004. "Wage Discrimination over the Life Course: A Comparison of Explanations." Social Problems 51:505–527.

McAdam, Doug. 1986. "Recruitment to High-Risk Activism." American Journal of Sociology 92:64–90.

McAdam, Doug, and Paulsen, Ronnelle. 1993. "Specifying the Relationship between Social Ties and Activism." American Journal of Sociology 99:640–667.

McAdam, Doug, and Snow, David A. 1997. Social Movements: Readings on their Emergence, Mobilization, and Dynamics. Los Angeles: Roxbury Publishing Company.

McBrier, Debra Branch. 2003. "Gender and Career Dynamics Within a Segmented Professional Labor Market: the Case of Law Academia." Social Forces 81:1201–1266.

McCarthy, Bill. 2002. "New Economics of Sociological Criminology." Annual Review of Sociology 28:417–442.

McClain, Dylan Loeb. 2005. "Richer Than Ever, but Watch Out for Missing Costs." New York Times. December 5:C6.

McCoy, Alfred W. 2006. A Question of Torture: CIA Interrogation, from the Cold War to the War on Terror. New York: Henry Holt.

McDermott, Monica, and Samson, Frank L. 2005. "White Racial and Ethnic Identity in the United States." Annual Review of Sociology 31:245–261.

McDill, Edward L., Natriello, Gary, and Pallas, Aaron. 1986. "A Population at Risk: Potential Consequences of Tougher School Standards for School Dropouts." American Journal of Education 94:135–181.

McGinn, Robert E. 1991. Science, Technology, and Society. Englewood Cliffs, NJ: Prentice Hall.

McGranahan, David A. 2003. "How People Make a Living in Rural America." In David L. Brown and Louis E. Swanson (eds.), Challenges for Rural American in the Twenty-First Century. University Park, PA: Pennsylvania State University Press, pp. 135–151.

McGuffey, C. Shawn, and Rich, B. Lindsay. 1999. "Playing in the Gender Transgression Zone: Race, Class, and Hegemonic Masculinity in Middle Childhood." Gender and Society 13:608–627.

McKinlay, John B. 1994. "A Case for Refocusing Upstream: The Political Economy of Illness." In Peter Conrad and Rachelle Kern (eds.), The Sociology of Health and Illness. New York: St. Martin's, pp. 509–530.

McKinlay, John B., and McKinlay, Sonja J. 1977. "The Questionable Effect of Medical Measures on the Decline of Mortality in the United States in the Twentieth Century." Milbank Memorial Fund Quarterly 55:405–428.

McLellan, David. 2006. Karl Marx: A Biography (4th ed.) London, UK: Palgrave Macmillan.

McPherson, J. Miller, Popielarz, Pamela A., and Drobnic, Sonja. 1992. "Social Networks and Organizational Dynamics." American Sociological Review 57:153–170.

McPherson, J. Miller, and Smith-Lovin, Lynn. 2002. "Cohesion and Membership Duration: Linking Groups, Relations and Individuals in an Ecology of Affiliation." Advances in Group Processes 19:1–36.

McPherson, J. Miller, Smith-Lovin, Lynn, and Brashears, Matthew E. 2006. "Social Isolation in America: Changes in Core Discussion Networks over Two Decades." American Sociological Review 71:353–375.

Mead, George Herbert. 1934. Mind, Self, and Society: From the Standpoint of a Social Behaviorist (Charles W. Morris, ed.). Chicago: University of Chicago Press.

Mead, Lawrence M. 1986. Beyond Entitlement: The Social Obligations of Citizenship. New York: Free Press.

———. 1992. The New Politics of Poverty: The Nonworking Poor in America. New York: Basic Books.

Mead, Sara. 2006. The Truth About Boys and Girls. Washington, DC: Education Sector.

Mellon, Margaret. 1993. "Altered Traits." Nucleus, Fall:4–6, 12.

Menjívar, Cecilia. 2000. Fragmented Ties. Berkeley, Calif.: University of California Press.

Merton, Robert. 1957. Social Theory and Social Structure (2nd ed.). New York: Free Press.

Messerschmidt, James W. 1993. Masculinities and Crime: Critique and Reconceptualization of Theory. Lanhan, MN: Rowman & Littlefield.

Messner, Steven F. 1989. "Economic Discrimination and Societal Homicide Rates: Further Evidence on the Cost of Inequality." American Sociological Review 54:597–611.

Meyer, David S. 2004. "Protest and Political Opportunities." Annual Review of Sociology 30:125–145.

Meyer, David S., and Staggenborg, Suzanne. 1996. "Movements, Countermovements, and the Structure of Political Opportunity." American Journal of Sociology 101:1628–1660.

Meyer, Pamela A., Pivetz, Timothy, Dignam, Timothy A., Homa, David M., Schoonover, Jaime, and Brody, Debra. 2003. "Surveillance for Elevated Blood Lead Levels Among Children—United States, 1997-2001." Morbidity and Mortality Weekly Report 2003:52(No. SS-10):1–21.

Meyerowitz, Joanne. 2002. How Sex Changed: A History of Transsexuality in the United States. Cambridge, MA: Harvard University Press.

Michalowski, Raymond J., and Kramer, Ronald C. 1987. "The Space Between Laws: The Problem of Corporate Crime in a Transnational Context." Social Problems 34:34–53.

Milkie, Melissa A. 1999. "Social Comparisons, Reflected Appraisals, and Mass Media: The Impact of Pervasive Beauty Images on Black and White Girls' Self Concepts." Social Psychology Quarterly 62:190–210.

Miller, Alan S., and Hoffmann, John P. 1999. "The Growing Divisiveness: Culture Wars or a War of Words." Social Forces 78:721–752.

Mills, C. Wright. 1940. "Situated Actions and Vocabularies of Motives." American Sociological Review 5:904–913.

———. 1956. The Power Elite. New York: Oxford University Press.

———. 1959. The Sociological Imagination. Oxford, England: Oxford University Press.

Minino, Arialdi M. 2002. "Deaths: Final Statistics for 2000." National Vital Statistics Report 50(15):1–120.

Mirowsky, John, and Ross, Catherine E. 1995. "Sex Differences in Distress: Real or Artifact?" American Sociological Review 60:449–468.

Mizruchi, Mark. 1989. "Similarity of Political Behavior Among Large American Corporations." American Journal of Sociology 95:401–424.

———. 1990. "Determinants of Political Opposition among Large American Corporations." Social Forces 68:1065–1088.

Moe, Richard, and Wilkie, Carter. 1997. Changing Places: Rebuilding Community in the Age of Sprawl. New York: Henry Holt.

Moen, Phyllis, Dempster-McClain, Donna, and Williams, Robin. 1989. "Social Integration and Longevity: An Event-History Analysis of Women's Roles and Resilience." American Sociological Review 54:635–647.

Mokdad, Ali H., Marks, James S., Stroup, Donna F. and Gerberding, Julie L. 2004. "Actual Causes of Death in the United States, 2000." Journal of the American Medical Association 291:1238–1245.

Molm, Linda D. 2003. "Theoretical Comparisons of Forms of Exchange." Sociological Theory 21:1–17.

Molm, Linda D., and Cook, Karen S. 1995. "Social Exchange Theory." In Karen S. Cook, Gary A. Fine, and James S. House (eds.), Sociological Perspectives on Social Psychology. New York: Allyn & Bacon, pp. 209–235.

Monroe, Charles R. 1995. World Religions: An Introduction. Amherst, NY: Prometheus.

Moore, Helen A., and Whitt, Hugh P. 1986. "Multiple Dimensions of the Moral Majority Platform: Shifting Interest Group Coalitions." Sociological Quarterly 27(3):423–439.

Morgeanthau, Tom. 1995. "What Color Is Black?" Newsweek. February 13:63–70.

Morris, Aldon D., and Mueller, Carol. 1992. Frontiers in Social Movement Theory. New Haven, CT: Yale University Press.

Morris, Martina, and Western, Bruce. 1999. "Inequality in Earnings at the Close of the Twentieth Century." Annual Review of Sociology 25:623–657.

Morton, Lois Wright. 2003. In David L. Brown and Louis E. Swanson (eds.), Challenges for Rural America in the Twenty-First Century. University Park, PA: Pennsylvania State University Press, pp. 290–304.

Moshoeshoe II. 1993. "Return to Self-Reliance: Balancing the African Condition and the Environment." In Pablo Piacetini (ed.), Story Earth: Native Voices on the Environment. San Francisco, CA: Mercury House, pp. 158–170.

Mulcahy, Aogan. 1995. "Claims-Making and the Construction of Legitimacy: Press Coverage of the 1981 Northern Irish Hunger Strike." Social Problems 42:449–467.

Munch, Allison, McPherson, J. Miller, and Smith-Lovin, Lynn. 1997. "Gender, Children, and Social Contact: The Effects of Childrearing for Men and Women." American Sociological Review 62:509–520.

Murphy, Caryle. 1994. "Egypt: An Uneasy Portent of Change." Current History 93:78–82.

Murr, Andrew, and Morganthau, Tom. 2001. "Burning Suburbia." Newsweek. January 15:32–33.

Murray, Charles A. 1984. Losing Ground: American Social Policy 1950–1980. New York: Basic Books.

Mutran, Elizabeth, and Reitzes, Donald C. 1984. "Intergenerational Support Activities and Well-Being Among the Elderly: A Convergence of Exchange and Symbolic Interaction Perspectives." American Sociological Review 49:117–130.

"NAFTA and Workers' Rights and Jobs." 2003. Public Citizen: Global Trade Watch. http://www.citizen.org/trade/nafta/ jobs/. Accessed June 2003.

Nagel, Joane. 1994. "Constructing Ethnicity: Creating and Recreating Ethnic Identity and Culture." Social Problems 41:152–176.

Nardi, Peter. 1992. Men's Friendships: Research on Men and Masculinities. Newbury Park, CA: Sage.

National Adoption Information Clearinghouse. Voluntary Relinquishment for Adoption: Numbers and Trends. http://naic .acf.hhs.gov/pubs/s_place.cfm.

National Alliance for Caregiving. 2003. Caregiving in the United States: Findings from a National Survey. Bethesda, MD: National Alliance for Caregiving.

National Center for Health Statistics. 2005. Health, United States, 2005. Hyattsville, MD.

National Coalition for the Homeless. 2006. "How Many People Experience Homelessness?" NCH Fact Sheet #2. http://www.nationalhomeless.org/publications/ facts/ How_Many.pdf. Accessed June 2006.

National Gay and Lesbian Task Force. 2006. States, Cities and Counties with Civil Rights Ordinances, Policies or Proclamations Prohibiting Discrimination on the Basis of Sexual Orientation. New York: National Gay and Lesbian Task Force.

National Highway Traffic Safety Administration. 2005. Traffic Safety Facts: Motorcycle Helmet Use Laws. Washington, DC: US Department of Transportation.

Nelson, Alan R., Smedley, Brian D. and Stith, Adrienne Y. 2002. Unequal Treatment: Confronting Racial and Ethnic Disparities in Health Care. Washington, DC: Institute of Medicine, National Academy Press.

Newman, Katherine S. 1999a. Falling from Grace: Downward Mobility in the Age of Affluence (rev. ed.). Berkeley: University of California Press.

———. 1999b. No Shame in My Game: The Working Poor in the Inner City. New York: Knopf.

Newman, Katherine S., and Massengill, Rebekah Peeples. 2006. The Texture of Hardship: Qualitative Sociology of Poverty, 1995–2005. Annual Review of Sociology 32:1–24.

NHSDA Report. 2003. Substance Use among American Indians or Alaska Natives. Washington D.C.: Substance Abuse and Mental Health Services Administration.

NICHD Early Child Care Research Network. 2005. Early Child Care and Children's Development in the Primary Grades: Follow-Up Results from the NICHD Study of Early Child Care. American Educational Research Journal 43:537–570.

Norrish, Barbara R., and Rundall, Thomas G. 2001. "Hospital Restructuring and the Work of Registered Nurses." Milbank Quarterly 79:55–79.

Ogletree, Charles, Jr., and Sarat, Austin. 2006. From Lynch Mobs to the Killing State: Race And the Death Penalty in America. New York: New York University Press.

Oldenburg, Ray. 1997. The Great Good Place: Cafés, Coffee Shops, Community Centers, Beauty Parlors, General Stores, Bars, Hangouts, and How They Get You Through the Day. New York: Marlowe.

Oleksyn, Veronika. 2006. "Birth of a Notion: Incentives Offered for Having More Kids." Seattle Times. July 22:A3.

Olsen, Gregg M. 1996. "Re-modeling Sweden: The Rise and Demise of the Compromise in a Global Economy." Social Problems 43:1–20.

Ontario Consultants on Religious Tolerance. 2006. "Religious Affiliation." http://www.religioustolerance.org/compuswrld.htm. Accessed June 2006.

Orenstein, Peggy. 1994. School Girls: Young Women, Self-Esteem, and the Confidence Gap. New York: Doubleday.

Osgood, D. Wayne, Wilson, Janet K., O'Malley, Patrick M., Bachman, Jerald G., and Johnston, Lloyd D. 1996. "Routine Activities and Individual Deviant Behavior." American Sociological Review 61:635–55.

Ouichi, William G., and Wilkins, Alan L. 1985. "Organizational Culture." Annual Review of Sociology 11:457–483.

"Out of Sight, Out of Mind." 2000. Economist. May 20:27–28.

Owens, Timothy J., Stryker, Sheldon, and Goodman, Norman. 2001. Extending Self-Esteem Research: Sociological and Psychological Currents. New York: Cambridge University Press.

Park, Kristin. 2005. "Choosing Childlessness: Weber's Typology of Action and Motives of the Voluntarily Childless." Sociological Inquiry 75:372–402.

Parreñas, Rhacel Salazar. 2000. "Migrant Filipina Domestic Workers and the International Division of Reproductive Labor." Gender and Society 14:560–581.

Parsons, Talcott. 1964. "The School Class as a Social System: Some of Its Functions in American Society." In Talcott Parsons (ed.), Social Structure and Personality. New York: Free Press.

Passos, Nikos, and Agnew, Robert (eds.). 1997. The Future of Anomie Theory. Boston: Northeastern University Press.

Paternoster, Raymond. 1989. "Absolute and Restrictive Deterrence in a Panel of Youth: Explaining the Onset, Persistence/Desistance, and Frequency of Delinquent Offending." Social Problems 36:289–309.

Paules, Greta Foff. 1991. Dishing It Out: Power and Resistance among Waitresses in a New Jersey Restaurant. Philadelphia: Temple University Press.

Paz, Juan J. 1993. "Support of Hispanic Elderly." In Harriett Pipes McAdoo (ed.), Family Ethnicity: Strength in Diversity. Newbury Park, CA: Sage, pp. 177–183.

Peabody, John W. 1996. "Economic Reform and Health Sector Policy: Lessons from Structural Adjustment Programs." Social Science and Medicine 43:823–835.

People for the American Way. 2004. Shattering the Myth: An Initial Snapshot of Voter Disenfranchisement in the 2004 Elections. Washington, DC: People for the American Way.

Pérez-Peña, Richard. 2004. "Study Says 50% of Children Enter Shelters with Asthma." New York Times. March 2:A23.

Perrow, Charles. 1984. Normal Accidents: Living with High-Risk Technologies. New York: Basic Books.

Perrucci, Robert, and Wysong, Earl. 2002. The New Class Society: Goodbye American Dream? (2nd ed.). Lanham, MD: Rowman & Littlefield.

Petersilia, Joan. 1999. "A Decade of Experimenting with Intermediate Sanctions: What Have We Learned?" Justice Research and Policy 1:9–23.

Peterson, Nicolas. 1993. "Demand Sharing: Reciprocity and the Pressure for Generosity among Foragers." American Anthropologist 95:860–874.

Peterson, Peter G. 1999. Gray Dawn: How the Coming Age Wave Will Transform America—and the World. New York: Times Books.

Peterson, Richard A., and Simkus, Albert. 1992. "How Musical Tastes Mark Occupational Status Groups." In Michelle Lamont and Marcel Fournier (eds.). Cultivating Differences: Symbolic Boundaries and the Making of Inequality. Chicago: University of Chicago Press, pp. 152–186.

Pew Research Center. 2003b. Religion and Politics: Contention and Consensus. Washington, D.C.: Pew Research Center.

———. 2005. Huge Racial Divide over Katrina and Its Consequences. Washington, D.C.: Pew Research Center.

———. 2006. Less Opposition to Gay Marriage, Adoption and Military Service. Washington, D.C.: Pew Research Center.

Piaget, Jean. 1954. The Construction of Reality in the Child. New York: Basic Books.

Piliavin, Irving, Gartner, Rosemary, Thornton, Craig, and Matsueda, Ross. 1986. "Crime, Deterrence, and Rational Choice." American Sociological Review 51:101–119.

Piven, Frances Fox, and Cloward, Richard A. 1988. Why Americans Don't Vote. New York: Pantheon.

———. 2000. Why Americans Still Don't Vote: And Why Politicians Want It That Way. Boston: Beacon.

Plummer, Gayle. 1985. "Haitian Migrants and Backyard Imperialism." Race and Class 26: 35–43.

Polling Report. 2006. "Religion." http://www.Pollingreport.com/religion.htm. Accessed June 2006.

Pollock, Philip H., III. 1982. "Organizations and Alienation. The Mediation Hypothesis Revisited." Sociological Quarterly 23:143–155.

Population Reference Bureau. 2006. World Population Data Sheet. Washington D.C.: Population Reference Bureau.

Porter, Eduardo. 2006. "Cost of Illegal Immigration May be Less than Meets the Eye." New York Times. April 16:B3.

Post, Tom. 1993. "Sailing into Big Trouble." Newsweek. November 1:34–35.

Poston, Dudley, Jr. 2000. "Social and Economic Development and the Fertility Transitions in Mainland China and Taiwan." Population and Development Review 26(Supplement):40–60.

Preimsberger, Duane. 1996. "Cops and Space Scientists: New Crime-Fighting Partners." Police Chief 63:108–114.

Prestby, J. E., Wandersman, A., Florin, P., Rich, R., and Chavis, D. M. 1990. "Benefits, Costs, Incentive Management and Participation in Voluntary Associations: A Means to Understanding and Promoting Empowerment." American Journal of Community Psychology 18:117–150.

Proctor, Bernadette, and Dalaker, Joseph. 2002. "Poverty in the United States, 2001." Current Population Reports P60–219. Washington, D.C.: U.S. Census Bureau.

Project on Student Debt. 2006. High Hopes, Big Debts. Berkeley, CA: Project on Student Debt.

Prokos, Anastasia, and Padavic, Irene. 2005. "An Examination of Competing Explanations for the Pay Gap among Scientists and Engineers." Gender & Society 19:523–543.

Public Citizen Water Privatization Overview. http://www.citizen.org/cmep/Water/general. Accessed April 2006.

Putnam, Robert D. 2000. Bowling Alone: The Collapse and Revival of American Community. New York: Simon & Schuster.

Qian, Zhenchao. 1999. "Who Intermarries? Education, Nativity, Region, and Interracial Marriage, 1980 and 1990." Journal of Comparative Family Studies 30:579–597.

Quadagno, Jill. 2002. Aging and the Life Course: An Introduction to Social Gerontology (2nd ed.). New York: McGraw Hill.

———. 2005. One Nation Uninsured: Why The U.S. Has No National Health Insurance. New York: Oxford University Press.

Quart, Alissa. 2003. Branded: The Buying and Selling of Teenagers. Cambridge, MA: Perseus.

Quillian, Lincoln. 1996. "Group Threat and Regional Change in Attitudes toward African-Americans." American Journal of Sociology 102:816–860.

Quinney, Richard. 1980. Class, State, and Crime (2nd ed.). New York: Longman.

Raley, J. Kelly. 1996. "A Shortage of Marriageable Men? A Note on the Role of Cohabitation in Black-White Differences in Marriage Rates." American Sociological Review 61:973–983.

Ramirez, Roberto R., and de la Cruz, Patrica. 2003. "The Hispanic Population in the U.S.: March 2002." Current Population Reports, Series P20–545. Washington, D.C.: U.S. Government Printing Office.

Rank, Mark R. 2004. One Nation, Underprivileged: Why American Poverty Affects Us All. New York: Oxford University Press.

Rankin, Susan R. 2003. Campus Climate for Gay, Lesbian, Bisexual, and Transgender People: A National Perspective. New York: National Gay and Lesbian Taskforce.

Reichman, Nancy. 1989. "Breaking Confidences: Organizational Influences on Insider Trading." Sociological Quarterly 30:185–204.

Reid, Lori L. 2002. "Occupational Segregation, Human Capital, and Motherhood: Black Women's Higher Exit Rates from Full-Time Employment." Gender and Society 16:728–747.

Reiman, Jeffrey. 1998. The Rich Get Richer and the Poor Get Prison: Ideology, Class, and Criminal Justice (5th ed.). Boston: Allyn & Bacon.

Reskin, Barbara. 1989. "Women Taking 'Male' Jobs Because Men Leave Them." IlliniWeek. July 20:7.

Ricento, Thomas, and Burnaby, Barbara (eds.). 1998. Language and Politics in the United States and Canada: Myths and Realities. Mahwah, NJ: Erlbaum.

Rich, Paul. 1999. "American Voluntarism, Social Capital, and Political Culture." Annals of the American Academy of Political and Social Science 565:15–34.

Rich, Robert. 1977. The Sociology of Law. Washington, D.C.: University Press of America.

Ridgeway, Cecilia L., and Smith-Lovin, Lynn. 1999. "The Gender System and Interaction." Annual Review of Sociology 25:191–216.

Riedmann, Agnes. 1987. "Ex-Wife at the Funeral: Keyed Anti-Structure." Free Inquiry in Sociology 16:123–129.

———. 1993. Science That Colonizes. Philadelphia: Temple University Press.

Rieker, Patricia R., and Bird, Chloe E. 2000. "Sociological Explanations of Gender Differences in Mental and Physical Health." In Chloe E. Bird, Peter Conrad, and Allan Fremont (eds.), Handbook of Medical Sociology. New York: Prentice-Hall, pp. 98–113.

Rietschlin, John. 1998. "Voluntary Association Membership and Psychological Distress." Journal of Health and Social Behavior 39:348–355.

Risman, Barbara J. 1998. Gender Vertigo: American Families in Transition. New Haven, CT: Yale University Press.

Ritzer, George. 1996. The McDonaldization of Society (rev. ed.). Thousand Oaks, CA: Pine Forge Press.

Robert, Christopher, and Carnevale, Peter J. 1997. "Group Choice in Ultimatum Bargaining." Organizational Behavior and Human Decision Processes 72:256–279.

Robert, Stephanie A., and House, James S. 2000. "Socioeconomic Inequalities in Health: Integrating, Individual-, Community-, and Societal-Level Theory and Research." In Gary L. Albrecht, Ray Fitzpatrick, and Susan C. Scrimshaw (eds.), Handbook of Social Studies in Health and Medicine. Thousand Oaks, CA: Sage, pp. 115–135.

Romaine, Suzanne. 2000. Language in Society: An Introduction to Sociolinguistics (2nd ed.). Oxford, England: Oxford University Press.

Romero, Simon, and Elder, Janet. 2003. "Hispanics in U.S. Report Optimism." New York Times. August 6:11.

Roschelle, Anne R., and Kaufman, Peter. 2004. "Fitting In and Fighting Back: Stigma Management Strategies Among Homeless Kids." Symbolic Interaction 27:23–46.

Rose, Peter. 1981. They and We: Racial and Ethnic Relations in the United States (3rd ed.). New York: Random House.

Rosen, Jeffrey. 2000. The Unwanted Gaze: The Destruction of Privacy in America. New York: Random House.

Rosenthal, Carolyn J. 1985. "Kinkeeping in the Familial Division of Labor." Journal of Marriage and the Family 47:965–974.

Ross, Catherine E., and Mirowsky, John. 1999. "Refining the Association between Education and Health: The Effects of Quantity, Credential, and Selectivity." Demography 36:445–460.

Ross, Stephen L., and Turner, Margery A. 2005. "Housing Discrimination in Metropolitan America: Explaining Changes between 1989 and 2000." Social Problems 52:152–180.

Rossman, Gabriel. 2004. "Elites, Masses, and Media Blacklists: The Dixie Chicks Controversy." Social Forces 83:61–67.

Rothman, Barbara Katz. 2000. Recreating Motherhood: Ideology and Technology in a Patriarchal Society. New Brunswick, NJ: Rutgers University Press.

———. 2005. Weaving a Family: Untangling Race and Adoption. New York: Beacon.

Rothman, David J. 1997. Beginnings Count: The Technological Imperative in American Health Care. New York: Oxford University Press.

Rotow, Thomas. 2000. "A Time to Join, A Time to Quit: The Influence of Life Cycle Transitions on Voluntary Association Membership." Social Forces 78:1133–1161.

Rubin, Barry. 2002. Islamic Fundamentalism in Egyptian Politics (2nd ed.). New York: Macmillan.

Rutter, Michael, Anderson-Wood, Lucie, Beckett, Celia, Bredenkamp, Diana, Castle, Jenny, Groothues, Christine, Kreppner, Jana, Keaveney, Lisa, Lord, Catherine, O'Conner, Thomas G., and the English and Romanian Adoptees (ERA) Study Team. 1999. "Quasi-Autistic Patterns Following Severe Early Global Privation." Journal of Child Psychology 40:537–549.

Ryan, William. 1981. Equality. New York: Pantheon Books.

Sadker, Myra, and Sadker, David. 1994. Failing at Fairness: How Our Schools Cheat Girls. New York: Simon & Schuster.

Sampson, Robert, Morenoff, Jeffrey, and Earls, Felton. 1999. "Beyond Social Capital: Spatial Dynamics of Collective Efficacy for Children." American Sociological Review 64:633–660.

Sampson, Robert, and Raudenbush, Stephen W. 1999. "Systematic Social Observation Of Public Spaces: A New Look At Disorder In Urban Neighborhoods." American Journal of Sociology 105:603–651.

Sampson, Robert J., Morenoff, Jeffrey D., and Gannon-Rowley, Thomas. 2002. "Assessing Neighborhood Effects: Social Processes and New Directions in Research." Annual Review of Sociology 28:443–478.

Sanday, Peggy Reeves. 1990. Fraternity Gang Rape: Sex, Brotherhood, and Privilege on Campus. New York: New York University Press.

Saporito, Salvatore. 2003. "Private Choices, Public Consequences: Magnet School Choice and Segregation by Race and Poverty." Social Problems 50:181–203.

Sassia, Saskia. 2006. Sociology of Globalization. New York: Norton.

Schaefer, Kristin D., Hennessy, James J., and Ponterotto, Joseph G. 1999. "Race as a Variable in Imposing and

Carrying out the Death Penalty in the U.S." Journal of Offender Rehabilitation 30:35–45.

Scharff, Virginia. 1991. Taking the Wheel: Women and the Coming of the Motor Age. New York: Free Press.

Scheck, Barry, Neufeld, Peter, and Dwyer, Jim. 2000. Actual Innocence: Five Days to Execution and Other Dispatches from the Wrongfully Convicted. New York: Doubleday.

Scheff, Thomas. 1966. Being Mentally Ill: A Sociological Theory. Chicago: Aldine.

Schneider, Mark, Teske, Paul, and Marschall, Melissa. 2000. Choosing Schools: Consumer Choice and the Quality of American Schools. Princeton, NJ: Princeton University Press.

Schor, Juliet B. 1998. The Overspent American: Upscaling, Downshifting, and the New Consumer. New York: Basic Books.

Schudson, Michael. 1995. The Power of News. Cambridge, MA: Harvard University Press.

Schur, Edwin M. 1979. Interpreting Deviance: A Sociological Introduction. New York: Harper & Row.

Schwartz, Barry. 1983. "George Washington and the Whig Conception of Heroic Leadership." American Sociological Review 48:18–33.

Schwartz, Barry, Markus, Hazel Rose, and Snibbe, Alana Conner. 2006. "Is Freedom Just Another Word for Many Things to Buy?" New York Times Magazine. February 26:14–15.

Scott, Janny. 2006. "Cities Shed Middle Class, and Are Richer and Poorer for It" New York Times. July 23:A1+.

Scott, Marvin B., and Lyman, Stafford M. 1968. "Accounts." American Sociological Review 33:46–62.

Scott, W. Richard. 2004. "Reflections on a Half-Century of Organizational Sociology." Annual Review of Sociology 30:1–21.

Seccombe, Karen, and Warner, Rebecca L. 2004. Marriage and Families: Relationships in Context. Belmont, CA: Wadsworth.

Sedlak, Andrea, and Broadhurst, Diane D. 1996. Third National Incidence Study of Child Abuse and Neglect. Washington, D.C.: U.S. Department of Health and Human Services.

Seltzer, Judith A. 1994. "Consequences of Marital Dissolution for Children." Annual Review of Sociology 20:235–266.

Semmerling, Tim Jon. 2006. "Evil" Arabs in American Popular Film: Orientalist Fear. Austin, TX: University of Texas Press.

Sen, Amartya. 1999. Development as Freedom. New York: Knopf.

Sen, Gita, and Grown, Caren. 1987. Development, Crises, and Alternative Visions. New York: Monthly Review Press.

"Separate and Unequal for Gypsies." 2006. New York Times. March 11:A14.

Settersten, Richard A. Jr., Furstenberg, Frank F. Jr., and Rumbaut, Rubén G. (eds.) 2006. On the Frontier of Adulthood: Theory, Research, and Public Policy. Chicago: University of Chicago Press.

Shapiro, Laura. 1994. "A Tomato with a Body that Just Won't Quit." Newsweek. June 6:80–82.

Shapiro, Thomas M. 2004. The Hidden Cost of Being African American: How Wealth Perpetuates Inequality. New York: Oxford University Press.

Sheler, Jeffrey L. 1995. "Keeping Faith in His Time." U.S. News and World Report. October 9:72–77.

Shenon, Philip. 2003. "Report on U.S. Antiterrorism Law Identifies Accusations of Abuses." New York Times. July 20:A11.

Sherif, Muzafer. 1936. The Psychology of Social Norms. New York: Harper & Row.

Sherkat, Darren E., and Ellison, Christopher G. 1999. "Recent Developments and Current Controversies in the Sociology of Religion." Annual Review of Sociology 25:363–394.

Shkilnyk, Anastasia M. 1985. A Poison Stronger Than Love: The Destruction of an Ojibwa Community. New Haven, CT: Yale University Press.

Shortt, Samuel. 1996. "Is Unemployment Pathogenic? A Review of Current Concepts with Lessons for Policy Planners." International Journal of Health Services 26:569–589.

Shover, Neal, and Wright, John Paul (eds.). 2000. Crimes of Privilege: Readings in White-Collar Crime. Oxford: Oxford University Press.

Siegel, Larry J. 1995. Criminology (5th ed.). Minneapolis, MN: West.

Simons, Ronald, and Gray, Phyllis. 1989. "Perceived Blocked Opportunity as an Explanation of Delinquency Among Lower-Class Black Males." Journal of Research on Crime and Delinquency 26:90–101.

Simpson, Richard L. 1985. "Social Control of Occupations and Work." Annual Review of Sociology 11:415–436.

Singh, Karan. 1993. "Let No Enemy Ever Wish Us Ill: The Hindu Vision of the Environment." In Pablo Piacentini (ed.), Story Earth: Native Voices on the Environment. San Francisco: Mercury House, pp. 146–156.

Skocpol, Theda. 1996. Boomerang: Clinton's Health Security Effort and the Turn Against Government in U.S. Politics. New York: Norton.

Sluka, Jeffrey A. (ed.). 2000. Death Squad: The Anthropology of State Terror. Philadelphia: University of Pennsylvania Press.

Smaje, Chris. 2000. Natural Hierarchies: The Historical Sociology of Race and Caste. Malden, MA: Blackwell.

Small, Mario Luis, and Newman, Katherine. 2002. "Urban Poverty After 'The Truly Disadvantaged': The Rediscovery of the Family, the Neighborhood, and Culture." Annual Review of Sociology 27:23–45.

Smith, Christian. 1991. The Emergence of Liberation Theology: Radical Religion and Social Movement Theory. Chicago: University of Chicago Press.

Smith, Jane I. 1999. Islam in America. New York: Columbia University Press.

Smith, Page. 1995. Democracy on Trial: The Japanese American Evacuation and Relocation in World War II. New York: Simon & Schuster.

Smith, Ryan A. 1997. "Race, Income, and Authority at Work: A Cross-Temporal Analysis of Black and White Men (1972–1994)." Social Problems 44:19–37.

Smock, Pamela J. 2000. "Cohabitation in the United States: An Appraisal of Research Themes, Findings, and Implications." Annual Review of Sociology 26:1–20.

Smolensky, Eugene, and Gootman, Jennifer Appleton (eds.). Working Families and Growing Kids: Caring for Children and Adolescents. Washington: National Academies Press.

Snow, David A., and Anderson, Leon. 1987. "Identity Work Among the Homeless: The Verbal Construction and Avowal of Personal Identities." American Journal of Sociology 92:1336–1371.

Snow, David A., Rochford, E. Burke, Jr., Worden, Steven K., and Benford, Robert D. 1986. "Frame Alignment Processes, Micromobilization, and Movement Participation." American Sociological Review 51:464–481.

Sohoni, Neera Kuckreja. 1994. "Where Are the Girls?" Ms. Magazine. July/August:96.

Southwell, Priscilla Lewis, and Everest, Marcy Jean. 1998. "The Electoral Consequences of Alienation: Nonvoting and Protest Voting in the 1992 Presidential Race." Social Science Journal 35:53–51.

Spring, Joel. 2004. Deculturalization and the Struggle for Equality: A Brief History of the Education of Dominated Cultures in the United States (4th ed.). New York: McGraw-Hill.

Staiger, Annegret. 2004. "Whiteness as Giftedness: Racial Formation at an Urban High School." Social Problems 51:161–181.

Stanglin, Douglas. 1992. "Toxic Wasteland." U.S. News and World Report. April 13:40–46.

Stark, Rodney, and Finke, Roger. 2000. Acts of Faith. Berkeley: University of California Press.

Starr, Paul. 1982. The Social Transformation of American Medicine. New York: Basic Books.

Steffensmeier, Darrell, and Allan, Emilie. 1996. "Gender and Crime: Toward a Gendered Theory of Female Offending." Annual Review of Sociology 22:459–487.

Steffensmeier, Darrell J., Allan, Emilie, Harer, Miles, and Streifel, Cathy. 1989. "Age and the Distribution of Crime." American Journal of Sociology 94:803–831.

Stern, Jessica. 2003. Terror in the Name of God: Why Religious Militants Kill. New York: HarperCollins.

Stiglitz, Joseph E. 2003. Globalization and Its Discontents. New York: Norton.

Stokes, Randall, and Anderson, Andy. 1990. "Disarticulation and Human Welfare in Less Developed Countries." American Sociological Review 55:63–74.

Stolte, John F., Fine, Gary Alan, and Cook, Karen S. 2001. "Sociological Miniaturism: Seeing the Big Through the Small in Social Psychology." Annual Review of Sociology 27:387–412.

Straus, Murray, and Gelles, Richard. 1986. "Societal Change and Change in Family Violence from 1975 to 1985 as Revealed by Two National Surveys." Journal of Marriage and the Family 48:465–479.

Street, Marc D. 1997. "Groupthink: An Examination of Theoretical Issues, Implications, and Future Research Suggestions." Small Group Research 28:72–93.

Stretesky, Paul, and Hogan, Michael J. 1998. "Environmental Justice: An Analysis of Superfund Sites in Florida." Social Problems 45:268–287.

Stroebe, Margaret S., and Stroebe, Wolfgang. 1983. "Who Suffers More? Sex Differences in Health Risks of the Widowed." Psychological Bulletin 93:279–301.

Sullivan, Deborah. 2001. Cosmetic Surgery: The Cutting Edge of Commercial Medicine in America. New Brunswick, NJ: Rutgers University Press.

Sullivan, Teresa A., Warren, Elizabeth, and Westbrook, Jay Lawrence. 2000. The Fragile Middle Class: Americans in Debt. New Haven, CT: Yale University Press.

Suomi, S. J., Harlow, H. H., and McKinney, W. T. 1972. "Monkey Psychiatrists." American Journal of Psychiatry 128(February): 927–932.

Sutherland, Edwin H. 1961. White-Collar Crime. New York: Holt, Reinhart & Winston.

Suzuki, Bob. 1989. "Asian Americans as the Model Minority." Change. November– December:12–20.

Swidler, Ann. 1986. "Culture in Action: Symbols and Strategies." American Sociological Review 51:273–286.

Takamura, Jeanette. 2002. Social Policy Issues and Concerns in a Diverse Aging Society. Generations 26(3):33–38.

Tannen, Deborah. 1990. You Just Don't Understand. New York: Morrow.

———. 1994. Talking from 9 to 5: How Women's and Men's Conversational Styles Affect Who Gets Heard, Who Gets Credit, and What Gets Done at Work. New York: Morrow.

Teachman, Jay. 2002. Stability Across Cohorts in Divorce Risk Factors. Demography 39:331–351.

Teachman, Jay D., Paasch, Kathleen M., Day, Randal D., and Carver, Karen P. 1997. "Poverty during Adolescence and Subsequent Educational Attainment." In G.J. Duncan and J. Brooks-Gunn (eds.), Consequences of Growing Up Poor. New York: Russell Sage Foundation, pp. 382–418.

Teachman, Jay D., Tedrow, Lucky M., and Crowder, Kyle D. 2000. "The Changing Demography of America's Families." Journal of Marriage and the Family 62:1234–1246.

Tedeschi, James T., and Riess, Marc. 1981. "Identities, the Phenomenal Self, and Laboratory Research." In J. T. Tedeschi (ed.), Impression Management Theory and Social Psychological Research. Orlando, FL: Academic.

Teitelbaum, Michael. 1975. "Relevance of Demographic Transition Theory to Developing Countries." Science 188(May 2):420–425.

Thomas, Darwin L., and Cornwall, Marie. 1990. "Religion and Family in the 1980's." In Alan Booth (ed.), Contemporary Families. Minneapolis, MN: National Council on Family Relations, pp. 265–274.

Thomas, W. I., and Thomas, Dorothy. 1928. The Child in America: Behavior Problems and Programs. New York: Knopf.

Thompson, Kevin. 1989. "Gender and Adolescent Drinking Problems: The Effects of Occupational Structure." Social Problems 36:30–47.

Thompson, Linda, and Walker, Alexis J. 1989. "Gender in Families: Women and Men in Marriage, Work and Parenthood." Journal of Marriage and the Family 51:845–872.

Thornberry, Terence P., and Farnworth, Margaret. 1982. "Social Correlates of Criminal Involvement: Further Evidence on the Relationship Between Social Status and Criminal Behavior." American Sociological Review 47:505–518.

Tillman, Robert, and Pontell, Henry N. 1992. "Is Justice Collar-Blind?: Punishing Medicaid Provider Fraud." Criminology 30:547–574.

Tilly, Charles. 1998. Durable Inequality. Berkeley: University of California Press.

———. 2004. Social Movements, 1768–2004. Boulder, CO, Paradigm Publishers.

Tjaden, Patricia, and Thoennes, Nancy. 1998. "Prevalence, Incidence, and Consequences of Violence against Women: Findings from the National Violence against Women Survey." National Institute of Justice Research in Brief. November.

———. 2000. Nature and Consequences of Intimate Partner Violence. Research Report 181867. Washington, D.C.: U.S. Department of Justice, National Institute of Justice.

Tomaskovich-Devey, Donald, and Skaggs, Sheryl. 2002. Sex Segregation, Labor Process Organization, and Gender Earnings Inequality. American Journal of Sociology 108:102–128.

Troeltsch, Ernst. 1931. The Social Teaching of the Christian Churches. New York: Macmillan.

Trudgill, Peter. 2000. Sociolinguistics: An Introduction to Language and Society. New York: Penguin.

Turner, Jonathan, and Beeghley, Leonard. 1981. The Emergence of Sociological Theory. Homewood, IL: Dorsey.

Turner, Jonathan, and Musick, David. 1985. American Dilemmas. New York: Columbia University Press.

Turner, R. Jay, and Avison, William R. 2003. "Status variations in stress exposure: Implications for the interpretation of research on race, socioeconomic status, and gender." Journal of Health and Social Behavior 44:488–505.

Turner, R. Jay, Wheaton, Blair, and Lloyd, Donald A. 1995. "The Epidemiology of Social Stress." American Sociological Review 60:104–125.

Twenge, Jean, Campbell, M. W. Keith, and Foster, Craig A.. 2003. "Parenthood and Marital Satisfaction: A Meta-Analytic Review." Journal of Marriage and Family 65:574–583.

Uehara, Edwina S. 1995. Reciprocity Reconsidered: Gouldner's Moral Norm of Reciprocity and Social Support. Journal of Social and Personal Relationships 12:483–502.

Ugger, Christopher, and Jeff Manza. 2002. "Democratic Contraction?: Political Consequences of Felon Disenfranchisement in the United States." American Sociological Review 67:777–803.

Umberson, Debra, Williams, Kristi, Powers, Daniel A., Chen, Meichu D., and Campbell, Anna M. 2005. "As Good as it Gets? A Life Course Perspective on Marital Quality." Social Forces 84:487–506.

UNAIDS/WHO. 2002. AIDS Epidemic Update: December 2002. Geneva, Switzerland: World Health Organization.

United Food and Commercial Workers. "Wal- Martization and Wages." http://www.ufcw.org. Accessed July 2006.

United Nations Development Programme. 2003. Human Development Report 2003. New York: United Nations.

———. 2005. Human Development Report 2005. New York: United Nations.

United Nations Environment Programme. 2002a. Global Environmental Outlook 3. New York: United Nations Environment Programme.

———. 2002b. Vital Water Graphics: Problems Related to Freshwater Resources. http://www .unep.org/vitalwater/21.htm. Accessed June 2003.

United Nations High Commissioner for Refugees. 2006. Refugees by Numbers: 2005. New York: United Nations.

United Nations Population Fund. 2000. State of World Population 2000. New York: United Nations Population Fund.

United Nations Statistics Division. 2003. The World's Women 2000: Trends and Statistics. http://unstats.un .org/unsd/demographic/ ww2000/table6a.htm.

U.S. Bureau of the Census. 1975. Historical Statistics of the United States: Colonial Times to 1970 (Bicentennial ed., Part 1). Washington, D.C.: U.S. Government Printing Office.

———. 2000a. Social and Demographic Characteristics of the U.S. Population, 1998. Issued July 2000 (CD-ROM).

———. 2000b. Statistical Abstract of the United States: 2000. Washington, D.C.: U.S. Government Printing Office.

———. 2002a. The Population Profile of the United States: 2000 (Internet Release). http://www.census.gov/population/www/ pop-profile/profile2000.html#cont. Accessed June 2003.

———. 2002b. Table 8: Income in 2001 by Educational Attainment for People 18 Years Old and Over, by Age, Sex, Race, and Hispanic Origin: March 2002. http://www.census.gov/population/socdemo/education/ppl-169/tab08 .pdf. Accessed June 2003.

———.2003a. Asset Ownership of Households: 2000. http://www.census.gov/hhes/www/wealth/1998_2000/wlth00-1.html. Accessed June 2003.

———. 2003b. Statistical Abstract of the United States: 2003. Washington, D.C.: U.S. Government Printing Office.

———. 2004. Poverty 2004 Tables. http://www .census.gov/hhes/www/poverty/poverty04/tables04 .html. Accessed June 2006.

———. 2005. Historical Income Tables—Households. http://www.census.gov/hhes/www/income/histinc/h01ar.html. Accessed June 2006.

———. 2006. Statistical Abstract of the United States: 2003. Washington, D.C.: U.S. Government Printing Office.

U.S. Bureau of Labor Statistics. 2002. Occupational Outlook Quarterly. Winter 2001–02.

———. 2005. Highlight of Women's Earnings in 2004. Report 987. Washington D.C.: U.S. Government Printing Office.

———. 2006a. Occupational Outlook Handbook 2006-07. Washington D.C.: U.S. Government Printing Office.

———. 2006b. Current Population Survey, Table 11. Employed Persons by detailed Occupation, Sex, Race, and Hispanic or Latino Ethnicity. www.bls.gov/cps. Accessed May 2006.

U.S. Committee for Refugees. 2002. World Refugee Survey 2001. Washington, D.C.: U.S. Committee for Refugees.

U.S. Department of Education. 2006. Digest of Education Statistics 2005. Washington D.C.: U.S. Government Printing Office.

U.S. Department of Health and Human Services. 2005. Health, United States, 2005. Washington D.C.: U.S. Government Printing Office.

U.S. Department of Justice. 1995. Crime in the United States: Uniform Crime Reports, 1995. Bureau of Justice Statistics. Washington D.C.: U.S. Government Printing Office.

———. 1998. Violence by Intimates: Analysis of Data on Crimes by Current or Former Spouses, Boyfriends, and Girlfriends. Report no. NCJ-167-237. Washington D.C.: U.S. Government Printing Office.

———. 2001. Sourcebook of Criminal Justice Statistics, 2001. Bureau of Justice Statistics. Washington D.C.: U.S. Government Printing Office.

———. 2002. Sourcebook of Criminal Justice Statistics, 2002. Bureau of Justice Statistics. Washington D.C.: U.S. Government Printing Office.

U.S. Department of Labor. 2002. Current Population Survey. Bureau of Labor Statistics. http://www.bls.gov/cps/cpsa2002.pdf. Accessed June 2003.

U.S. Department of State. 2005. Country Report on Human Rights Practices: India. Washington, D.C.: U.S. Government Printing Office.

———. 2006. Background note: Sudan. http://www.state .gov/r/pa/ei/bgn/5424.htm. Accessed June 2006.

U.S. General Accounting Office. 1996. "Death Penalty Sentencing: Research Indicates Pattern of Racial Disparities." In H. A. Bedau (ed.), The Death Penalty in America: Current Controversies. New York: Oxford University Press, pp. 268–272.

Useem, Bert, and Goldstone, Jack A. 2002. "Forging Social Order and Its Breakdown: Riot and Reform in U.S. Prisons." American Sociological Review 67: 499–525.

Vago, Steven. 1989. Law and Society (2nd ed.). Englewood Cliffs, NJ: Prentice-Hall.

Vallas, Steven P. 1987. "White Collar Proletarians? The Structure of Clerical Work and Levels of Class Consciousness." Sociological Quarterly 28:523–540.

Vallas, Steven P., and Yarrow, Michael. 1987. "Advanced Technology and Worker Alienation." Working and Occupations 14(February): 126–142.

van de Walle, Etienne, and Knodel, John. 1980. "Europe's Fertility Transition." Population Bulletin 34(6):1–43.

van Vugt, Mark, and Snyder, Mark (eds.). 2002. Special Issue: Cooperation in Society: Fostering Community Action and Civic Participation. American Behavioral Scientist. 45:769–782.

Vannini, Phillip. 2004. "The Meanings of a Star: Interpreting Music Fans' Reviews." Symbolic Interaction 27:47–69.

Vares, Tiina, and Braun, Virginia. 2006. "Spreading the Word, but What Word is That? Viagra and Male Sexuality in Popular Culture." Sexualities 9:315–332.

Varner, Gary E. 1994. "The Prospects for Consensus and Convergence in the Animal Rights Debate." Hastings Center Report 24:24–28.

Vaughan, Diane. 1996. The Challenger Launch Decision: Risky Technology, Culture, and Deviance at NASA. Chicago: University of Chicago Press.

Veblen, Thorstein. 1919. The Vested Interests and the State of the Industrial Arts. New York: Huebsch.

Wagner, Dennis. 2006. "Meth Lays Siege to Indian Country." USA Today. March 30:3A.

Wald, Kenneth D. 1987. Religion and Politics in the United States. New York: St. Martin's.

Waldman, Steven. 1992. "Benefits 'R' Us." Newsweek. August 10:56–58.

Walker, Iain, and Smith, Heather J. 2002. Relative Deprivation: Specification, Development, and Interpretation. New York: Cambridge University Press.

Wallace, Walter. 1969. Sociological Theory. Hawthorne, NY: Aldine.

Wallerstein, Judith, and Kelly, J. 1980. Surviving the Breakup. New York: Basic Books.

Waltman, Jerold. 2000. The Politics of the Minimum Wage. Urbana, IL: University of Illinois Press.

Washburn, Jennifer. 2006. University, Inc.: The Corporate Corruption of American Higher Education. New York: Basic Books.

Watamura, Sarah E., Donzella, Bonny, Alwin, Jan, and Gunnar, Megan R. 2003. "Morning-to-Afternoon Increases in Cortisol Concentrations for Infants and Toddlers at Child Care: Age Differences and Behavioral Correlates." Child Development 74:1006–1020.

Waters, Mary C., and Jimenez, Tomas R. 2005. "Assessing Immigrant Assimilation: New Empirical and Theoretical Challenges." Annual Review of Sociology 31:105–125.

Weber, Max. 1954. Law in Economy and Society. (Max Rheinstein, ed., Edward Shils and Max Reinstein, trans.) Cambridge, MA: Harvard University Press. (Originally published 1914.)

———. 1958. The Protestant Ethic and the Spirit of Capitalism. (Talcott Parsons, trans.) New York: Scribner's. (Originally published 1904–1905.)

———. 1970a. "Bureaucracy." In H. H. Gerth and C. Wright Mills (trans.), From Max Weber: Essays in Sociology. New York: Oxford University Press. (Originally published 1910.)

———. 1970b. "Class, Status, and Party." In H. H. Gerth and C. Wright Mills (trans.), From Max Weber: Essays in Sociology. New York: Oxford University Press. (Originally published 1910.)

———. 1970c. "Religion." In H. H. Gerth and C. Wright Mills (trans.), From Max Weber: Essays in Sociology. New York: Oxford University Press. (Originally published 1910).

Wechsler, David. 1958. The Measurement and Appraisal of Adult Intelligence (4th ed.). Baltimore, MD: Williams & Wilkins.

Weil, Frederick. 1985. "The Variable Effects of Education on Liberal Attitudes." American Sociological Review 50:458–474.

———. 1989. "The Sources and Structure of Legitimation in Western Democracies." American Sociological Review 54:682–706.

Weitz, Rose. 2001. "Women and Their Hair: Seeking Power Through Resistance and Accommodation." Gender & Society 15:667–686.

———. 2004. Rapunzel's Daughters: What Women's Hair Tells Us About Women's Lives. New York: Farrar, Straus, and Giroux.

———. 2007. The Sociology of Health, Illness, and Health Care: A Critical Approach (4th ed.). Belmont, CA: Wadsworth.

Wellman, Barry (ed.). 1999. Networks in the Global Village: Life in Contemporary Communities. Boulder, CO: Westview.

Wellman, Barry, and Berkowitz, S. D. (eds.) 1988. Social Structures: A Network Approach. New York: Cambridge University Press.

Wellman, Barry, Salaff, Janet, Dimitrova, Dimitrina, Garton, Laura, and Gulia, Milena. 1996. "Computer Networks as Social Networks: Collaborative Work, Telework, and Virtual Community." Annual Review of Sociology 22:213–238.

Wellman, Barry, and Wortley, Scot. 1990. "Different Strokes from Different Folks: Community Ties and Social Support." American Journal of Sociology 96:558–588.

Welsh, Sandy. 1998. "Gender and Sexual Harassment." Annual Review of Sociology 25:169–190.

West, Candace, and Zimmerman, Don H. 1987. "Doing Gender." Gender and Society 1:125–151.

Weston, Kath. 1991. Families We Choose: Lesbians, Gays, Kinship. New York: Columbia University Press.

Whisenant, Warren, Miller, John, and Pedersen, Paul M. 2005. "Systemic Barriers in Athletic Administration: An Analysis of Job Descriptions for Interscholastic Athletic Directors." Sex Roles 53:911–918.

White, Lynn, Jr. 1967. "The Historical Roots of Our Ecologic Crisis." Science 155:1203–1207.

White, Lynn K. 1994. "Co-residence and Leaving Home: Young Adults and Their Parents." Annual Review of Sociology 20:81–102.

Whorf, Benjamin L. 1956. Language, Thought, and Reality. Cambridge, MA: MIT Press.

Wilentz, Amy. 1993. "Love and Haiti." New Republic 209:18–19.

Wilkinson, Richard G. 1996. Unhealthy Societies: The Afflictions of Inequality. London: Routledge.

———. 2005. The Impact of Inequality. London: New Press.

Williams, Christine L. 1992. "The Glass Escalator: Hidden Advantages for Men in the 'Female' Professions." Social Problems 39:253–267.

Williams, David R. 1998. "African-American Health: The Role of the Social Environment." Journal of Urban Health: Bulletin of the New York Academy of Medicine 75:300–321.

Williams, David R., and Jackson, Pamela Braboy. 2005. "Social Sources of Racial Disparities in Health." Health Affairs 24:325–35.

Williams, Kirk, and Drake, Susan. 1980. "Social Structure, Crime, and Criminalization: An Empirical Examination of the Conflict Perspective." Sociological Quarterly 21:563–575.

Williams, Marian R., and Holcolm, Jefferson E. 2001. "Racial Disparity and Death Sentences in Ohio." Journal of Criminal Justice 29:207–218.

Williams, Rhys H. 1995. "Constructing the Public Good: Social Movements and Cultural Resources." Social Problems 42:124–144.

Wilson, Edward O. 1978. "Introduction: What Is Sociobiology?" In Michael S. Gregory, Anita Silvers, and Diane Sutch (eds.), Sociobiology and Human Nature. San Francisco: Jossey-Bass.

Wilson, George. 1997. "Pathways to Power: Racial Differences in the Determinants of Job Authority." Social Problems 44:38–52.

Wilson, James Q. 1992. "Crime, Race, and Values." Society 30:90–93.

Wilson, Thomas C. 1986. "Interregional Migration and Racial Attitudes." Social Forces 65:177–186.

———. 1991. "Urbanism, Migration, and Tolerance: A Reassessment." American Sociological Review 56:117–123.

Wilson, William J. 1978. The Declining Significance of Race. Chicago: University of Chicago Press.

———. 1987. The Truly Disadvantaged. Chicago: University of Chicago Press.

———. 1996. When Work Disappears: the World of the New Urban Poor. New York: Knopf.

Wimberly, Dale. 1990. "Investment Dependence and Alternative Explanations of Third World Mortality: A Cross-National Study." American Sociological Review 55:75–91.

Winders, Bill. 1999. "The Roller Coaster of Class Conflict: Class Segments, Mass Mobilization, and Voter Turnout in the U.S., 1840–1996." Social Forces 77:833–860.

Winerip, Michael. 2003. "Rigidity in Florida and its Consequences." New York Times. July 23:A15.

Winkler, Anne E., McBride, Timothy D., and Andrews, Courtney. 2005. "Wives Who Outearn Their Husbands: A Transitory Or Persistent Phenomenon For Couples?" Demography 42:523–53.

Winnick, Terri A. 2005. "From Quackery to 'Complementary' Medicine: The American Medical Profession Confronts Alternative Therapies." Social Problems 52:38–61.

Wirth, Louis. 1938. "Urbanism as a Way of Life." American Journal of Sociology 44(1):1–24.

Wiseman, Claire V., Gray, James J., Mosimann, James E., and Ahrens, Anthony H. 1992. "Cultural Expectations of Thinness in Women: An Update." International Journal of Eating Disorders 11:85–89.

Wolf, Martin. 2005. Why Globalization Works (2nd ed.). New Haven, CT: Yale University Press.

Woodberry, Robert D., and Smith, Christian S. 1998. "Fundamentalism et al: Conservative Protestants in America." Annual Review of Sociology 24:25–56.

World Bank. 2003. World Development Indicators 2003. Washington, D.C.: The World Bank.

World Health Organization. 1998. Fact Sheet No. 154: Tobacco Epidemic: Health Dimensions. Geneva, Switzerland: World Health Organization.

———. 2000. Fact Sheet No. 241: Female Genital Mutilation. Geneva, Switzerland: World Health Organization.

"World Population Beyond 6 Billion." 1999. Population Bulletin 54(1).

Wright, Erik O. 1985. Classes. London: Verso.

Wright, Victoria Clay, Chang, Jeani, Jeng, Gary, and Macaluso, Maurizio. 2006. "Assisted Reproductive Technology Surveillance, United States, 2003." Morbidity and Mortality Weekly Report 55(SS04):1–22.

Wrong, Dennis. 1961. "The Oversocialized Conception of Man in Modern Sociology." American Sociological Review 26(April):183–193.

———. 1979. Power. New York: Harper & Row.

Wuthnow, Robert. 1988. The Restructuring of American Religion. Princeton, NJ: Princeton University Press.

Yamane, David. 1994. "Professional Socialization for What?" Footnotes 22(March):7.

Yount, Kristin R. 1991. "Ladies, Flirts, and Tomboys: Strategies for Managing Sexual Harassment in an Underground Coal Mine." Contemporary Journal of Ethnography 19:396–422.

Zhou, Min, and Gatewood, James V. 2000. "Mapping the Terrain: Asian American Diversity and the Challenges of the Twenty-First Century." Asian American Policy Review 9:5–29.

Zuo, JiPing, and Benford, Robert. 1995. "Mobilization Processes and the 1989 Chinese Democracy Movement." Sociological Quarterly 36:131–156.

Zweigenhaft, Richard L., and Domhoff, G. William. 1998. Diversity in the Power Elite: Have Women and Minorities Reached the Top? New Haven, CT: Yale University Press.

Photo Credits

This page constitutes an extension of the copyright page. We have made every effort to trace the ownership of all copyrighted material and to secure permission from copyright holders. In the event of any question arising as to the use of any material, we will be pleased to make the necessary corrections in future printings. Thanks are due to the following authors, publishers, and agents for permission to use the material indicated.

Chapter 1. 1: © Royalty-Free/CORBIS 3: © Tony Freeman/PhotoEdit 5-7: all, © Brown Brothers 8: top left, © Brown Brothers 8: bottom right, © Ed Kashi 9: center right, © Bettmann Archive/CORBIS 9: bottom, © Bettmann Archive/CORBIS 11: © AP/Wide World Photos 13: © Jeff Greenberg/PhotoEdit 15: © Joel Gordon 22: © David Young Wolff/PhotoEdit 23: © Marc and Evelyne Bernheim/Woodfin Camp & Associates. 25: © AP/Wide World Photos

Chapter 2. 31: © AP/Wide World Photos 34: right, © Dagmar Fabricus/Stock, Boston Inc. 34: left, © Harold Schultz/FPG 35: © AP/Wide World Photos 40: © David Young Wolff/PhotoEdit 44: © Bob Daemmrich/The Image Works 45: © Joel Gordon 46: © David Austen/Woodfin Camp & Associated 51: © AP/Wide World Photos

Chapter 3. 54: © David Young Wolff/PhotoEdit 57: © Martin Rogers/Stock, Boston Inc. 59: © Richard Hutchings/PhotoEdit 64: © Joel Gordon 67: © Sean Sprague/The Image Works 68: © O'Brien Productions/CORBIS 70: © R. Maiman/CORBIS Sygma

Chapter 4. 74: © Robert Caputo/Stock, Boston Inc. 78: © Catherine Karnow/Woodfin Camp & Associates 80: © Anthony Bannister/Gallo Images/CORBIS 84: © Bruce Lee Smith/Gamma Liaison 85: © David Young-Wolff/PhotoEdit 88: © Jose Luiz Pelaez/CORBIS 90: © Patrick Ward/Stock, Boston Inc. 91: © Royalty-Free/CORBIS

Chapter 5. 97: © Tony Freeman/PhotoEdit 100: © M. Greenlar/The Image Works 102: © Left Lane Productions/CORBIS 103: © Bob Daemmrich/Stock, Boston Inc. 106: © Joel Gordon 106: © Joel Gordon 108: © David Brinkerhoff 113: © Ariel Skelley/PictureQuest 114: © Mark Peterson/CORBIS 116: © Charles Gupton/CORBIS

Chapter 6. 120: © AP/Wide World Photos 122: © Michael Newman/PhotoEdit 125: © Royalty-Free/CORBIS 127: © Alon Reininger/Contact Press Images 128: © Bob Daemmrich/Stock, Boston Inc. 137: © Richard Lord/PhotoEdit 138: © A. Ramey/Woodfin Camp & Associates 140: © Joel Gordon

Chapter 7. 143: © AP/Wide World Photos 147: © Andrew Lichtenstein/The Image Works 149: left, © Paul Barton/CORBIS 149: right, © Stephanie Maze/Woodfin Camp & Associates 151: © Michael Dwyer/Stock, Boston Inc. 154: © AP/Wide World Photos 157: © Gerd Ludwig/Woodfin Camp & Associates 160: © Nathan Benn/CORBIS 164: © AFP/Getty 168: © Les Stone/The Image Works 170: © AP/Wide World Photos

Chapter 8. 174: © Cleo Photography/PhotoEdit 180: © Peter Tumely/CORBIS 182: © Ted Spiegel/CORBIS 183: © Sandy Felsenthal/CORBIS 188: © Michael Newman/ PhotoEdit 191: © Ed Kashi/CORBIS 193: © RON HAVIV/VII/AP/Wide World Photos

Chapter 9. 197: © Royalty-Free/CORBIS 202: © Tony Freeman/PhotoEdit 205: © Tom & Dee Ann McCarthy/CORBIS 207: © A. Ramey/Stock, Boston Inc. 212: © Cleve Bryant/ PhotoEdit 216: © Kim Kulish/CORBIS 220: © Royalty-Free/CORBIS

Chapter 10. 222: © Richard T. Nowitz/CORBIS 224: © Richard Wood/Index Stock 225: © Bob Daemmrich / The Image Works 229: © AP/Wide World Photos 232: © AP/Wide World Photos 234: © Joel Gordon 237: © Francis Hogan/Electronic Publishing Services Inc., N.Y.C 240: © Catherine Ursillo/ Photo Researchers Inc. 242: © Earl & Nazima Kowall/CORBIS

Chapter 11. 245: © John & Yva Momatiuk/The Image Works 248: © AP/Wide World Photos 249: © Shelley Gazin/CORBIS 250: Gerd Ludwig/Woodfin Camp & Associates 253: © Jim Craigmyle/CORBIS 255: © Eastcott-Momatiuk/Woodfin Camp & Associates 257: © Peter Beck/CORBIS 258: © Mark Peterson/CORBIS 260: © Big Cheese Photo LLC/Alamy 267: © Richard Hutchings/Photo Researchers Inc.

Chapter 12. 271: © Will & Deni McIntyre/CORBIS 274: © Charles Gupton/Stock, Boston Inc. 275: © Bob Daemmrich/The Image Works 277: © AP/Wide World Photos 283: bottom, © James Shaffer/PhotoEdit 283: top, © Mark Harmel/FPG 284: © AP/Wide World Photos 290: © Vince Streano/CORBIS 292: © Robert Azzi/Woodfin Camp & Associates 294: © Henry Francis du Pont Winterthur Museum 297: © Mark Harmel/FPG

Chapter 13. 300: © Royalty-Free/CORBIS 302: © AP / Wide World Photos 305: © 1994 Tom Muscionico/Contact Press Images 309: © Paul Conklin/PhotoEdit 315: © Joe Rodriguez/Black Star Publishing 317: © Michael Justice 319: © Sharon G. Henry 320: © Michael Newman/PhotoEdit 321: © AP/Wide World Photos 323: top right, © Peter Menzel 323: bottom, © Seth Resnick

Name Index

Subject Index

Abortion rights movement, 375
Abuse. *See* Domestic violence
Accidents, normal, 385–386
Accommodation, intergroup, 178
Accounts, 88–89
Acculturation, intergroup, 178
Achieved status, 75–76, 144
Addams, Jane, 9
Adolescence
    delinquency, 133
    homeless during, 90–91
    motherhood in, 259
    socialization, 66, 69
    U. S. norms, 251
Adoption, 260–261
Adulthood
    early, 252
    middle age, 254
    socialization, 68
    transitions to, 251–252
Afghanistan, 292
Africa, 192–193, 343
African Americans
    attitudes toward, 185
    capital convictions, 140
    college enrollment, 280
    crime rates, 135–136
    educational attainment, 185
    female-headed families, 185–186
    imprisoned, 138
    income, 185
    Islamic, 292
    life expectancy, 231–232
    media stereotypes, 47
    mortality rates, 344–345
    murder by, 129
    poor, 160–161
    population percentage, 184
    segregation, 182–183
    social position, 184–185
    status of, 76, 193–194
African National Congress, 370
Age
    crime correlate, 133–134
    divorce rate and, 263
    gender and, 220–221
    life expectancy and, 232
    voting and, 309–310
Ageism, 179
Aging, 219–220, 234
Agricultural societies, 80–81
AIDS. *See* HIV/AIDS
Al Qaeda, 169
Alcoholism, 83–84
Alienation, bureaucratic, 117
Alienation, work, 326
AMA. *See* American Medical Association (AMA)
American Civil Liberties Union, 372
American Dream, 155–155
American Medical Association (AMA), 236–237
American Sign Language, 43
Amish church, 291
Amnesty International, 301
Anomie, 123
Anticipatory socialization, 66
Arab Americans
    Christians, 190
    educational achievement, 190
    9/11 attacks and, 190–191
    population percentages, 190
Ascribed status, 76, 144
Asian Americans
    discrimination against, 188–189
    educational achievement, 189
    profile, 188
Asians, 169, 231

ASL. *See* American Sign Language
Assimilation. *See* Social integration
Australian aborigines, 89
Authoritarian systems, 304
Authoritarianism, 181
Authority, 301–302
Automobile, 46

Battered women's syndrome, 11–12
Behavior
    collective, 366–367
    cultural view, 33–36
    patterns of, 8
    sociobiological view, 36–37
Berlin Wall, 383
Bias, history text, 273
Bias, social-desirability, 22
bin Laden, Osama, 169, 369
Birth rate, 336
Black Muslims, 292
Blacklisting, 308
Blackouts, electrical, 385–396
Blacks. *See* African Americans
Blame avoidance, 88
Blue-collar workers, 323–324
Body builders, 125
Boot camps, prison, 70–71
Bowling Along (Putnam), 112
Bragging, 90
BRCA1 gene, 49
Breast cancer, 49
Bride killings, 200
Bureaucracies
    characteristics, 115–116
    corporate, 318
    critiques of, 117–118
    defined, 115
    ideal model, 116
    organizational culture, 116–117